Random matrices arise from, and have important applications to, number theory, probability, combinatorics, representation theory, quantum mechanics, solid-state physics, quantum field theory, quantum gravity, and many other areas of physics and mathematics.

This volume of surveys and research results, based largely on lectures given at the Spring 1999 MSRI program of the same name, covers broad areas such as topologic and combinatorial aspects of random matrix theory; scaling limits, universalities, and phase transitions in matrix models; universalities for random polynomials; and applications to integrable systems. Its stress on the interaction between physics and mathematics will make it a welcome addition to the shelves of graduate students and researchers in both fields, as will its expository emphasis.

Mathematical Sciences Research Institute
Publications

40

Random Matrix Models and Their Applications

Mathematical Sciences Research Institute Publications

Vol. 1   Freed and Uhlenbeck: *Instantons and Four-Manifolds*, second edition
Vol. 2   Chern (editor): *Seminar on Nonlinear Partial Differential Equations*
Vol. 3   Lepowsky, Mandelstam, and Singer (editors): *Vertex Operators in Mathematics and Physics*
Vol. 4   Kac (editor): *Infinite Dimensional Groups with Applications*
Vol. 5   Blackadar: *K-Theory for Operator Algebras*, second edition
Vol. 6   Moore (editor): *Group Representations, Ergodic Theory, Operator Algebras, and Mathematical Physics*
Vol. 7   Chorin and Majda (editors): *Wave Motion: Theory, Modelling, and Computation*
Vol. 8   Gersten (editor): *Essays in Group Theory*
Vol. 9   Moore and Schochet: *Global Analysis on Foliated Spaces*
Vol. 10–11   Drasin, Earle, Gehring, Kra, and Marden (editors): *Holomorphic Functions and Moduli I, II*
Vol. 12–13   Ni, Peletier, and Serrin (editors): *Nonlinear Diffusion Equations and Their Equilibrium States I, II*
Vol. 14   Goodman, de la Harpe, and Jones: *Coxeter Graphs and Towers of Algebras*
Vol. 15   Hochster, Huneke, and Sally (editors): *Commutative Algebra*
Vol. 16   Ihara, Ribet, and Serre (editors): *Galois Groups over $\mathbb{Q}$*
Vol. 17   Concus, Finn, and Hoffman (editors): *Geometric Analysis and Computer Graphics*
Vol. 18   Bryant, Chern, Gardner, Goldschmidt, and Griffiths: *Exterior Differential Systems*
Vol. 19   Alperin (editor): *Arboreal Group Theory*
Vol. 20   Dazord and Weinstein (editors): *Symplectic Geometry, Groupoids, and Integrable Systems*
Vol. 21   Moschovakis (editor): *Logic from Computer Science*
Vol. 22   Ratiu (editor): *The Geometry of Hamiltonian Systems*
Vol. 23   Baumslag and Miller (editors): *Algorithms and Classification in Combinatorial Group Theory*
Vol. 24   Montgomery and Small (editors): *Noncommutative Rings*
Vol. 25   Akbulut and King: *Topology of Real Algebraic Sets*
Vol. 26   Judah, Just, and Woodin (editors): *Set Theory of the Continuum*
Vol. 27   Carlsson, Cohen, Hsiang, and Jones (editors): *Algebraic Topology and Its Applications*
Vol. 28   Clemens and Kollar (editors): *Current Topics in Complex Algebraic Geometry*
Vol. 29   Nowakowski (editor): *Games of No Chance*
Vol. 30   Grove and Petersen (editors): *Comparison Geometry*
Vol. 31   Levy (editor): *Flavors of Geometry*
Vol. 32   Cecil and Chern (editors): *Tight and Taut Submanifolds*
Vol. 33   Axler, McCarthy, and Sarason (editors): *Holomorphic Spaces*
Vol. 34   Ball and Milman (editors): *Convex Geometric Analysis*
Vol. 35   Levy (editor): *The Eightfold Way*
Vol. 36   Gavosto, Krantz, and McCallum (editors): *Contemporary Issues in Mathematics Education*
Vol. 37   Schneider and Siu (editors): *Several Complex Variables*
Vol. 38   Billera, Björner, Green, Simion, and Stanley (editors): *New Perspectives in Geometric Combinatorics*
Vol. 39   Haskell, Pillay, and Steinhorn (editors): *Model Theory, Algebra, and Geometry*

Volumes 1–4 and 6–27 are available from Springer-Verlag

Random Matrix Models and Their Applications

Edited by

Pavel Bleher
*Indiana University–Purdue University
Indianapolis*

Alexander Its
*Indiana University–Purdue University
Indianapolis*

CAMBRIDGE
UNIVERSITY PRESS

Pavel Bleher
Department of Mathematics
Indiana University–
  Purdue University (IUPUI)
402 North Blackford Street
Indianapolis, IN 46202-3216
United States

Alexander Its
Department of Mathematics
Indiana University
  Purdue University (IUPUI)
402 North Blackford Street
Indianapolis, IN 46202-3216
United States

Mathematical Sciences
  Research Institute
1000 Centennial Drive
Berkeley, CA 94720
United States

Series Editor
Silvio Levy

MSRI Editorial Committee
Joe P. Buhler (chair)
Alexandre Chorin
Silvio Levy
Jill Mesirov
Robert Osserman
Peter Sarnak

The Mathematical Sciences Research Institute wishes to acknowledge support by the National Science Foundation.

PUBLISHED BY THE PRESS SYNDICATE OF THE UNIVERSITY OF CAMBRIDGE
The Pitt Building, Trumpington Street, Cambridge, United Kingdom

CAMBRIDGE UNIVERSITY PRESS
The Edinburgh Building, Cambridge CB2 2RU, UK
40 West 20th Street, New York, NY 10011-4211, USA
10 Stamford Road, Oakleigh, Melbourne 3166, Australia
Ruiz de Alarcón 13, 28014 Madrid, Spain
Dock House, The Waterfront, Cape Town 8001, South Africa

http://www.cambridge.org

© Mathematical Sciences Research Institute 2001

Printed in the United States of America

A catalogue record for this book is available from the British Library.

Library of Congress Cataloging in Publication data:

Random matrix models and their applications / edited by Pavel M. Bleher, Alexander R. Its.
  p. cm. (Mathematical Sciences Research Institute publications ; 40)
Includes bibliographical references and index.
ISBN 0-521-80209-1
1. Random matrices. I. Bleher, Pavel M. (Pavel Maximovich), 1947– II. Its, Alexander R. III. Series.

QA188.R34 2001
512.9'434–dc21   2001025395

Contents

Preface	ix
Symmetrized Random Permutations JINHO BAIK AND ERIC M. RAINS	1
Hankel Determinants as Fredholm Determinants ESTELLE L. BASOR, YANG CHEN, AND HAROLD WIDOM	21
Universality and Scaling of Zeros on Symplectic Manifolds PAVEL BLEHER, BERNARD SHIFFMAN, AND STEVE ZELDITCH	31
z-Measures on Partitions, Robinson–Schensted–Knuth Correspondence, and $\beta = 2$ Random Matrix Ensembles ALEXEI BORODIN AND GRIGORI OLSHANSKI	71
Phase Transitions and Random Matrices GIOVANNI M. CICUTA	95
Matrix Model Combinatorics: Applications to Folding and Coloring PHILIPPE DI FRANCESCO	111
Interrelationships Between Orthogonal, Unitary and Symplectic Matrix Ensembles PETER J. FORRESTER AND ERIC M. RAINS	171
Dual Isomonodromic Tau Functions and Determinants of Integrable Fredholm Operators J. HARNAD	209
Functional Equations and Electrostatic Models for Orthogonal Polynomials MOURAD E. H. ISMAIL	225
Random Words, Toeplitz Determinants, and Integrable Systems I ALEXANDER R. ITS, CRAIG A. TRACY, AND HAROLD WIDOM	245
Random Permutations and the Discrete Bessel Kernel KURT JOHANSSON	259

Solvable Matrix Models VLADIMIR KAZAKOV	271
The τ-Function for Analytic Curves I. K. KOSTOV, I. KRICHEVER, M. MINEEV-WEINSTEIN, P. B. WIEGMANN, AND A. ZABRODIN	285
Integration over Angular Variables for Two Coupled Matrices G. MAHOUX, M. L. MEHTA, AND J.-M. NORMAND	301
Integrable Lattices: Random Matrices and Random Permutations PIERRE VAN MOERBEKE	321
SL(2) and z-Measures ANDREI OKOUNKOV	407
Some Matrix Integrals Related to Knots and Links PAUL ZINN-JUSTIN	421

Preface

This volume represents the most recent trends in the random matrix theory with a special emphasis on the exchange of ideas between physical and mathematical communities. The main topics include:

- random matrix theory and combinatorics
- scaling limits; universalities and phase transitions in matrix models
- topologico-combinatorial aspects of the theory of random matrix models
- scaling limit of correlations between zeros on complex and symplectic manifolds

Most contributions are based on talks and series of lectures given by the authors during the MSRI semester "Random Matrix Models and Their Applications" in Spring 1999, and have an expository or pedagogical style.

One of the basic ideas of the MSRI semester was to bring together the leading experts, both physicists and mathematicians, to discuss the latest results in the theory of matrix models and its applications. The book follows this line: it is divided roughly in half between physics and mathematics. The papers by physicists (G. Cicuta; Ph. Di Francesco; V. Kazakov; G. Mahoux, M. Mehta, J.-M. Normand; P. Zinn-Justin) give an overview of different physical problems in which the random matrix theory plays a decisive role, along with a rich variety of methods and ideas used to solve the problems. This includes enumeration of Feynman graphs on Riemann surfaces in the context of two-dimensional quantum gravity, spin systems on random surfaces, "meander problem" and random foldings, enumeration of knots and links, phase transitions and critical phenomena in random matrix models, interacting matrix models, etc.

The papers by mathematicians are devoted to recent breakthrough results on the statistics of longest increasing subsequence in random permutations and related problems of representation theory (J. Baik, E. Rains; A. Borodin, G. Olshanski; A. Its, C. Tracy, H. Widom; K. Johansson; A. Okounkov), universality of correlations between zeros on complex and symplectic manifolds (P. Bleher, B. Shiffman, S. Zelditch), applications of Hankel matrices to the theory of random matrices (E. Basor, Y. Chen, H. Widom), orthogonal polynomials (M. Ismail), interpolation properties of the ensembles of random matrices (P. Forrester, E. Rains), and integrable systems in the theory of random matrix models (J. Harnad and P. van Moerbeke). The paper of I. Kostov, I. Krichever, M. Mineev-Vainstein, P. Wiegmann, and A. Zabrodin is written by physicists and mathematicians and it relates conformal maps to integrable systems and matrix models.

We would like to express our gratitude to the MSRI Director, David Eisenbud, and the Deputy Directors, Hugo Rossi and Joe Buhler, for their help and support during the semester. We thank the series editor, Silvio Levy, for suggesting the publication of this volume and for his careful editing.

Our work in organizing the MSRI semester "Random Matrix Models" and the present volume was partially supported by the School of Science of Indiana University – Purdue University Indianapolis and through NSF Grants DMS-9970625 (Bleher) and DMS-9801608 (Its). We gratefully acknowledge this support.

Pavel Bleher
Alexander Its

Symmetrized Random Permutations

JINHO BAIK AND ERIC M. RAINS

ABSTRACT. Selecting N random points in a unit square corresponds to selecting a random permutation. Placing symmetry restrictions on the points, we obtain special kinds of permutations: involutions, signed permutations and signed involutions. We are interested in the statistics of the length (in numbers of points) of the longest up/right path in each symmetry type as the number of points increases to infinity. The limiting distribution functions are expressed in terms of a Painlevé II equation. In addition to the Tracy–Widom distributions of random matrix theory, we also obtain two new classes of distribution functions interpolating between the GOE and GSE, and between the GUE and GOE^2 Tracy–Widom distribution functions. Applications to random vicious walks and site percolation are also discussed

1. Introduction

Suppose that we are selecting n points, $p_1, p_2, \ldots, p_n$, at random in a rectangle, say $R = [0,1] \times [0,1]$ (see Figure 1). We denote by π the configuration of n random points. With probability 1, no two points have same x-coordinates nor y-coordinates. An up/right path of π is a collection of points $p_{i_1}, p_{i_2}, \ldots, p_{i_k}$ such that $x(p_{i_1}) < x(p_{i_2}) < \cdots < x(p_{i_k})$ and $y(p_{i_1}) < y(p_{i_2}) < \cdots < y(p_{i_k})$. The length of such a path is defined by the number of the points in the path. Now we denote by $l_n(\pi)$ the length of the longest up/right path of a random points configuration π.

As one can see from Figure 1, a configuration of n points gives rise to a permutation. For the example at hand, the corresponding permutation is $\left(\begin{smallmatrix} 1 & 2 & 3 & 4 & 5 \\ 5 & 1 & 3 & 2 & 4 \end{smallmatrix}\right)$. Therefore we can identity random points in R and random permutations, and we use the same notation π. In this identification, $l_n(\pi)$ is the length of the longest increasing subsequence of a random permutation.

The longest increasing subsequence has been of great interest for a long time (see [AD2], [OR], [BDJ1], for example). Especially as $n \to \infty$, it is known that $E(l_n) \sim 2\sqrt{n}$ [LS], [VK1; VK2] (also [AD1; Se; Jo2]) and $\text{Var}(l_n) \sim c_0 n^{1/3}$

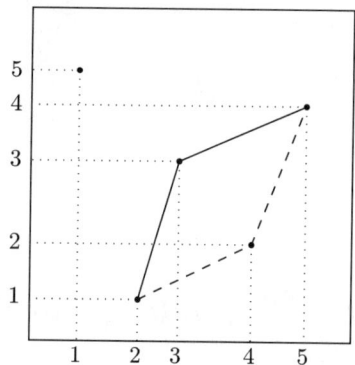

Figure 1. Random points in a rectangle.

[BDJ1] with some numerical constant $c_0 = 0.8132\cdots$. Moreover, the limiting distribution of l_n after proper scaling is obtained in [BDJ1] in terms of the solution to the Painlevé II equation (see Section 3 for precise statement). An interesting feature is that the limiting distribution function above is also the limiting distribution of the (scaled) largest eigenvalue of a random GUE matrix [TW1], the so-called "GUE Tracy–Widom distribution" F_2. In other words, properly centered and scaled, the length of the longest increasing subsequence of a random permutation behaves statistically for large n like the largest eigenvalue of a random GUE matrix. There have been many papers concerning the relations on combinatorics and random matrix theory: we refer the reader to [Re; Ge; Ke; Ra; Jo2; BDJ1; BDJ2; TW3; Bo; Jo1; Jo4; Ok; BOO; TW4; Jo3; BR1; BR2; ITW; St; PS2; PS1; Ba; BR3]. The purpose of this paper is to survey the analytic results of the recent papers [BR1; BR2] and discuss related topics.

In random matrix theory, three ensembles play important roles, GUE, GOE and GSE (see [Me], for example). Since random permutation is related to GUE, it would be interesting to ask which object in combinatorics is related to GOE and GSE. For this purpose, we consider symmetrized permutations. In terms of random points, 5 symmetry types of the rectangle R are considered, denoted by the symbols □, ◸, ◺, ⊡, and ⊠. Throughout this paper (and also in [BR1; BR2]), the symbol ⊛ is used to denote an arbitrary choice of the five possibilities above. Let $\delta = \{(t,t) : 0 \leq t \leq 1\}$, the diagonal line, and $\delta^t = \{(t, 1-t) : 0 \leq t \leq 1\}$, the anti-diagonal line. Consider the following random points selections:

□ select n points in R at random.

◸ select n points in $R \setminus \delta$ and m points in δ at random, and add their reflection images about δ.

◺ select n points in $R \setminus \delta^t$ and m points in δ^t at random, and add their reflection images about δ^t.

⊡ select n points at random in R, and add their rotational images about the center $(1/2, 1/2)$.

⊠ select n points in $R \setminus \delta$, m_+ points in δ and m_- points in δ^t at random, and add their reflection images about both δ and δ^t.

Define the map ι on S_n by $\iota(x) = n+1-x$. Let $\text{fp}(\pi)$ denote the number of points satisfying $\pi(x) = x$, and $\text{fpi}(\pi)$ denote the number of points satisfying $\pi(x) = \iota(x)$ (fpi represents negated points: see Remark 3 below). Each of the processes above corresponds to picking a random permutation from each of the following ensembles:

$$S_n^\square = S_n,$$
$$S_{n,m}^{\boxtimes} = \{\pi \in S_{2n+m} : \pi = \pi^{-1},\ \text{fp}(\pi) = m\},$$
$$S_{n,m}^{\boxtimes} = \{\pi \in S_{2n+m} : \pi = \iota\pi^{-1}\iota,\ \text{fpi}(\pi) = m\},$$
$$S_n^{\square} = \{\pi \in S_{2n} : \pi = \iota\pi\iota\},$$
$$S_{n,m_+,m_-}^{\boxtimes} = \{\pi \in S_{4n+2m_++2m_-} : \pi = \pi^{-1},\ \pi = \iota\pi^{-1}\iota,\ \text{fp}(\pi) = 2m_+,\ \text{fpi}(\pi) = 2m_-\}.$$

We denote the length of the longest increasing subsequence (equivalently, the longest up/right path) of π in each of the ensemble respectively by

$$L_n^\square,\quad L_{n,m}^{\boxtimes},\quad L_{n,m}^{\boxtimes},\quad L_n^{\square},\quad L_{n,m_+,m_-}^{\boxtimes}.$$

REMARK 1. The map $\pi \mapsto \iota^{-1}\pi$ gives a bijection between $S_{n,m}^{\boxtimes}$ and $S_{n,m}^{\boxtimes}$. Thus $L_{n,m}^{\boxtimes}$ has the same statistics with the length of the longest *decreasing* subsequence of a random involution with m fixed points taken from $S_{n,m}^{\boxtimes}$. From the definition, $L_{n,m}^{\boxtimes}$ is the random variable describing the length of the longest *increasing* subsequence of a random involution taken from the same ensemble.

REMARK 2. We may identify S_{2n} with the set of bijections from

$$\{-n,\ldots,-2,-1,1,2,\ldots,n\}$$

onto itself. In this identification, $S_n^{\square}$ becomes the set of signed permutations; $\pi(x) = -\pi(-x)$. The longest increasing subsequence problem of a random signed permutation is considered in [TW3] and [Bo].

REMARK 3. Under the identification in Remark 2, $S_{n,m_+,m_-}^{\boxtimes}$ becomes the set of signed involutions with m_+ fixed points and m_- negated points (we call x a negated point if $\pi(x) = -x$.)

In this paper, we are interested in the statistics of $L^{\circledast}$ as $n \to \infty$. Especially for ⌀, ⌁ and ⊠, we are interested in the cases when $m = [\sqrt{2n}\alpha]$ for ⌀, $m = [\sqrt{2n}\beta]$ for ⌁, and $m_+ = [\sqrt{n}\alpha]$ and $m_- = [\sqrt{n}\beta]$ for ⊠ with fixed $\alpha, \beta \geq 0$ where $[k]$ denotes the largest integer less than or equal to k. Then for most cases, the expected values have the same asymptotics. Namely, if we set $N = n, 2n+m, 2n+m, 2n, 4n+2m_++2m_-$ for each of $\square, \boxtimes, \boxtimes, \square, \boxtimes$ case respectively, we have

$$\lim_{N \to \infty} \frac{E(L^{\circledast})}{\sqrt{N}} = 2,$$

when $0 \leq \alpha \leq 1$ and $\beta \geq 0$ are fixed for ▨, ◨ and ⊠. When $\alpha > 1$, we have different expected value in the limit (see Section 3.)

On the other hand, the variance behaves asymptotically like $c_0 N^{1/3}$ but now with different constant c_0 depending on the symmetry type. It is because each symmetry type has different limiting distribution: $L_n^\square$ has GUE fluctuation, $L_{n,m}^\boxtimes$ GOE fluctuation and $L_n^\boxdot$ GUE^2 fluctuation (see Section 3 below for precise statements). Here GUE^2 denotes the statistics of a superimposition of eigenvalues of two random GUE matrices. Similarly for GOE^2. The cases of ▨ and ⊠ show more interesting features. For ▨, the limiting distribution function changes depending on the value of $\alpha = m/\sqrt{2n}$. The fluctuation is GSE when $\alpha < 1$, GOE when $\alpha = 1$ and Gaussian when $\alpha > 1$. By taking suitable scaling limit $\alpha \to 1$, we can find a certain smooth transition between GSE and GOE. For ⊠, the value $\alpha = m_+/\sqrt{n}$ determines the limiting distribution ; the value m_- plays no role in the transition. The fluctuation is GUE when $\alpha < 1$, GOE^2 when $\alpha = 1$, and Gaussian when $\alpha > 1$.

In Section 2, we define the Tracy–Widom distributions for GUE, GOE and GSE as well as new classes of distribution functions describing the transition around $\alpha = 1$. Main results are stated in Section 3, and Section 4 includes some applications and the related problems. Most of the results in this article are taken from [BR1; BR2]. Theorems 3.1 and 3.5 for □ were first proved in [BDJ1], and Theorem 3.1 for ⊡ was first obtained in [TW4; Bo]. The only new result is Theorem 4.2.

2. Tracy–Widom Distribution Functions

Let $u(x)$ be the solution of the Painlevé II (PII) equation,

$$u_{xx} = 2u^3 + xu, \tag{2-1}$$

with the boundary condition

$$u(x) \sim -\operatorname{Ai}(x) \quad \text{as} \quad x \to +\infty, \tag{2-2}$$

where Ai is the Airy function. The proof of the (global) existence and the uniqueness of the solution was first established in [HM]: the asymptotics as $x \to -\infty$ are (see [HM; DZ2], for example)

$$u(x) = -\operatorname{Ai}(x) + O\left(\frac{e^{-(4/3)x^{3/2}}}{x^{1/4}}\right), \quad \text{as } x \to +\infty, \tag{2-3}$$

$$u(x) = -\sqrt{\frac{-x}{2}}\left(1 + O\left(\frac{1}{x^2}\right)\right), \quad \text{as } x \to -\infty. \tag{2-4}$$

Recall that $\mathrm{Ai}(x) \sim \dfrac{e^{-(2/3)x^{3/2}}}{2\sqrt{\pi}x^{1/4}}$ as $x \to +\infty$. Define

$$v(x) := \int_\infty^x u(s)^2\, ds, \qquad (2\text{--}5)$$

so that $v'(x) = u(x)^2$.

We introduce the Tracy–Widom distributions. (Note that $q := -u$, which Tracy and Widom used in their papers, solves the same differential equation with the boundary condition $q(x) \sim +\mathrm{Ai}(x)$ as $x \to \infty$.)

DEFINITION (TRACY–WIDOM DISTRIBUTION FUNCTIONS). Set

$$F(x) := \exp\left(\frac{1}{2}\int_x^\infty v(s)\,ds\right) = \exp\left(-\frac{1}{2}\int_x^\infty (s-x)u(s)^2\,ds\right),$$

$$E(x) := \exp\left(\frac{1}{2}\int_x^\infty u(s)\,ds\right),$$

and set

$$F_2(x) := F(x)^2 = \exp\left(-\int_x^\infty (s-x)u(s)^2\,ds\right),$$

$$F_1(x) := F(x)E(x) = F_2(x)^{1/2} e^{\frac{1}{2}\int_x^\infty u(s)\,ds},$$

$$F_4(x) := F(x)\bigl(E(x)^{-1} + E(x)\bigr)/2 = F_2(x)^{1/2}\,\frac{e^{-\frac{1}{2}\int_x^\infty u(s)\,ds} + e^{\frac{1}{2}\int_x^\infty u(s)\,ds}}{2}.$$

In [TW1; TW2], Tracy and Widom proved that under proper centering and scaling, the distribution of the largest eigenvalue of a random GUE/GOE/GSE matrix converges to $F_2(x)$ / $F_1(x)$ / $F_4(x)$ as the size of the matrix becomes large. We note that from the asymptotics (2–3) and (2–4), for some positive constant c,

$$F(x) = 1 + O\bigl(e^{-cx^{3/2}}\bigr) \qquad \text{as } x \to +\infty, \qquad (2\text{--}6)$$

$$E(x) = 1 + O\bigl(e^{-cx^{3/2}}\bigr) \qquad \text{as } x \to +\infty, \qquad (2\text{--}7)$$

$$F(x) = O\bigl(e^{-c|x|^3}\bigr) \qquad \text{as } x \to -\infty, \qquad (2\text{--}8)$$

$$E(x) = O\bigl(e^{-c|x|^{3/2}}\bigr) \qquad \text{as } x \to -\infty. \qquad (2\text{--}9)$$

Hence in particular, $\lim_{x\to+\infty} F_\beta(x) = 1$ and $\lim_{x\to-\infty} F_\beta(x) = 0$, $\beta = 1,2,4$. Monotonicity of $F_\beta(x)$ follows from the fact that $F_\beta(x)$ is the limit of a sequence of distribution functions. Therefore $F_\beta(x)$ is indeed a distribution function.

As indicated in Introduction, we need new classes of distribution functions to describe the phase transitions from GSE to GOE and from GUE to GOE². First we consider the Riemann–Hilbert problem (RHP) for the Painlevé II equation

[FN; JMU]. Let Γ be the real line $\mathbb{R}$, oriented from $+\infty$ to $-\infty$. Let $m(\,\cdot\,;x)$ be the solution of the following RHP:

$$\begin{cases} m(z;x) \text{ is analytic in } z \in \mathbb{C} \setminus \Gamma, \\ m_+(z;x) = m_-(z;x)\begin{pmatrix} 1 & -e^{-2i(\frac{4}{3}z^3+xz)} \\ e^{2i(\frac{4}{3}z^3+xz)} & 0 \end{pmatrix} & \text{for } z \in \Gamma, \quad (2\text{-}10) \\ m(z;x) = I + O(\frac{1}{z}) \text{ as } z \to \infty. \end{cases}$$

Here $m_+(z;x)$ and m_- are the limits of $m(z';x)$ as $z' \to z$ from the left and right of the contour Γ: $m_\pm(z;x) = \lim_{\varepsilon \downarrow 0} m(z \mp i\varepsilon; x)$. Relation (2-10) corresponds to the RHP for the PII equation with the special monodromy data $p = -q = 1, r = 0$ (see [FN; JMU], also [FZ; DZ2]). In particular if the solution is expanded at $z = \infty$,

$$m(z;x) = I + \frac{m_1(x)}{z} + O\left(\frac{1}{z^2}\right), \qquad \text{as } z \to \infty, \qquad (2\text{-}11)$$

we have

$$2i(m_1(x))_{12} = -2i(m_1(x))_{21} = u(x),$$
$$2i(m_1(x))_{22} = -2i(m_1(x))_{11} = v(x),$$

where $u(x)$ and $v(x)$ are defined in Equations (2-1) to (2-5). Therefore the Tracy–Widom distributions above are expressed in terms of the residue at ∞ of the solution to the RHP (2-10). It is noteworthy that the new distributions below which interpolate the Tracy–Widom distributions require additional information of the solution of RHP.

DEFINITION. Let $m(z;x)$ be the solution of RHP (2-10) and denote by $m_{jk}(z;x)$ the (jk)-entry of $m(z;x)$. For $w > 0$, define

$$F^{\boxtimes}(x;w) := F(x)$$
$$\times \Big(\big(m_{22}(-iw;x) - m_{12}(-iw;x)\big)E(x)^{-1} + \big(m_{22}(-iw;x) + m_{12}(-iw;x)\big)E(x) \Big)/2,$$

and for $w < 0$, define

$$F^{\boxtimes}(x;w) := e^{\frac{8}{3}w^3 - 2xw} F(x)$$
$$\times \Big(\big(-m_{21}(-iw;x) + m_{11}(-iw;x)\big)E(x)^{-1} - \big(m_{21}(-iw;x) + m_{11}(-iw;x)\big)E(x) \Big)/2.$$

Also define

$$F^{\boxtimes}(x;w) := m_{22}(-iw;x)F_2(x), \qquad w > 0,$$
$$F^{\boxtimes}(x;w) := -e^{\frac{8}{3}w^3 - 2xw} m_{21}(-iw;x)F_2(x), \qquad w < 0.$$

First $F^{\boxtimes}(x;w)$ and $F^{\boxtimes}(x;w)$ are real from Lemma 2.1(i) below. Note that $F^{\boxtimes}(x;w)$ and $F^{\boxtimes}(x;w)$ are continuous at $w = 0$ since at $z = 0$, the jump condition of the RHP (2-10) implies

$$(m_{12})_+(0;x) = -(m_{11})_-(0;x),$$
$$(m_{22})_+(0;x) = -(m_{21})_-(0;x).$$

In fact, $F^{\square}(x;w)$ and $F^{\boxtimes}(x;w)$ are entire in $w \in \mathbb{C}$ from the RHP (2–10). From (2–6)–(2–9) and Lemma 2.1(ii) below, we see that

$$\lim_{x \to +\infty} F^{\square}(x;w), F^{\boxtimes}(x;w) = 1, \qquad \lim_{x \to -\infty} F^{\square}(x;w), F^{\boxtimes}(x;w) = 0$$

for any fixed $w \in \mathbb{R}$. Also Theorem 3.3 below shows that $F^{\square}(x;w)$ and $F^{\boxtimes}(x;w)$ are limits of distribution functions, implying that they are monotone in x. Therefore, $F^{\square}(x;w)$ and $F^{\boxtimes}(x;w)$ are indeed distribution functions for each $w \in \mathbb{R}$.

We close this section summarizing some properties of $m(-iw;x)$ in the following lemma. In particular the lemma implies that $F^{\square}(x;w)$ interpolates between $F_4(x)$ and $F_1(x)$, and $F^{\boxtimes}(x;w)$ interpolates between $F_2(x)$ and $F_1(x)^2$ (see Corollary 2.2).

LEMMA 2.1. *Let* $\sigma_3 = \begin{pmatrix} 1 & 0 \\ 0 & -1 \end{pmatrix}$, $\sigma_1 = \begin{pmatrix} 0 & 1 \\ 1 & 0 \end{pmatrix}$, *and set* $[a,b] = ab - ba$.

(i) *For real* w, $m(-iw;x)$ *is real.*

(ii) *For fixed* $w \in \mathbb{R}$, *we have*

$$m(-iw;x) = \left(I + e^{-cx^{3/2}}\right)\begin{pmatrix} 1 & -e^{\frac{8}{3}w^3 - 2xw} \\ 0 & 1 \end{pmatrix}, \qquad w > 0,\ x \to +\infty,$$

$$m(-iw;x) = \left(I + e^{-cx^{3/2}}\right)\begin{pmatrix} 1 & 0 \\ -e^{-\frac{8}{3}w^3 + 2xw} & 1 \end{pmatrix}, \qquad w < 0,\ x \to +\infty,$$

$$m(-iw;x) \sim \frac{1}{\sqrt{2}}\begin{pmatrix} 1 & -1 \\ 1 & 1 \end{pmatrix} e^{(-\frac{4}{3}w^3 + xw)\sigma_3} e^{(\frac{\sqrt{2}}{3}(-x)^{3/2} + \sqrt{2}w^2(-x)^{1/2})\sigma_3},$$
$$w > 0,\ x \to -\infty,$$

$$m(-iw;x) \sim \frac{1}{\sqrt{2}}\begin{pmatrix} 1 & 1 \\ -1 & 1 \end{pmatrix} e^{(-\frac{4}{3}w^3 + xw)\sigma_3} e^{(-\frac{\sqrt{2}}{3}(-x)^{3/2} - \sqrt{2}w^2(-x)^{1/2})\sigma_3},$$
$$w < 0,\ x \to -\infty.$$

(iii) *For any* x, *we have*

$$\lim_{w \to 0+} m(-iw;x) = \lim_{w \to 0-} \sigma_1 m(-iw;x)\sigma_1$$

$$= \begin{pmatrix} \frac{1}{2}(E(x)^2 + E(x)^{-2}) & -E(x)^2 \\ \frac{1}{2}(-E(x)^2 + E(x)^{-2}) & E(x)^2 \end{pmatrix}. \qquad (2\text{--}12)$$

(iv) *For fixed* $w \in \mathbb{R} \setminus \{0\}$, $m(-iw;x)$ *solves the differential equation*

$$\frac{d}{dx}m = w[m,\sigma_3] + u(x)\sigma_1 m,$$

where $u(x)$ *is the solution of the PII equation* (2–1), (2–2).

COROLLARY 2.2. *We have*

$$F^{\square}(x;0) = F_1(x), \qquad \lim_{w \to \infty} F^{\square}(x;w) = F_4(x), \qquad \lim_{w \to -\infty} F^{\square}(x;w) = 0,$$

$$F^{\boxtimes}(x;0) = F_1(x)^2, \qquad \lim_{w \to \infty} F^{\boxtimes}(x;w) = F_2(x), \qquad \lim_{w \to -\infty} F^{\boxtimes}(x;w) = 0.$$

PROOF. The values at $w = 0$ follow from (2–12). For $w \to \pm\infty$, note that from the RHP (2–10), we have $\lim_{z \to \infty} m(z;x) = I$. □

3. Main Results

As in the Introduction, let N denote $n, 2n+m, 2n+m, 2n, 4n+2m_+ + 2m_-$ for each of $\square, \boxslash, \boxbslash, \boxdot, \boxtimes$ case respectively. We scale the random variables: for permutations and involutions,

$$\chi_n^\square = \frac{L_n^\square - 2\sqrt{N}}{N^{1/6}}, \quad \chi_{n,m}^{\boxslash} = \frac{L_{n,m}^{\boxslash} - 2\sqrt{N}}{N^{1/6}}, \quad \chi_{n,m}^{\boxbslash} = \frac{L_{n,m}^{\boxbslash} - 2\sqrt{N}}{N^{1/6}},$$

and for signed permutations and signed involutions,

$$\chi_n^{\boxdot} = \frac{L_n^{\boxdot} - 2\sqrt{N}}{2^{2/3} N^{1/6}}, \quad \chi_{n,m_+,m_-}^{\boxtimes} = \frac{L_{n,m_+,m_-}^{\boxtimes} - 2\sqrt{N}}{2^{2/3} N^{1/6}}.$$

All the results in this section are taken from [BR2] which utilizes the algebraic work of [BR1].

First, we state the results for random permutations and random signed permutations. The result for random permutations was first obtained in [BDJ1], and the result for random signed permutations in [TW4; Bo].

THEOREM 3.1. *For fixed $x \in \mathbb{R}$,*

$$\lim_{n\to\infty} \Pr(\chi_n^\square \leq x) = F_2(x),$$
$$\lim_{n\to\infty} \Pr(\chi_n^{\boxdot} \leq x) = F_2(x)^2.$$

For the involution cases, we have the following limits.

THEOREM 3.2. *For each fixed α and β, and for fixed $x \in \mathbb{R}$, we have: for $\boxslash$,*

$$\lim_{n\to\infty} \Pr(\chi_{n,[\sqrt{2n\alpha}]}^{\boxslash} \leq x) = F_4(x), \qquad 0 \leq \alpha < 1,$$
$$\lim_{n\to\infty} \Pr(\chi_{n,[\sqrt{2n}]}^{\boxslash} \leq x) = F_1(x),$$
$$\lim_{n\to\infty} \Pr(\chi_{n,[\sqrt{2n\alpha}]}^{\boxslash} \leq x) = 0, \qquad \alpha > 1;$$

for $\boxbslash$,

$$\lim_{n\to\infty} \Pr(\chi_{n,[\sqrt{2n\beta}]}^{\boxbslash} \leq x) = F_1(x), \qquad \beta \geq 0;$$

and for $\boxtimes$,

$$\lim_{n\to\infty} \Pr(\chi_{n,[\sqrt{n\alpha}],[\sqrt{n\beta}]}^{\boxtimes} \leq x) = F_2(x), \qquad 0 \leq \alpha < 1, \ \beta \geq 0,$$
$$\lim_{n\to\infty} \Pr(\chi_{n,[\sqrt{n}],[\sqrt{n\beta}]}^{\boxtimes} \leq x) = F_1(x)^2, \qquad \beta \geq 0,$$
$$\lim_{n\to\infty} \Pr(\chi_{n,[\sqrt{n\alpha}],[\sqrt{n\beta}]}^{\boxtimes} \leq x) = 0, \qquad \alpha > 1, \ \beta \geq 0.$$

This theorem shows that for $\boxslash$ and $\boxtimes$, the limiting distributions differ depending on the value of α. As indicated earlier in the Introduction, as $\alpha \to 1$ at a certain rate, we obtain smooth transitions. From Corollary 2.2, the following results are consistent with Theorem 3.2.

THEOREM 3.3. *For fixed $w \in \mathbb{R}$, $\beta \geq 0$ and $x \in \mathbb{R}$,*

$$\lim_{n\to\infty} \Pr(\chi^{\boxslash}_{n,m} \leq x) = F^{\boxslash}(x;w), \qquad m = [\sqrt{2n} - 2w(2n)^{1/3}],$$

$$\lim_{n\to\infty} \Pr(\chi^{\boxtimes}_{n,m_+,m_-} \leq x) = F^{\boxtimes}(x;w), \qquad m_+ = [\sqrt{n} - 2wn^{1/3}], \ m_- = [\sqrt{n}\beta].$$

When $\alpha > 1$, Theorem 3.2 shows that we have used inappropriate scaling. In a proper scaling, we obtain normal distribution $N(0,1)$.

THEOREM 3.4. *For fixed $\alpha > 1$ and $\beta \geq 0$, as $n \to \infty$,*

$$\frac{L^{\boxslash}_{n,[\sqrt{2n\alpha}]} - (\alpha + 1/\alpha)\sqrt{N}}{\sqrt{(1/\alpha - 1/\alpha^3)}N^{1/4}} \to N(0,1) \qquad \text{in distribution,}$$

$$\frac{L^{\boxtimes}_{n,[\sqrt{n\alpha}],[\sqrt{n}\beta]} - (\alpha + 1/\alpha)\sqrt{N}}{\sqrt{(1/\alpha - 1/\alpha^3)}N^{1/4}} \to N(0,1) \qquad \text{in distribution.}$$

All the results above are on the convergence in distribution. We also have convergence of moments for all the cases. The case $\square$ was first obtained in [BDJ1].

THEOREM 3.5. *For each case of the preceding theorems, all the moments of the random variable converge to the moments of the corresponding limiting distribution.*

From this result, we can obtain the asymptotics of variances. Especially for $\square$, the variance is

$$\lim_{N\to\infty} \frac{\mathrm{Var}(l^{\square}_n)}{N^{1/3}} = \int_{-\infty}^{\infty} x^2 dF_2(x) - \left(\int_{-\infty}^{\infty} x dF_2(x)\right)^2.$$

Evaluating the integrals numerically, we obtain $0.8132\ldots$ (see [TW1]).

The outline of the proofs is as follows. First we consider the Poisson generating function. It is to let the number of points be Poisson. For example, we define

$$P^{\square}_l(t) = e^{-t^2} \sum_{n=0}^{\infty} \frac{t^{2n}}{n!} \Pr(L^{\square}_n \leq l).$$

The de-Poissonization lemma [Jo2] tells us that in the limit $n \to \infty$, we have

$$\Pr(L^{\square}_n \leq l) \sim P^{\square}_l(n^2),$$

hence it is enough to obtain the asymptotics of the Poisson generating function. The crucial point is that there is a determinantal formula for each Poisson generating function. For the cases $\square$ and $\boxdot$, the determinant is of Toeplitz type [Ge; Ra], while for the rest, it is of Hankel type [BR1]. In fact, as in [Ge], there are general identities between sum of Schur functions and determinantal formulae (see [BR1] for details), which can be used to consider other type of Young tableaux problems (see Section 4.4 below). Now general theory connects Toeplitz/Hankel determinants and orthogonal polynomials. It turns out that to analyze all the cases above, only one set of orthogonal polynomials are needed, namely the orthogonal polynomials on the unit circle with respect to the weight $e^{2t\cos\theta}d\theta/(2\pi)$.

Now following Fokas, Its and Kitaev [FIK], there is a Riemann–Hilbert representation for orthogonal polynomials. Let Σ be the unit circle in the complex plane oriented counterclockwise. Let $Y(z)$ be a 2×2 matrix-valued function satisfying

$$Y(z) \text{ is analytic in } \mathbb{C}\setminus\Sigma,$$

$$Y_+(z) = Y_-(z)\begin{pmatrix} 1 & z^{-k}e^{t(z+z^{-1})} \\ 0 & 1 \end{pmatrix} \quad \text{for } z\in\Sigma,$$

$$Y(z)\begin{pmatrix} z^{-k} & 0 \\ 0 & z^k \end{pmatrix} = I + O(z^{-1}) \quad \text{as } z\to\infty,$$

where $Y_+(z)$ and $Y_-(z)$ denote the limit of $Y(z')$ as $z'\to z$ satisfying $|z'|<1$ and $|z'|>1$, respectively. Then one finds that for example, the 11 entry of $Y(z)$ is the k-th monic orthogonal polynomial with respect to the weight $e^{2t\cos\theta}d\theta/(2\pi)$. Once we have a Riemann–Hilbert representation, we can employ the steepest-descent method (Deift–Zhou method) developed by Deift and Zhou [DZ1] to find asymptotics as parameters become large (or small). For our case, the parameters are t and k, and taking proper scaling, we obtain precise asymptotics which eventually yield the convergence in distribution and convergence of moments. We also mention that in our analysis, equilibrium measures play a crucial role as in the papers [DKMVZ; DKMVZ2; DKMVZ3]. The asymptotic analysis of $Y(z)$ was first obtained in [BDJ1], and further extended in [BR2].

4. Applications and Related Topics

4.1. Random Involutions and Random Signed Involutions. The ensemble $S_{n,m}^{\boxtimes}$ is the set of involutions with n 2-cycles and m 1-cycles. In the previous section, we considered the limiting statistics when n and m are related by $m = [\sqrt{2n}\alpha]$ with α being finite ; either fixed or $\alpha \to 1$ with certain rate. It is of interest to consider the whole set of involutions without constraints on the number of fixed points. Similarly, the signed involutions without constraint on the number of fixed points and negated points is also of interest. We define the ensembles of involutions and signed involutions

$$\tilde{S}_n = \{\pi \in S_n : \pi = \pi^{-1}\},$$

$$\tilde{S}_n^{\boxtimes} = \{\pi \in S_{2n} : \pi = \pi^{-1}, \ \pi = \iota\pi\iota\},$$

and denote by $\tilde{L}_n(\pi)$ and $\tilde{L}_n^{\boxtimes}(\pi)$ the length of the longest increasing subsequence of $\pi \in \tilde{S}_n$ and that of $\pi \in \tilde{S}_n^{\boxtimes}$, respectively.

Noting $\tilde{S}_n = \bigcup_{2k+m=n} S_{k,m}^{\boxtimes}$, we have

$$\Pr(\tilde{L}_n \leq l) = \frac{1}{|\tilde{S}_n|}\sum_{2k+m=n}\Pr(L_{k,m}^{\boxtimes}\leq l)|S_{k,m}|. \qquad (4\text{-}1)$$

It is not difficult to check that (see [Kn, pp. 66-67]) as $n \to \infty$, the main contribution to the sum $|\tilde{S}_n| = \sum_{2k+m=n} |S_{k,m}|$ comes from $\sqrt{2k} - (2k)^{\varepsilon+1/4} \leq m \leq \sqrt{2k} + (2k)^{\varepsilon+1/4}$. Comparing with the scaling $m = [\sqrt{2k} - 2w(2k)^{1/3}]$ in Theorem 3.3, the main contribution to the sum (4–1) comes from when $w = 0$, or $\alpha = 1$. Thus we obtain GOE Tracy–Widom distribution function in the limit. Similarly, we can obtain the convergence of moments. The signed involution case is analogous.

THEOREM 4.1. *For fixed $x \in \mathbb{R}$,*

$$\lim_{n \to \infty} \Pr\left(\frac{\tilde{L}_n - 2\sqrt{n}}{n^{1/6}} \leq x\right) = F_1(x),$$

$$\lim_{n \to \infty} \Pr\left(\frac{\tilde{L}_n^{\boxtimes} - 2\sqrt{n}}{2^{2/3} n^{1/6}} \leq x\right) = F_1(x)^2.$$

We also have convergence of all the moments.

This result should be compared with the results on random permutation and random signed permutation where the limiting distribution was $F_2(x)$ and $F_2(x)^2$ under the same scaling of the above (see [BDJ1], [TW3], [Bo] and Theorem 3.1).

REMARK. As in the permutation case, the length of the longest *increasing* subsequence and the length of the longest *decreasing* subsequence of random involutions have the same statistics. This can be seen by noting that there is a bijection (called the Robinson–Schensted correspondence; see [Kn], for example) between the set $\tilde{S}_n$ of involutions of n letters and the set of standard Young tableaux of size n, and the rows and the columns of standard Young tableaux have the same statistics under the push forward of the uniform probability distribution on $\tilde{S}_n$ under this bijection.

4.2. β-Plancherel Measure on the Set of Young Diagrams. Let Y_n be the set of Young diagrams, or equivalently partitions, of size n. Given a partition $\lambda = (\lambda_1, \lambda_2, \ldots) \vdash n$, let d_λ denote the number of standard Young tableaux of shape λ. We introduce the β-Plancherel measure M_n^β on Y_n defined by

$$M_n^\beta(\lambda) = \frac{d_\lambda^\beta}{\sum_{\mu \vdash n} d_\mu^\beta}, \qquad \lambda \in Y_n.$$

When $\beta = 2$, this is the Plancherel measure which arises in the representation theory. We are interest in the typical shape and the fluctuation of λ where λ is taken randomly from the probability space Y_n with M_n^β.

A motivation introducing the measure above is the result of Regev [Re], who proved that for fixed $\beta > 0$ and fixed l, as $n \to \infty$,

$$\sum_{\substack{\lambda \vdash n \\ \lambda_1 \leq l}} (d_\lambda)^\beta \sim$$

$$\left(\frac{l^{l^2/2} l^n}{(\sqrt{2\pi})^{(l-1)/2} n^{(l-1)(l+2)/4}} \right)^\beta \frac{n^{(l-1)/2}}{l!} \int_{\mathbb{R}^l} e^{-\frac{1}{2}\beta l \sum_j x_j^2} \prod_{j<k} |x_j - x_k|^\beta d^l x.$$

The multiple integral on the right hand side is the Selberg integral which can be computed exactly for each β. When $\beta = 1, 2, 4$, this integral is the normalization constant of the probability density of eigenvalues in GOE, GUE and GSE, respectively (see [Me], for example). So the basic question is if the β in the definition of the β-Plancherel measure corresponds to the β in the random matrix theory.

The well-known Robinson–Schensted correspondence [Sc] establishes a bijection between S_n and the pairs of standard Young tableaux with the same shape of size n, RS : $\pi \mapsto (P(\pi), Q(\pi))$. Especially, we obtain $\sum_{\mu \vdash n} d_\mu^2 = |S_n| = n!$. Moreover, under RS, the length of the longest increasing subsequence of $\pi \in S_n$ is equal to the number of boxes in the first row of $P(\pi)$ (or equally of $Q(\pi)$). Therefore under RS, the Plancherel measure M_n^2 is simply the push forward of the uniform probability measure on S_n to Y_n, and the number of boxes in the first row of a random Young diagram and the length of the longest increasing subsequence of a random permutation have the same statistics: GUE ($\beta = 2$) fluctuation in the limit, by Theorem 3.1.

If RS(π) = (P, Q), then RS(π^{-1}) = (Q, P) (see [Kn], for example). Therefore the set of involutions $\tilde{S}_n$ is bijective to the set of (single) standard Young tableaux, and the number of boxes in the first row of a random Young diagram taken under the probability M_n^1 has the same statistics with the length of the longest increasing subsequence of a random involution: GOE ($\beta = 1$) fluctuation in the limit, by Theorem 4.1.

In fact, it is shown in [Ok; BOO; Jo1] that, for the case of $\beta = 2$, in the large n limit, the number of boxes in the k-th of a random Young diagram corresponds to the k-th largest eigenvalue of a random GUE matrix. Also the typical shape of λ is obtained in [LS] and [VK1] which is related to Wigner's semicircle law [Ke]. On the other hand, the second row of a random Young diagram for $\beta = 1$ is discussed in [BR2] implying that it corresponds to the second eigenvalue of a random GOE matrix. It would be interesting to obtain similar results for general row for $\beta = 1$ and also for general β.

4.3. Random Turn Vicious Walker Model.

Random permutations and random involutions arise also in certain random walk process in a one dimensional integer lattice. We call a particle left-movable (resp. right-movable) if its left (resp. right) site is vacant. Initially, p particles are located at the points $1, 2, 3, \ldots, p$. At each time step t ($t = 1, 2, \ldots$), one particle among the left-movable particles is selected at random, and is moved to its left site (so at $t = 1$, the leftmost particle is moved.) Suppose this process is repeated for n time steps. It is found that [Fo] there is a bijection between all the possible configurations and the set of Young tableaux with at most p rows of size n: simply, in the k-th row, write the times when the particle is originally at the point k (see Figure 2).

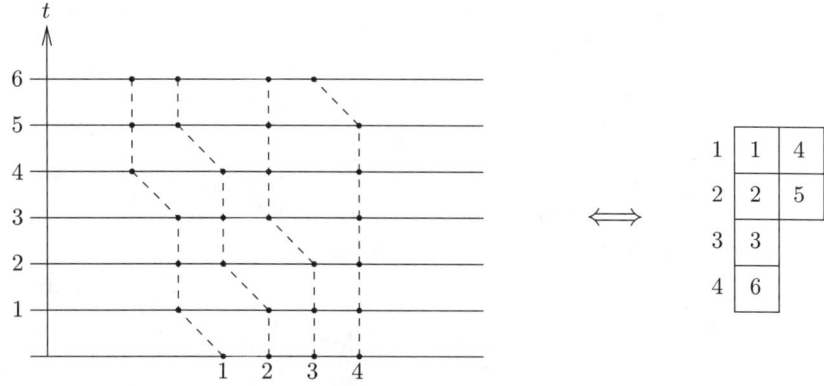

Figure 2. Random turn vicious walkers and Young tableaux.

In this correspondence the number of moves made by the k-th particle (counted from the left) is the number of boxes in the k-th row of the corresponding Young tableau. If we take the limit $p \to \infty$, we remove the constraint on the number of rows. In other words, we start with countably many particles located at $1, 2, 3, \ldots$ and at each time step, we move a randomly selected left-movable particle to its left site. Then the statistics of the number of moves made by the leftmost particle in n time steps is identical to that of the number of boxes in the first row of a random Young diagram under the probability M_n^1 (or the length of the longest increasing subsequence of a random involution). Hence, by Theorem 4.1, we obtain GOE fluctuation in the large n limit.

Now suppose after n steps of left moves, by taking n steps of right moves, we want the particles brought back to their original positions. The first n steps give a Young tableau and the next n steps give another Young tableau. So we are in the situation of pairs of Young tableaux, the case $\beta = 2$. Especially if the number of particles are infinite, the statistics of the moves made by the leftmost particle is identical to the length of the longest increasing subsequence of a random permutation: GUE fluctuation [Fo] in the large n limit.

4.4. Symmetrized Versions of Johansson's Model. In [Jo1], Johansson introduced a certain model which has several probabilistic interpretations, as a randomly growing Young diagram, a totally asymmetric one dimensional exclusion process, a certain zero-temperature directed polymer in a random environment or as a kind of first-passage site percolation model. Here we consider the symmetrized versions. The limiting distributions are parallel to the results of Section 3 depending on the symmetry type.

Each model is defined on $\mathbb{M}$, a subset of $\mathbb{Z}^2$, and at each site $(i,j) \in \mathbb{M}$, we define a random variable $w(i,j)$ of geometric distribution. We are interested in "the length of the longest up/right path". Let $g(q)$ denote the geometric distribution with parameter q, so that $\Pr\bigl(g(q)=k\bigr) = (1-q)q^k$ for $k=0,1,2,\ldots$. Let $g'(\beta,q)$ be the probability distribution defined by

$$\Pr(g'(\beta,q) = k) = \frac{1-q^2}{1+\beta q} \beta^{k \bmod 2} q^k.$$

Take $0 < q < 1$, and let $\alpha, \beta \geq 0$ such that $\alpha\sqrt{q}, \beta\sqrt{q} < 1$. Otherwise stated, the random variables $w(i,j)$ are independent each other. We denote by $(j,k) \nearrow (j',k')$ the set of up/right paths π from (j,k) to (j',k').

(i) Let $\mathbb{M} = \mathbb{Z}_+^2$. Let
$$w(i,j) \sim g(q).$$

Define
$$G^{\square}(N) = \max\left\{ \sum_{(i,j) \in \pi} w(i,j) : \pi \in (1,1) \nearrow (N,N) \right\}.$$

(ii) Let $\mathbb{M} = \mathbb{Z}^2$. Let $w(i,j) = 0$ if $i=0$ or $j=0$, and for $i,j \neq 0$, let
$$w(i,j) = w(-i,-j) \sim g(q).$$

Define
$$G^{\square}(N) = \max\left\{ \sum_{(i,j) \in \pi} w(i,j) : \pi \in (-N,-N) \nearrow (N,N) \right\}.$$

(iii) Let $\mathbb{M} = \mathbb{Z}_+^2$. Let
$$w(i,j) = w(j,i) \sim g(q), \quad i \neq j,$$
$$w(i,i) \sim g(\alpha\sqrt{q}).$$

Define
$$G^{\boxtimes}(N) = \max\left\{ \sum_{(i,j) \in \pi} w(i,j) : \pi \in (1,1) \nearrow (N,N) \right\}.$$

(iv) Let $\mathbb{M} = \mathbb{Z}_+ \times \mathbb{Z}_-$. Let
$$w(i,-j) = w(j,-i) \sim g(q), \quad i \neq -j,$$
$$w(i,-i) \sim g'(\beta\sqrt{q}).$$

Define
$$G^{\boxtimes}(N) = \max\left\{ \sum_{(i,j) \in \pi} w(i,j) : \pi \in (1,-N) \nearrow (N,-1) \right\}.$$

(v) Let $\mathbb{M} = \mathbb{Z}^2$. Let $w(i,j) = 0$ if $i = 0$ or $j = 0$. Otherwise

$$w(i,j) = w(-i,-j) \sim g(q), \quad |i| \neq |j|,$$
$$w(i,i) = w(-i,-i) \sim g(\alpha\sqrt{q}),$$
$$w(i,-i) = w(-i,i) \sim g'(\beta\sqrt{q}).$$

Define

$$G^{\boxtimes}(N) = \max\left\{\sum_{(i,j)\in\pi} w(i,j) : \pi \in (-N,-N) \nearrow (N,N)\right\}.$$

We are interested in the limiting distributions of $G^{\circledast}(N)$.

THEOREM 4.2. *Set*

$$\eta(q) = \frac{2\sqrt{q}}{1-\sqrt{q}}, \qquad \rho(q) = \frac{q^{1/6}(1+\sqrt{q})^{1/3}}{1-\sqrt{q}}.$$

Then:

(i)
$$\lim_{N\to\infty} \Pr\left(\frac{G^{\square}(N;q) - \eta(q)N}{\rho(q)N^{1/3}} \leq x\right) = F_2(x).$$

(ii)
$$\lim_{N\to\infty} \Pr\left(\frac{G^{\boxdot}(N;q) - \eta(q)(2N)}{2^{2/3}\rho(q)(2N)^{1/3}} \leq x\right) = F_2(x)^2.$$

(iii)
$$\lim_{N\to\infty} \Pr\left(\frac{G^{\boxslash}(N;q,\alpha) - \eta(q)N}{\rho(q)N^{1/3}} \leq x\right) = F_4(x), \qquad 0 \leq \alpha < 1,$$
$$\lim_{N\to\infty} \Pr\left(\frac{G^{\boxslash}(N;q,\alpha) - \eta(q)N}{\rho(q)N^{1/3}} \leq x\right) = F^{\boxslash}(x;w), \qquad \alpha = 1 - \frac{2w}{\rho(q)N^{1/3}},$$
$$\lim_{N\to\infty} \Pr\left(\frac{G^{\boxslash}(N;q,\alpha) - \eta(q)N}{\rho(q)N^{1/3}} \leq x\right) = 0, \qquad \alpha > 1.$$

(iv)
$$\lim_{N\to\infty} \Pr\left(\frac{G^{\boxbslash}(N;q,\beta) - \eta(q)N}{\rho(q)N^{1/3}} \leq x\right) = F_1(x), \qquad 0 \leq \beta.$$

(v)
$$\lim_{N\to\infty} \Pr\left(\frac{G^{\boxtimes}(N;q,\alpha,\beta) - \eta(q)(2N)}{2^{2/3}\rho(q)(2N)^{1/3}} \leq x\right) = F_2(x), \quad 0 \leq \alpha < 1, \ \beta \geq 0,$$

$$\lim_{N\to\infty} \Pr\left(\frac{G^{\boxtimes}(N;q,\alpha,\beta) - \eta(q)(2N)}{2^{2/3}\rho(q)(2N)^{1/3}} \leq x\right) = F^{\boxtimes}(x;w),$$
$$\alpha = 1 - \frac{2w}{\rho(q)(2N)^{1/3}}, \ \beta \geq 0,$$

$$\lim_{N\to\infty} \Pr\left(\frac{G^{\boxtimes}(N;q,\alpha,\beta) - \eta(q)(2N)}{2^{2/3}\rho(q)(2N)^{1/3}} \leq x\right) = 0, \quad \alpha > 1, \ \beta \geq 0.$$

REMARK. The result in part (i) is obtained in [Jo1] in the more general case of $(1,1) \nearrow (M,N)$, $M \neq N$.

The proof of the theorem is analogous to that of the result in Section 3. There again is a Toeplitz/Hankel determinantal formula for each case. In fact, the only change with the symmetrized permutation case is that we have $(1 + q + 2\sqrt{q}\cos\theta)^N d\theta/(2\pi)$ instead of $e^{2t\cos\theta} d\theta/(2\pi)$ for the weight. In analyzing the asymptotics of orthogonal polynomials, we need different scaling, but once we scale, the analysis is parallel. One important common property of the two weights above is that both of the supports of their equilibrium measures change from the full circle to a part of the circle with one gap depending on $2t/k < 1$ and $2t/k > 1$, and $N/k < \eta(q)^{-1}$ and $N/k > \eta(q)^{-1}$. And also in the one gap case, the equilibrium measures decay like a square root at the ends of the supports.

Acknowledgments

The authors thank the organizers of the workshop on Random Matrix Models and their Applications for their invitations.

References

[AD1] D. Aldous and P. Diaconis, "Hammersley's interacting particle process and longest increasing subsequences", *Prob. Th. and Rel. Fields* **103** (1995), 199–213.

[AD2] D. Aldous and P. Diaconis, "Longest increasing subsequences: from patience sorting to the Baik-Deift-Johansson theorem", *Bull. Amer. Math. Soc. (N.S.)* **36**:4 (1999), 413–432.

[Ba] J. Baik, "Random vicious walks and random matrices", *Comm. Pure Appl. Math* **53**:11 (2000), 1385–1410.

[BDJ1] J. Baik, P. Deift, and K. Johansson, "On the distribution of the length of the longest increasing subsequence of random permutations", *J. Amer. Math. Soc.* **12**:4 (1999), 1119–1178.

[BDJ2] J. Baik, P. Deift, and K. Johansson, "On the distribution of the length of the second row of a Young disgram under Plancherel measure", *Geom. Funct. Anal.* **10**:4 (2000), 702–731.

[BR1] J. Baik and E. M. Rains, "Algebraic aspects of increasing subsequences", to appear in *Duke J. Math.* See http://xxx.lanl.gov/abs/math.CO/9905083.

[BR2] J. Baik and E. M. Rains, "The asymptotics of monotone subsequences of involutions", to appear in *Duke J. Math.* See http://xxx.lanl.gov/abs/math.CO/9905084.

[BR3] J. Baik and E. M. Rains, "Limiting distributions for a polynuclear growth model with external sources", *J. Stat. Phys.*, 100(3/4) (2000), 523–541.

[Bo] A. Borodin, "Longest increasing subsequences of random colored permutations", *Electron. J. Combin.* **6**:1:R13, 1999.

[BOO] A. Borodin, A. Okounkov, and G. Olshanski, "On asymptotics of Plancherel measures for symmetric groups", *J. Amer. Math. Soc.* **13**:3 (2000), 481–515.

[DKMVZ] P. Deift, T. Kriecherbauer, K. McLaughlin, S. Venakides, and X. Zhou, "Asymptotics for polynomials orthogonal with respect to varying exponential weights", *Internat. Math. Res. Notices* **16** (1997), 759–782.

[DKMVZ2] P. Deift, T. Kriecherbauer, K. McLaughlin, S. Venakides, and X. Zhou, "Strong asymptotics of orthogonal polynomials with respect to exponential weights", *Comm. Pure Appl. Math.* **52**:12 (1999), 1491–1552.

[DKMVZ3] P. Deift, T. Kriecherbauer, K. McLaughlin, S. Venakides, and X. Zhou, "Uniform asymptotics for polynomials orthogonal with respect to varying exponential weights and applications to universality questions in random matrix theory", *Comm. Pure Appl. Math.* **52**:11 (1999), 1335–1425.

[DZ1] P. Deift and X. Zhou, "A steepest descent method for oscillatory Riemman-Hilbert problems; asymptotics for the MKdV equation", *Ann. of Math.* **137** (1993), 295–368.

[DZ2] P. Deift and X. Zhou, "Asymptotics for the Painlevé II equation", *Comm. Pure Appl. Math.* **48** (1995), 277–337.

[FN] H. Flaschka and A. Newell, "Monodromy and spectrum preserving deformations, I", *Comm. Math. Phys.* **76**:1 (1980), 67–116.

[FIK] A. Fokas, A. Its, and V. Kitaev, "Discrete Painlevé equations and their appearance in quantum gravity", *Comm. Math. Phys.* **142** (1991), 313–344.

[FZ] A. Fokas and X. Zhou, "On the solvability of Painlevé II and IV", *Comm. Math. Phys.* **144** (1992), 601–622.

[Fo] P. Forrester, "Random walks and random permutations", preprint, 1999. See http://xxx.lanl.gov/abs/math.CO/9907037.

[Ge] I. Gessel, "Symmetric functions and P-recursiveness", *J. Combin. Theory Ser. A* **53** (1990), 257–285.

[HM] S. Hastings and J. McLeod, "A boundary value problem associated with the second Painlevé transcendent and the Korteweg de Vries equation", *Arch. Rational Mech. Anal.* **73** (1980), 31–51.

[ITW] A. Its, C. Tracy, and H. Widom, "Random words, Toeplitz determinants and integrable systems, I", preprint, 1999. See http://xxx.lanl.gov/abs/math.CO/9909169.

[JMU] M. Jimbo, T. Miwa, and K. Ueno, "Monodromy preserving deformations of linear ordinary differential equations with rational coefficients, I. General theory and τ-function", *Phys. D* **2**:2 (1981), 306–352.

[Jo1] K. Johansson, "Shape fluctuations and random matrices", *Comm. Math. Phys.* **209**:2 (1999), 437–476.

[Jo2] K. Johansson, "The longest increasing subsequence in a random permutation and a unitary random matrix model", *Math. Res. Lett.*, 5(1-2) (1998), 63–82.

[Jo3] K. Johansson, "Discrete orthogonal polynomial ensembles and the Plancherel measure", to appear in *Ann. Math.* See http://xxx.lanl.gov/abs/math.CO/9906120.

[Jo4] K. Johansson, "Transversal fluctuations for increasing subsequences on the plane", *Probab. Theory Related Fields* 116(4) (2000), 445–456.

[Ke] S. Kerov, "Transition probabilities of continual Young diagrams and the Markov moment problem", *Funct. Anal. Appl* **27** (1993), 104–117.

[Kn] D. E. Knuth, *The art of computer programming*, v.3: *Sorting and searching*, Addison Wesley, Reading, MA, 2nd ed., 1973.

[LS] B. Logan and L. Shepp, "A variational problem for random Young tableaux", *Adv. in Math.* **26** (1977), 206–222.

[Me] M. Mehta, *Random matrices*, "Academic press, San Diago, second edition, 1991.

[OR] A. Odlyzko and E. Rains, "On longest increasing subsequences in random permutations", *Technical report, AT&T Labs*, 1998.

[Ok] A. Okounkov, "Random matrices and random permutations", preprint, 1999. See http://xxx.lanl.gov/abs/math.CO/9903176.

[PS1] M. Prähofer and H. Spohn, "Universal distributions for growth processes in $1+1$ dimensions and random matrices", *Phys. Rev. Lett.* **84** (2000), 4882.

[PS2] M. Prähofer and H. Spohn, "Statistical self-similarity of one-dimensional growth processes", *Physica A* **279** (2000), 342.

[Ra] E. M. Rains, "Increasing subsequences and the classical groups", *Electron. J. Combin.* **5**:1:R12, 1998.

[Re] A. Regev, "Asymptotic values for degrees associated with strips of Young diagrams", *Adv. in Math.* **41** (1981), 115–136.

[Sc] C. Schensted, "Longest increasing and decreasing subsequences", *Canad. J. Math.* **13** (1961), 179–191.

[Se] T. Seppäläinen, "A microscopic model for the burgers equation and longest increasing subsequences", *Electron. J. Prob.* **1**:5, 1995.

[St] R. Stanley, "Generalized riffle shuffles and quasisymmetric functions", preprint, 1999. See http://xxx.lanl.gov/abs/math.CO/9912025.

[TW1] C. Tracy and H. Widom, "Level-spacing distributions and the Airy kernel", *Comm. Math. Phys.* **159** (1994), 151–174.

[TW2] C. Tracy and H. Widom, "On orthogonal and symplectic matrix ensembles", *Comm. Math. Phys.* **177** (1996), 727–754.

[TW3] C. Tracy and H. Widom, "Random unitary matrices, permutations and Painlevé", *Comm. Math. Phys.* **207**:3 (1999), 665–685.

[TW4] C. Tracy and H. Widom, "On the distribution of the lengths of the longest monotone subsequences in random words", preprint, 1999. See http://xxx.lanl.gov/abs/math.CO/9904042.

[VK1] A. Vershik and S. Kerov, "Asymptotics of the Plancherel measure of the symmetric group and the limiting form of Young tables", *Soviet Math. Dokl.* **18** (1977), 527–531.

[VK2] A. Vershik and S. Kerov, "Asymptotic behavior of the maximum and generic dimensions of irreducible representations of the symmetric group", *Funct. Anal. Appl* **19** (1985), 21–31.

JINHO BAIK
DEPARTMENT OF MATHEMATICS
PRINCETON UNIVERSITY
PRINCETON, NJ 08544
UNITED STATES

INSTITUTE FOR ADVANCED STUDY
PRINCETON, NJ 08540
UNITED STATES
    jbaik@math.princeton.edu
    http://www.math.princeton.edu/~jbaik

ERIC M. RAINS
AT&T RESEARCH
FLORHAM PARK, NJ 07932
UNITED STATES
    rains@research.att.com

Hankel Determinants as Fredholm Determinants

ESTELLE L. BASOR, YANG CHEN, AND HAROLD WIDOM

> ABSTRACT. Hankel determinants which occur in problems associated with orthogonal polynomials, integrable systems and random matrices are computed asymptotically for weights that are supported in an semi-infinite or infinite interval. The main idea is to turn the determinant computation into a random matrix "linear statistics" type problem where the Coulomb fluid approach can be applied.

1. Introduction

Let w be a weight function supported on L (a subset of $\mathbb{R}$) that has finite moments of all orders

$$\mu_n = \int_L x^n w(x)\, dx.$$

With $w(x)$ we associate the Hankel matrix (μ_{i+j}), where $i, j = 0, \ldots, n-1$. The purpose of this paper is the determination of

$$D_n[w] := \det(\mu_{i+j})_{i,j=0}^{n-1}$$

for large n with suitable conditions on w. If L is a single interval, say $[-1, 1]$, then the asymptotic form of such determinants was computed by Szegö [1918] and later by Hirschmann [1966] for quite general w.

Our main result is as follows. Suppose we replace $w(x)$ by a function given in the form $w_0(x)U(x)$ where $w_0(x)$ is the weight $e^{-x}x^\nu$. Then for appropriate functions w, the determinants are given asymptotically as $n \to \infty$ by

$$D_n[w] = \exp\bigl(c_1 n^2 \log n + c_2 n^2 + c_3 n \log n + c_4 n + c_5 n^{1/2} + c_6 \log n + c_7 + o(1)\bigr) \quad (1)$$

Basor was supported in part by NSF Grants DMS-9623278 and DMS-9970879. Widom was supported in part by NSF Grant DMS-9732687 and EPSRC Visiting Fellowship Grant GR/M16580.

where

$$c_1 = 1, \quad c_2 = -3/2, \quad c_3 = \nu, \quad c_4 = -\nu + \log 2\pi,$$

$$c_5 = \frac{2}{\pi} \int_0^\infty \log(U(x^2))\, dx, \quad c_6 = \nu^2/2 - 1/6,$$

$$c_7 = 4/3 \log G(1/2) + (1/3 + \nu/2) \log \pi + (\nu/2 - 1/18) \log 2$$
$$- \log G(1+\nu) - (\nu/2) \log U(0) + \frac{1}{2\pi^2} \int_0^\infty x S(x)^2\, dx,$$

$$S(x) = \int_0^\infty \cos(xy) \log U(y^2)\, dy,$$

and G is the Barnes function (see Section 2).

In Section 2 we establish an identity relating $D_n[w]$, $D_n[w_0]$, and a certain Fredholm determinant and a description of the Fredholm determinant from a "linear statistics" point of view. A computation of $D_n[w_0]$ is also included. Then in Section 3 the Coulomb fluid approach is used to compute the asymptotics of the Fredholm determinant. This, along with the computation of $D_n[w_0]$ allows us to give a heuristic, Coulomb fluid derivation of the formula. A rigorous proof based on operator theory techniques developed in [Basor 1997; Tracy and Widom 1994] will appear in a forthcoming paper. In Section 3 the Hermite case is also included.

2. Preliminaries

Let $p_i(x)$ be polynomials orthonormal with respect to $w_0(x)$ (the reference weight function) over L

$$\int_L p_i(x) p_j(x) w_0(x)\, dx = \delta_{i,j},$$

with strictly positive leading coefficients. For later convenience we also write $\phi_j = \sqrt{w_0} p_j$ as the orthonormal functions. Consider the determinant,

$$\det\left(\int_L p_i(x) p_j(x) w(x)\, dx \right)_{i,j=0}^{n-1}.$$

If $p_i(x) = \sum_{j=0}^i c_{ij} x^j$, then

$$\det\left(\int_L p_i p_j w\, dx \right)_{i,j=0}^{n-1} = \det\left(\int_L \sum_{k=0}^i \sum_{l=0}^j c_{ik} c_{jl} x^{k+l} w\, dx \right)_{i,j=0}^{n-1}$$

$$= \left(\prod_{i=0}^{n-1} c_{ii} \right)^2 \det\left(\int_L x^{j+k} w\, dx \right)_{j,k=0}^{n-1}$$

$$= \left(\prod_{i=0}^{n-1} c_{ii} \right)^2 \det(\mu_{j+k})_{j,k=0}^{n-1}. \tag{2}$$

If $w = w_0$ then the left side of (2) is 1. So,

$$\det\left(\int_L p_i p_j w\, dx\right)_{i,j=0}^{n-1} = \frac{\det(\mu_{i+j})_{i,j=0}^{n-1}}{\det(\mu_{i+j}^0)_{i,j=0}^{n-1}} = \frac{D_n[w]}{D_n[w_0]}, \qquad (3)$$

where

$$\mu_i^0 := \int_L x^i w_0\, dx,$$

are the moments of the reference weight w_0, and

$$D_n[w_0] = \frac{1}{\left(\prod_{i=0}^{n-1} c_{ii}\right)^2}.$$

We now express the left side of (3) as a Fredholm determinant.

$$\det\left(\int_L p_i p_j w\, dx\right) = \det\left(\int_L \phi_i \phi_j \left(1 - \left(1 - \frac{w}{w_0}\right)\right) dx\right)$$
$$= \det(\delta_{i,j} - M_{i,j}),$$

where

$$M_{i,j} := \int_L \phi_i \phi_j (1 - w/w_0)\, dx =: \int_L \phi_i \phi_j F\, dx = \int_L \phi_i \phi_j (1 - U)\, dx.$$

We have the standard expansion

$$-\log\det(\delta_{i,j} - M_{i,j}) = \sum_{p=1}^{\infty} \frac{\operatorname{tr} \mathbf{M}^p}{p},$$

where the matrix $\mathbf{M}$ has elements $M_{i,j}$. To compute $\operatorname{tr}\mathbf{M}^p$, we first look at the simpler case of $p = 3$. We see that $\operatorname{tr}\mathbf{M}^3$ equals

$$\sum_{i,j,k=0}^{n-1} M_{i,j} M_{j,k} M_{k,i}$$

$$= \int dX \sum_{i,j,k=0}^{n-1} \phi_i(x_1)\phi_j(x_1)F(x_1)\phi_j(x_2)\phi_k(x_2)F(x_2)\phi_k(x_3)\phi_i(x_3)F(x_3)$$

$$= \int dX\, F(x_1)F(x_2)F(x_3) \sum_{i=0}^{n-1} \phi_i(x_3)\phi_i(x_1) \sum_{j=0}^{n-1} \phi_j(x_1)\phi_j(x_2) \sum_{k=0}^{n-1} \phi_k(x_2)\phi_k(x_3)$$

$$= \int dX\, K_n(x_1,x_2)K_n(x_2,x_3)K_n(x_3,x_1)F(x_1)F(x_2)F(x_3),$$

where $\int dX$ stands for $\int_L \cdots \int_L dx_1\, dx_2\, dx_3$ and where

$$K_n(x,y) := \sum_{i=0}^{n-1} \phi_i(x)\phi_i(y) = a_n \frac{\phi_n(x)\phi_{n-1}(y) - \phi_n(y)\phi_{n-1}(x)}{x - y}. \qquad (4)$$

The last equality of (4) is the Christoffel–Darboux formula [Szegő 1975] and a_n are the off diagonal recurrence coefficients of p_i. The generalization to integer p is obvious and we find,
$$\operatorname{tr} \mathbf{M}^p = \operatorname{tr}(K_n F)^p,$$
where the operator $K_n F$ has kernel $K_n(x,y)F(y)$. So
$$\det(\delta_{i,j} - M_{i,j}) = \det(I - K_n F), \tag{5}$$
and I in (5) has kernel $\delta(x-y)$. We now come to the linear statistics.

If x_i, for $i = 1, \ldots, n$, are random variables with the joint probability density function
$$p(x_1, \ldots, x_n) \propto \prod_{i=1}^{n} w_0(x_i) \prod_{1 \leq j, k \leq n} |x_j - x_k|^2, \quad x_i \in L, \tag{6}$$
then $Q = \sum_{i=1}^{n} f(x_i)$ (the linear statistics) is also a random variable. Consider the generating function of Q, $\langle \exp(-Q) \rangle$, where
$$\langle (\ldots) \rangle := \frac{\int_L \cdots \int_L (\ldots) p(x_1, \ldots, x_n) \, dx_1 \ldots dx_n}{\int_L \cdots \int_L p(x_1, \ldots, x_n) \, dx_1 \ldots dx_n}.$$
Recall the Heine formula [Szegő 1975] for Hankel determinants:
$$\det(\mu_{i+j})_{i,j=0}^{n-1} = \frac{1}{n!} \int_L \cdots \int_L dx_1 \ldots dx_n \prod_{i=1}^{n} w(x_i) \prod_{1 \leq j, k \leq n} |x_j - x_k|^2, \tag{7}$$
and the analogous one with μ_{i+j} replaced by μ_{i+j}^0 and w replaced by w_0. If we write $w = \exp(-v)$, $w_0 = \exp(-v_0)$ and $v = v_0 + f$ then
$$D_n[w] = \det(\mu_{i+j}^0)_{i,j=0}^{n-1} \langle \exp(-Q) \rangle = D_n[w_0] \langle \exp(-Q) \rangle. \tag{8}$$
In this notation $f(x) = -\log U(x)$. So our strategy is to choose a suitable w_0 for which there is exact result for $D_n[w_0]$, and compute $\langle \exp(-Q) \rangle$ as a Fredholm determinant for n large. In the next section we will use a heuristic method to give an indication how the results for $\langle \exp(-Q) \rangle$ for large n can be found. If we take $w_0(x) = x^\nu \exp(-x)$, $\nu > -1$ and $L = [0, \infty)$ then $p_i(x)$ are the orthonormal Laguerre polynomials. It is well known [Szegő 1975] that
$$c_{ii}^2 = \frac{1}{\Gamma(1+i+\nu)\Gamma(1+i)}.$$
So
$$D_n[w_0] = \prod_{i=0}^{n-1} \Gamma(1+i+\nu)\Gamma(1+i) = \frac{G(n+\nu+1)}{G(\nu+1)} \frac{G(n+1)}{G(1)}, \tag{9}$$
where the Barnes function G [Barnes 1900; Whittaker and Watson 1962] satisfies the functional equation $G(z+1) = \Gamma(z)G(z)$, with the initial condition $G(1) = 1$. The asymptotics of the Barnes function are computed in [Whittaker and Watson 1962] and since $G(1+a+n)$ is asymptotic to
$$n^{(n+a)^2/2 - 1/12} e^{-3/4 n^2 - an} (2\pi)^{(n+a)/2} G^{2/3}(1/2) \pi^{1/6} 2^{-1/36}$$

we can directly apply this formula with $a = 0$ and $a = \nu$ to obtain asymptotically

$$D_n[w_0] = \exp\left\{d_1 n^2 \log n + d_2 n^2 + d_3 n \log n + d_4 n + d_5 \log n + d_6 + o(1)\right\} \quad (10)$$

where

$$d_1 = 1, \quad d_2 = -3/2, \quad d_3 = \nu, \quad d_4 = -\nu + \log 2\pi, \quad d_5 = \nu^2/2 - 1/6,$$
$$d_6 = (4/3)\log G(1/2) + (1/3 + \nu/2)\log \pi + (\nu/2 - 1/18)\log 2 - \log G(1+\nu).$$

3. The Coulomb Fluid Method

For suitably chosen f, $\langle \exp(-Q) \rangle$ for large n was computed in [Chen and Lawrence 1998] starting from the Heine formula. An alternative and shorter derivation is given here. Now if Q is in some sense "small" then by expanding up to Q^2, we have, $\langle \exp(-Q) \rangle \approx 1 - \langle Q \rangle + \frac{1}{2}\langle Q^2 \rangle$. This can be reproduced by expanding

$$\exp\left(-\langle Q \rangle - \frac{(\langle Q \rangle^2 - \langle Q^2 \rangle)}{2}\right),$$

up to $\langle Q^2 \rangle$ and $\langle Q \rangle^2$. With the introduction of the microscopic density $\varrho(x) := \sum_{i=1}^{n} \delta(x - x_i)$, one finds

$$\langle Q \rangle = \int_L f(x) \langle \varrho(x) \rangle \, dx,$$
$$\langle Q \rangle^2 - \langle Q^2 \rangle = \int_L \int_L f(x) \left(\langle \varrho(x) \rangle \langle \varrho(y) \rangle - \langle \varrho(x)\varrho(y) \rangle\right) f(y) \, dx \, dy.$$

In the Coulomb fluid approach, expected to be valid for large n, we replace $\langle \varrho(x) \rangle$ by the equilibrium density $\sigma(x)$, which is supposed to be supported in a single interval (a, b). It is then a simple exercise to show that the correlation function

$$\langle \varrho(x) \rangle \langle \varrho(y) \rangle - \langle \varrho(x)\varrho(y) \rangle$$

is replaced by

$$\frac{1}{2\pi^2 \sqrt{(b-x)(x-a)}} \frac{\partial}{\partial y} \left(\frac{\sqrt{(b-y)(y-a)}}{x-y}\right).$$

Therefore,

$$\langle \exp(-Q) \rangle \sim \exp(-S_1 - S_2), \quad (11)$$

where

$$S_1 = \frac{1}{4\pi^2} \int_a^b \int_a^b \frac{f(x)}{\sqrt{(b-x)(x-a)}} \frac{\partial}{\partial y}\left(\frac{\sqrt{(b-y)(y-a)}}{x-y}\right) f(y) \, dx \, dy,$$
$$S_2 = \int_a^b \sigma(x) f(x) \, dx.$$

The end points of the interval, a and b, are determined by the normalization condition $\int_a^b \sigma(x)\,dx = n$ and a supplementary condition [Chen and Lawrence 1998].

So the constants c_i, $i = 1, \ldots, 6$ and part of c_7 are obtained from the asymptotic expansion of $D_n[w_0]$ while c_5 and the last two terms of c_7 are obtained from the large n behaviour of S_2 and S_1 respectively. For $w_0(x) = x^\nu \exp(-x)$, $x \geq 0$ it is known that $a = 0$, $b = 4n + 2\nu$ and

$$\sigma(x) = -\nu \delta_+(x) + \frac{1}{2\pi}\sqrt{\frac{b-x}{x}}, \quad 0 \leq x < b.$$

So,

$$S_2 = -\frac{\nu}{2}f(0) + \frac{1}{2\pi}\int_0^b f(x)\sqrt{\frac{b-x}{x}}\,dx$$
$$\to \frac{\nu}{2}\log U(0) - \frac{2n^{1/2}}{\pi}\int_0^\infty \log U(x^2)\,dx, \quad \text{as } n \to \infty. \tag{12}$$

As $n \to \infty$, S_1 tends to

$$\frac{1}{4\pi^2}\int_0^\infty\int_0^\infty \frac{f(x)}{\sqrt{x}}f(y)\frac{\partial}{\partial y}\frac{\sqrt{y}}{x-y}\,dx\,dy.$$

Changing the integration variables $x = s^2$, $y = t^2$ and noting

$$-\frac{1}{2}\int_{-\infty}^\infty |x|\exp[-ixt]\,dx = \frac{1}{t^2},$$

we find

$$S_1 \to -\frac{1}{2\pi^2}\int_0^\infty x\left(\int_0^\infty \log U(s^2)\cos(xs)\,ds\right)^2 dx$$
$$= -\frac{1}{2\pi^2}\int_0^\infty xS(x)^2\,dx. \tag{13}$$

Therefore (1) follows from (10), (12) and (13).

We can also make use of the above information to determine the recurrence coefficients of the monic polynomials $P_j(x)$, orthogonal with respect to $x^\nu \exp(-tx)U(x)$. The parameter t is introduced for later convenience. The recurrence relations reads

$$xP_n(x) = P_{n+1}(x) + \alpha_n(t)P_n(x) + \beta_n(t)P_{n-1}(x).$$

From the basic properties of the Hankel determinant generated by the weight $w(x,t) = x^\nu \exp(-tx)U(x)$, one finds

$$\alpha_n(t) = -\frac{d}{dt}\ln\frac{D_{n+1}(t)}{D_n(t)},$$

$$\beta_n(t) = \frac{D_{n+1}(t)D_{n-1}(t)}{(D_n(t))^2},$$

where $D_n(t) = D_n[w(\,\cdot\,,t)]$. With the asymptotics, we find, as $n \to \infty$, that

$$\alpha_n(1) = 2n + \nu + 1 - \left(\frac{1}{\pi}\int_0^\infty \log U(x^2)\,dx\right) n^{-1/2} + o(1)$$

$$\beta_n(1) = n^2 + \nu n - \left(\frac{1}{2\pi}\int_0^\infty \log U(x^2)\,dx\right) n^{1/2} + o(1),$$

are the recurrence coefficients of those monic polynomials orthogonal with respect to $w(x) = x^\nu \exp(-x)U(x)$. As a further application of the asymptotic formula, we study the short noise generating function of an n-channel disordered conductor [Muttalib and Chen 1996] where the f of the linear statistics is

$$f(x) := M\ln\frac{x+z}{x+1}, \quad |z| = 1.$$

As $n \to \infty$, S_1 tends to

$$-M^2 \log \frac{\sqrt{z}+1}{2z^{1/4}},$$

while S_2 tends to

$$\frac{\nu M}{2}\log z + 2\sqrt{n}M(1 - \sqrt{z}).$$

This generalises the results of [Muttalib and Chen 1996] to $\nu \neq 0$.

Now suppose w is supported in $(-\infty, \infty)$. We adopt the same strategy to determine the large n behaviour of the associated Hankel determinant:

$$D_n[w] = D_n[w_0]\frac{D_n[w]}{D_n[w_0]},$$

where the "reference" Hankel determinant is generated by the Hermite weight $w_0(x) = \exp(-x^2)$, where $x \in (-\infty, \infty)$. Now $a = -b = -\sqrt{2n}$ and $\sigma(x) = \frac{1}{\pi}\sqrt{x^2 - b^2}$. Thus, as $n \to \infty$,

$$\frac{D_n[u]}{D_n[w_0]} \sim \exp(-S_1 - S_2),$$

where

$$\begin{aligned} S_1 &= -\frac{1}{8\pi^2}\int_{-\infty}^\infty |k|\hat{f}(-k)\hat{f}(k)\,dk \\ S_2 &= \frac{\sqrt{2n}}{\pi}\int_{-\infty}^\infty f(x)\,dx. \end{aligned} \quad (14)$$

Here $\hat{f}(k) = \int_{-\infty}^\infty \exp(ikx)f(x)\,dx$. Equations (14) are essentially those found by Kac and by Akhiezer [Akhiezer 1964]. Therefore the large n behaviour of $D_n[w]$ follows from

$$D_n[w_0] = (2\pi)^{n/2}2^{-n^2/2}G(n+1),$$

the asymptotics of the Barnes function, and Equation (14).

Acknowledgement

Part of this work was performed when the authors were participating in the programme on Random Matrices and their Applications at the MSRI, Berkeley. The authors thank Prof. D. Eisenbud for the hospitality of the Institute, and the organizers, Profs. P. Bleher and A. Its, for their invitation, which made this collaboration possible.

References

[Akhiezer 1964] N. I. Akhiezer, "Continual analogue to some theorems on Toeplitz matrices", *Ukrain. Mat. Zh.* **16** (1964), 445–462. In Russian; translated in *Amer. Soc. Trans. Series* (2) **50** (1964), 295–316.

[Barnes 1900] E. W. Barnes, "The theory of the G-function", *Quart. J. Pure and Appl. Math.* **31** (1900), 264–313.

[Basor 1997] E. L. Basor, "Distribution functions for random variables for ensembles of positive Hermitian matrices", *Comm. Math. Phys.* **188**:2 (1997), 327–350.

[Chen and Lawrence 1998] Y. Chen and N. Lawrence, "On the linear statistics of Hermitian random matrices", *J. Phys. A* **31**:4 (1998), 1141–1152.

[Hirschman 1966] I. I. Hirschman, Jr., "The strong Szegö limit theorem for Toeplitz determinants", *Amer. J. Math.* **88** (1966), 577–614.

[Muttalib and Chen 1996] K. A. Muttalib and Y. Chen, "Distribution function for shot noise in Wigner-Dyson ensembles", *Internat. J. Modern Phys. B* **10**:16 (1996), 1999–2006.

[Szegő 1918] G. Szegő, "A Hankel-féle formákról", *Mathematikai és Természettudományi Értesétő* **36** (1918), 497–538. Translated as "Hankel forms", *Amer. Math. Soc. Transl.* (2) **108** (1977), 1–36; reprinted as pp. 113–148 in his *Collected papers*, v. 1, 1915–1927, edited by R. Askey, Birkhäuser, Boston, 1982.

[Szegő 1975] G. Szegő, *Orthogonal polynomials*, 4th ed., Colloquium Publications **23**, Amer. Math. Soc., Providence, 1975.

[Tracy and Widom 1994] C. A. Tracy and H. Widom, "Fredholm determinants, differential equations and matrix models", *Comm. Math. Phys.* **163**:1 (1994), 33–72.

[Whittaker and Watson 1962] E. T. Whittaker and G. N. Watson, *A course of modern analysis*, 4th ed., Cambridge University Press, New York, 1962.

ESTELLE L. BASOR
DEPARTMENT OF MATHEMATICS
CALIFORNIA POLYTECHNIC STATE UNIVERSITY
SAN LUIS OBISPO, CA 93407
UNITED STATES
   ebasor@calpoly.edu

YANG CHEN
DEPARTMENT OF MATHEMATICS
IMPERIAL COLLEGE
180 QUEEN'S GATE
LONDON SW7 2BZ
UNITED KINGDOM
   y.chen@ic.ac.uk

HAROLD WIDOM
DEPARTMENT OF MATHEMATICS
UNIVERSITY OF CALIFORNIA
SANTA CRUZ, CA 96054
UNITED STATES
   widom@cats.ucsc.edu

Universality and Scaling of Zeros on Symplectic Manifolds

PAVEL BLEHER, BERNARD SHIFFMAN, AND STEVE ZELDITCH

ABSTRACT. This article is concerned with random holomorphic polynomials and their generalizations to algebraic and symplectic geometry. A natural algebro-geometric generalization involves random holomorphic sections $H^0(M, L^N)$ of the powers of any positive line bundle $L \to M$ over any complex manifold. Our main interest is in the statistics of zeros of k independent sections (generalized polynomials) of degree N as $N \to \infty$. We fix a point P and focus on the ball of radius $1/\sqrt{N}$ about P. Magnifying the ball by the factor $\sqrt{N}$, we found in a prior work that the statistics of the configurations of simultaneous zeros of random k-tuples of sections tend to a universal limit independent of P, M, L. We review this result and generalize it further to the case of pre-quantum line bundles over almost-complex symplectic manifolds (M, J, ω). Following earlier work of Shiffman and Zelditch, we replace $H^0(M, L^N)$ in the complex case with the "asymptotically holomorphic" sections defined by Boutet de Monvel and Guillemin and (from another point of view) by Donaldson and Auroux. We then give a generalization to an m-dimensional setting of the Kac–Rice formula for zero correlations, which we use together with earlier results to prove that the scaling limits of the n-point correlation functions for zeros of random k-tuples of asymptotically holomorphic sections belong to the same universality class as in the complex case. In our prior work, we showed that the limit correlations are short range; here we show further that the limit "connected correlations" decay exponentially with respect to the square of the maximum distance between points.

1. Introduction

A well-known theme in random matrix theory (RMT), zeta functions, quantum chaos, and statistical mechanics, is the universality of scaling limits of correlation functions. In RMT, the relevant correlation functions are for eigenvalues of random matrices (see [De; TW; BZ; BK; So] and their references). In the case of zeta functions, the correlations are between the zeros [KS]. In

---

Research partially supported by NSF grants #DMS-9970625 (Bleher), #DMS-9800479 (Shiffman), #DMS-9703775 (Zelditch).

quantum dynamics, they are between eigenvalues of 'typical' quantum maps whose underlying classical maps have a specified dynamics. In the 'chaotic case' it is conjectured that the correlations should belong to the universality class of RMT, while in integrable cases they should belong to that of Poisson processes. The latter has been confirmed for certain families of integrable quantum maps, scattering matrices and Hamiltonians (see [Ze2; RS; Sa; ZZ] and their references). In statistical mechanics, there is a large literature on universality of critical exponents [Car]; other rigorous results include analysis of universal scaling limits of Gibbs measures at critical points [Sin]. In this article we are concerned with a somewhat new arena for scaling and universality, namely that of RPT (random polynomial theory) and its algebro-geometric generalizations [Han; Hal; BBL; BD; BSZ1; BSZ2; SZ1; NV]. The focus of these articles is on the configurations and correlations of zeros of random polynomials and their generalizations, which we discuss below. Random polynomials can also be used to define random holomorphic maps to projective space, but we leave that for the future. Our purpose here is partly to review the results of [SZ1; BSZ1; BSZ2] on universality of scaling limits of correlations between zeros of random holomorphic sections on complex manifolds. More significantly, we give an improved version of our formula from [BSZ2] for determining zero correlations from joint probability distributions, and we apply this formula together with results in [SZ2] to extend our limit zero correlation formulas to the case of almost-complex symplectic manifolds.

Notions of universality depend on context. In RMT, one fixes a set of matrices (for example, a group $U(N)$ or a symmetric space such as $\text{Sym}(N)$, the $N \times N$ real symmetric matrices) and endows it with certain kinds of probability measures μ_N. These measures bias towards certain types of matrices and away from others, and one may ask how the eigenvalue correlations depend on the $\{\mu_N\}$ in the large N limit. In RPT one could similarly endow spaces of polynomials of degree N with a variety of measures, and ask how correlations between zeros depend on them in the large N limit. However, the version of universality which concerns us in this article and in [BSZ1; BSZ2] lies in another direction. We are interested in very general notions of *polynomial* that arise in geometry, and in how the statistics of zeros depends on the geometric setting in which these polynomials live. We will always endow our generalized polynomials with Gaussian measures (or with essentially equivalent spherical measures).

The generalized polynomials studied in [SZ1; BSZ1; BSZ2] were holomorphic sections $H^0(M, L^N)$ of powers of a positive line bundle $L \to M$ over a compact Kähler manifold (M, ω) of a given dimension m. Such sections form the Hilbert space of quantum wave functions which quantize (M, ω) in the sense of geometric quantization [At; Wo]. Recall that geometric quantization begins with a symplectic manifold (M, ω) such that $\frac{1}{\pi}[\omega] \in H^2(M, \mathbb{Z})$. There then exists a complex Hermitian line bundle $(L, h) \to M$ and a Hermitian connection ∇ with curvature ω. To obtain a Hilbert space of sections, one needs additionally to fix a *polarization* of M, i.e. a Lagrangian sub-bundle $\mathcal{L}$ of TM, and we define *polarized sections* to be those satisfying $\nabla_v s = 0$ when v is tangent to $\mathcal{L}$. In the Kähler case, one takes $\mathcal{L} = T^{1,0}M$, the holomorphic sub-bundle. Thus, polarized sections are holomorphic sections. The power N plays the role of the inverse Planck constant, so that the high power $N \to \infty$ limit is the semiclassical limit.

When $M = \mathbb{CP}^m$ and $L = \mathcal{O}(1)$ (the hyperplane section line bundle), holomorphic sections of $L^N = \mathcal{O}(N)$ are just homogeneous holomorphic polynomials of degree N. (In general, one may embed $M \subset \mathbb{CP}^d$ for some d, and then holomorphic sections $s \in H^0(M, L^N)$ may be identified with restrictions of polynomials on $\mathbb{CP}^d$ to M, for $N \gg 0$ by the Kodaira vanishing theorem.) We equip L with a Hermitian metric h and endow M with the volume form dV induced by the curvature ω of L. The pair (h, dV) determine an $\mathcal{L}^2$ norm on $H^0(M, L^N)$ and hence a Gaussian probability measure μ_N. All probabilistic notions such as expectations or correlations are with reference to this measure. The basic theme of the results of [BSZ1; BSZ2] was that in a certain scaling limit, the correlations between zeros are universal in the sense of being independent of M, L, ω, and other details of the setting.

The geometric setting was extended even further by two of the authors in [SZ2] by allowing (M, ω) to be any compact symplectic manifold with integral symplectic form, i.e. $\frac{1}{\pi}[\omega] \in H^2(M, \mathbb{Z})$. Complex line bundles with $c_1(L) = \frac{1}{\pi}[\omega]$ are known in this context as 'pre-quantum line bundles' (cf. [Wo]). It has been known for some time [BG] that there are good analogues of holomorphic sections of powers of such line bundles in this context. Interest in symplectic analogues of holomorphic line bundles and their holomorphic sections has grown recently because of Donaldson's [Do1] use of asymptotically holomorphic sections of powers of pre-quantum line bundles over symplectic manifolds in constructing embedded symplectic submanifolds, Lefschetz pencils and other constructions of an algebro-geometric nature [Do1; Do2; Au1; Au2; AK; BU1; BU2; Sik]. Given an almost complex structure J on M which is compatible with ω, we follow Boutet de Monvel and Guillemin [BG] (see also [GU] and [BU1; BU2]) in defining spaces $H_J^0(M, L^N)$ of *almost holomorphic* sections of the pre-quantum line bundle $L \to M$ with curvature ω. A Hermitian metric h on L and ω determine an $\mathcal{L}^2$ norm and hence a Gaussian measure μ_N on $H_J^0(M, L^N)$.

Our main concern is with the zeros Z_s of k-tuples $s = (s_1, \ldots, s_k)$ of holomorphic or almost-holomorphic sections. We let $|Z_s|$ denote Riemannian $(2m - 2k)$-volume on Z_s, regarded as a measure on M:

$$(|Z_s|, \varphi) = \int_{Z_s} \varphi \, d\mathrm{Vol}_{2m-2k} .$$

As in [BSZ1; BSZ2], we introduce the punctured product

$$M_n = \{(z^1, \ldots, z^n) \in \underbrace{M \times \cdots \times M}_{n} : z^p \neq z^q \text{ for } p \neq q\} \qquad (1\text{-}1)$$

and consider the product measures on M_n,

$$|Z_s|^n := \big(\underbrace{|Z_s| \times \cdots \times |Z_s|}_{n}\big).$$

The expectation $\mathbf{E}\,|Z_s|^n$ is called the *n-point zero correlation measure*. We write

$$\mathbf{E}\,|Z_s|^n = K_{nk}^N(z^1,\ldots,z^n)\,dz,$$

where dz denotes the product volume form on M_n. The generalized function $K_{nk}^N(z^1,\ldots,z^n)$ is called the *n-point zero correlation function*.

The main points are first to express these correlation measures in terms of the joint probability distribution

$$\widetilde{\mathbf{D}}_{(z^1,\ldots,z^n)}^N = \widetilde{D}^N(x^1,\ldots,x^n,\xi^1,\ldots,\xi^n;z^1,\ldots,z^n)\,dx\,d\xi$$

of the random variables $s(z^1),\ldots,s(z^n),\nabla s(z^1),\ldots,\nabla s(z^n)$, and secondly to prove that the latter has a universal scaling limit. Here dx denotes volume measure on $L_{z^1}^N \oplus \cdots \oplus L_{z^n}^N$, and $d\xi$ is volume measure on $(T_M^* \otimes L^N)_{z^1} \oplus \ldots \oplus (T_M^* \otimes L^N)_{z^n}$. For more details and precise definitions, see §4.1. As for the first point, we have the following formula for the correlation measures in terms of the joint probability distribution:

THEOREM 1.1. *The n-point zero correlation function for random almost holomorphic sections of $L^N \to M$ is given by*

$$K_{nk}^N(z) = \int d\xi\, \widetilde{D}_n^N(0,\xi,z) \prod_{p=1}^n \sqrt{\det(\xi^p \xi^{p*})}.$$

One of our main results is Theorem 4.3, which gives a general form of Theorem 1.1 with $H_J^0(M,L^N)$ replaced by a finite dimensional space of sections of an arbitrary vector bundle over a Riemannian manifold. Theorem 4.3 is a generalization of the Kac–Rice formula [Kac; Ri] (see also [BD; EK; Hal; SSm]) to higher dimensions. A special case of Theorem 4.3 was given by J. Neuheisel [Ne] in a parallel study of correlations of nodal sets (zero sets of eigenfunctions of the Laplacian) on spheres.

The correlations of course depend heavily on the geometry of the bundle. For instance, it was shown in [SZ1] that $Z_{s_N} \to \omega$ for almost every sequence $\{s_N\}$ of holomorphic sections of L^N. That is, zeros tend almost surely to congregate in highly (positively) curved regions. To find universal quantities, we scale around a point $z_0 \in M$. The most vivid case is where $k = m$ so that almost surely the simultaneous zeros of the k-tuple of sections form a discrete set. The density of zeros in a unit ball $B_1(z_0)$ around z_0 then grows like N^m, so we rescale the zeros in the $1/\sqrt{N}$ ball $B_{1/\sqrt{N}}(z_0)$ by a factor of $\sqrt{N}$ to get configurations of zeros with a constant density as $N \to \infty$. Our problem is whether the statistics of these configurations tend to a limit and whether the limit is universal. In [BSZ2, Th. 3.4] (see also [BSZ1]), it was shown that when L is a positive holomorphic line bundle over a complex manifold M, the scaled n-point correlation functions $K_{nk}^N(\frac{z^1}{\sqrt{N}},\ldots,\frac{z^1}{\sqrt{N}})$ converge in the high power limit to a universal correlation function $K_{nkm}^\infty(z^1,\ldots,z^n)$ on the punctured product $(\mathbb{C}^m)_n$ depending only on the dimension m of the manifold and the codimension k of the zero set. Our main application of Theorem 1.1 is that this universality law for the scaling limits of the zero correlation functions extends to the general symplectic case:

THEOREM 1.2. *Let L be the pre-quantum line bundle over a $2m$-dimensional compact integral symplectic manifold (M,ω). Let $z_0 \in M$ and choose complex local coordinates $\{z_j\}$ centered at z_0 so that $\omega|_{z_0} = \frac{i}{2}\sum dz_j \wedge d\bar{z}_j$ and $(\partial/\partial z_j)|_{z_0} \in T^{1,0}M$ $(1 \leq j \leq m)$. Let $\mathcal{S} = H_J^0(M, L^N)^k$ $(k \geq 1)$, and give $\mathcal{S}$ the standard Gaussian measure μ. Then*

$$\frac{1}{N^{nk}} K_{nk}^N \left(\frac{z^1}{\sqrt{N}},\ldots,\frac{z^n}{\sqrt{N}}\right) \to K_{nkm}^\infty(z^1,\ldots,z^n)$$

(weakly in $\mathcal{D}'((\mathbb{C}^m)_n)$), where $K_{nkm}^\infty(z^1,\ldots,z^n)$ is the universal scaling limit in the Kähler setting.

The proof of this result is similar to the holomorphic case [BSZ2]. Using Theorem 1.1, we reduce the scaling limit of $K_n(z)$ to that of the joint probability density $\widetilde{\mathbf{D}}_{(z^1,\ldots,z^n)}^N$. It was shown in [SZ2, Theorem 5.4] that the latter has a universal scaling limit:

$$\widetilde{\mathbf{D}}_{(z^1/\sqrt{N},\ldots,z^n/\sqrt{N})}^N \longrightarrow \mathbf{D}_{(z^1,\ldots,n^n)}^\infty, \qquad (1\text{--}2)$$

where $\mathbf{D}_{(z^1,\ldots,z^n)}^\infty$ is a universal Gaussian measure supported on the holomorphic 1-jets, and $\{z_j\}$ are the complex local coordinates of Theorem 1.2.

Let us say a few words on the proof of (1–2). Recall that a Gaussian measure on $\mathbb{R}^p$ is a measure of the form

$$\gamma_\Delta = \frac{e^{-\frac{1}{2}\langle \Delta^{-1}x,x\rangle}}{(2\pi)^{p/2}\sqrt{\det \Delta}} dx_1 \cdots dx_p, \qquad (1\text{--}3)$$

where Δ is a positive definite symmetric $p \times p$ matrix. Since $\widetilde{\mathbf{D}}_{(z^1,\ldots,z^n)}^N$ is the push-forward of a Gaussian measure, we have $\widetilde{\mathbf{D}}_{(z^1,\ldots,z^n)}^N = \gamma_{\Delta^N}$ where Δ^N is the covariance matrix of the random variables $(s(z^p), \nabla s(z^p))$. The main step in the proof in [SZ2] was to show that the covariance matrices Δ^N underlying $\widetilde{\mathbf{D}}^N$ tend in the scaling limit to a semi-positive matrix Δ^∞. To deal with singular measures, we introduced a class of generalized Gaussians whose covariance matrices are only semi-positive definite. A generalized Gaussian is simply a Gaussian supported on the subspace corresponding to the positive eigenvalues of the covariance matrix. It followed that the scaled distributions $\widetilde{\mathbf{D}}^N$ tend to a generalized Gaussian γ_{Δ^∞} 'vanishing in the $\bar{\partial}$-directions.' To prove that $\Delta^N \to \Delta^\infty$, we expressed Δ^N in terms of the *Szegő kernel* $\Pi_N(x,y)$ and its derivatives. The Szegő kernel is essentially the orthogonal projection from $\mathcal{L}^2(M, L^N) \to H_J^0(M, L^N)$. Since it is more convenient to deal with scalar kernels than sections, we pass from

$L \to M$ to the associated principal S^1 bundle $X \to M$. Sections s of L^N are then canonically identified with equivariant functions $\hat{s}$ on X transforming by $e^{iN\theta}$ under the S^1 action. The space $H^0_J(M, L^N)$ then corresponds to a space $\mathcal{H}^2_N(X)$ of equivariant functions. In the holomorphic case, these functions are CR functions; i.e., they satisfy the tangential Cauchy–Riemann equations $\bar{\partial}_b \hat{s} = 0$. In the symplectic almost-complex case they are 'almost CR' functions in a sense defined by Boutet de Monvel and Guillemin. The scalar Szegő kernels are then the orthogonal projections $\Pi_N : \mathcal{L}^2(X) \to \mathcal{H}^2_N(X)$. The main ingredient in the proof of (1–2) was the *scaling asymptotics* of the Szegő kernels $\Pi_N(x, y)$. In 'preferred' local coordinates (z, θ) on X (see §3.2), the scaling asymptotics read:

$$\Pi_N\left(z_0 + \frac{u}{\sqrt{N}}, \frac{\theta}{N}, z_0 + \frac{v}{\sqrt{N}}, \frac{\varphi}{N}\right)$$

$$\sim e^{i(\theta-\varphi)} e^{u \cdot \bar{v} - \frac{1}{2}(|u|^2 + |v|^2)} \left\{1 + \frac{1}{\sqrt{N}} p_1(u, v; z_0) + \cdots \right\}.$$

The universal limit correlation functions $K^\infty_{nkm}(z^1, \ldots, z^n)$ are described in [BSZ2] (see also [BSZ1]). They are given in terms of the level 1 Szegő kernel for the (reduced) Heisenberg group (see §2.3),

$$\Pi^{\mathbf{H}}_1(z, \theta; w, \varphi) = \frac{1}{\pi^m} e^{i(\theta-\varphi) + i \operatorname{Im}(z \cdot \bar{w}) - \frac{1}{2}|z-w|^2} = \frac{1}{\pi^m} e^{i(\theta-\varphi) + z \cdot \bar{w} - \frac{1}{2}(|z|^2 + |w|^2)},$$

and its first and second derivatives at the points $(z, w) = (z^p, z^{p'})$. Indeed, the correlation functions are universal rational functions in $z^p_q, \bar{z}^p_q, e^{z^p \cdot \bar{z}^{p'}}$, and are smooth functions on $(\mathbb{C}^m)_n$. We let

$$\widetilde{K}_{nkm}(z^1, \ldots, z^n) := (K^\infty_{1km})^{-n} K_{nkm}(z^1, \ldots, z^n)$$

denote the "normalized" n-point limit correlation function, where

$$K^\infty_{1km} = \frac{m!}{\pi^k (m-k)!}$$

is the expected volume density of the zero set. For example [BSZ1; BSZ2],

$$\widetilde{K}^\infty_{21m}(z^1, z^2) = \frac{\left[\frac{1}{2}(m^2+m) \sinh^2 t + t^2\right] \cosh t - (m+1) t \sinh t}{m^2 \sinh^3 t} + \frac{(m-1)}{2m},$$

$$\text{for } t = \frac{|z^1 - z^2|^2}{2}. \quad (1\text{--}4)$$

Formula (1–4) with $m = 1$ agrees with the scaling limit pair correlation function of Hannay [Han] (see also [BBL]) for zeros of polynomials in one complex variable, i.e. for $M = \mathbb{CP}^1$ and $L = \mathcal{O}(1)$.

The correlations are "short range" in the sense that $\widetilde{K}_{nkm}(z^1, \ldots, z^n) = 1 + O(r^4 e^{-r^2})$, where r is the minimum distance between the points z^p [BSZ2]. We show in §5.3 that in fact the "connected n-point correlations" are $o(e^{-R^2/n})$, where R is the maximum distance between the points.

2. Line Bundles on Complex Manifolds

We begin with some notation and basic properties of sections of holomorphic line bundles, their zero sets, and Szegő kernels. We also provide two examples that will serve as model cases for studying correlations of zeros of sections of line bundles in the high power limit.

2.1. Sections of Holomorphic Line Bundles. Let $L \to M$ be a holomorphic line bundle over a compact complex manifold. Thus, at each $z \in M$, $L_z \simeq \mathbb{C}$ is a complex line and locally, over a sufficiently small open set $U \subset M$, $L \simeq U \times \mathbb{C}$. For background on line bundles and other objects of complex geometry, we refer to [GH].

A key notion is that of *positive* line bundle. By definition, this means that there exists a smooth Hermitian metric h on L with positive curvature form

$$\Theta_h = -\partial\bar{\partial}\log\|e_L\|_h^2,$$

where e_L denotes a local holomorphic frame (= nonvanishing section) of L over an open set $U \subset M$, and $\|e_L\|_h = h(e_L, e_L)^{1/2}$ denotes the h-norm of e_L. A basic example is the hyperplane bundle $\mathcal{O}(1) \to \mathbb{CP}^m$, the dual of the tautological line bundle. When $m = 1$, its square is the holomorphic tangent bundle $T\mathbb{CP}^1$. Its positivity is equivalent to the positivity of the curvature of $\mathbb{CP}^1$ in the usual sense of differential geometry. Hyperbolic surfaces $\mathbf{H}^2/\Gamma$ have negatively curved tangent bundles, but their cotangent bundles $T^*(\mathbf{H}^2/\Gamma)$ are positively curved. In the case of complex tori $\mathbb{C}/\Lambda$ (where $\Lambda \subset \mathbb{C}$ is a lattice), both the tangent and cotangent bundles are flat. The positive 'pre-quantum' line bundle there is the bundle whose sections are theta functions.

Intuitively speaking, positive curvature at w creates a potential well which traps a particle near z. On the quantum level, this particle is a wave function (holomorphic section) $\Pi_N(z, w)$ which is concentrated at w. This wave function is known to mathematicians as the 'Szegő kernel', and to physicists as the 'coherent state' centered at w. The simplest (but non-compact) case is where $M = \mathbb{C}^m$ and where $\Theta_h = \sum_{j=1}^m dz_j \wedge d\bar{z}_j$ (cf. [Do1]). We note that $\Theta_h = dA$, where $A = \frac{1}{2}\sum_{j=1}^m z_j d\bar{z}_j - \bar{z}_j dz_j$ is a connection form on the trivial bundle $L = \mathbb{C}^m \times \mathbb{C} \to \mathbb{C}^m$. The associated covariant derivative on sections is given by $\bar{\partial}_A f = \bar{\partial}f + A^{0,1}f$, where $A^{0,1}$ is the $(0,1)$ component of A. Then $\bar{\partial}_A e^{-|z|^2/2} = 0$, i.e. there is a Gaussian holomorphic section concentrated at $w = 0$. As will be seen in §2.3, it is essentially the Szegő kernel of the Heisenberg group.

According to the above intuition, positive line bundles should have a plentiful supply of global holomorphic sections. Indeed, the space $H^0(M, L^N)$ of holomorphic sections of $L^N = L \otimes \cdots \otimes L$ is a complex vector space of dimension $d_N = \frac{c_1(L)}{m!}N^m + \cdots$ given by the Hilbert polynomial ([Kl; Na]; see [SSo, Lemma 7.6]). It is in part because the dimension d_N increases so rapidly with N that probabilities and correlations simplify so much as $N \to \infty$.

To define the term 'Szegő kernel' we need to define a Hilbert space structure on $H^0(M, L^N)$: We give M the Hermitian metric corresponding to the Kähler form $\omega = \frac{\sqrt{-1}}{2}\Theta_h$ and the induced Riemannian volume form

$$dV_M = \frac{1}{m!}\omega^m. \tag{2-1}$$

Since $\frac{1}{\pi}\omega$ is a de Rham representative of the Chern class $c_1(L) \in H^2(M, \mathbb{R})$, it follows from (2-1) that $\mathrm{Vol}(M) = \frac{\pi^m}{m!}c_1(L)^m$.

The metric h induces Hermitian metrics h^N on L^N given by $\|s^{\otimes N}\|_{h^N} = \|s\|_h^N$. We give $H^0(M, L^N)$ the Hermitian inner product

$$\langle s_1, s_2\rangle = \int_M h^N(s_1, s_2)\, dV_M \qquad (s_1, s_2 \in H^0(M, L^N)). \tag{2-2}$$

We first define the Szegő kernels as the orthogonal projections

$$\Pi_N : \mathcal{L}^2(M, L^N) \to H^0(M, L^N).$$

The projections Π_N can be given in terms of orthonormal bases $\{S_j^N\}$ of sections of $H^0(M, L^N)$ by

$$\Pi_N(z, w) = \sum_{j=1}^{d_N} S_j^N(z) \otimes \overline{S_j^N(w)},$$

so that

$$(\Pi_N s)(w) = \int_M h_z^N\bigl(s(z), \Pi_N(z, w)\bigr) dV_M(z), \quad s \in \mathcal{L}^2(M, L^N).$$

Since we are studying the asymptotics of the Π_N as $N \to \infty$, we find it useful to instead view the Szegő kernels as projections on the same space of functions. We show how this is accomplished below.

2.2. Lifting the Szegő Kernel. As in [BG; Ze1; SZ1; BSZ2; SZ2] and elsewhere, we analyze the $N \to \infty$ limit by lifting the analysis of holomorphic sections over M to a certain S^1 bundle $X \to M$. We let L^* denote the dual line bundle to L, and we consider the circle bundle $X = \{\lambda \in L^* : \|\lambda\|_{h^*} = 1\}$, where h^* is the norm on L^* dual to h. Let $\pi : X \to M$ denote the bundle map; if $v \in L_z$, then $\|v\|_h = |(\lambda, v)|$, $\lambda \in X_z = \pi^{-1}(z)$. Note that X is the boundary of the disc bundle $D = \{\lambda \in L^* : \rho(\lambda) > 0\}$, where $\rho(\lambda) = 1 - \|\lambda\|_{h^*}^2$. The disc bundle D is strictly pseudoconvex in L^*, since Θ_h is positive, and hence X inherits the structure of a strictly pseudoconvex CR manifold. Associated to X is the contact form $\alpha = -i\partial\rho|_X = i\bar{\partial}\rho|_X$. We also give X the volume form

$$dV_X = \frac{1}{m!}\alpha \wedge (d\alpha)^m = \alpha \wedge \pi^* dV_M. \tag{2-3}$$

The setting for our analysis of the Szegő kernel is the Hardy space $\mathcal{H}^2(X) \subset \mathcal{L}^2(X)$ of square-integrable CR functions on X, i.e., functions that are annihilated by the Cauchy–Riemann operator $\bar{\partial}_b$ (see [St, pp. 592–594]) and are $\mathcal{L}^2$

with respect to the inner product

$$\langle F_1, F_2 \rangle = \frac{1}{2\pi} \int_X F_1 \overline{F_2} dV_X\,, \quad F_1, F_2 \in \mathcal{L}^2(X)\,. \tag{2-4}$$

Equivalently, $\mathcal{H}^2(X)$ is the space of boundary values of holomorphic functions on D that are in $\mathcal{L}^2(X)$. We let $r_\theta x = e^{i\theta} x$ ($x \in X$) denote the S^1 action on X and denote its infinitesimal generator by $\frac{\partial}{\partial \theta}$. The S^1 action on X commutes with $\bar{\partial}_b$; hence $\mathcal{H}^2(X) = \bigoplus_{N=0}^\infty \mathcal{H}_N^2(X)$ where $\mathcal{H}_N^2(X) = \{F \in \mathcal{H}^2(X) : F(r_\theta x) = e^{iN\theta} F(x)\}$. A section s_N of L^N determines an equivariant function $\hat{s}_N$ on L^* by the rule

$$\hat{s}_N(\lambda) = \left(\lambda^{\otimes N}, s_N(z) \right)\,, \quad \lambda \in L_z^*\,, \; z \in M\,,$$

where $\lambda^{\otimes N} = \lambda \otimes \cdots \otimes \lambda$. We henceforth restrict $\hat{s}$ to X and then the equivariance property takes the form $\hat{s}_N(r_\theta x) = e^{iN\theta} \hat{s}_N(x)$. The map $s \mapsto \hat{s}$ is a unitary equivalence between $H^0(M, L^N)$ and $\mathcal{H}_N^2(X)$. (This follows from (2–3)–(2–4) and the fact that $\alpha = d\theta$ along the fibers of $\pi : X \to M$.)

We now define the (lifted) Szegő kernel to be the orthogonal projection $\Pi_N : \mathcal{L}^2(X) \to \mathcal{H}_N^2(X)$. It is defined by

$$\Pi_N F(x) = \int_X \Pi_N(x, y) F(y)\, dV_X(y)\,, \quad F \in \mathcal{L}^2(X)\,. \tag{2-5}$$

As above, it can be given as

$$\Pi_N(x, y) = \sum_{j=1}^{d_N} \widehat{S}_j^N(x) \overline{\widehat{S}_j^N(y)}\,, \tag{2-6}$$

where $S_1^N, \ldots, S_{d_N}^N$ form an orthonormal basis of $H^0(M, L^N)$. Note that although the Szegő kernel Π_N is defined on X, its absolute value is well-defined on M. In particular, on the diagonal we have

$$\Pi_N(z, z) = \Pi_N(z, \theta; z, \theta) = \sum_{j=1}^{d_N} \|S_j^N(z)\|_{h^N}^2\,.$$

2.3. Model Examples.

The Szegő kernels and their derivatives were worked out explicitly in [BSZ2] for two model cases, namely for the hyperplane section bundle over $\mathbb{CP}^m$ and for the Heisenberg bundle over $\mathbb{C}^m$, i.e. the trivial line bundle with curvature equal to the standard symplectic form on $\mathbb{C}^m$. These cases are important, since by universality, the scaling limits of correlation functions for all line bundles coincide with those of the model cases.

In fact, the two models are locally equivalent in the CR sense. In the case of $\mathbb{CP}^m$, the circle bundle X is the $2m+1$ sphere S^{2m+1}, which is the boundary of the unit ball $B^{2m+2} \subset \mathbb{C}^{m+1}$. In the case of $\mathbb{C}^m$, the circle bundle is the reduced Heisenberg group $\mathbf{H}_{\text{red}}^m$, which is a discrete quotient of the simply connected Heisenberg group $\mathbb{C}^m \times \mathbb{R}$.

We summarize here the formulas for the Szegő kernels from [BSZ2] in these model cases; for further details see [BSZ2, §1.3]. For the first example (see also [SZ1, §4.2]), $M = \mathbb{CP}^m$ and L is the hyperplane section bundle $\mathcal{O}(1)$. Sections $s \in H^0(\mathbb{CP}^m, \mathcal{O}(1))$ are linear functions on $\mathbb{C}^{m+1}$, so that the zero divisors Z_s are projective hyperplanes. The line bundle $\mathcal{O}(1)$ carries a natural metric h_{FS} given by

$$\|s\|_{h_{\mathrm{FS}}}([w]) = \frac{|(s,w)|}{|w|}, \qquad w = (w_0, \ldots, w_m) \in \mathbb{C}^{m+1},$$

for $s \in \mathbb{C}^{m+1*} \equiv H^0(\mathbb{CP}^m, \mathcal{O}(1))$, where $|w|^2 = \sum_{j=0}^m |w_j|^2$ and $[w] \in \mathbb{CP}^m$ denotes the complex line through w. The Kähler form on $\mathbb{CP}^m$ is the Fubini–Study form

$$\omega_{\mathrm{FS}} = \frac{\sqrt{-1}}{2} \Theta_{h_{\mathrm{FS}}} = \frac{\sqrt{-1}}{2} \partial \bar\partial \log |w|^2.$$

The dual bundle $L^* = \mathcal{O}(-1)$ is the affine space $\mathbb{C}^{m+1}$ with the origin blown up, and $X = S^{2m+1} \subset \mathbb{C}^{m+1}$. The N-th tensor power of $\mathcal{O}(1)$ is denoted $\mathcal{O}(N)$. An orthonormal basis for the space $H^0(\mathbb{CP}^m, \mathcal{O}(N))$ of homogeneous polynomials on $\mathbb{C}^{m+1}$ of degree N is the set of monomials:

$$s_J^N = \left[\frac{(N+m)!}{\pi^m j_0! \cdots j_m!}\right]^{\frac{1}{2}} z^J, \quad z^J = z_0^{j_0} \cdots z_m^{j_m}, \qquad J = (j_0, \ldots, j_m), \ |J| = N$$

Hence the Szegő kernel for $\mathcal{O}(N)$ is given by

$$\Pi_N(x,y) = \sum_J \frac{(N+m)!}{\pi^m j_0! \cdots j_m!} x^J \bar y^J = \frac{(N+m)!}{\pi^m N!} \langle x, y \rangle^N.$$

Note that

$$\Pi(x,y) = \sum_{N=1}^\infty \Pi_N(x,y) = \frac{m!}{\pi^m}(1 - \langle x, y \rangle)^{-(m+1)},$$

which is the classical Szegő kernel for the $(m+1)$-ball.

The second example is the linear model $\mathbb{C}^m \times \mathbb{C} \to \mathbb{C}^m$ for positive line bundles $L \to M$ over Kähler manifolds and their associated Szegő kernels. Its associated principal S^1 bundle $\mathbb{C}^m \times S^1 \to \mathbb{C}^m$, which may be identified with the boundary of the disc bundle $D \subset L^*$ in the dual line bundle, is the *reduced Heisenberg group* $\mathbf{H}^m_{\mathrm{red}}$. Let us summarize its definition and properties. We start with the usual (simply connected) Heisenberg group $\mathbf{H}^m = \mathbb{C}^m \times \mathbb{R}$ with group law

$$(\zeta, t) \cdot (\eta, s) = (\zeta + \eta, t + s + \mathrm{Im}(\zeta \cdot \bar\eta)).$$

The identity element is $(0,0)$ and $(\zeta, t)^{-1} = (-\zeta, -t)$. The Lie algebra of $\mathbf{H}_m$ is spanned by elements $Z_1, \ldots, Z_m, \bar Z_1, \ldots, \bar Z_m, T$ satisfying the canonical commutation relations $[Z_j, \bar Z_k] = -i\delta_{jk}T$ (all other brackets are zero). Below we will select such a basis of left invariant vector fields.

UNIVERSALITY AND SCALING OF ZEROS ON SYMPLECTIC MANIFOLDS 41

We can regard $\mathbf{H}^m$ as a strictly convex CR manifold which may be embedded in $\mathbb{C}^{m+1}$ as the boundary of a strictly pseudoconvex domain, namely the upper half space $\mathcal{U}^m := \{z \in \mathbb{C}^{m+1} : \operatorname{Im} z_{m+1} > \frac{1}{2}\sum_{j=1}^{m}|z_j|^2\}$. $\mathbf{H}^m$ acts simply transitively on $\partial \mathcal{U}^m$ (cf. [St], XII), and we get an identification of $\mathbf{H}^m$ with $\partial \mathcal{U}^m$ by:
$$[\zeta, t] \to (\zeta, t + i|\zeta|^2) \in \partial \mathcal{U}^m.$$

The linear model for the principal S^1 bundle is the reduced Heisenberg group $\mathbf{H}_{\text{red}}^m = \mathbf{H}^m/\{(0, 2\pi k) : k \in \mathbb{Z}\} = \mathbb{C}^m \times S^1$ with group law
$$(\zeta, e^{it}) \cdot (\eta, e^{is}) = (\zeta + \eta, e^{i[t+s+\operatorname{Im}(\zeta \cdot \bar{\eta})]}).$$

It is the principal S^1 bundle over $\mathbb{C}^m$ associated to the line bundle $L_\mathbf{H} = \mathbb{C}^m \times \mathbb{C}$. The metric on $L_\mathbf{H}$ with curvature $\Theta = \sum dz_q \wedge d\bar{z}_q$ is given by setting $h_\mathbf{H}(z) = e^{-|z|^2}$; i.e., $|f|_{h_\mathbf{H}} = |f|e^{-|z|^2/2}$. The reduced group $\mathbf{H}_{\text{red}}^m$ may be viewed as the boundary of the dual disc bundle $D \subset L_\mathbf{H}^*$ and hence is a strictly pseudoconvex CR manifold.

We then define the Hardy space $\mathcal{H}^2(\mathbf{H}_{\text{red}}^m)$ of CR holomorphic functions to be the functions in $\mathcal{L}^2(\mathbf{H}_{\text{red}}^m)$ satisfying the left-invariant Cauchy–Riemann equations $\bar{Z}_q^L f = 0$ ($1 \leq q \leq m$) on $\mathbf{H}_{\text{red}}^m$. Here, $\{\bar{Z}_q^L\}$ denotes a basis of the left-invariant anti-holomorphic vector fields on $\mathbf{H}_{\text{red}}^m$. Let us recall their definition: we first equip $\mathbf{H}_{\text{red}}^m$ with its left-invariant connection form $\alpha^L = \frac{1}{2}(\sum_q (u_q dv_q - v_q du_q) + d\theta)$ ($\zeta = u + iv$), whose curvature equals the symplectic form $\omega = \sum_q du_q \wedge dv_q$. The left-invariant (CR-) holomorphic vector fields Z_q^L and anti-holomorphic vector fields $\bar{Z}_q^L$ are the horizontal lifts of the vector fields $\frac{\partial}{\partial z_q}$ and $\frac{\partial}{\partial \bar{z}_q}$, respectively, with respect to α^L. They span the left-invariant CR structure of $\mathbf{H}_{\text{red}}^m$ and are given by
$$Z_q^L = \frac{\partial}{\partial z_q} + \frac{i}{2}\bar{z}_q \frac{\partial}{\partial \theta}, \quad \bar{Z}_q^L = \frac{\partial}{\partial \bar{z}_q} - \frac{i}{2}z_q \frac{\partial}{\partial \theta}.$$

The vector fields $\{\frac{\partial}{\partial \theta}, Z_q^L, \bar{Z}_q^L\}$ span the Lie algebra of $\mathbf{H}_{\text{red}}^m$ and satisfy the canonical commutation relations above.

For $N = 1, 2, \ldots$, we define $\mathcal{H}_N^2 \subset \mathcal{H}^2(\mathbf{H}_{\text{red}}^m)$ as the (infinite-dimensional) Hilbert space of square-integrable CR functions f such that $f \circ r_\theta = e^{iN\theta} f$ as before. The Szegő kernel $\Pi_N^\mathbf{H}(x, y)$ is the orthogonal projection to $\mathcal{H}_N^2$. It is given by
$$\Pi_N^\mathbf{H}(x, y) = \frac{1}{\pi^m} N^m e^{iN(t-s)} e^{N(\zeta \cdot \bar{\eta} - \frac{1}{2}|\zeta|^2 - \frac{1}{2}|\eta|^2)}, \quad x = (\zeta, t), \ y = (\eta, s).$$

The Szegő kernels $\Pi_N^\mathbf{H}$ are Heisenberg dilates of the level 1 kernel $\Pi_1^\mathbf{H}$:
$$\Pi_N^\mathbf{H}(x, y) = N^m \Pi_1^\mathbf{H}(\delta_{\sqrt{N}} x, \delta_{\sqrt{N}} y),$$
where the Heisenberg dilations (or scalings) δ_r are the automorphisms of $\mathbf{H}^m$
$$\delta_r(z, \theta) = (rz, r^2 \theta), \quad r \in \mathbb{R}^+.$$

(The dilation $\delta_{\sqrt{N}}$ descends to a homomorphism of $\mathbf{H}^m_{\mathrm{red}}$.)

REMARK. The group $\mathbf{H}^m_{\mathrm{red}}$ acts by left translation on $\mathcal{H}^2_1$. The generators of this representation are the right-invariant vector fields $Z^R_q, \bar{Z}^R_q$ together with $\frac{\partial}{\partial \theta}$. They are horizontal with respect to the right-invariant contact form $\alpha^R = \frac{1}{2}(\sum_q(u_q dv_q - v_q du_q)) - d\theta)$ and are given by:

$$Z^R_q = \frac{\partial}{\partial z_q} - \frac{i}{2}\bar{z}_q \frac{\partial}{\partial \theta}, \quad \bar{Z}^R_q = \frac{\partial}{\partial \bar{z}_q} + \frac{i}{2}z_q \frac{\partial}{\partial \theta}.$$

In physics terminology, Z^R_q is known as an annihilation operator and $\bar{Z}^R_q$ is a creation operator.

The representation $\mathcal{H}^2_1$ is irreducible and may be identified with the Bargmann–Fock space of entire holomorphic functions on $\mathbb{C}^n$ which are square integrable relative to $e^{-|z|^2}$. The identification goes as follows: the function $\varphi_o(z,\theta) := e^{i\theta}e^{-|z|^2/2}$ is CR-holomorphic and is also the ground state for the right invariant annihilation operator; i.e., it satisfies

$$\bar{Z}^L_q \varphi_o(z,\theta) = 0 = Z^R_q \varphi_o(z,\theta).$$

In the physics terminology, the level 1 Szegő kernel $\Pi^{\mathbf{H}}_1$, which is the left translate of φ_o by $(-w, -\varphi)$, is the coherent state associated to the phase space point w. Any element $F(z,\theta)$ of $\mathcal{H}^2_1$ may be written in the form $F(z,\theta) = f(z)\varphi_o$. Then $\bar{Z}^L_q F = (\frac{\partial}{\partial \bar{z}_q}f)\varphi_o$, so that F is CR if and only if f is holomorphic. Moreover, $F \in \mathcal{L}^2(\mathbf{H}^m_{\mathrm{red}})$ if and only if f is square integrable relative to $e^{-|z|^2}$.

3. Almost-Complex Symplectic Manifolds

In [SZ2], the study of the Szegő kernel was extended to almost-complex symplectic manifolds, and parametrices and resulting off-diagonal asymptotics for the Szegő kernel were obtained in this general setting. We now summarize the basic geometric and analytic constructions of [SZ2] for the almost-complex symplectic case.

We denote by (M,ω) a compact symplectic manifold such that $[\frac{1}{\pi}\omega]$ is an integral cohomology class. We also fix a compatible almost complex structure J satisfying $\omega(v, Jv) > 0$. We denote by $T^{1,0}M$, respectively $T^{0,1}M$, the holomorphic (respectively anti-holomorphic) sub-bundles of the complex tangent bundle, i.e. $J = i$ on $T^{1,0}$ and $J = -i$ on $T^{0,1}$. It is well known (see [Wo, Prop. 8.3.1]) that there exists a Hermitian line bundle $(L, h) \to M$ and a metric connection ∇ on L whose curvature Θ_L satisfies $\frac{i}{2}\Theta_L = \omega$. The 'quantization' of (M,ω) at Planck constant $1/N$ should be a Hilbert space of polarized sections of the N-th tensor power L^N of L ([GS, p. 266]). In the complex case, polarized sections are simply holomorphic sections. The notion of polarized sections is problematic in the non-complex symplectic setting, since the Lagrangian subspaces $T^{1,0}M$ defining the complex polarization are not integrable and there usually are no 'holomorphic' sections. A subtle but compelling replacement for the notion of polarized section has been proposed by Boutet de Monvel and Guillemin [BG], and it is this notion which was used in [SZ2].

To define these polarized sections, we work as above on the associated principal S^1 bundle $X \to M$ with $X = \{v \in L^* : |v|_h = 1\}$. We let α be the connection 1-form on X given by ∇; we then have $\frac{1}{\pi}d\alpha = \pi^*\omega$, and thus α is a contact form on X, i.e., $\alpha \wedge (d\alpha)^m$ is a volume form on X. In the complex case, X was a CR manifold. In the general almost-complex symplectic case it is an almost CR manifold. The almost CR structure is defined as follows: The kernel of α defines a horizontal hyperplane bundle $H \subset TX$. Using the projection $\pi : X \to M$, we may pull back J to an almost complex structure on H. We denote by $H^{1,0}$, respectively $H^{0,1}$ the eigenspaces of eigenvalue i, respectively $-i$, of J. The splitting $TX = H^{1,0} \oplus H^{0,1} \oplus \mathbb{C}\frac{\partial}{\partial\theta}$ defines the almost CR structure on TX. We also define local orthonormal frames $Z_1, \ldots, Z_n$ of $H^{1,0}$, respectively $\bar{Z}_1, \ldots, \bar{Z}_m$ of $H^{0,1}$, and dual orthonormal coframes $\vartheta_1, \ldots, \vartheta_m$, respectively $\bar{\vartheta}_1, \ldots, \bar{\vartheta}_m$. On the manifold X we have $d = \partial_b + \bar{\partial}_b + \frac{\partial}{\partial\theta} \otimes \alpha$, where $\partial_b = \sum_{j=1}^m \vartheta_j \otimes Z_j$ and $\bar{\partial}_b = \sum_{j=1}^m \bar{\vartheta}_j \otimes \bar{Z}_j$. Note that for an $\mathcal{L}^2$ section s^N of L^N, we have

$$(\nabla_{L^N} s^N)\hat{} = d^h \hat{s}^N,$$

where $d^h = \partial_b + \bar{\partial}_b$ is the horizontal derivative on X.

3.1. The $\bar{D}$ Complex and Szegő Kernels.
In the complex case, a holomorphic section s of L^N lifts to a function $\hat{s} \in \mathcal{L}^2_N(X)$ satisfying $\bar{\partial}_b \hat{s} = 0$. The operator $\bar{\partial}_b$ extends to a complex satisfying $\bar{\partial}_b^2 = 0$, which is a necessary and sufficient condition for having a maximal family of CR holomorphic coordinates. In the non-integrable case $\bar{\partial}_b^2 \neq 0$, and there may be no solutions of $\bar{\partial}_b f = 0$. To define polarized sections and their equivariant lifts, Boutet de Monvel and Guillemin [Bou; BG] defined a complex $\bar{D}_j$, which is a good replacement for $\bar{\partial}_b$ in the non-integrable case. Their main result is:

THEOREM 3.1 [BG, Lemma 14.11 and Theorem A 5.9]. *There exists an S^1-invariant complex of first order pseudodifferential operators $\bar{D}_j$ over X*

$$0 \to C^\infty(\Lambda_b^{0,0}) \xrightarrow{\bar{D}_0} C^\infty(\Lambda_b^{0,1}) \xrightarrow{\bar{D}_1} \cdots \xrightarrow{\bar{D}_{m-1}} C^\infty(\Lambda_b^{0,m}) \to 0,$$

where $\Lambda_b^{0,j} = \Lambda^j(H^{0,1}X)^*$, such that:

(i) $\sigma(\bar{D}_j) = \sigma(\bar{\partial}_b)$ *to second order along* $\Sigma := \{(x, r\alpha_x) : x \in X, r > 0\} \subset T^*X$;
(ii) *The orthogonal projector* $\Pi : \mathcal{L}^2(X) \to \mathcal{H}^2(X)$ *onto the kernel of $\bar{D}_0$ is a complex Fourier integral operator which is microlocally equivalent to the Cauchy–Szegő projector of the holomorphic case;*
(iii) $(\bar{D}_0, \frac{\partial}{\partial\theta})$ *is jointly elliptic.*

We refer to the kernel $\mathcal{H}^2(X) = \ker \bar{D}_0 \cap \mathcal{L}^2(X)$ as the Hardy space of square-integrable 'almost CR functions' on X. The $\mathcal{L}^2$ norm is with respect to the inner product (2–4) as in the holomorphic case. Since the S^1 action on X commutes with $\bar{D}_0$, we have as before the decomposition $\mathcal{H}^2(X) = \bigoplus_{N=0}^{\infty} \mathcal{H}_N^2(X)$, where $\mathcal{H}_N^2(X)$ denotes the almost CR functions on X that transform by the factor $e^{iN\theta}$ under the action r_θ. By property (iii) above, they are smooth functions. We denote by $H_J^0(M, L^N)$ the space of sections corresponding to $\mathcal{H}_N^2(X)$ under the map $s \mapsto \hat{s}$. Elements of $H_J^0(M, L^N)$ are the 'almost holomorphic sections' of L^N. (Note that products of almost holomorphic sections are not necessarily almost holomorphic.) We henceforth identify $H_J^0(M, L^N)$ with $\mathcal{H}_N^2(X)$. By the Riemann–Roch formula of [BG, Lemma 14.14], the dimension of $H_J^0(M, L^N)$ (or $\mathcal{H}_N^2(X)$) is given by $d_N = \frac{c_1(L)}{m!} N^m + \cdots$ (for N sufficiently large), as before. (The estimate $d_N \sim \frac{c_1(L)}{m!} N^m$ also follows from [SZ2, §4.2].)

As before, we let $\Pi_N : \mathcal{L}^2(X) \to \mathcal{H}_N^2(X)$ denote the orthogonal projection. The level N Szegő kernel $\Pi_N(x, y)$ is given as in the holomorphic case by (2–5) or (2–6), using an orthonormal basis $S_1^N, \ldots, S_{d_N}^N$ of $H_J^0(N, L^N) \equiv \mathcal{H}_N^2(X)$.

3.2. Scaling Limit of the Szegő Kernel.

Our analysis is based on the near-diagonal scaling asymptotics of the Szegő kernel from [SZ2]. These asymptotics are given in terms of the Heisenberg dilations $\delta_{\sqrt{N}}$, using local 'Heisenberg coordinates' at a point $x_0 \in X$. These coordinates are given in terms of *preferred coordinates* at $P_0 = \pi(x_0)$ and a *preferred frame* at P_0. A coordinate system $(z_1, \ldots, z_m)$ on a neighborhood U of P_0 is said to be preferred if

$$(g - i\omega)|_{P_0} = \sum_{j=1}^{m} dz_j \otimes d\bar{z}_j \big|_0 .$$

Here g denotes the Riemannian metric $g(v, w) := \omega(v, Jw)$ induced by the symplectic form ω. Preferred coordinates satisfy the following three (redundant) conditions:

(i) $\partial/\partial z_j|_{P_0} \in T^{1,0}(M)$, for $1 \le j \le m$,
(ii) $\omega(P_0) = \omega_0$,
(iii) $g(P_0) = g_0$,

where ω_0 is the standard symplectic form and g_0 is the Euclidean metric:

$$\omega_0 = \frac{i}{2} \sum_{j=1}^{m} dz_j \wedge d\bar{z}_j = \sum_{j=1}^{m} (dx_j \otimes dy_j - dy_j \otimes dx_j), \quad g_0 = \sum_{j=1}^{m} (dx_j \otimes dx_j + dy_j \otimes dy_j).$$

A *preferred frame* for $L \to M$ at P_0 is a local frame (=nonvanishing section) e_L on U such that

(i) $\|e_L\|_{P_0} = 1$;
(ii) $\nabla e_L|_{P_0} = 0$;
(iii) $\nabla^2 e_L|_{P_0} = -(g + i\omega) \otimes e_L|_{P_0} \in T_M^* \otimes T_M^* \otimes L.$

A preferred frame can be constructed by multiplying an arbitrary frame by a function with specified 2-jet at P_0; any two such frames necessarily agree to third order at P_0.

DEFINITION. A *Heisenberg coordinate chart* at a point x_0 in the principal bundle X is a coordinate chart $\rho : U \approx V$ with $0 \in U \subset \mathbb{C}^m \times \mathbb{R}$ and $\rho(0) = x_0 \in V \subset X$ of the form
$$\rho(z_1, \ldots, z_m, \theta) = e^{i\theta} h(z)^{1/2} e_L^*(z),$$
where e_L is a preferred local frame for $L \to M$ at $P_0 = \pi(x_0)$, and $(z_1, \ldots, z_m)$ are preferred coordinates centered at P_0. We require that P_0 have coordinates $(0, \ldots, 0)$ and $e_L^*(P_0) = x_0$.

The following near-diagonal asymptotics of the Szegő kernel is the key analytical result on which our analysis of the scaling limit for correlations of zeros is based.

THEOREM 3.2 [SZ2, Theorem 2.3]. *Let $P_0 \in M$ and choose a Heisenberg coordinate chart about P_0. Then*

$$N^{-m} \Pi_N \left(\frac{u}{\sqrt{N}}, \frac{\theta}{N}; \frac{v}{\sqrt{N}}, \frac{\varphi}{N} \right)$$
$$= \Pi_1^{\mathbf{H}}(u, \theta; v, \varphi) \left[1 + \sum_{r=1}^{K} N^{-r/2} b_r(P_0, u, v) + N^{-(K+1)/2} R_K(P_0, u, v, N) \right],$$

where $\|R_K(z_0, u, v, N)\|_{\mathcal{C}^j(\{|u|\leq\rho,\ |v|\leq\rho\})} \leq C_{K,j,\rho}$ *for* $j \geq 0$, $\rho > 0$ *and* $C_{K,j,\rho}$ *is independent of the point z_0 and choice of coordinates.*

This asymptotic formula has several applications to symplectic geometry, in addition to our result on zero correlations. For example, Theorem 3.2 is used in [SZ2] to obtain symplectic versions of the following results in complex geometry:

(i) the asymptotic expansion theorem of [Ze1],
(ii) the Tian almost isometry theorem [Ti],
(iii) the Kodaira embedding theorem (see [GH] or [SSo]).

The symplectic forms of these theorems are based on the symplectic Kodaira maps $\Phi_N : M \to PH_J^0(M, L^N)^*$, which are defined as in the holomorphic case by $\Phi_N(z) = \{s^N : s^N(z) = 0\}$. Equivalently, we choose an orthonormal basis $S_1^N, \ldots, S_{d_N}^N$ of $H_J^0(M, L^N)$ and write

$$\Phi_N : M \to \mathbb{CP}^{d_N - 1}, \qquad \Phi_N(z) = \left(S_1^N(z) : \ldots : S_{d_N}^N(z) \right).$$

We now state the symplectic generalizations of the above three theorems:

THEOREM 3.3 [SZ2, Theorems 3.1–3.2]. *Let $L \to (M, \omega)$ be the pre-quantum line bundle over a $2m$-dimensional symplectic manifold, and let $\{\Phi_N\}$ be its Kodaira maps. Then:*

(i) *There exists a complete asymptotic expansion:*
$$\Pi_N(z,z) = a_0 N^m + a_1(z) N^{m-1} + a_2(z) N^{m-2} + \cdots$$
for certain smooth coefficients $a_j(z)$ with $a_0 = \pi^{-m}$. Hence, the maps Φ_N are well-defined for $N \gg 0$.

(ii) *Let ω_{FS} denote the Fubini–Study form on $\mathbb{CP}^{d_N-1}$. Then*
$$\left\| \frac{1}{N} \Phi_N^*(\omega_{FS}) - \omega \right\|_{\mathcal{C}^k} = O\left(\frac{1}{N}\right)$$
for any k.

(iii) *For N sufficiently large, Φ_N is an embedding.*

For proofs we refer to [SZ2]. (See also [BU2] for a proof of a similar Kodaira embedding theorem.)

4. Correlations of Zeros

In §5, we shall use Theorem 3.2 and the methods of [BSZ2] to extend the results of [BSZ1; BSZ2] on the universality of the scaling limit of the n-point zero correlations to the case of almost complex symplectic manifolds. The basis for our argument is Theorem 2.2 from [BSZ2], which generalizes a formula of Kac [Kac] and Rice [Ri] for zeros of functions on $\mathbb{R}^1$, and of [Hal] for zeros of (real) Gaussian vector fields (see also [BD; EK; Ne; SSm]). However, we shall need to consider the case where the joint probability distributions are singular, and hence we give below a complete proof of a more general result (Theorem 4.3) on the correlations of zeros of sections of $\mathcal{C}^\infty$ vector bundles.

4.1. General Formula for Zero Correlations. For our general setting, we let (V, h) be a $\mathcal{C}^\infty$ (real) vector bundle over an oriented $\mathcal{C}^\infty$ Riemannian manifold (M, g). (Here, h denotes a $\mathcal{C}^\infty$ metric on V.) Suppose that $\mathcal{S}$ is a finite dimensional subspace of the space $\mathcal{C}^\infty(M, V)$ of global $\mathcal{C}^\infty$ sections of V, and let $d\mu$ be a probability measure on $\mathcal{S}$ given by a semi-positive $\mathcal{C}^0$ 'rapidly decaying' volume form. We say that a $\mathcal{C}^0$ volume form $\psi dx_1 \wedge \cdots dx_d$ on $\mathbb{R}^d$ is *rapidly decaying* if $\psi(x) = o(\|x\|^{-N})$ for all $N \in \mathbb{Z}^+$. (In this paper, we are primarily interested in the case where $d\mu$ is a Gaussian measure.) The purpose of this section is to study the zero set Z_s of a random section $s \in \mathcal{S}$ and to obtain formulas for the expected value and n-point correlations of the volume measure $|Z_s|$. We shall later apply our results to the case where $V = L^N \oplus \cdots \oplus L^N$, for a complex line bundle L over a compact almost complex symplectic manifold M and where $\mathcal{S} = \mathcal{H}_N^2 \oplus \cdots \oplus \mathcal{H}_N^2$. (Recall that $\mathcal{H}_N^2$ is the space of almost holomorphic sections of L^N.) Then the zero sets Z_s are the simultaneous zeros of (random) k-tuples of almost holomorphic sections.

Our formulation involving general vector bundles also allows us to reduce the study of n-point correlations to the case $n = 1$, i.e., to expected densities (or volumes) of zero sets. We first describe the formula (Theorem 4.2) for this expected zero density. This formula is given in terms of the 'joint probability density,' which is a measure on the space $J^1(M,V)$ of 1-jets of sections of V.

Recall that we have the exact sequence of vector bundles

$$0 \longrightarrow T_M^* \otimes V \xrightarrow{\iota} J^1(M,V) \xrightarrow{\pi_V} V \to 0. \tag{4-1}$$

We let

$$\mathcal{E}: M \times \mathcal{S} \to V, \quad \mathcal{E}(z,s) = s(z)$$

denote the evaluation map, and we say that $\mathcal{S}$ *spans* V if $\mathcal{E}$ is surjective, i.e., if $\{s(z): s \in \mathcal{S}\}$ spans V_z for all $z \in M$. We are mainly interested in the jet map

$$\mathcal{J}: M \times \mathcal{S} \to J^1(M,V), \quad \mathcal{J}(z,s) = J_z^1 s = \text{ the 1-jet of } s \text{ at } z.$$

Note that $\mathcal{E} = \pi_V \circ \mathcal{J}$.

Note that a *measure* on an N-dimensional manifold Y is a current $\nu \in \mathcal{D}^0(Y)' = \mathcal{D}'^N(Y)$ of order 0. We can write $\nu = f d\text{Vol}_Y$, where $f \in \mathcal{D}'^0(Y)$. (Recall that $\mathcal{D}^p(Y)$ denotes the space of compactly supported $\mathcal{C}^\infty$ p-forms on Y, and $\mathcal{D}'^p(Y) = \mathcal{D}^{N-p}(Y)'$.) Some authors refer to f as a measure, but to keep the distinction, we shall call elements of $\mathcal{D}'^0(Y)$ *generalized functions*.

To describe the induced volume forms on the total spaces of the bundles in (4-1), we write $g(z) = \sum g_{qq'}(z) du_q \otimes du_{q'}$, $h_{jj'} = h(e_j, e_{j'})$, where $\{u_1, \ldots, u_m\}$ are local coordinates in M and $\{e_1, \ldots, e_k\}$ is a local frame in V (here $m = \dim M$, $k = \text{rank } V$). We let $G = \det(g_{qq'})$, $H = \det(h_{jj'})$. We further let $dz = \sqrt{G} du_1 \wedge \cdots \wedge du_m$ denote Riemannian volume in M, and we write

$$x = \sum_j x_j e_j(z) \in V_z, \quad dx = \sqrt{H(z)} dx_1 \wedge \cdots \wedge dx_k,$$

$$\xi = \sum_{j,q} \xi_{jq} du_q \otimes e_j|_z \in (T_M^* \otimes V)_z, \quad d\xi = G(z)^{-k/2} H(z)^{m/2} \prod_{j,q} d\xi_{jq}.$$

The induced volume measures on V and $T_M^* \otimes V$ are given by $dx\, dz$ and $d\xi\, dz$ respectively. We give V a connection that preserves h; its covariant derivative provides a splitting $\nabla: J^1(M,V) \to T_M^* \otimes V$ of (4-1), and hence $dx\, d\xi\, dz$ provides a volume form on $J^1(M,V)$.

DEFINITION. The *1-jet density* of μ is the measure

$$\mathbf{D} := \mathcal{J}_*(dz \times \mu)$$

on the space $J^1(M,V)$ of 1-jets. We write

$$\mathbf{D} = D(x, \xi, z)\, dx\, d\xi\, dz \quad D(x, \xi, z) \in \mathcal{D}'^0(J^1(M,V)).$$

We let ρ_ε denote a $\mathcal{C}^\infty$ 'approximate identity' on V of the form

$$\rho_\varepsilon(v) = \varepsilon^{-k}\rho(\varepsilon^{-1}v),$$

with

$$\rho \in \mathcal{C}^\infty(V), \quad \int_{V_z} \rho(x,z)\,dx = 1, \quad \rho \geq 0, \quad \rho(v) = 0 \text{ for } \|v\| \geq 1.$$

We let $\tilde{\rho}_\varepsilon \in \mathcal{C}^\infty(J^1(M,V))$ be given by

$$\tilde{\rho}_\varepsilon(x,\xi,z) = \rho_\varepsilon(x,z).$$

or formally, $\tilde{\rho}_\varepsilon = \rho_\varepsilon \circ \pi_V$.

LEMMA 4.1. *Suppose that $\mathcal{E}$ spans V. Then there exists a unique positive measure $\mathbf{D}^0$ on $T_M^* \otimes V$ such that*

$$\iota_* \mathbf{D}^0 = \lim_{\varepsilon \to 0} \tilde{\rho}_\varepsilon \mathbf{D}.$$

Moreover, $\mathbf{D}^0$ is independent of the choice of local frame $\{e_j\}$, connection ∇, and approximate identity ρ_ε.

PROOF. The surjectivity of $\mathcal{E} = \pi_V \circ \mathcal{J}$ guarantees that the normal bundle N_ι is disjoint from the wave front set of $D(x,\xi,z)$ and hence $\iota^* D(x,\xi,z)$ is well-defined (see [Hö, Th. 8.2.4]). Thus we can define

$$\mathbf{D}^0 := \iota^* D(x,\xi,z)\,d\xi\,dz. \tag{4-2}$$

To verify the equation of the lemma, it suffices by the continuity of ι^* to consider the case where $D(x,\xi,z) \in \mathcal{C}^\infty$. In this case, $\mathbf{D}^0 = D(0,\xi,z)\,d\xi\,dz$, and hence

$$\tilde{\rho}_\varepsilon \mathbf{D} \to \delta_0(x) D(0,\xi,z)\,dx\,d\xi\,dz = \iota_*\bigl(D(0,\xi,z)\,d\xi\,dz\bigr) = \iota_* \mathbf{D}^0.$$

Since dx and $d\xi$ are intrinsic volume forms, it follows that $\mathbf{D}^0$ is independent of the choice of local frame $\{e_j\}$ (and local coordinates). To show that $D(0,\xi,z)$ does not depend on the choice of connection on V, write $s = \sum x_j e_j$, $\nabla s = \sum \xi_{jq}\,dz_q \otimes e_j$, $\xi_{jq} = \frac{\partial x_j}{\partial z_q} + \sum_k x_k \theta_{jq}^k$. Then if we consider the flat connection $\nabla' s = \sum \xi'_{jq} dz_q \otimes e_j$, $\xi'_{jq} = \frac{\partial x_j}{\partial z_q}$, we have

$$\frac{\partial(\xi_{jq}, x_j)}{\partial(\xi'_{jq}, x_j)} = 1.$$

Hence, $dx\,d\xi' = dx\,d\xi$ so that $D'(0,\xi,z) = D(0,\xi,z)$. □

We note that

$$(J_{z_0}^1)_* \mu = D(x,\xi,z_0)\,dx\,d\xi,$$

so that $D(x,\xi,z_0)\,dx\,d\xi$ is the joint probability distribution of the random variables $X_j^{z_0}$, $\Xi_{jq}^{z_0}$ on $\mathcal{S}$ given by

$$X_j^{z_0}(s) = x_j(z_0), \quad \Xi_{jq}^{z_0}(s) = \xi_{jq}(z_0) \quad (1 \leq j \leq k,\ 1 \leq q \leq m).$$

UNIVERSALITY AND SCALING OF ZEROS ON SYMPLECTIC MANIFOLDS 49

This is a special case of the *n-point joint probability distribution* defined below.

For a vector-valued 1-form $\xi \in T^*_{M,z} \otimes V_z = \text{Hom}(T_{M,z}, V_z)$, we let $\xi^* \in \text{Hom}(V_z, T_{M,z})$ denote the adjoint to ξ (i.e., $\langle \xi^* v, t \rangle = \langle v, \xi t \rangle$). We consider the endomorphism $\xi\xi^* \in \text{Hom}(V_z, V_z)$, and we write

$$|||\xi||| = \sqrt{\det(\xi\xi^*)}.$$

(Note that $||| \cdot |||$ is not a norm.) In terms of a local frame $\{e_j\}$,

$$|||\xi||| = \sqrt{H} \, \|\xi_1 \wedge \ldots \wedge \xi_k\|, \quad \xi = \sum_j \xi_j \otimes e_j. \quad (4\text{--}3)$$

To verify (4–3), write

$$\xi_j = \sum_{q=1}^m \xi_{jq} du_q;$$

then

$$\xi^* = \sum_{j,q} \xi^*_{jq} \frac{\partial}{\partial u_q} \otimes e^*_j, \quad \xi^*_{jq} = \sum_{j',q'} h_{jj'} \gamma_{q'q} \bar{\xi}_{j'q'},$$

where $(\gamma_{qq'}) = (g_{qq'})^{-1}$; hence we have

$$\xi\xi^* = \sum_{j,j',j'',q,q'} h_{j'j''} \xi_{jq} \gamma_{q'q} \bar{\xi}_{j''q'} \, e_j \otimes e^*_{j'}.$$

Its determinant is given by

$$\det(\xi\xi^*) = H \det \left(\sum_{q,q'} \xi_{jq} \gamma_{q'q} \bar{\xi}_{j'q'} \right)_{1 \leq j,j' \leq k} = H \det \langle \xi_j, \xi_{j'} \rangle = H \, \|\xi_1 \wedge \ldots \wedge \xi_k\|^2,$$

which gives (4–3).

Let us assume that $\mathcal{S}$ spans V. Then the incidence set $I := \{(z,s) \in M \times \mathcal{S} : s(z) = 0\}$ is a smooth submanifold and hence by Sard's theorem applied to the projection $I \to \mathcal{S}$, the zero set

$$Z_s = \{z \in M : z(s) = 0\}$$

is a smooth $(m-k)$-dimensional submanifold of M for almost all s. (In the holomorphic case, this is called Bertini's Theorem.) We let $|Z_s|$ denote Riemannian $(m-k)$-volume on Z_s, regarded as a measure on M:

$$(|Z_s|, \varphi) = \int_{Z_s} \varphi \, d\text{Vol}_{m-k} \quad \text{for a.a. } s \in \mathcal{S}.$$

Its expected value is the positive measure $\mathbf{E}\,|Z_s|$ given by

$$(\mathbf{E}\,|Z_s|, \varphi) = \mathbf{E}\,(|Z_s|, \varphi) = \int_{\mathcal{S}} d\mu(s) \int_{Z_s} \varphi \, d\text{Vol}_{m-k} \leq +\infty,$$

for $\varphi \in \mathcal{C}^0(M)$ and $\varphi \geq 0$. (Recall that $\mathbf{E}$ denotes expectation.) In fact the following general density formula tells us that $(\mathbf{E}|Z_s|, \varphi) < +\infty$ if the test function φ has compact support.

THEOREM 4.2. *Let $M, V, \mathcal{S}, d\mu$ be as above, and suppose that $\mathcal{S}$ spans V. Then*

$$\mathbf{E}|Z_s| = \pi_*\big(\sqrt{\det(\xi\xi^*)}\,\mathbf{D}^0\big) \in \mathcal{D}'^m(M), \qquad (4\text{--}4)$$

where $\pi : T_M^* \otimes V \to M$ is the projection.

Note that although $\mathbf{D}^0$ depends on the metric h on V, the measure $\sqrt{\det(\xi\xi^*)}\mathbf{D}^0$ is independent of h. In the case where $D(x,\xi,z) \in \mathcal{C}^0$, (4–4) becomes

$$\mathbf{E}|Z_s| = K_1(z)\,dz, \quad K_1(z) = \int D(0,\xi,z)\sqrt{\det(\xi\xi^*)}\,d\xi. \qquad (4\text{--}5)$$

Before proceeding further, we first give a heuristic explanation of (4–5). Suppose that $D \in \mathcal{C}^0$ and fix a point $z_0 \in M$. Let us consider the case where $\operatorname{rank} V = \dim M = m$ so that the zeros are discrete. Then the probability of finding a zero in a small ball $\mathbb{B}_r = \mathbb{B}_r(z_0)$ of radius r about z_0 is approximately $K_1(z_0)\operatorname{Vol}(\mathbb{B}_r)$. If the radius r is very small, we can suppose that the sections $s \in \mathcal{S}$ are approximately linear:

$$s(z) \approx X^{z_0} + \Xi^{z_0} \cdot (z - z_0), \qquad (4\text{--}6)$$

where we have written s in terms of a local frame for V and local coordinates in M. Here, $X^{z_0} = X^{z_0}(s) = \big(X_j^{z_0}(s)\big)$, respectively $\Xi^{z_0} = \Xi^{z_0}(s) = \big(\Xi_{jq}^{z_0}(s)\big)$, is a vector-valued, respectively matrix-valued, random variable on $\mathcal{S}$. Then the probability that the linearized section s given by (4–6) has a zero in $\mathbb{B}_r$ is given by

$$\mu\{s \in \mathcal{S} : X^{z_0} \in \Xi^{z_0}(\mathbb{B}_r)\} = \int_{\mathbb{R}^{m^2}} \int_{\xi(\mathbb{B}_r)} D(x,\xi,z_0)\,dx\,d\xi$$

$$\approx \int \operatorname{Vol}(\xi(\mathbb{B}_r)) D(0,\xi,z_0)\,d\xi.$$

Since $\operatorname{Vol}(\xi(\mathbb{B}_r)) = |||\xi|||\operatorname{Vol}(\mathbb{B}_r)$, we have

$$K_1(z_0) \approx \frac{\mu\{s \in \mathcal{S} : X^{z_0} \in \Xi^{z_0}(\mathbb{B}_r)\}}{\operatorname{Vol}(\mathbb{B}_r)} \approx \int D(0,\xi,z_0)|||\xi|||\,d\xi.$$

The linear approximation (4–6) leads to a similar explanation in the case where $\operatorname{rank} V < \dim M$; we leave this to the reader.

Before embarking on the proof of Theorem 4.2, we show how the theorem provides a generalization of Theorem 1.1 on the correlations between zeros. Let us first review the definition of these correlations.

DEFINITION. Let $M, V, \mathcal{S}, d\mu$ be as above, and suppose that $\mathcal{S}$ spans V. Let M_n denote the punctured product (1–1). The *n-point zero correlation measure* is the expectation $\mathbf{E}|Z_s|^n$, where

$$|Z_s|^n = \big(\underbrace{|Z_s| \times \cdots \times |Z_s|}_{n}\big),$$

which is a well-defined measure on M_n for almost all $s \in \mathcal{S}$. We write

$$\mathbf{E}|Z_s|^n = K_n(z^1, \ldots, z^n)\, dz.$$

The generalized function $K_n(z^1, \ldots, z^n)$ is called the *n-point zero correlation function*.

We suppose $n \geq 2$ and write

$$\tilde{s}(z) = (s(z^1), \ldots, s(z^n))$$

for $(z^1, \ldots, z^n) \in M_n$, regarded as a section of the vector bundle

$$V_n := \bigoplus_{p=1}^{n} \pi_p^* V \longrightarrow M_n,$$

where $\pi_p : M_n \to M$ denotes the projection onto the p-th factor. We then have the evaluation map

$$\mathcal{E}_n : M_n \times \mathcal{S} \to V, \quad \mathcal{E}_n(z, s) = \tilde{s}(z),$$

and the jet map

$$\mathcal{J}_n : M_n \times \mathcal{S} \to J^1(M_n, V_n), \quad \mathcal{J}(z,s) = J_z^1 \tilde{s} = (J_{z^1}^1 s, \ldots, J_{z^n}^1 s).$$

We also write

$$x = (x^1, \ldots, x^n) \in V_n,$$
$$\xi = (\xi^1, \ldots, \xi^n) \in (T_M^* \otimes V)_{z^1} \oplus \cdots \oplus (T_M^* \otimes V)_{z^n} \subset (T_{M_n}^* \otimes V_n)_z,$$
$$dx = dx^1 \cdots dx^n, \quad d\xi = d\xi^1 \cdots d\xi^n, \quad dz = dz^1 \cdots dz^n.$$

DEFINITION. The *n-point density* at $(z^1, \ldots, z^n) \in M_n$ is the probability measure

$$\mathbf{D}_n := D_n(x, \xi, z)\, dx\, d\xi\, dz = \mathcal{J}_{n*}(dz \times \mu)$$

on the space $J^1(M_n, V_n)$. Note that this measure is supported on the sub-bundle

$$\pi_1^*(T_M^* \otimes V) \oplus \ldots \oplus \pi_n^*(T_M^* \otimes V) \subset T_{M_n}^* \otimes V_n.$$

The *(n-point) joint probability distribution* at $(z^1, \ldots, z^n)$ is the joint probability distribution $D_n(x, \xi, z)\, dx\, d\xi = (J_z^1)_* \mu$ of the (complex) random variables

$$X_{jp}^z(s) := x_j(z^p), \quad \Xi_{jpq}^z(s) := \xi_{jq}(z^p) \qquad (1 \leq j \leq k,\ 1 \leq p \leq n,\ 1 \leq q \leq m).$$

If the evaluation map $\mathcal{E}_n$ is surjective, we also write as before

$$\mathbf{D}_n^0 = \iota^* D(x,\xi,z)\,d\xi\,dz,$$

so that

$$\iota_* \mathbf{D}_n^0 = \lim_{\varepsilon \to 0} \tilde{\rho}_\varepsilon^n \mathbf{D}_n.$$

Thus, Theorem 4.2 applied to $V_n \to M_n$ yields our general formula for the n-point correlations of zeros:

THEOREM 4.3. *Let $V \to M$ be a $\mathcal{C}^\infty$ vector bundle over an oriented Riemannian manifold. Consider the ensemble $(\mathcal{S}, \mu)$, where $\mathcal{S}$ is a finite-dimensional subspace of $\mathcal{C}^\infty(M, V)$ and μ is given by a $\mathcal{C}^0$ rapidly decaying volume form on $\mathcal{S}$. Suppose that $\mathcal{S}$ spans V_n, where n is a positive integer. Then*

$$\mathbf{E}\,|Z_s|^n = \pi_* \left(\sqrt{\prod_{p=1}^n \det(\xi^p \xi^{p*})}\, \mathbf{D}^0 \right). \tag{4-7}$$

In the case where $D_n(x,\xi,z) \in \mathcal{C}^0$, (4-7) becomes

$$\mathbf{E}\,|Z_s|^n = K_n(z)\,dz, \quad K_n(z) = \int d\xi\, D_n(0,\xi,z) \prod_{p=1}^n \sqrt{\det(\xi^p \xi^{p*})}.$$

Our proof of Theorem 4.2 uses the following *coarea formula* of Federer:

LEMMA 4.4 [Fe, 3.2.12]. *Let $f : Y \to \mathbb{R}^k$ be a $\mathcal{C}^\infty$ map, where Y is an oriented m-dimensional Riemannian manifold. For $\gamma \in \mathcal{L}^1(Y)$, we have*

$$\int_{\mathbb{R}^k} dx_1 \cdots dx_k \int_{f^{-1}(x)} \gamma\, d\mathrm{Vol}_{m-k} = \int_Y \gamma\, \|df_1 \wedge \cdots \wedge df_k\|\, d\mathrm{Vol}_Y.$$

Recall that by Sard's theorem, $f^{-1}(x)$ is an $(m-k)$-dimensional submanifold for almost all $x \in \mathbb{R}^k$.

As a consequence of Lemma 4.4, for $\psi \in \mathcal{C}^0(\mathbb{R}^k)$ we have

$$\int_{\mathbb{R}^k} \psi(x) |f^{-1}(x)|\, dx_1 \cdots dx_k = (\psi \circ f) \|df_1 \wedge \cdots \wedge df_k\|\, d\mathrm{Vol}_Y \in \mathcal{D}'^m(Y), \tag{4-8}$$

where $|f^{-1}(x)|$ denotes $(m-k)$-dimensional volume measure on $f^{-1}(x)$.

REMARK. Federer's coarea formula, which is actually valid for Lipschitz maps, can be regarded as in integrated form of the *Leray formula*

$$|f^{-1}(x)| = \|df_1 \wedge \cdots \wedge df_k\| \left. \frac{d\mathrm{Vol}_Y}{df_1 \wedge \cdots \wedge df_k} \right|_{f^{-1}(x)}.$$

PROOF OF THEOREM 4.2. We restrict to a neighborhood U of an arbitrary point $z_0 \in M$. Since $\mathcal{S}$ spans V, we can choose U so that there exist sections $e_1, \ldots, e_k \in \mathcal{S}$ that form a local frame for V over U. Since $\mathbf{D}^0$ is independent of the connection, we can further assume that $\nabla|_U$ is the flat connection $\nabla s = \sum ds_j \otimes e_j$.

For a section $s \in \mathcal{S}$, we write $s(z) = \sum_{j=1}^{k} s_j(z) e_j(z)$ $(z \in U)$ and we let $\hat{s} = (s_1, \ldots, s_k) : U \to \mathbb{R}^k$. Then

$$|||\nabla s||| = \sqrt{H} \, \|ds_1 \wedge \cdots \wedge ds_k\|.$$

Thus by (4–8),

$$\int_{\mathbb{R}^k} \rho_\varepsilon(x) |\hat{s}^{-1}(x)| \, dx = (\rho_\varepsilon \circ s) \, |||\nabla s||| \, dz \in \mathcal{D}'^m(U), \qquad (4\text{–}9)$$

where we write, as before, $dx = \sqrt{H(z)} dx_1 \cdots dx_k$.

Let π_U, π' denote the projections given in the commutative diagram:

$$\begin{array}{ccccc}
U \times \mathcal{S} & \xrightarrow{\partial} & J^1(U,V) & \xleftarrow{\iota} & T_U^* \otimes V \\
& \searrow \pi_U & \downarrow \pi' & \swarrow \pi & \\
& & U & &
\end{array}$$

Integrating (4–9) over $\mathcal{S}$, we obtain

$$\int_{\mathbb{R}^k} \rho_\varepsilon(x) \mathbf{E} \, |\hat{s}^{-1}(x)| \, dx = \pi_{U*}(\rho_\varepsilon \circ s \, |||\nabla s||| \, dz \times \mu) = \pi'_* \big(\rho_\varepsilon(x) \, |||\xi||| \, \mathbf{D}\big)$$
$$\to \pi'_*\big(|||\xi||| \, \iota_* \mathbf{D}^0\big) = \pi_*(|||\xi||| \, \mathbf{D}^0). \qquad (4\text{–}10)$$

To complete the proof of Theorem 4.2, it suffices to show that the map

$$\Psi : \mathbb{R}^k \to \mathcal{D}'^m(U), \quad \Psi(x) = \mathbf{E} \, |\hat{s}^{-1}(x)|$$

is continuous; i.e., for all test functions $\varphi \in \mathcal{D}(U)$, the map $x \mapsto \mathbf{E}\,(|\hat{s}^{-1}(x)|, \varphi)$ is continuous. Indeed, if Ψ is continuous, then

$$\left(\int_{\mathbb{R}^k} \rho_\varepsilon(x) \mathbf{E} \, |\hat{s}^{-1}(x)| \, dx, \varphi\right) = \int \mathbf{E}\,(|\hat{s}^{-1}(x)|, \varphi) \rho_\varepsilon(x) \, dx$$
$$\to \mathbf{E}\,(|\hat{s}^{-1}(0)|, \varphi) = \mathbf{E}\,(|Z_s|, \varphi),$$

and (4–4) follows from (4–10).

To verify the continuity of Ψ, we extend $\{e_1, \ldots, e_k\}$ to a basis $\{e_1, \ldots, e_k, \ldots, e_d\}$ of $\mathcal{S}$, and we write

$$d\mu(s) = \psi(c_1, \ldots, c_d) \, dc, \quad s = \sum_{i=1}^{d} c_i e_i.$$

We note that

$$\hat{s}^{-1}(x_1, \ldots, x_k) = Z_{[s - \sum_{1}^{k} x_j e_j]},$$

and therefore

$$\Psi(x_1, \ldots, x_k) = \int |Z_s| \psi(c_1 + x_1, \ldots, c_k + x_k, c_{k+1}, \ldots, c_d) \, dc.$$

Write $c + x = (c_1 + x_1, \ldots, c_k + x_k, c_{k+1}, \ldots, c_d)$. We let $\tau : I \to \mathbb{R}^d$ denote the projection given by
$$\tau\left(z, \sum c_i e_i\right) = (c_1, \ldots, c_d).$$
For a test function $\varphi \in \mathcal{D}(U)$, we have
$$(\Psi(x), \varphi) = \int_{\mathbb{R}^d} (|Z_s|, \varphi) \psi(c+x) \, dc = \int_{\mathbb{R}^d} \left(|\tau^{-1}(c)|, \varphi(z)\right) \psi(c+x) \, dc$$
$$= \int_I \varphi(z) \psi(c+x) \|dc_1 \wedge \cdots \wedge dc_d\|_I \, d\operatorname{Vol}_I(z, c), \qquad (4\text{–}11)$$
where the last equality is by the coarea formula (4–8) applied to τ.

Suppose that $x^\nu \to x^0 \in \mathbb{R}^k$. In order to use (4–11) to show that $(\Psi(x^\nu), \varphi) \to (\Psi(x^0), \varphi)$, we note that $\|dc_1 \wedge \cdots \wedge dc_d\|_I \leq 1$ and hence
$$\varphi(z) \psi(c + x^\nu) \|dc_1 \wedge \cdots \wedge dc_d\|_I \leq \varphi(z) \gamma(\|c\| - R),$$
where
$$\gamma(r) = \sup_{\|c\| \geq r} \psi(c), \qquad R = \sup_\nu \|x^\nu\|.$$
We let $I(r) = \{(z, \sum c_i e_i) \in I : \|c\| = r\}$ denote the sphere bundle of radius r in the vector bundle $I \to M$. Then
$$\int_I \varphi(z) \gamma(\|c\| - R) \, d\operatorname{Vol}_I(z, c) = \int_0^{+\infty} dr \, \gamma(r - R) \int_{I(r)} \varphi(z) \, d\operatorname{Vol}_{I(r)}$$
$$\leq C \int_0^{+\infty} dr \, \gamma(r - R) r^{d-1} .$$
Since by hypothesis $\gamma(r) = o(r^{-d-1})$, we conclude that the integral is finite and thus the Lebesgue dominated convergence theorem implies that $(\Psi(x^\nu), \varphi) \to (\Psi(x^0), \varphi)$. □

4.2. Zero Correlations on Complex Manifolds. We now describe the jet density **D** in the case where (V, h) is a complex Hermitian vector bundle. In this case, we choose a complex local frame $\{e_1, \ldots, e_k\}$ and we let $H_\mathbb{C} = \det(h_{jj'})$, $h_{jj'} = h(e_j, e_{j'})$. We write
$$x = \sum_j x_j e_j, \quad \xi = \sum_{j,q} \xi_{jq} du_q \otimes e_j = \sum_j \xi_j \otimes e_j,$$
where ξ_{jq}, x_j are complex. We then have
$$\mathbf{D} = D(x, \xi, z) \, dx \, d\xi \, dz,$$
where this time
$$dx = H_\mathbb{C}(z) \prod_j d\operatorname{Re} x_j \, d\operatorname{Im} x_j, \quad d\xi = G(z)^{-k/2} H_\mathbb{C}(z)^m \prod_{j,q} d\operatorname{Re} \xi_{jq} \, d\operatorname{Im} \xi_{jq}.$$
We also have
$$\|\|\xi\|\| = H_\mathbb{C} \|\xi_1 \wedge \cdots \wedge \xi_k \wedge \bar{\xi}_1 \wedge \cdots \wedge \bar{\xi}_k\|.$$

We can now specialize Theorems 4.2–4.3 to the case where V is a holomorphic line bundle over a complex manifold M and the sections in $\mathcal{S}$ are holomorphic. If we now let $\{z_q\}$ denote complex local coordinates, we can write

$$\xi = \xi' + \xi'' = \sum_{j,q} (\xi'_{jq} \, dz_q + \xi''_{jq} d\bar{z}_q) \otimes e_j \,.$$

Since $\bar{\partial}s = 0$ for all $s \in \mathcal{S}$, the support of the measure $\mathbf{D}$ is contained in $V \oplus (T_M^{*h} \otimes V)$, i.e., those (x,ξ) with $\xi''_{jq} \equiv 0$ (using a holomorphic frame $\{e_j\}$ and a connection ∇ of type (1,0)). Hence on the support of $\mathbf{D}$, we have $\xi_j \in T_M^{*h}$, and hence

$$\||\xi\|| = H \, \|\xi_1 \wedge \cdots \wedge \xi_k\|^2 = \det(\xi\xi^*)_{\mathbb{C}} \,,$$

where $(\xi\xi^*)_{\mathbb{C}} \in \mathrm{Hom}_{\mathbb{C}}(V_z, V_z)$ denotes the complex endomorphism. Hence as a special case of Theorem 4.3, we obtain:

THEOREM 4.5. *Let $V \to M$ be a holomorphic line bundle over a complex manifold M and let $\mathcal{S}$ be a finite dimensional complex subspace of $H^0(M,V)$. We give $\mathcal{S}$ a semi-positive rapidly decaying volume form μ. If $\mathcal{S}$ spans V_n, then*

$$\mathbf{E}\,|Z_s|^n = \pi_* \left(\textstyle\prod_{p=1}^n \det(\xi\xi^*)_{\mathbb{C}} \, \mathbf{D}^0 \right) . \tag{4-12}$$

In the case where the image of $\mathcal{J}_n$ contains all the holomorphic 1-jets, we can write $\mathbf{D}_n = D_n(x, \xi', z) \, dx \, d\xi' \, dz$, $D_n(x, \xi, z) \in \mathcal{C}^0$. Then (4-12) yields the following result from [BSZ2, Th. 2.1]:

$$\mathbf{E}\,|Z_s|^n = K_n(z)\,dz\,, \quad K_n(z) = \int d\xi\, D_n(0,\xi,z) \prod_{p=1}^n \det(\xi^p \xi^{p*})_{\mathbb{C}} \,. \tag{4-13}$$

5. Universality of the Scaling Limit of the Correlations

We return to our complex Hermitian line bundle (L,h) on a compact almost complex $2m$-dimensional symplectic manifold M with symplectic form $\omega = \frac{i}{2}\Theta_L$, where Θ_L is the curvature of L with respect to a connection ∇. Theorem 1.1 follows from Theorem 4.3 applied to the vector bundle

$$V = \underbrace{L^N \oplus \cdots \oplus L^N}_{k}$$

and the (finite-dimensional) space of sections

$$\mathcal{S} = H^0_J(M, L^N)^k \subset \mathcal{C}^\infty(M, V)\,.$$

5.1. Gaussian Measures.

Recalling (2–2), we consider the Hermitian inner product on $H^0_J(M, L^N)$:

$$\langle s_1, s_2 \rangle = \int_M h^N(s_1, s_2) \frac{1}{m!} \omega^m \quad (s_1, s_2 \in H^0_J(M, L^N)).$$

We give S the Gaussian probability measure $\mu_N = \nu_N \times \cdots \times \nu_N$, where ν_N is the 'normalized' complex Gaussian measure on $H^0_J(M, L^N)$:

$$\nu_N(s) = \left(\frac{d_N}{\pi}\right)^{d_N} e^{-d_N |c|^2} dc, \quad s = \sum_{j=1}^{d_N} c_j S^N_j.$$

Here $\{S^N_j\}$ is an orthonormal basis for $H^0_J(M, L^N)$ (with respect to the Hermitian inner product (2–2)) and dc is $2d_N$-dimensional Lebesgue measure. The normalization is chosen so that $\mathbf{E}\langle s, s\rangle = 1$. This Gaussian is characterized by the property that the $2d_N$ real variables $\operatorname{Re} c_j$, $\operatorname{Im} c_j$ ($j = 1, \ldots, d_N$) are independent identically distributed (i.i.d.) random variables with mean 0 and variance $\frac{1}{2d_N}$; i.e.,

$$\mathbf{E} c_j = 0, \quad \mathbf{E} c_j c_k = 0, \quad \mathbf{E} c_j \bar{c}_k = \frac{1}{d_N} \delta_{jk}.$$

Picking a random element of S means picking k sections of $H^0_J(M, L^N)$ independently and at random.

REMARK. Since we are interested in the zero sets Z_s, which do not depend on constant factors, we could just as well suppose our sections lie in the unit sphere $SH^0_J(M, L^N)$ with respect to the Hermitian inner product (2–2), and pick random sections with respect to the spherical measure. This gives the same expectations for $|Z_s|^n$ as the Gaussian measure on $H^0_J(M, L^N)$.

We now review the concept of 'generalized Gaussian measures' from [SZ2], which is one of the ingredients in obtaining the (universal) scaling limit of the joint probability distribution, which in turn yields the universality of the scaling limit of the correlation of zeros on symplectic manifolds. (For further details and related results, see [SZ2, §5.1].) To begin, a (non-singular) Gaussian measure γ_Δ on $\mathbb{R}^p$ given by (1–3) has second moments

$$\langle x_j x_k \rangle_{\gamma_\Delta} = \Delta_{jk}.$$

The measure γ_Δ is characterized by its Fourier transform

$$\widehat{\gamma_\Delta}(t_1, \ldots, t_p) = e^{-\frac{1}{2} \sum \Delta_{jk} t_j t_k}. \tag{5–1}$$

The push-forward of a Gaussian measure by a surjective linear map is also Gaussian. Since we need to push forward Gaussian measures (on the spaces $H^0_J(M, L^N)$) by linear maps that are sometimes not surjective, we shall consider the case where Δ is positive semi-definite. In this case, we can still use (5–1) to define a measure γ_Δ, which we call a *generalized Gaussian*. If Δ has null eigenvalues, then γ_Δ is a Gaussian measure on the subspace $\Lambda_+ \subset \mathbb{R}^p$ spanned by the positive eigenvectors. If γ is a generalized Gaussian on $\mathbb{R}^p$ and $L:\mathbb{R}^p \to \mathbb{R}^q$ is a (not necessarily surjective) linear map, then $L_*\gamma$ is a generalized Gaussian on $\mathbb{R}^q$. By studying the Fourier transform, it is easy to see that the map $\Delta \mapsto \gamma_\Delta$ is a continuous map from the positive semi-definite matrices to the space of positive measures on $\mathbb{R}^p$ (with the weak topology).

5.2. Densities and the Szegő Kernel.
We now consider the n-point joint probability distribution of a (Gaussian) random almost holomorphic section $s \in H^0_J(M, L^N)$ having prescribed values $s(z^p) = x^p$ and prescribed derivatives $\nabla s(z^p) = \xi^p$ (for $1 \leq p \leq n$). We denote this density by $\widetilde{D}^N_n(x, \xi, z)\, dx\, d\xi$ as in [SZ2], where $z = (z^1, \ldots, z^n)$. Having equipped $H^0_J(M, L^N)$ with the Gaussian measure ν_N, and recalling that the joint probability distribution

$$\widetilde{\mathbf{D}}^N_z := \widetilde{D}^N_n(x, \xi, z)\, dx\, d\xi = (J^1_z)_* \nu_N,$$

is the push-forward of ν_N by a linear map, we conclude that the joint probability distribution is a generalized Gaussian measure on the complex vector space of 1-jets of sections:

$$\widetilde{\mathbf{D}}^N_z = \gamma_{\Delta^N(z)}.$$

To be more precise, we consider the $n(2m+1)$ complex-valued random variables X_p, Ξ_{pq} ($1 \leq p \leq n$, $1 \leq q \leq 2m$) on $\mathcal{H}^2_N(X) \equiv H^0_J(M, L^N)$ given by

$$X_p(s) = s(z^p, 0), \quad \Xi_{pq}(s) = (\nabla_q)s(z^p, 0),$$

where

$$\nabla_q = \frac{1}{\sqrt{N}} \frac{\partial^h}{\partial z_q}, \quad \nabla_{m+q} = \frac{1}{\sqrt{N}} \frac{\partial^h}{\partial \bar{z}_q} \quad (1 \leq q \leq m),$$

for $s \in \mathcal{H}^2_N(X)$. Here, $\partial^h/\partial z_q$ denotes the horizontal lift to X of the tangent vector $\partial/\partial z_q$ on M. The covariance matrix $\Delta^N(z)$ is given by the Szegő kernel and its covariant derivatives, as follows:

$$\Delta^N(z) = \begin{pmatrix} A^N & B^N \\ B^{N*} & C^N \end{pmatrix},$$

$$(A^N)^p_{p'} = \mathbf{E}\left(X_p \bar{X}_{p'}\right) = \frac{1}{d_N} \Pi_N(z^p, 0; z^{p'}, 0),$$

$$(B^N)^p_{p'q'} = \mathbf{E}\left(X_p \bar{\Xi}_{p'q'}\right) = \frac{1}{d_N} \overline{\nabla}^2_{q'} \Pi_N(z^p, 0; z^{p'}, 0),$$

$$(C^N)^{pq}_{p'q'} = \mathbf{E}\left(\Xi_{pq} \bar{\Xi}_{p'q'}\right) = \frac{1}{d_N} \nabla^1_q \overline{\nabla}^2_{q'} \Pi_N(z^p, 0; z^{p'}, 0),$$

for $p, p' = 1, \ldots, n$ and $q, q' = 1, \ldots, 2m$. Here, ∇^1_q and ∇^2_q denote the differential operator on $X \times X$ given by applying ∇_q to the first and second factors, respectively. (We note that A^N, B^N, C^N are $n \times n$, $n \times 2mn$, $2mn \times 2mn$ matrices, respectively; p, q index the rows, and p', q' index the columns.) In [BSZ2] we proved that the joint probability density has a universal scaling limit, and in [SZ1] this result was extended to the symplectic case:

THEOREM 5.1 [SZ2, Theorem 5.4]. *Let L be a pre-quantum line bundle over a $2m$-dimensional compact integral symplectic manifold (M, ω). Choose Heisenberg coordinates $\{z_j\}$ about a point $P_0 \in M$. Then*

$$\widetilde{\mathbf{D}}^N_{(z^1/\sqrt{N},\ldots,z^n/\sqrt{N})} \longrightarrow \mathbf{D}^\infty_{(z^1,\ldots,n^n)} = \gamma_{\Delta^\infty(z)}$$

where $\mathbf{D}^\infty_{(z^1,\ldots,z^n)}$ is a universal Gaussian measure supported on the holomorphic 1-jets, and $\Delta^N(z/\sqrt{N}) \to \Delta^\infty(z)$.

Theorem 1.2 then follows immediately from Theorems 1.1 and 5.1. In fact, we have the error estimate

$$\left(\frac{1}{N^{nk}} K^N_{nk}\left(\frac{z^1}{\sqrt{N}},\ldots,\frac{z^n}{\sqrt{N}}\right), \varphi\right) = \left(K^\infty_{nkm}(z^1,\ldots,z^n), \varphi\right) + O\left(\frac{1}{\sqrt{N}}\right),$$

for all $\varphi \in \mathcal{D}^{mn}((\mathbb{C}^m)_n)$.

A technically interesting novelty in the proof of Theorem 5.1 is the role of the $\bar{\partial}$ operator. In the holomorphic case, $\widetilde{\mathbf{D}}^N_{(z^1,\ldots,z^n)}$ is supported on the subspace of sections satisfying $\bar{\partial}s = 0$. In the almost complex case, sections do not satisfy this equation, so $\widetilde{\mathbf{D}}^N_{(z^1,\ldots,z^n)}$ is a measure on a higher-dimensional space of jets. However, Theorem 5.1 says that the mass in the '$\bar{\partial}$-directions' shrinks to zero as $N \to \infty$.

An alternate statement of Theorem 5.1 involves equipping the unit spheres $H^0(M, L^N)$ with Haar probability measure, and letting $\mathbf{D}^N_{(z^1,\ldots,z^n)}$ be the corresponding joint probability distribution on $SH^0_J(M, L^N)$. In [SZ1, Theorem 0.2], it was shown that these non-Gaussian measures $\mathbf{D}^N$ also have the same scaling limit $\mathbf{D}^\infty$.

The matrix Δ^∞ is given in terms of the Szegő kernel for the Heisenberg group:

$$\Delta^\infty(z) = \frac{m!}{c_1(L)^m} \begin{pmatrix} A^\infty(z) & B^\infty(z) \\ B^\infty(z)^* & C^\infty(z) \end{pmatrix}, \qquad (5\text{--}2)$$

where

$$A^\infty(z)^p_{p'} = \Pi^{\mathbf{H}}_1(z^p, 0; z^{p'}, 0),$$

$$B^\infty(z)^p_{p'q'} = \begin{cases} (z^{p'}_{q'} - z^p_{q'})\Pi^{\mathbf{H}}_1(z^p, 0; z^{p'}, 0) & \text{for } 1 \leq q \leq m, \\ 0 & \text{for } m+1 \leq q \leq 2m, \end{cases}$$

$$C^\infty(z)^{pq}_{p'q'} = \begin{cases} \left(\delta_{qq'} + (\bar{z}^{p'}_q - \bar{z}^p_q)(z^p_{q'} - z^{p'}_{q'})\right)\Pi^{\mathbf{H}}_1(z^p, 0; z^{p'}, 0) & \text{for } 1 \leq q, q' \leq m, \\ 0 & \text{otherwise.} \end{cases}$$

For details, see [SZ2].

Equation (5–2) says that the variances in the anti-holomorphic directions vanish. If we remove the rows and columns of the matrices corresponding to $m + 1 \leq q \leq 2m$, then we get the covariance matrix

$$\Delta^\infty_h(z) = \frac{m!}{c_1(L)^m} \begin{pmatrix} A^\infty(z) & B^\infty_h(z) \\ B^\infty_h(z)^* & C^\infty_h(z) \end{pmatrix} \qquad (5\text{--}3)$$

for the joint probability distribution in the holomorphic case. (Here A^∞, B_h^∞, and C_h^∞ are $n \times n$, $n \times mn$, and $mn \times mn$ matrices, respectively.) In [BSZ2], we used (4–13) and (5–3) to obtain formulas for the scaling limit zero correlations K_{nkm}^∞. We briefly summarize here how it was done: Let us write

$$\mathbf{D}^\infty_{(z^1,\ldots,z^n)} = \mathbf{D}^\infty_z = D^\infty(x,\xi,z)\,dx\,d\xi\,.$$

The function $D^\infty(0,\xi,z)$ is Gaussian in ξ, but is not normalized as a probability density. It is given by

$$D^\infty(0,\xi,z)\,d\xi = \frac{1}{\pi^n \det A^\infty(z)} \gamma_{\Lambda^\infty(z)}\,, \qquad (5\text{–}4)$$

where

$$\Lambda^\infty(z) = C_h^\infty(z) - B_h^\infty(z)^* A^\infty(z)^{-1} B_h^\infty(z)\,. \qquad (5\text{–}5)$$

We first consider the $k=1$ case of the limit correlation function for the zero divisor (complex hypersurface) of one random section. By (4–13), (5–4), and the identity $\det \Delta_h^\infty = \det \Lambda^\infty \det A^\infty$, we obtain

$$K_{n1m}^\infty(z^1,\ldots,z^n) = \frac{1}{\pi^n \det A^\infty(z)} \int_{\mathbb{C}^{mn}} \prod_{p=1}^n \left(\sum_{q=1}^m |\xi_q^p|^2 \right) d\gamma_{\Lambda^\infty(z)}(\xi)\,. \qquad (5\text{–}6)$$

The integral in (5–6) is a sum of $(2n)$-th moments of the Gaussian measure $\gamma_{\Lambda^\infty(z)}$, and can be evaluated using the Wick formula. Indeed, in the pair correlation case $n=2$, (5–6) yields the explicit formula (1–4).

For the case of random k-tuples $s = (s^1,\ldots,s^k) \in \mathcal{S} = H_J^0(M,L^N)^k$ (where the zero sets are of codimension k), the 1-jets $J_z^1 s^1, \ldots, J_z^1 s^k$ are i.i.d. random vectors, and we have

$$K_{nkm}^\infty(z^1,\ldots,z^n)$$
$$= \frac{1}{[\pi^n \det A^\infty(z)]^k} \int_{\mathbb{C}^{kmn}} \prod_{p=1}^n \det_{1 \le j, j' \le k} \left(\sum_{q=1}^m \xi_{jq}^p \bar{\xi}_{j'q}^p \right) d\gamma_{I_k \otimes \Lambda^\infty(z)}(\xi)\,, \qquad (5\text{–}7)$$

where I_k denotes the $k \times k$ identity matrix; i.e.,

$$\left(I_k \otimes \Lambda^\infty(z)\right)_{j'p'q'}^{jpq} = \delta_{j'}^j \Lambda^\infty(z)_{p'q'}^{pq}\,.$$

For further details and explicit formulas, see [BSZ2] for the case $k=n=2$, and see [BSZ3] for the point pair correlation case $n=2$, $k=m$. Indeed, we show in [BSZ3] that for small values of $r := |z^1 - z^2|$, we have

$$\widetilde{K}_{2mm}(z^1, z^2) = \frac{m+1}{4} r^{4-2m} + O(r^{8-2m})\,, \qquad m = 1, 2, 3, \ldots\,.$$

5.3. Decay of Correlations.

Let us define the *normalized n-point scaling limit zero correlation function*

$$\widetilde{K}^\infty_{nkm}(z) = (K^\infty_{1km})^{-n} K^\infty_{nkm}(z) = \left(\frac{\pi^k(m-k)!}{m!}\right)^n K^\infty_{nkm}(z). \tag{5-8}$$

In [BSZ2], we showed that the limit correlations are "short range" in the following sense:

THEOREM 5.2 [BSZ2, Theorem 4.1]. *The correlation functions satisfy the estimate*

$$\widetilde{K}^\infty_{nkm}(z^1,\ldots,z^n) = 1 + O(r^4 e^{-r^2}) \quad \text{as } r \to \infty, \quad r = \min_{p \neq p'} |z^p - z^{p'}|.$$

PROOF. We review here the proof of this estimate. Writing

$$A = \pi^m A^\infty, \quad B = \pi^m B^\infty_h, \quad C = \pi^m C^\infty_h, \quad \Lambda = \pi^m \Lambda^\infty,$$

we have:

$$A^p_{p'} = e^{i\,\mathrm{Im}(z^p \cdot \bar z^{p'})} e^{-\frac{1}{2}|z^p - z^{p'}|^2},$$

$$B^p_{p'q'} = (z^p_{q'} - z^{p'}_{q'}) A^p_{p'},$$

$$C^{pq}_{p'q'} = \left[\delta_{qq'} + (\bar z^{p'}_q - \bar z^p_q)(z^p_{q'} - z^{p'}_{q'})\right] A^p_{p'}.$$

This implies that

$$\begin{aligned}
A &= I + O(e^{-r^2/2}), & A^p_p &= 1, \\
B &= O(r e^{-r^2/2}), & & \\
C &= I + O(r^2 e^{-r^2/2}) \quad \text{as } r \to \infty, & C^{pq}_{pq} &= 1.
\end{aligned} \tag{5-9}$$

Recalling (5–5), we have

$$\Lambda = I + O(r^2 e^{-r^2/2}), \quad \Lambda^{pq}_{pq} = 1 + O(r^2 e^{-r^2}), \quad \text{as } r \to \infty. \tag{5-10}$$

We now use the Wick formula to evaluate the integral in (5–7). (Formula (5–7) is homogeneous of order 0 in the matrix entries, so is not affected when A^∞, Λ^∞ are multiplied by π^m.) Note that the Wick formula involves terms that are products of diagonal elements of Λ, and products that contain at least two off-diagonal elements of Λ. The former terms are of the form $1 + O(r^2 e^{-r^2})$, and the latter are $O(r^4 e^{-r^2})$. Similarly, $\det A = 1 + O(e^{-r^2})$, and the estimate follows. □

The theorem can be extended to estimates of the connected correlation functions (called also truncated correlation functions, cluster functions, or cumulants), as follows. The *n*-point connected correlation function is defined as (see, e.g., [GJ, p. 286])

$$\widetilde{T}^\infty_{nkm}(z^1,\ldots,z^n) = \sum_G (-1)^{l+1}(l-1)! \prod_{j=1}^{l} \widetilde{K}^\infty_{n_j km}(z^{p_{j1}},\ldots,z^{p_{jn_j}}), \tag{5-11}$$

where the sum is taken over all partitions $G = (G_1, \ldots, G_l)$ of the set $(1, \ldots, n)$ and $G_j = (p_{j1}, \ldots, p_{jn_j})$. In particular, recalling that $\widetilde{K}^\infty_{1km} \equiv 1$,

$$\widetilde{T}^\infty_{1km}(z^1) = \widetilde{K}^\infty_{1km}(z^1) = 1,$$

$$\widetilde{T}^\infty_{2km}(z^1, z^2) = \widetilde{K}^\infty_{2km}(z^1, z^2) - \widetilde{K}^\infty_{1km}(z^1)\widetilde{K}^\infty_{1km}(z^2) = \widetilde{K}^\infty_{2km}(z^1, z^2) - 1,$$

$$\widetilde{T}^\infty_{3km}(z^1, z^2, z^3) = \widetilde{K}^\infty_{3km}(z^1, z^2, z^3) - \widetilde{K}^\infty_{2km}(z^1, z^2)\widetilde{K}^\infty_{1km}(z^3)$$
$$- \widetilde{K}^\infty_{2km}(z^1, z^3)\widetilde{K}^\infty_{1km}(z^2) - \widetilde{K}^\infty_{2km}(z^2, z^3)\widetilde{K}^\infty_{1km}(z^1)$$
$$+ 2\widetilde{K}^\infty_{1km}(z^1)\widetilde{K}^\infty_{1km}(z^2)\widetilde{K}^\infty_{1km}(z^3)$$
$$= \widetilde{K}^\infty_{3km}(z^1, z^2, z^3) - \widetilde{K}^\infty_{2km}(z^1, z^2)$$
$$- \widetilde{K}^\infty_{2km}(z^1, z^3) - \widetilde{K}^\infty_{2km}(z^2, z^3) + 2,$$

and so on. The inverse of (5–11) is

$$\widetilde{K}^\infty_{nkm}(z^1, \ldots, z^n) = \sum_G \prod_{j=1}^l \widetilde{T}^\infty_{n_j km}(z^{p_{j1}}, \ldots, z^{p_{jn_j}}) \qquad (5\text{–}12)$$

(Möbius theorem). The advantage of the connected correlation functions is that they go to zero if at least one of the distances $|z^i - z^j|$ goes to infinity (see Corollary 5.8 below). In our case the connected correlation functions can be estimated as follows. Define

$$d(z^1, \ldots, z^n) = \max_{\mathcal{G}} \prod_{l \in L} |z^{i(l)} - z^{f(l)}|^2 e^{-|z^{i(l)} - z^{f(l)}|^2/2}. \qquad (5\text{–}13)$$

where the maximum is taken over all oriented connected graphs $\mathcal{G} = (V, L)$ "with zero boundary" such that $V = (z^1, \ldots, z^n)$. Here V denotes the set of vertices of $\mathcal{G}$, L the set of edges, and $i(l)$ and $f(l)$ stand for the initial and final vertices of the edge l, respectively. The graph $\mathcal{G}$ is said to be have zero boundary if $\sum\{l : l \in L\}$ is a 1-cycle; i.e., for each vertex $z^p \in V$, the number of edges beginning at z^p equals the number ending at z^p. (There must be at least one edge beginning at each vertex, since $\mathcal{G}$ is assumed to be connected. Graphs may have any number of edges connecting the same two vertices.) Observe that the maximum in (5–13) is achieved at some graph $\mathcal{G}$, because $te^{-t/2} \leq 2/e < 1$ and therefore the product in (5–13) is at most $(2/e)^{|L|}$.

THEOREM 5.3. *The connected correlation functions satisfy the estimate*

$$\widetilde{T}^\infty_{nkm}(z^1, \ldots, z^n) = O(d(z^1, \ldots, z^n)),$$

provided that $\min_{p \neq q} |z^p - z^q| \geq c > 0$.

To prove the theorem, let us introduce the n-point functions

$$\widehat{K}_n(z^1, \ldots, z^n) = \det A_{nkm}(z^1, \ldots, z^n)\, \widetilde{K}^\infty_{nkm}(z^1, \ldots, z^n) \qquad (5\text{–}14)$$

$$= [(m-k)!/m!]^n \int \prod_{p=1}^n {\det}_{1 \leq j, j' \leq k} \left(\sum_{q=1}^m \xi^p_{jq} \bar{\xi}^p_{j'q} \right) d\gamma_{I_k \otimes \Lambda(z)}(\xi)$$

where $A_{nkm} = I_k \otimes A$, an $nk \times nk$ matrix. (Note that $\det A_{nkm} = (\det A)^k$. It was shown in [BSZ2, Lemma 3.3] that $\det A > 0$ at distinct points z^p.) We also consider the corresponding "connected functions"

$$\widehat{T}_n(z^1,\ldots,z^n) = \sum_G (-1)^{l+1}(l-1)! \prod_{j=1}^l \widehat{K}_{n_j}(z^{p_{j1}},\ldots,z^{p_{jn_j}}), \qquad (5\text{--}15)$$

and we note that the Möbius inversion formula applies to $\widehat{K}_n, \widehat{T}_n$.

Observe that we can rewrite $\widehat{K}_n(z^1,\ldots,z^n)$ as a sum over Feynman diagrams. Namely, each term in the Wick sum for the integral in (5–14) corresponds to a graph $\mathcal{F} = (V,L)$ (Feynman diagram) such that $V = (z^1,\ldots,z^n)$ and the edges $l \in L$ connect the paired variables $\xi_{jq}^{i(l)}, \bar{\xi}_{jq'}^{f(l)}$ in the given Wick term. We have that

$$\widehat{K}_n(z^1,\ldots,z^n) = [(m-k)!/m!]^n \sum_{\mathcal{F}} W_{\mathcal{F}}(z^1,\ldots,z^n), \qquad (5\text{--}16)$$

where the function $W_{\mathcal{F}}(z^1,\ldots,z^n)$ is the sum over all terms in the Wick sum corresponding to the Feynman diagram $\mathcal{F}$. (In other words, to get $W_{\mathcal{F}}(z^1,\ldots,z^n)$ we fix the indices p,p' of the pairings $(\xi_{jq}^p, \bar{\xi}_{jq'}^{p'})$ prescribed by $\mathcal{F}$ and sum up in the Wick formula over all indices j,q at every z^p.) Note that each graph $\mathcal{F}$ in the sum (5–16), having arisen from a term in the Wick sum, has zero boundary.

A remarkable property of the "connected functions" is that they are represented by the sum over connected Feynman diagrams (see, e.g., [GJ]):

$$\widehat{T}_n(z^1,\ldots,z^n) = [(m-k)!/m!]^n \sum_{\mathcal{F}}{}^{\text{conn}} W_{\mathcal{F}}(z^1,\ldots,z^n).$$

We conclude from (5–10) that for all connected Feynman diagrams $\mathcal{F}$,

$$W_{\mathcal{F}}(z^1,\ldots,z^n) = O(d(z^1,\ldots,z^n)), \quad \text{provided that } \min_{p \neq q}|z^p - z^q| \geq c > 0.$$

Summing over $\mathcal{F}$, we obtain the following estimate:

LEMMA 5.4. $\widehat{T}_n(z^1,\ldots,z^n) = O(d(z^1,\ldots,z^n))$, provided that $\min_{p \neq q}|z^p - z^q| \geq c > 0$.

It remains to relate $\widetilde{T}_{nkm}^\infty(z^1,\ldots,z^n)$ to $\widehat{T}_n(z^1,\ldots,z^n)$. To do this, we introduce the functions

$$Q_n(z^1,\ldots,z^n) = \sum_G (-1)^{l+1}(l-1)! \prod_{j=1}^l \det A_{n_jkm}(z^{p_{j1}},\ldots,z^{p_{jn_j}}),$$

which are the connected functions for $\det A_{nkm}(z^1,\ldots,z^n)$, and

$$R_n(z^1,\ldots,z^n) = \sum_G (-1)^{l+1}(l-1)! \prod_{j=1}^l \frac{1}{\det A_{n_jkm}(z^{p_{j1}},\ldots,z^{p_{jn_j}})},$$

UNIVERSALITY AND SCALING OF ZEROS ON SYMPLECTIC MANIFOLDS   63

which are the connected functions for $\dfrac{1}{\det A_{nkm}(z^1,\ldots,z^n)}$. Recall the Möbius inversion formula

$$\frac{1}{\det A_{nkm}(z^1,\ldots,z^n)} = \sum_G \prod_{j=1}^{l} R_{n_j}(z^{p_{j1}},\ldots,z^{p_{jn_j}}). \qquad (5\text{-}17)$$

We have the following relation between $\widetilde{T}^{\infty}_{nkm}(z^1,\ldots,z^n)$ and $\widehat{T}_n(z^1,\ldots,z^n)$.

LEMMA 5.5.
$$\widetilde{T}^{\infty}_{nkm}(z^1,\ldots,z^n)$$
$$= \sum_{G,H}{}^{\text{conn}} \prod_{j=1}^{l} \widehat{T}_{n_j}(z^{p_{j1}},\ldots,z^{p_{jn_j}}) \prod_{j=1}^{l'} R_{m_j}(z^{p'_{j1}},\ldots,z^{p'_{jm_j}}), \qquad (5\text{-}18)$$

where the sum is taken over all pairs $\{G = (G_1,\ldots,G_l), H = (H_1,\ldots,H_{l'})\}$ of partitions of the set $(1,\ldots,n)$ which are "mutually connected" in the sense that there is no proper subset S of the set $(1,\ldots,n)$ such that S is a union of some subsets G_j and is also a union of some subsets H_j. In (5-18), $G_j = (p_{j1},\ldots,p_{jn_j})$ and $H_j = (p'_{j1},\ldots,p'_{jm_j})$.

PROOF. The proof is by induction on n. From (5-12),

$$\widetilde{T}^{\infty}_{nkm}(z^1,\ldots,z^n) = \widetilde{K}^{\infty}_{nkm}(z^1,\ldots,z^n) - \sum_F{}' \prod_{j=1}^{l} \widetilde{T}^{\infty}_{n_jkm}(z^{p_{j1}},\ldots,z^{p_{jn_j}}) \qquad (5\text{-}19)$$

where the summation goes over all partitions $F = (F_1,\ldots,F_l)$ with at least two elements in the partition (i.e., $l \geq 2$). From (5-15) and (5-17), we have

$$\widetilde{K}^{\infty}_{nkm}(z^1,\ldots,z^n) = \widehat{K}_n(z^1,\ldots,z^n)\frac{1}{\det A_{nkm}(z^1,\ldots,z^n)} \qquad (5\text{-}20)$$
$$= \sum_{G,H} \prod_{j=1}^{l} \widehat{T}_{n_j}(z^{p_{j1}},\ldots,z^{p_{jn_j}}) \prod_{j=1}^{l'} R_{m_j}(z^{p'_{j1}},\ldots,z^{p'_{jm_j}}),$$

where the sum is taken over all pairs $\{G = (G_1,\ldots,G_l), H = (H_1,\ldots,H_{l'})\}$ of partitions of the set $(1,\ldots,n)$. If we use the inductive assumption that (5-18) holds when the number of points is less than n and apply that assumption to $\widetilde{T}^{\infty}_{n_jkm}(z^{p_{j1}},\ldots,z^{p_{jn_j}})$ in (5-19), we obtain that

$$\sum_F{}' \prod_{j=1}^{l} \widetilde{T}^{\infty}_{n_jkm}(z^{p_{j1}},\ldots,z^{p_{jn_j}})$$
$$= \sum_{G,H}{}^{\text{disconn}} \prod_{j=1}^{l} \widehat{T}_{n_j}(z^{p_{j1}},\ldots,z^{p_{jn_j}}) \prod_{j=1}^{l'} R_{m_j}(z^{p'_{j1}},\ldots,z^{p'_{jm_j}}), \qquad (5\text{-}21)$$

where the sum on the right is taken over all partitions $G = (G_1,\ldots,G_l)$ of the set $(1,\ldots,n)$ and all partitions $H = (H_1,\ldots,H_p)$ of the set $(1,\ldots,n)$ which are "mutually disconnected" in the sense that there is a proper subset S of the set

$(1, \ldots, n)$ that is simultaneously a union of some subsets G_j and a union of some subsets H_j. When we substitute (5–20) and (5–21) into (5–19) and take the difference on the right of (5–19), disconnected pairs $\{G, H\}$ will be cancelled out and we will be left with mutually connected $\{G, H\}$. This proves the lemma. $\square$

LEMMA 5.6. *The functions* $Q_n(z^1, \ldots, z^n)$ *satisfy the estimate*
$$Q_n(z^1, \ldots, z^n) = O(d(z^1, \ldots, z^n)), \tag{5-22}$$
provided that $\min_{p \neq q} |z^p - z^q| \geq c > 0$.

PROOF. By the determinant formula,
$$\det A_{nkm}(z^1, \ldots, z^n) = (\det A)^k = \sum_\pi (-1)^{\sigma(\pi)} \prod_{j=1}^{k} \prod_{p=1}^{n} A_p^{\pi_j(p)},$$
where the sum is over all k-tuples $\pi = (\pi_1, \ldots, \pi_k)$ of permutations of $(1, \ldots, n)$. We claim that
$$Q_n(z^1, \ldots, z^n) = \sum_\pi{}^{\text{conn}} (-1)^{\sigma(\pi)} \prod_{j=1}^{k} \prod_{p=1}^{n} A_p^{\pi_j(p)}, \tag{5-23}$$
where the summation on the right goes over the set of k-tuples $\pi = (\pi_1, \ldots, \pi_k)$ such that no proper subset of $(1, \ldots, n)$ is invariant under the group generated by the π_j. (Each such π corresponds to a connected graph consisting of edges beginning at p and ending at $\pi_j(p)$, for all p, j.) Indeed,
$$Q_n(z^1, \ldots, z^n) = \det A_{nkm}(z^1, \ldots, z^n) - \sum_F{}' \prod_{j=1}^{l} Q_{n_j}(z^{p_{j1}}, \ldots, z^{p_{jn_j}}),$$
where the summation on the right goes over all partitions $F = (F_1, \ldots, F_l)$ with $l \geq 2$. Using this equation, we prove (5–23) by induction (cf. the proof of Lemma 5.5). The estimate (5–22) now follows from (5–23) and (5–9). $\square$

LEMMA 5.7. *The functions* $R_n(z^1, \ldots, z^n)$ *satisfy the estimate*
$$R_n(z^1, \ldots, z^n) = O(d(z^1, \ldots, z^n)), \tag{5-24}$$
provided that $\min_{p \neq q} |z^p - z^q| \geq c > 0$.

PROOF. We have the identity
$$0 = \sum_{G,H}{}^{\text{conn}} \prod_{j=1}^{l} Q_{n_j}(z^{p_{j1}}, \ldots, z^{p_{jn_j}}) \prod_{j=1}^{l'} R_{m_j}(z^{p'_{j1}}, \ldots, z^{p'_{jm_j}}), \quad n \geq 2. \tag{5-25}$$
The proof of this identity is the same as that of Lemma 5.5. Indeed, the connected functions of
$$\det A_{nkm} \frac{1}{\det A_{nkm}} = 1$$

UNIVERSALITY AND SCALING OF ZEROS ON SYMPLECTIC MANIFOLDS 65

are equal to 0 (except that the 1-point connected function equals 1); hence (5–25) follows.

The identity (5–25) can be rewritten as

$\det A_{nkm}(z^1,\ldots,z^n)\, R_n(z^1,\ldots,z^n)$

$$= -\sum_{G,H}^{\text{conn}'} \prod_{j=1}^{l} Q_{n_j}(z^{p_{j1}},\ldots,z^{p_{jn_j}}) \prod_{j=1}^{l'} R_{m_j}(z^{p'_{j1}},\ldots,z^{p'_{jm_j}}), \quad (5\text{–}26)$$

where the summation on the right goes over all mutually connected pairs of partitions $\{G,H\}$ with at least two elements in H (i.e., $l' \geq 2$). Now the estimate (5–24) follows by induction from Lemma 5.6 and identity (5–26). □

Theorem 5.3 follows from Lemmas 5.4, 5.5 and 5.7. The theorem yields the following more explicit estimate:

COROLLARY 5.8. *The connected correlation functions satisfy the estimate*

$$\widetilde{T}^\infty_{nkm}(z^1,\ldots,z^n) = o(e^{-R^2/n}), \quad R = \max_{p,q}|z^p - z^q|,$$

provided that $\min_{p\neq q}|z^p - z^q| \geq c > 0.$

PROOF. We must show that

$$d(z^1,\ldots,z^n) \leq o(e^{-R^2/n}). \quad (5\text{–}27)$$

Assume without loss of generality that $|z^1 - z^n| = R$. Let $\mathcal{G} = (V,L)$ be an oriented connected graph with zero boundary as in the definition of $d(z^1,\ldots,z^n)$. Since z^1 and z^n are connected by a chain of loops in $\mathcal{G}$, we can choose disjoint sets of edges $L', L'' \subset L$ such that L' forms a path starting at z^1 and ending at z^n, and L'' forms a path starting at z^n and ending at z^1. This means that there is a sequence $z^1 = z^{i_1}, z^{i_2}, \ldots, z^{i_{n'}} = z^n$ such that $L' = \{l_1,\ldots,l_{n'-1}\}$, where l_j begins at z^{i_j} and ends at $z^{i_{j+1}}$. By removing any loops in L', we can assume that the z^{i_j} are distinct and thus $n' \leq n$. A similar description holds for L''. Let $r_j = |z^{i_j} - z^{i_{j+1}}|$. We note that

$$R \leq \sum r_j \leq \left((n'-1)\sum r_j^2\right)^{1/2},$$

where the second inequality is by Cauchy–Schwarz. We then have

$$\prod_{l\in L'}|z^{i(l)} - z^{f(l)}|^2 e^{-|z^{i(l)}-z^{f(l)}|^2/2} = \prod_{j=1}^{n'-1} r_j^2 e^{-r_j^2/2}$$

$$\leq R^{2n'-2} e^{-\frac{1}{2}\sum r_j^2} \leq R^{2n'-2} e^{-R^2/(2n'-2)},$$

and hence

$$\prod_{l\in L'}|z^{i(l)} - z^{f(l)}|^2 e^{-|z^{i(l)}-z^{f(l)}|^2/2} \leq R^{2n-2} e^{-R^2/(2n-2)}, \quad R \geq 1.$$

The same inequality also holds for the product over the path L''. Since each term of the product in (5-13) is less than 1, we then have

$$\prod_{l \in L} |z^{i(l)} - z^{f(l)}|^2 e^{-|z^{i(l)} - z^{f(l)}|^2/2} \leq \prod_{l \in L' \cup L''} |z^{i(l)} - z^{f(l)}|^2 e^{-|z^{i(l)} - z^{f(l)}|^2/2}$$
$$\leq o(e^{-R^2/n}).$$

Taking the supremum over all graphs, we obtain (5-27). □

REMARK. The preceding proof gives the bound

$$d(z^1, \ldots, z^n) \leq R^{4n-4} e^{-R^2/(n-1)}, \quad R \geq 1.$$

Hence we actually have the estimate

$$\widetilde{T}_{nkm}^{\infty}(z^1, \ldots, z^n) = O\big(R^{4n-4} e^{-R^2/(n-1)}\big), \quad \text{provided that } \min_{p \neq q} |z^p - z^q| \geq c > 0.$$

This implies Theorem 5.2 because of the inversion formula (5-12).

References

[At] M. F. Atiyah, *The geometry and physics of knots*, Lezioni Lincee, Cambridge Univ. Press, Cambridge, 1990.

[Au1] D. Auroux, "Asymptotically holomorphic families of symplectic submanifolds", *Geom. Funct. Anal.* **7** (1997), 971–995.

[Au2] D. Auroux, "Symplectic 4-manifolds as branched coverings of $\mathbb{CP}^2$", *Invent. Math.* **139** (2000), 551–602.

[AK] D. Auroux and L. Katzarkov, "Branched coverings of $\mathbb{CP}^2$ and invariants of symplectic 4-manifolds", preprint, 1999.

[BFG] M. Beals, C. Fefferman and R. Grossman, "Strictly pseudoconvex domains in $\mathbb{C}^n$", *Bull. Amer. Math. Soc.* **8** (1983), 125–322.

[BD] P. Bleher and X. Di, "Correlations between zeros of a random polynomial", *J. Stat. Phys.* **88** (1997), 269–305.

[BSZ1] P. Bleher, B. Shiffman and S. Zelditch, "Poincaré–Lelong approach to universality and scaling of correlations between zeros", *Comm. Math. Phys.* **208** (2000), 771–785.

[BSZ2] P. Bleher, B. Shiffman and S. Zelditch, "Universality and scaling of correlations between zeros on complex manifolds", *Invent. Math.* **142** (2000), 351–395.

[BSZ3] P. Bleher, B. Shiffman and S. Zelditch, "Correlations between zeros and supersymmetry", preprint. See http://www.arxiv.org/abs/math-ph/0011016.

[BBL] E. Bogomolny, O. Bohigas, and P. Leboeuf, "Quantum chaotic dynamics and random polynomials", *J. Stat. Phys.* **85** (1996), 639–679.

[BU1] D. Borthwick and A. Uribe, "Almost complex structures and geometric quantization", *Math. Res. Lett.* **3** (1996), 845–861.

[BU2] D. Borthwick and A. Uribe, "Nearly Kählerian embeddings of symplectic manifolds", preprint. See http://www.arxiv.org/abs/math.DG/9812041.

[Bch] T. Bouche, "Asymptotic results for Hermitian line bundles over complex manifolds: the heat kernel approach", pp. 67–81 in *Higher-dimensional complex varieties* (Trento, 1994), de Gruyter, Berlin, 1996.

[Bou] L. Boutet de Monvel, "Hypoelliptic operators with double characteristics and related pseudodifferential operators", *Comm. Pure and Appl. Math.* **27** (1974), 585–639.

[BG] L. Boutet de Monvel and V. Guillemin, *The spectral theory of Toeplitz operators*, Ann. Math. Studies **99**, Princeton Univ. Press, Princeton, 1981.

[BK] A. Boutet de Monvel and A. Khorunzhy, "On universality of the smoothed eigenvalue density of large random matrices", *J. Phys. A* **32** (1999), L413–L417.

[BS] L. Boutet de Monvel and J. Sjöstrand, "Sur la singularité des noyaux de Bergman et de Szegő", *Astérisque* **34–35** (1976), 123–164.

[BZ] E. Brezin and A. Zee, "Universality of the correlations between eigenvalues of large random matrices", *Nucl. Phys. B* **402** (1993), 613–627.

[Car] J. Cardy, *Scaling and renormalization in statistical physics*, Cambridge Lecture Notes in Physics **5**, Cambridge Univ. Press, Cambridge, 1996.

[Cat] D. Catlin, "The Bergman kernel and a theorem of Tian", in *Analysis and geometry in several complex variables*, edited by G. Komatsu and M. Kuranishi, Birkhäuser, Boston, 1999.

[De] P. A. Deift, *Orthogonal polynomials and random matrices: a Riemann–Hilbert approach*, Courant Lecture Notes in Mathematics, Courant Institute of Mathematical Sciences, New York, 1999.

[Do1] S. Donaldson, "Symplectic submanifolds and almost complex geometry", *J. Diff. Geom.* **44** (1996), 666–705.

[Do2] S. Donaldson, "Lefschetz fibrations in symplectic geometry", in *Proceedings of the International Congress of Mathematicians*, vol. II, Berlin, 1998.

[EK] A. Edelman and E. Kostlan, "How many zeros of a random polynomial are real?", *Bull. Amer. Math. Soc.* **32** (1995), 1–37.

[Fe] H. Federer, *Geometric measure theory*, Springer, New York, 1969.

[Fo] G. B. Folland, *Harmonic analysis in phase space*, Princeton Univ. Press, Princeton, 1989.

[GJ] J. Glimm and A. Jaffe, *Quantum physics: a functional integral point of view*, 2nd ed., Springer, New York, 1987.

[GH] P. Griffiths and J. Harris, *Principles of algebraic geometry*, Wiley, New York, 1978.

[GS] V. Guillemin and S. Sternberg, *Symplectic techniques in physics*, Cambridge Univ. Press, Cambridge, 1984.

[GU] V. Guillemin and A. Uribe, "Laplace operator on tensor powers of a line bundle", *Asympt. Anal.* **1** (1988), 105–113.

[Hal] B. I. Halperin, "Statistical mechanics of topological defects", in *Physics of Defects, Les Houches Session XXXV*, North-Holland, 1980.

[Han] J. H. Hannay, "Chaotic analytic zero points: exact statistics for those of a random spin state", *J. Phys. A* **29** (1996), 101–105.

[Hö] L. Hörmander, *The analysis of linear partial differential operators I*, Grund. Math. Wiss. **256**, Springer, New York, 1983.

[Kac] M. Kac, "On the average number of real roots of a random algebraic equation", *Bull. Amer. Math. Soc.* **49** (1943), 314–320.

[KS] N. Katz and P. Sarnak, *Random Matrices, "Frobenius eigenvalues, and monodromy*, AMS Colloq. Pub. **45**, Amer. Math. Soc, Providence, RI, 1998.

[Kl] S. Kleiman, "Toward a numerical theory of ampleness", *Ann. of Math.* **84** (1966), 293–344.

[Le] P. Lelong, "Intégration sur un ensemble analytique complexe", *Bull. Soc. Math. France* **85** (1957), 239–262.

[MM] V. A. Malyshev and R. A. Minlos, *Gibbs random fields: Cluster expansions*, Kluwer, Dordrecht, 1991.

[MelS] A. Melin and J. Sjöstrand, "Fourier integral operators with complex valued phase functions", pp. 120–233 in Lecture Notes in Math. **459**, Springer, New York, 1975.

[MenS] A. Menikoff and J. Sjöstrand, "On the eigenvalues of a class of hypoelliptic operators", *Math. Ann.* **235** (1978), 55–85.

[Na] Y. Nakai, "A criterion of an ample sheaf on a projective scheme", *Amer. J. Math.* **85** (1963), 14–26.

[Ne] J. Neuheisel, *The asymptotic distribution of nodal sets on spheres*, Ph.D. thesis, Johns Hopkins University, 2000.

[NV] S. Nonnenmacher and A. Voros, "Chaotic eigenfunctions in phase space", *J. Stat. Phys.* **92** (1998), 431–518.

[Ri] S. O. Rice, "Mathematical analysis of random noise", *Bell System Tech. J.* **23** (1944), 282–332 and **24** (1945), 46–156. Reprinted as 133–294 in *Selected papers on noise and stochastic processes*, Dover, New York, 1954.

[RS] Z. Rudnick and P. Sarnak, "The pair correlation function of fractional parts of polynomials", *Comm. Math. Phys.* **194** (1998), 61–70.

[Sa] P. Sarnak, "Arithmetic quantum chaos", pp. 183–236 in Israel Math. Conference Proc. **8**, 1995.

[SSo] B. Shiffman and A. J. Sommese, *Vanishing theorems on complex manifolds*, Progress in Math. **56**, Birkhäuser, Boston, 1985.

[SZ1] B. Shiffman and S. Zelditch, "Distribution of zeros of random and quantum chaotic sections of positive line bundles", *Comm. Math. Phys.* **200** (1999), 661–683.

[SZ2] B. Shiffman and S. Zelditch, "Random almost holomorphic sections of ample line bundles on symplectic manifolds", preprint. See http://www.arxiv.org/abs/math.SG/0001102.

[SSm] M. Shub and S. Smale, "Complexity of Bezout's theorem II: Volumes and probabilities", pp. 267–285 in *Computational algebraic geometry* (Nice, 1992), Progr. Math. **109**, Birkhäuser, Boston, 1993.

[Sik] J. C. Sikorav, "Construction de sous-variétés symplectiques (d'après S. K. Donaldson et D. Auroux)", pp. 231–253 in *Séminaire Bourbaki* 1997/98, Astérisque **252**, Soc. Math. de France, 1998.

[Sin] Ya. G. Sinai, *Theory of phase transitions: rigorous results*, Pergamon Press, New York, 1982.

[Sj] J. Sjöstrand, "Density of resonances for strictly convex analytic obstacles", *Can. J. Math.* **48** (1996), 397–447.

[So] A. Soshnikov, "Universality at the edge of the spectrum in Wigner random matrices", *Comm. Math. Phys.* **207** (1999), 697–733.

[St] E. M. Stein, *Harmonic analysis*, Princeton Univ. Press, Princeton, 1993.

[Ti] G. Tian, "On a set of polarized Kähler metrics on algebraic manifolds", *J. Diff. Geom.* **32** (1990), 99–130.

[TW] C. A. Tracy and H. Widom, "Universality of the distribution functions of random matrix theory II", preprint. See http://www.arxiv.org/abs/math-ph/9909001.

[Wo] N. M. J. Woodhouse, *Geometric quantization*, Clarendon Press, Oxford, 1992.

[Ze1] S. Zelditch, "Szegő kernels and a theorem of Tian", *Int. Math. Res. Notices* **6** (1998), 317–331.

[Ze2] S. Zelditch, "Level spacings for integrable quantum maps in genus zero", *Comm. Math. Phys.* **196** (1998), 289–318.

[ZZ] S. Zelditch and M. Zworski, "Spacings between phase shifts in a simple scattering problem", *Comm. Math. Phys.* **204** (1999), 709–729.

[Zi] P. Zinn-Justin, "Universality of correlation functions of Hermitian random matrices in an external field", *Comm. Math. Phys.* **194** (1998), 631–650.

PAVEL BLEHER
DEPARTMENT OF MATHEMATICAL SCIENCES
INDIANA UNIVERSITY–PURDUE UNIVERSITY INDIANAPOLIS
INDIANAPOLIS, IN 46202
UNITED STATES
bleher@math.iupui.edu

BERNARD SHIFFMAN
DEPARTMENT OF MATHEMATICS
JOHNS HOPKINS UNIVERSITY
BALTIMORE, MD 21218
UNITED STATES
shiffman@math.jhu.edu

STEVE ZELDITCH
DEPARTMENT OF MATHEMATICS
JOHNS HOPKINS UNIVERSITY
BALTIMORE, MD 21218
UNITED STATES
zelditch@math.jhu.edu

z-Measures on Partitions, Robinson–Schensted–Knuth Correspondence, and $\beta = 2$ Random Matrix Ensembles

ALEXEI BORODIN AND GRIGORI OLSHANSKI

ABSTRACT. We suggest a hierarchy of all the results known so far about the connection of the asymptotics of combinatorial or representation theoretic problems with "$\beta = 2$ ensembles" arising in the random matrix theory. We show that all such results are, essentially, degenerations of one general situation arising from so-called generalized regular representations of the infinite symmetric group.

Introduction

In last few years there appeared a number of papers indicating a strong connection of certain asymptotic problems of enumerative combinatorics and representation theory of symmetric groups with the random matrix theory; see [Baik et al. 1999a; 1999b; Baik and Rains 1999a; 1999b; Borodin 1998a; 1998b; 1999; $\geq$ 2001; Borodin and Olshanski 1998a; 1998b; 2000a; Borodin et al. 2000; Johansson 2000; 1999; Okounkov 1999b; 1999a; Olshanski 1998a; 1998b; Tracy and Widom 1998; 1999], for a partial list. Such a connection was also anticipated in earlier works [Regev 1981; Kerov 1993; 1994]. For other interesting connections see also [Borodin 2000b; Borodin and Okounkov 2000; Okounkov 2001].

In this paper we suggest a hierarchy of all the results known so far about the connection of the asymptotics of combinatorial or representation theoretic problems with so-called "$\beta = 2$ ensembles" arising in random matrix theory. (These ensembles are characterized by the property that their correlation functions have determinantal form with a scalar kernel; see below.) We show that all such results are, essentially, degenerations of one general situation arising from so-called generalized regular representations of the infinite symmetric group; see [Kerov et al. 1993] and Section 3 below.

Olshanski was supported by the Russian Foundation for Basic Research under grant 98–01–00303.

It is worth noting that though the hierarchy provides a clear understanding why this or that problem should have this or that asymptotics, the technical aspects of the proof are sometimes nontrivial and should not be underestimated.

Many claims cited below were recently proved by Kurt Johansson, we would like to thank him for keeping us informed about his work.

1. z-Measures

For $n = 1, 2, \ldots$, let $\mathbb{Y}_n$ denote the set of partitions of n, which will be identified with Young diagrams with n boxes. We agree that $\mathbb{Y}_0$ consists of a single element — the zero partition or the empty diagram $\varnothing$.

Given $\lambda \in \mathbb{Y}_n$, we write $|\lambda| = n$ and denote by $d = d(\lambda)$ the number of diagonal boxes in λ. We shall use the Frobenius notation [Macdonald 1995]

$$\lambda = (p_1, \ldots, p_d \mid q_1, \ldots, q_d).$$

Here $p_i = \lambda_i - i$ is the number of boxes in the i-th row of λ on the right of the i-th diagonal box; likewise, $q_i = \lambda'_i - i$ is the number of boxes in the i-th column of λ below the i-th diagonal box (λ' stands for the transposed diagram).

Note that

$$p_1 > \cdots > p_d \geq 0, \qquad q_1 > \cdots > q_d \geq 0, \qquad \sum_{i=1}^d (p_i + q_i + 1) = |\lambda|.$$

The numbers p_i, q_i are called the *Frobenius coordinates* of the diagram λ.

Let $b = (i, j)$ be a box of λ; here i, j are the row number and the column number of b. Recall the definition of the *content* and the *hook length* of b:

$$c(b) = j - i, \qquad h(b) = (\lambda_i - j) + (\lambda'_j - i) + 1.$$

We will consider two complex parameters z, z' such that the numbers $(z)_k (z')_k$ and $(-z)_k (-z')_k$ are real and strictly positive for any $k = 1, 2, \ldots$. Here and below

$$(a)_k = a(a+1) \ldots (a+k-1), \qquad (a)_0 = 1,$$

denotes the Pochhammer symbol.

The above assumption on z, z' means that one of the following two conditions holds:

- either $z' = \bar{z}$ and $z \in \mathbb{C} \setminus \mathbb{Z}$
- or $z, z' \in \mathbb{R}$ and there exists $m \in \mathbb{Z}$ such that $m < z, z' < m + 1$.

We set

$$t = zz'$$

and note that $t > 0$.

For a Young diagram λ let $\dim \lambda$ denote the number of the standard Young tableaux of shape λ. Equivalently, $\dim \lambda$ is the dimension of the irreducible representation (of the symmetric group of degree $|\lambda|$) corresponding to λ; see

[Macdonald 1995]. The well-known *hook formula* for $\dim \lambda$ has the following form (see [Macdonald 1995], for example):

$$\dim \lambda = \frac{|\lambda|!}{\prod_{b\in\lambda} h(b)}.$$

In the Frobenius coordinates this formula takes the form

$$\dim \lambda = \frac{|\lambda|!}{\prod_{i=1}^{d} p_i! q_i!} \det\left[\frac{1}{p_i + q_j + 1}\right].$$

We introduce a function on the Young diagrams depending on the parameters z, z':

$$M_{z,z'}(\lambda) = \frac{\prod_{b\in\lambda}(c(b)+z)(c(b)+z')}{(t)_{|\lambda|}} \cdot \frac{\dim^2 \lambda}{|\lambda|!}$$

$$= \frac{|\lambda|!}{(t)_{|\lambda|}} \cdot \prod_{b\in\lambda} \frac{(c(b)+z)(c(b)+z')}{h^2(b)}. \qquad (1\text{–}1)$$

We agree that $M_{z,z'}(\varnothing) = 1$. Thanks to our assumption on the parameters, $M_{z,z'}(\lambda) > 0$ for all λ.

PROPOSITION 1.1. *For any* $n = 0, 1, 2, \ldots$,

$$\sum_{\lambda \in \mathbb{Y}_n} M_{z,z'}(\lambda) = 1,$$

so that the restriction of $M_{z,z'}$ to $\mathbb{Y}_n$ is a probability distribution on $\mathbb{Y}_n$.

We shall denote this distribution by $M_{z,z'}^{(n)}$ and call it the *n-th level z-measure*.

Proposition 1.1 is an easy corollary of Proposition 3.1 below.

Let $\mathbb{Y} = \mathbb{Y}_0 \sqcup \mathbb{Y}_1 \sqcup \ldots$ denote the set of all Young diagrams. Consider the negative binomial distribution on the nonnegative integers, which depends on t and an additional parameter ξ, $0 < \xi < 1$:

$$\pi_{t,\xi}(n) = (1-\xi)^t \frac{(t)_n}{n!} \xi^n, \qquad n = 0, 1, \ldots.$$

For $\lambda \in \mathbb{Y}$ we set

$$M_{z,z',\xi}(\lambda) = M_{z,z'}(\lambda)\, \pi_{t,\xi}(|\lambda|).$$

By the construction, $M_{z,z',\xi}(\cdot)$ is a probability distribution on $\mathbb{Y}$, which can be viewed as a mixture of the finite distributions $M_{z,z'}^{(n)}$. From the formulas for $M_{z,z'}$ and $\pi_{t,\xi}$ we get an explicit expression for $M_{z,z',\xi}$:

$$M_{z,z',\xi}(\lambda) = (1-\xi)^t \xi^{|\lambda|} \prod_{b\in\lambda} \frac{(c(b)+z)(c(b)+z')}{h^2(b)}$$

$$= (1-\xi)^t \xi^{\sum_{i=1}^{d}(p_i+q_i+1)} t^d \prod_{i=1}^{d} \frac{(z+1)_{p_i}(z'+1)_{p_i}(-z+1)_{q_i}(-z'+1)_{q_i}}{p_i! p_i! q_i! q_i!} \det^2\left[\frac{1}{p_i+q_j+1}\right].$$

We shall call $M_{z,z',\xi}$ the *mixed z-measure*. Following a certain analogy with models of statistical physics (compare [Vershik 1996]) one may call $(\mathbb{Y}, M_{z,z',\xi})$ the *grand canonical ensemble*.

Let $\mathbb{Z}'$ denote the set of half-integers,

$$\mathbb{Z}' = \mathbb{Z} + \tfrac{1}{2} = \{\ldots, -\tfrac{3}{2}, -\tfrac{1}{2}, \tfrac{1}{2}, \tfrac{3}{2}, \ldots\},$$

and let $\mathbb{Z}'_+$ and $\mathbb{Z}'_-$ be the subsets of positive and negative half-integers, respectively. It will be sometimes convenient to identify both $\mathbb{Z}'_+$ and $\mathbb{Z}'_-$ with $\mathbb{Z}_+ = \{0, 1, 2, \ldots\}$ by making use of the correspondence $\pm(k + \tfrac{1}{2}) \leftrightarrow k$, where $k \in \mathbb{Z}_+$.

Denote by $\mathrm{Conf}(\mathbb{Z}')$ the space of all finite subsets of $\mathbb{Z}'$ which will be called *configurations*. We define an embedding $\lambda \mapsto X$ of the set $\mathbb{Y}$ of Young diagrams into the set $\mathrm{Conf}(\mathbb{Z}')$ of configurations in $\mathbb{Z}'$ as follows:

$$\lambda = (p_1, \ldots, p_d \,|\, q_1, \ldots, q_d) \mapsto X = \{p_1 + \tfrac{1}{2}, \ldots, p_d + \tfrac{1}{2}, -q_1 - \tfrac{1}{2}, \ldots, -q_d - \tfrac{1}{2}\}. \tag{1-2}$$

Under the identification $\mathbb{Z}' \simeq \mathbb{Z}_+ \sqcup \mathbb{Z}_+$, the map $\lambda \mapsto X$ is simply associating to λ the collection of its Frobenius coordinates. The image of the map consists exactly of the configurations X with the property $|X \cap \mathbb{Z}'_+| = |X \cap \mathbb{Z}'_-|$. We call such configurations *balanced*.

Under the embedding $\lambda \mapsto X$ the probability measure $M_{z,z',\xi}$ on $\mathbb{Y}$ turns into a probability measure on the balanced configurations in $\mathbb{Z}'$. According to the conventional terminology [Daley and Vere-Jones 1988], we get a point process on $\mathbb{Z}'$; let us denote it as $\mathcal{P}_{z,z',\xi}$.

The *n-th correlation function* $\rho_n^{(z,z',\xi)}(x_1, \ldots, x_n)$ of $\mathcal{P}_{z,z',\xi}$ is the probability that the random point configuration contains the points $x_1, \ldots, x_n$.

In [Borodin and Olshanski 2000a] we have computed all the correlation functions of $\mathcal{P}_{z,z',\xi}$. To state the result we need some notation.

Consider the following functions in u depending on z, z', ξ as parameters (compare [Borodin and Olshanski 2000a]):

$$\psi_\pm(u) = t^{1/2}\,\xi^{u+1/2}\,(1-\xi)^{\pm(z+z')}\,\frac{\Gamma(u+1\pm z)\Gamma(u+1\pm z')}{\Gamma(1\pm z)\Gamma(1\pm z')\Gamma(u+1)\Gamma(u+1)},$$

$$P_\pm(u) = (\psi_\pm(u))^{1/2}\, F\bigl(\mp z, \mp z'; u+1; \tfrac{\xi}{\xi-1}\bigr),$$

$$Q_\pm(u) = \frac{t^{1/2}\xi^{1/2}\,(\psi_\pm(u))^{1/2}}{1-\xi}\,\frac{F\bigl(1\mp z, 1\mp z'; u+2; \tfrac{\xi}{\xi-1}\bigr)}{u+1}.$$

Here $F(a,b;c;w)$ is the Gauss hypergeometric function.

THEOREM 1.2 [Borodin and Olshanski 2000a]. *The correlation functions of $\mathcal{P}_{z,z',\xi}$ have the form*

$$\rho_n^{(z,z',\xi)}(x_1, \ldots, x_n) = \det[K(x_i, x_j)]_{i,j=1}^n, \qquad x_1, \ldots, x_n \in \mathbb{Z}',$$

where
$$K(x,y) = \frac{F_1(x)G_1(y) + F_2(x)G_2(y)}{x-y}, \qquad (1\text{--}3)$$

with

$$\begin{aligned}
F_1(x) &= \begin{cases} -Q_+\!\left(x-\tfrac{1}{2}\right) & \text{for } x > 0, \\ P_-\!\left(-x-\tfrac{1}{2}\right) & \text{for } x < 0; \end{cases} \\
F_2(x) &= \begin{cases} P_+\!\left(x-\tfrac{1}{2}\right) & \text{for } x > 0, \\ Q_-\!\left(-x-\tfrac{1}{2}\right) & \text{for } x < 0; \end{cases} \\
G_1(x) &= \begin{cases} P_+\!\left(x-\tfrac{1}{2}\right) & \text{for } x > 0, \\ -Q_-\!\left(-x-\tfrac{1}{2}\right) & \text{for } x < 0; \end{cases} \\
G_2(x) &= \begin{cases} Q_+\!\left(x-\tfrac{1}{2}\right) & \text{for } x > 0, \\ P_-\!\left(-x-\tfrac{1}{2}\right) & \text{for } x < 0. \end{cases}
\end{aligned} \qquad (1\text{--}4)$$

We call $K(x,y)$ the *hypergeometric kernel*.

REMARKS 1.3. 1. The hypergeometric kernel has no singularity on the diagonal: the numerator of (1.3) vanishes if $x = y$.

2. The hypergeometric kernel satisfies the relation
$$K(x,y) = \operatorname{sgn}(x)\operatorname{sgn}(y) K(y,x). \qquad (1\text{--}5)$$

This shows that the kernel is Hermitian with respect to the indefinite inner product in $\ell^2(\mathbb{Z}') = \ell^2(\mathbb{Z}'_+) \oplus \ell^2(\mathbb{Z}'_-)$ given by the operator $\operatorname{id} \oplus (-\operatorname{id})$.

3. The restriction of the hypergeometric kernel to $\mathbb{Z}'_+$ has the form
$$\frac{P_+(x-\tfrac{1}{2})Q_+(y-\tfrac{1}{2}) - P_+(y-\tfrac{1}{2})Q_+(x-\tfrac{1}{2})}{x-y}.$$

Note that this kernel is symmetric. We will call it the *positive part* of the hypergeometric kernel.

4. Kernels with the symmetry (1.5) appeared before in works of mathematical physicists on solvable models of systems with positive and negative charged particles; see [Alastuey and Forrester 1995; Cornu and Jancovici 1987; 1989; Gaudin 1985; Forrester 1986; 1988; 1989] and references therein. The mixed z-measure can also be interpreted as a model for positive and negative particles on $\mathbb{Z}'$: positive particles may occupy locations in $\mathbb{Z}'_+$, negative — in $\mathbb{Z}'_-$. The square of the Cauchy determinant
$$\det{}^2\left[\frac{1}{p_i+q_j+1}\right] = \frac{\prod_{i<j}[(p_i-p_j)(q_i-q_j)]^2}{\prod_{i,j}(p_i+q_j+1)^2}$$

in the formula for $M_{z,z',\xi}$ above encodes the logarithmic interaction of the charged particles.

5. The papers [Okounkov 1999a; 2001] contain another derivation of Theorem 1.2 and a generalization of the mixed z-measures $M_{z,z',\xi}$.

2. Three Versions of the Robinson–Schensted–Knuth Correspondence

A description of the RSK algorithm can be found in [Fulton 1997; Sagan 1991].

We start with the "widest" version of the RSK correspondence due to Knuth [1970].

Denote by $B_{k,l}^n$ the set of 'bijections' between two sets of size n, the first set consists of (possibly repeated) numbers from 1 to k and the second set consists of (possibly repeated) numbers from 1 to l. Such bijections are in one-to-one correspondence with matrices of size $k \times l$ with nonnegative integral entries, total sum of entries equal to n: the (i,j)–entry shows how many times the element $i \in \{1, \ldots, k\}$ is associated with the element $j \in \{1, \ldots, l\}$. Clearly,

$$|B_{k,l}^n| = \binom{kl+n-1}{n} = \frac{kl(kl+1)\cdots(kl+n-1)}{n!}.$$

The RSK algorithm establishes a bijection of $B_{k,l}^n$ and the set of ordered pairs of semi-standard Young tableaux of the same shape with n boxes, the first tableau has entries from the set $\{1, \ldots, k\}$, while the second — from the set $\{1, \ldots, l\}$. (Recall that a semi-standard Young tableau is a tableau whose entries are weakly increasing along the rows and strictly increasing along the columns. In a standard tableau we have strictly increasing entries in both directions.)

As is well-known, the number of semi-standard Young tableaux of shape λ with entries from $\{1, \ldots, k\}$ is equal to the value of the Schur symmetric function $s_\lambda(1, 1, \ldots, 1, 0, 0, \ldots)$ where the number of 1's equals k. This value can be written in the following form (see [Macdonald 1995, I.3, Ex. 4], for example),

$$s_\lambda(\underbrace{1, 1, \ldots, 1}_{k}, 0, 0, \ldots) = \prod_{b \in \lambda} \frac{c(b) + k}{h(b)}.$$

Recall also that the number of standard Young tableaux of shape λ is $\dim \lambda$.

Hence, if we consider the uniform probability distribution on $B_{k,l}^n$, then, with respect to its image on the set of Young diagrams with n boxes, the probability of a Young diagram $\lambda \in \mathbb{Y}_n$ equals

$$\frac{n!}{kl(kl+1)\cdots(kl+n-1)} \prod_{b \in \lambda} \frac{(c(b)+k)(c(b)+l)}{h^2(b)}.$$

Comparing this with (1.1) we conclude that this distribution coincides with $M_{z,z'}^{(n)}$ for $z = k$, $z' = l$.

Note that these values of z, z' do not satisfy our conditions on the parameters imposed in Section 1. The reason is that for such z, z' the values of $M_{z,z'}^{(n)}$ can be zero, for example $M_{z,z'}^{(n)}(\lambda) = 0$ for all λ with length (number of nonzero parts) greater than $\min\{k, l\}$. However, all values of $M_{z,z'}^{(n)}$ remain nonnegative. We consider such situation as a specific degeneration of the regular picture (when the values $M_{z,z'}^{(n)}$ are strictly positive).

Two other (earlier) versions of the RSK correspondence are due to Robinson [1938] and Schensted [1961].

Denote by $B_{k,\infty}^n$ the set of words of length n built from the alphabet $\{1,\ldots,k\}$ (our notation will become clear soon). It is a subset of $B_{k,n}^n$ characterized by the property that the numbers in the second set are all distinct (they encode the order of letters $\{1,\ldots,k\}$ in the word). It means that in the corresponding matrices of size $k \times n$ every column has exactly one nonzero element which is equal to 1. Obviously, $|B_{k,\infty}^n| = k^n$.

In this case the RSK algorithm establishes a bijection of $B_{k,\infty}^n$ and the set of ordered pairs of Young tableaux of the same shape with n boxes; the first tableau is semi-standard and it is filled with numbers from 1 to k, and the second tableau is standard. This means that the probability of a Young diagram $\lambda \in \mathbb{Y}_n$ with respect to the image of the uniform distribution on $B_{k,\infty}^n$ equals

$$k^{-n} \prod_{b \in \lambda} \frac{c(b) + k}{h(b)} \cdot \dim \lambda.$$

It is easy to see from (1.1) that this is the limit of $M_{z,z'}^{(n)}$ for $z = k$ and $z' \to \infty$.

Finally, if we forbid for both sets in the definition of $B_{k,l}^n$ to have repetitions, then we get the symmetric group S_n. It would be logical to denote the symmetric group by $B_{\infty,\infty}^n$; see below. In the language of matrices, it means that we consider $n \times n$ matrices with 0's and 1's such that in each row and each column there is exactly one nonzero element. Clearly, $|S_n| = n!$.

The RSK algorithm provides a bijection of the set of permutations of n symbols and the set of ordered pairs of standard Young tableaux of the same shape with n boxes. Hence, the probability of a Young diagram $\lambda \in \mathbb{Y}_n$ with respect to the distribution coming from the uniform distribution on S_n equals $\dim^2 \lambda / n!$. This distribution on the Young diagrams is called the *Plancherel distribution*. The relation (1.1) easily implies that the Plancherel distribution is the limit of $M_{z,z'}^{(n)}$ as $z, z' \to \infty$.

For bijections from $B_{k,l}^n$, $B_{k,\infty}^n$, $B_{\infty,\infty}^n$ we define a weakly increasing subsequence to be a sequence of pairs of associated elements, first element is from the first set, second element is from the second set, which weakly increase in each element. Under the RSK correspondence the length of the longest weakly increasing subsequence of a bijection coincides with the length of the first row of the corresponding Young diagram in all three cases described above; see [Knuth 1970; Schensted 1961].

3. Harmonic Analysis on the Infinite Symmetric Group

For more detailed discussion of the material of this section see [Kerov et al. 1993; Vershik and Kerov 1981a; Olshanski 1998a].

We define the infinite symmetric group $S(\infty)$ as the inductive limit of the finite symmetric groups S_n with respect to natural embeddings $S_n \to S_{n+1}$. Equivalently, $S(\infty)$ is the group of *finite* permutations of the set $\{1,2,\ldots\}$.

By a *character* of $S(\infty)$ (in the sense of von Neumann) we mean any central, positive definite function χ on $S(\infty)$, normalized by the condition $\chi(e)=1$. We assign to χ a function $M(\lambda)$ on the set $\mathbb{Y} = \sqcup \mathbb{Y}_n$ of Young diagrams as follows: for any $n=1,2,\ldots$,

$$\chi\big|_{S_n} = \sum_{\lambda \in \mathbb{Y}_n} M(\lambda) \frac{\chi^\lambda}{\dim \lambda},$$

where χ^λ denotes the irreducible character of S_n (in the conventional sense), indexed by $\lambda \in \mathbb{Y}_n$, and $\dim \lambda = \chi^\lambda(e)$ is its dimension. Let $M^{(n)}$ stand for the restriction of the function M to $\mathbb{Y}_n$; this is a probability distribution on $\mathbb{Y}_n$. Conversely, let $M = \{M^{(n)}\}$ be a function on $\mathbb{Y}$ such that each $M^{(n)}$ is a probability distribution; then M corresponds to a character χ if (and only if) the distributions $M^{(n)}$ obey a natural coherence relation, which comes from the classical Young branching rule for the irreducible characters of the finite symmetric groups; see [Vershik and Kerov 1981a; Olshanski 1998a]. (Equivalently, the function $\varphi(\lambda) = M(\lambda)/\dim \lambda$ must be a *harmonic function on the Young graph* $\mathbb{Y}$ in the sense of Vershik and Kerov; see [Vershik and Kerov 1981a; Olshanski 1998a].)

PROPOSITION 3.1. *The z-measures $M^{(n)}_{z,z'}$ introduced in Section 1 satisfy the coherence relation mentioned above and, consequently, define a character $\chi_{z,z'}$ of $S(\infty)$.*

Several direct proofs of the proposition are known. A simple proof is given in [Olshanski 1998a, §7]. About generalizations, see [Kerov 2000; Borodin and Olshanski 2000b].

Note that the degenerations $M^{(n)}_{k,l}$, $M^{(n)}_{k,\infty}$, and $M^{(n)}_{\infty,\infty}$ of the z-measures also correspond to certain characters, which will be denoted as $\chi_{k,l}$, $\chi_{k,\infty}$, and $\chi_{\infty,\infty}$, respectively. The character $\chi_{\infty,\infty}$ is easily described: it takes value 1 at $e \in S(\infty)$ and vanishes at all other elements of the group.

By the very definition of the characters of $S(\infty)$, they form a convex set. The extreme points of that set are called the *indecomposable* characters, and the other points are called *decomposable* characters.

According to a remarkable theorem due to Thoma [1964] (see also [Vershik and Kerov 1981a; Wassermann 1981; Kerov et al. 1998]), the indecomposable characters of $S(\infty)$ are parametrized by the points of the infinite dimensional simplex

$$\Omega = \left\{ \alpha_1 \geq \alpha_2 \geq \ldots \geq 0,\ \beta_1 \geq \beta_2 \geq \ldots \geq 0 \ \bigg|\ \sum_{i=1}^\infty (\alpha_i + \beta_i) \leq 1 \right\},$$

which is called the *Thoma simplex*. Given a point $\omega = (\alpha, \beta) \in \Omega$, we denote by $\chi^{(\omega)}$ the corresponding indecomposable character.

The characters $\chi_{k,\infty}$ and $\chi_{\infty,\infty}$ are indecomposable: the former corresponds to the point ω with $\alpha_1 = \alpha_2 = \cdots = \alpha_k = 1/k$ (all other coordinates are zero), and the latter corresponds to the point $\omega = (0,0)$ (all coordinates are zero). The characters $\chi_{z,z'}$ (with z, z' satisfying the conditions of Section 1) and $\chi_{k,l}$ are decomposable.

Every character can be uniquely represented as a convex combination of the indecomposable ones,
$$\chi = \int_\Omega \chi^{(\omega)} P(d\omega).$$
Here P is a probability measure on Ω, which is called the *spectral measure* of the character χ. Moreover, any probability measure on Ω is a spectral measure of a character, so that the set of characters of $S(\infty)$ is isomorphic, as a convex set, to the set of probability measures on the Thoma simplex. Under this isomorphism, indecomposable characters correspond to delta measures on Ω.

Given a concrete decomposable character χ, a natural problem is to describe explicitly its spectral measure P. This will be referred to as the *problem of harmonic analysis*.

This problem is readily solved for the degenerate characters $\chi_{k,l}$:

PROPOSITION 3.2. *Let* $\chi = \chi_{k,l}$ *with* $k \le l$. *Set* $a = l - k$. *Then the spectral measure is concentrated on the* $(k-1)$-*dimensional subsimplex*
$$\alpha_1 + \cdots + \alpha_k = 1, \qquad \alpha_{k+1} = \alpha_{k+2} = \cdots = \beta_1 = \beta_2 \ldots = 0$$
of Ω *and has density*
$$\mathrm{const} \cdot \prod_{1 \le i < j \le k} (\alpha_i - \alpha_j)^2 \cdot \prod_{i=1}^k \alpha_i^a$$
with respect to the Lebesgue measure.

For the characters $\chi_{z,z'}$ with nonintegral parameters the problem of harmonic analysis is highly nontrivial and will be briefly discussed at the end of Section 8. One of the first results in this direction is as follows (recall that two measures are called disjoint if there exist disjoint Borel sets supporting them).

PROPOSITION 3.3. *Let* $P_{z,z'}$ *denote the spectral measure of* $\chi_{z,z'}$. *Except the obvious equality* $P_{z,z'} = P_{z',z}$, *the measures* $P_{z,z'}$ *are pairwise disjoint.*

Notice the following general result which relates the spectral measure P of a character χ to the finite probability distributions $M^{(n)}$. Let us embed $\mathbb{Y}_n$ into Ω by:
$$\lambda = (p_1, \ldots, p_d \mid q_1, \ldots, q_d) \in \mathbb{Y}_n$$
$$\mapsto \left\{ \frac{p_1 + \frac{1}{2}}{n}, \ldots, \frac{p_d + \frac{1}{2}}{n}, 0, 0, \ldots; \frac{q_1 + \frac{1}{2}}{n}, \ldots, \frac{q_d + \frac{1}{2}}{n}, 0, 0, \ldots \right\} \in \Omega. \quad (3\text{-}1)$$

PROPOSITION 3.4. *As $n \to \infty$, the push-forwards of the measures $M^{(n)}$ under these embeddings weakly converge to P.*

This is a special case of a more general result proved in [Kerov et al. 1998].

The characters of $S(\infty)$ can be related to representations in two ways.

The first way is rather evident. Each character χ is a positive definite function on $S(\infty)$, so that it determines a unitary representation of $S(\infty)$, which will be denoted as $\Pi(\chi)$. When χ is indecomposable, $\Pi(\chi)$ is a *factor* representation of finite type in the sense of von Neumann; see [Thoma 1984].

The second way is a bit more involved. Set $G = S(\infty) \times S(\infty)$ and let K denote the diagonal subgroup in G, which is isomorphic to $S(\infty)$. We interpret χ as a function on the first copy of $S(\infty)$, which is a subgroup of G, and then extend it to the whole group G by the formula

$$\psi(g_1, g_2) = \chi(g_1 g_2^{-1}), \qquad g_1, g_2 \in S(\infty).$$

Note that ψ is the only extension of χ that is a K-biinvariant function on G. The function ψ is also positive definite, so that one can assign to it a unitary representation in the canonical way. This representation of the group G will be denoted by $T(\chi)$. By the very construction, it possesses a distinguished K-invariant vector.

Note that $\Pi(\chi)$ coincides with the restriction of $T(\chi)$ to the first copy of $S(\infty)$. If χ is indecomposable, $\chi = \chi^{(\omega)}$, then $T(\chi) = T(\chi^{(\omega)})$ is irreducible. The representations of the form $T(\chi^{(\omega)})$ are exactly the irreducible unitary representations of the group G possessing a K-invariant vector (such a vector is unique, within a scalar factor). Thus, the Thoma simplex can be identified with the *spherical dual* to (G, K). It is worth noting that the irreducible representations of the form $T(\chi^{(\omega)})$ (except two trivial cases) are *not* tensor products of irreducible representations of the factors $S(\infty)$. For more details about the representations $T(\chi^{(\omega)})$, see [Vershik and Kerov 1981b; Wassermann 1981; Olshanski 1989].

The representation $T(\chi_{\infty,\infty})$ is readily described: it coincides with the natural representation of the group G realized in the Hilbert space $\ell^2(G/K)$. Note that G/K is identified with the group $S(\infty)$ on which G acts by left and right shifts, so that $T(\chi_{\infty,\infty})$ may be called the *regular* representation of G. As for $\Pi(\chi_{\infty,\infty})$, it provides a classical realization of the hyperfinite von Neumann factor of type II_1.

The representations $T(\chi_{z,z'})$ are called the *generalized regular* representations of G. The term is motivated by the fact that each $T(\chi_{z,z'})$ can be realized as the inductive limit of a chain of the form

$$\cdots \to (\mathrm{Reg}_n, v_n) \to (\mathrm{Reg}_{n+1}, v_{n+1}) \to \cdots,$$

where Reg_n stands for the (bi)regular representation of the group $S_n \times S_n$ in the space of functions on S_n and v_n is a certain vector in that space, depending on the parameters z, z' (v_n is given by a certain central function on S_n). When

$z' = \bar{z}$, the generalized regular representations admit a very nice realization in certain L^2 spaces of functions defined on a compactification of the group $S(\infty)$. We refer to [Kerov et al. 1993] for the exposition of this construction.

Finally, note that for any decomposable character χ, the decomposition of $T(\chi)$ into irreducible representations is governed by the spectral measure P:

$$T(\chi) = \int_\Omega T(\chi^{(\omega)}) P(d\omega).$$

4. Mixing

Theorem 1.2 computes the correlation functions of a point process obtained from the distributions $M_{z,z'}^{(n)}$ mixed together by the negative binomial distribution with parameters (t, ξ); see Section 1. In this section we consider the degenerations of the mixing procedure in the cases when z and z' are integers, when z is an integer and $z' \to \infty$, and when z and z' both tend to infinity while $\xi \to 1$.

If $z = k$ and $z' = l$ are positive integers then nothing interesting happens — we have to mix the corresponding measures on $\mathbb{Y}_n$'s by the negative binomial distribution with parameters (kl, ξ).

If $z = k$ is a positive integer and $z' \to \infty$, or $z \to \infty$ and $z' \to \infty$ then $t = zz'$ goes to ∞. If we keep $\theta = t\xi$ fixed (hence, $\xi \to 0$) then the negative binomial distribution degenerates to the Poisson distribution with parameter θ. The mixing procedure with Poisson distribution is called *poissonization*.

The degeneration $\xi \to 1$ is a bit more delicate. Let us embed $\mathbb{Z}'$ into the punctured line $\mathbb{R}^* = \mathbb{R} \setminus \{0\}$ and then rescale the lattice by multiplying the coordinates of its points by $(1-\xi)$. Then the coordinates of the point configuration in $\mathbb{R}^*$ that corresponds to $\lambda \in \mathbb{Y}_n$ (as defined in (1.2)) after rescaling differ from the coordinates of the image of λ in Ω (as defined in (3.1)) by the scaling factor $(1-\xi)n$.

The discrete distribution on the positive semiaxis with

$$\text{Prob}\{(1-\xi)n\} = (1-\xi)^t \frac{(t)_n}{n!} \xi^n, \quad n = 0, 1, 2, \ldots,$$

which depends on the parameter $\xi \in (0,1)$, converges, as $\xi \to 1$, to the gamma distribution with parameter t

$$\gamma(ds) = \frac{s^{t-1}}{\Gamma(t)} e^{-s} ds.$$

This brings us to the following construction. Consider the space $\widetilde\Omega = \Omega \times \mathbb{R}_+$ with the probability measure

$$\widetilde P_{z,z'} = P_{z,z'} \otimes \frac{s^{t-1}}{\Gamma(t)} e^{-s} ds.$$

Let us embed $\mathbb{Y} = \mathbb{Y}_0 \sqcup \mathbb{Y}_1 \sqcup \mathbb{Y}_2 \sqcup \ldots$ into $\widetilde\Omega = \Omega \times \mathbb{R}_+$ by sending a Young diagram $\lambda \in \mathbb{Y}_n$ to the pair consisting of its image in Ω and the number $(1-\xi)n$.

PROPOSITION 4.1. *The push-forwards of $M_{z,z',\xi}$ under the embeddings described above converge, as $\xi \to 1$, to $\widetilde{P}_{z,z'}$.*

Exact claims with a detailed description of this convergence will appear in [Borodin and Olshanski $\geq$ 2001a].

5. Ensembles

We introduce several terms that will be used below.

By the word *ensemble* throughout this paper we will mean a stochastic point process (i.e, a probability measure on the space of point configurations) whose correlation functions $\rho_n(x_1, \ldots, x_n)$ are given by determinantal formulas of the form

$$\rho_n(x_1, \ldots, x_n) = \det[K(x_i, x_j)]_{i,j=1}^n,$$

where $K(x, y)$ is a certain kernel. We will call $K(x, y)$ the *correlation kernel*.

The process $\mathcal{P}_{z,z',\xi}$ is an example; the ensemble lives on $\mathbb{Z}'$ and the correlation kernel is the hypergeometric kernel; see Theorem 1.2. We will call it the *discrete z-ensemble*.

In all our examples the points of the ensembles will vary in discrete or continuous subsets of the real line. Such a subset will be called the *phase space* of the corresponding ensemble. For example, $\mathbb{Z}'$ is the phase space of $\mathcal{P}_{z,z',\xi}$.

There is a class of *orthogonal polynomial ensembles* characterized by the condition of having a fixed finite number of points, say k, the joint probability distribution of which has the density

$$\text{const} \cdot \prod_{1 \leq i,j \leq k} (x_i - x_j)^2 \prod_{i=1}^k w(x_i)$$

with respect to either the Lebesgue measure, if the phase space is continuous, or counting measure, if the phase space is discrete. A standard argument due to Dyson [1962] (see also [Mehta 1991]) shows that the correlation kernel is the Christoffel–Darboux kernel of order k for orthogonal polynomials on the phase space with respect to the weight function $w(x)$. If a_n denotes the top degree coefficient of p_n,

$$p_n(x) = a_n x^n + \{\text{lower degree terms}\},$$

and $h_n = \|p_n\|^2$ then the kernel has the form

$$K(x, y) = \frac{a_{k-1}}{a_k h_{k-1}} \frac{p_k(x) p_{k-1}(y) - p_{k-1}(x) p_k(y)}{x - y} \sqrt{w(x) w(y)}.$$

Below we will consider the following orthogonal polynomial ensembles:

- *Laguerre ensemble*: phase space $\mathbb{R}_+$, weight function $w(x) = x^a e^{-x}$, $a > -1$;
- *Hermite ensemble*: phase space $\mathbb{R}$, weight function $w(x) = e^{-x^2}$;
- *Charlier ensemble*: phase space $\mathbb{Z}_+$, weight function $w(x) = \theta^x/x!$, $\theta > 0$;

- *Meixner ensemble*: phase space $\mathbb{Z}_+$, weight function $w(x) = (a+1)_x \xi^x / x!$, $a > -1$, $\xi \in (0,1)$.

Corresponding normalizing constants for the orthogonal polynomials can be found in [Erdélyi 1953b; Koekoek and Swarttouw 1998; Nikiforov et al. 1991]. The Christoffel–Darboux kernels for these ensembles will be called *Laguerre, Hermite, Charlier,* and *Meixner kernels*, respectively.

We will also deal with the *Airy ensemble* (see [Forrester 1993; Tracy and Widom 1994a]): the phase space is $\mathbb{R}$, the correlation kernel is

$$\frac{A(x)A'(y) - A'(x)A(y)}{x-y}$$

where $A(x)$ is the Airy function.

Two other ensembles that we will need are the ensemble arising from poissonized Plancherel distributions for symmetric groups (see [Borodin et al. 2000] and Section 6 below) with the phase space $\mathbb{Z}'$ and the kernel of the form (1.3), (1.4) where

$$P_\pm(x) = \theta^{\frac{1}{4}} J_x(2\sqrt{\theta}), \quad Q_\pm(x) = \theta^{\frac{1}{4}} J_{x+1}(2\sqrt{\theta}), \tag{5-1}$$

$\theta > 0$ is a parameter, $J_\nu(x)$ is the Bessel function; and the ensemble arising from the problem of harmonic analysis on $S(\infty)$ described in Section 3 (see [Borodin and Olshanski 1998a] and Section 6 below) with the phase space $\mathbb{R}^*$ and the kernel of the form (1.3) where

$$F_1(x) = \begin{cases} -\mathcal{Q}_+(x) & \text{for } x > 0, \\ \mathcal{P}_-(-x) & \text{for } x < 0; \end{cases}$$

$$F_2(x) = \begin{cases} \mathcal{P}_+(x) & \text{for } x > 0, \\ \mathcal{Q}_-(-x) & \text{for } x < 0; \end{cases}$$

$$G_1(x) = \begin{cases} \mathcal{P}_+(x) & \text{for } x > 0, \\ -\mathcal{Q}_-(-x) & \text{for } x < 0; \end{cases}$$

$$G_2(x) = \begin{cases} \mathcal{Q}_+(x) & \text{for } x > 0, \\ \mathcal{P}_-(-x) & \text{for } x < 0, \end{cases}$$

where we have defined

$$\mathcal{P}_\pm(x) = \frac{(zz')^{1/4}}{(\Gamma(1\pm z)\Gamma(1\pm z')\,x)^{1/2}} \, W_{\frac{\pm(z+z')+1}{2},\frac{z-z'}{2}}(x),$$

$$\mathcal{Q}_\pm(x) = \frac{(zz')^{3/4}}{(\Gamma(1\pm z)\Gamma(1\pm z')\,x)^{1/2}} \, W_{\frac{\pm(z+z')-1}{2},\frac{z-z'}{2}}(x),$$

z, z' satisfy the assumptions stated in Section 1, and $W_{\kappa,\mu}(x)$ is the Whittaker function; see [Erdélyi 1953a]. We will call these ensembles the *Plancherel ensemble* and the *continuous z-ensemble*, respectively. The kernel for the first one will be called the *Plancherel kernel*, for the second one — the *Whittaker kernel*.

When an ensemble lives on $\mathbb{R}^*$ or $\mathbb{Z}'$, one may single out its *positive part* — the restriction to $\mathbb{R}_+ \subset \mathbb{R}^*$ or $\mathbb{Z}'_+ \subset \mathbb{Z}'$, respectively. The correlation kernel of the positive part is the corresponding restriction of the correlation kernel of the initial ensemble. We will use the term "positive part of the kernel" for such restrictions.

The positive part of the Plancherel kernel has been independently found in [Johansson 1999] where it was called the *discrete Bessel kernel*; see Section 9.

The random point configurations of the discrete z-ensemble and of the Plancherel ensemble are finite with probability 1. The random point configuration of the Airy ensemble is, with probability 1, infinite, bounded from above, unbounded from below, and has no finite accumulation points. The random point configuration of the continuous z-ensemble is, with probability 1, infinite, bounded from above and below, and has zero (which is *not* in the phase space) as its only accumulation point.

6. Correlations After Mixing

In accordance with the notation of Section 2, it is natural to denote the measures on Young diagrams with n boxes coming from $B_{k,l}^n$, $B_{k,\infty}^n$, $B_{\infty,\infty}^n$ as $M_{k,l}^{(n)}$, $M_{k,\infty}^{(n)}$, $M_{\infty,\infty}^{(n)}$, respectively, and the corresponding mixtures (i.e., measures on the set of all Young diagrams) as $M_{k,l,\xi}$, $M_{k,\infty,\theta}$, $M_{\infty,\infty,\theta}$. We want to see how the hypergeometric kernel will behave in these degenerate cases.

We start with the case when z, z' are positive integers, say, $z = k$, $z' = l$, $k \leq l$. Set $a = l - k$.

PROPOSITION 6.1 [Borodin and Olshanski 2000a], [Johansson 2000]. *Let $\lambda = (p_1, \ldots, p_d \mid q_1, \ldots, q_d) \in \mathbb{Y}$ be distributed according to $M_{k,l,\xi}$. Then the distribution of points $\{k + p_1, \ldots, k + p_d\}$ coincides with the restriction of the k-point Meixner ensemble with parameters (a, ξ) to the set $\{k, k+1, \ldots\}$.*

This claim corresponds to the fact that the hypergeometric functions entering the hypergeometric kernel become Meixner polynomials if z or z' is integral; see [Borodin and Olshanski 2000a]. Furthermore, the positive part of the hypergeometric kernel becomes the Christoffel–Darboux kernel for Meixner polynomials (shifted by k).

Now we pass to $M_{k,\infty,\theta}$.

PROPOSITION 6.2 [Johansson 1999]. *Let $\lambda = (p_1, \ldots, p_d \mid q_1, \ldots, q_d) \in \mathbb{Y}$ be distributed according to $M_{k,\infty,\theta}$. Then the distribution of points $\{k + p_1, \ldots, k + p_d\}$ coincides with the restriction of the k-point Charlier ensemble with parameter θ to the set $\{k, k+1, \ldots\}$.*

The easiest way to see this is to examine the degeneration of Meixner polynomials with parameters (a, ξ) to Charlier polynomials with parameter θ when $a \to \infty$, $\theta = k(k+a)\xi$ is fixed.

Next, consider the situation when z and z' both go to ∞.

PROPOSITION 6.3 [Borodin et al. 2000]. *Let $\lambda = (p_1, \ldots, p_d \,|\, q_1, \ldots, q_d) \in \mathbb{Y}$ be distributed according to $M_{\infty,\infty,\theta}$. Then the random point configuration*

$$\{p_1 + \tfrac{1}{2}, \ldots, p_d + \tfrac{1}{2}, -q_1 - \tfrac{1}{2}, \ldots, -q_d - \tfrac{1}{2}\}$$

forms the Plancherel ensemble with parameter θ.

This claim corresponds to the degeneration of the hypergeometric function to the Bessel J-function when first two parameters go to infinity and the argument goes to zero so that the product of these three numbers is fixed (and equals θ).

As for the representation theoretic picture, we have the following claim.

PROPOSITION 6.4 [Borodin and Olshanski 1998a; Borodin ≥ 2001]. *Let*

$$((\alpha, \beta), s) \in \widetilde{\Omega} = \Omega \times \mathbb{R}_+$$

be distributed according to $\widetilde{P}_{z,z'}$. Then the random point configuration

$$(s\alpha_1, s\alpha_2, \ldots, -s\beta_1, -s\beta_2, \ldots)$$

forms the continuous z-ensemble.

REMARK 6.5. When one of the parameters z, z' becomes integral, say, $z = k \in \{1, 2, \ldots\}$, and $z' = z + a$, $a > -1$, the Whittaker kernel degenerates to the Laguerre kernel of order k with parameter a. Then Proposition 6.4 implies that the measure $\widetilde{P}_{z,z'}$ gets concentrated on the finite-dimensional subset of $\widetilde{\Omega} = \Omega \times \mathbb{R}_+$ where $\alpha_{k+1} = \alpha_{k+2} = \cdots = \beta_1 = \beta_2 = \cdots = 0$, and on this subset, in the new coordinates $x_i = s\alpha_i$ (s is the coordinate on $\mathbb{R}_+$), it equals

$$\text{const} \cdot \prod_{1 \leq i < j \leq k} (x_i - x_j)^2 \prod_{i=1}^{k} x_i^a e^{-x_i} dx_i.$$

See [Borodin and Olshanski 1998b, Remark 2.4]. This agrees with Proposition 3.2.

7. Asymptotics When Mixing Parameters Tend to a Limit

We start with $M_{k,l,\xi}$. Assume that $a = l - k \geq 0$. Then Proposition 3.4, the degeneration of the hypergeometric kernel to the Whittaker kernel (Proposition 6.4) and the coincidence of the Whittaker kernel with the Laguerre kernel when at least one parameter is integral (Remark 6.5) imply the following claim.

PROPOSITION 7.1. *Let $\lambda \in \mathbb{Y}$ be distributed according to $M_{k,l,\xi}$. Then the random point configuration $\{(1-\xi)\lambda_1, \ldots, (1-\xi)\lambda_k\}$ converges, as $\xi \to 1$, to the Laguerre ensemble.*

Now we pass to $M_{k,\infty,\theta}$. The fact that the character of $S(\infty)$ corresponding to $M_{k,\infty}^{(n)}$ is indecomposable and corresponds to the point $\alpha_1 = \cdots = \alpha_k = 1/k$ in Ω (see Section 3) leads to the following statement.

Consider the embedding of the set of Young diagrams with length at most k into $\mathbb{R}_+^k$ defined by dividing the lengths of rows of a Young diagram by θ.

PROPOSITION 7.2. *Under the embeddings described above $M_{k,\infty,\theta}$ weakly converges to the delta measure at the point $(1/k, \ldots, 1/k)$ as $\theta \to \infty$.*

One can also ask about fluctuations of $M_{k,\infty,\theta}$ around the limit delta measure. Johansson proved the following statement.

PROPOSITION 7.3 [Johansson 1999]. *Let $\lambda \in \mathbb{Y}$ be distributed according to $M_{k,\infty,\theta}$. Then the random point configuration*

$$\left\{\frac{\lambda_1 - \theta/k}{\sqrt{2\theta/k}}, \ldots, \frac{\lambda_k - \theta/k}{\sqrt{2\theta/k}}\right\}$$

converges, as $\theta \to \infty$, to the k-point Hermite ensemble.

The convergence of distribution of the first point of the random configuration from Proposition 7.3 was proved by Tracy and Widom, [Tracy and Widom 1999].

Propositions 7.2 and 7.3 correspond to a certain degeneration of Charlier polynomials to Hermite polynomials which follows from a more general degeneration of Laguerre polynomials with large argument and parameter to Hermite polynomials; see [Temme 1990].

The most interesting case is $M_{\infty,\infty,\theta}$. The reason is simple: the number of points (rows of Young diagrams) is unbounded in this case. One can look at at least two different regimes when $\theta \to \infty$: "in the bulk of spectrum" or "at the edge of spectrum".

PROPOSITION 7.4 [Borodin et al. 2000], [Johansson 1999]. *Let $\lambda \in \mathbb{Y}$ be distributed according to $M_{\infty,\infty,\theta}$. Then the random point configuration*

$$\left\{\frac{\lambda_1 - 2\sqrt{\theta}}{\theta^{\frac{1}{6}}}, \frac{\lambda_2 - 2\sqrt{\theta}}{\theta^{\frac{1}{6}}}, \ldots\right\}$$

converges, as $\theta \to +\infty$, to the Airy ensemble.

The convergence of distributions of the first and the second points of the random configuration from Proposition 7.4 was proved earlier in [Baik et al. 1999a; 1999b].

Proposition 7.4 is the result of degeneration of the Bessel functions (5.1) to the Airy function and its derivative.

For the results on the asymptotics "in the bulk of spectrum", see [Borodin et al. 2000]. These results correspond to the degeneration of the Plancherel kernel to the *discrete sine kernel*

$$\frac{\sin(a(x-y))}{\pi(x-y)}, \qquad x, y \in \mathbb{Z}, \quad 0 < a \leq \frac{\pi}{2}.$$

It was also mentioned in [Johansson 1999] that under a certain limit procedure the Plancherel kernel degenerates to the conventional sine kernel $\sin(\pi(x-y))/(\pi(x-y))$ on $\mathbb{R}$.

One can also consider "double limits" of $M_{k,l,\xi}$ and $M_{k,\infty,\theta}$ (or, equivalently, Meixner and Charlier ensembles) when at least two parameters tend to critical values. Then the scaling procedure must involve at least two large parameters. For $M_{k,\infty,\theta}$ the asymptotics looks as follows.

PROPOSITION 7.5 [Johansson 1999]. *Let $\lambda \in \mathbb{Y}$ be distributed according to $M_{k,\infty,\theta}$. Then the random point configuration*

$$\left\{ \frac{\lambda_1 - \theta/k - 2\sqrt{\theta}}{(1+\sqrt{\theta}/k)^{\frac{2}{3}}\theta^{\frac{1}{6}}}, \frac{\lambda_2 - \theta/k - 2\sqrt{\theta}}{(1+\sqrt{\theta}/k)^{\frac{2}{3}}\theta^{\frac{1}{6}}}, \ldots \right\}$$

converges, as $k \to \infty$ and $\theta \to \infty$, to the Airy ensemble.

The result corresponds to a degeneration of Charlier polynomials to the Airy function [Johansson 1999].

For $M_{k,l,\xi}$ a similar result was proved for $k, l \to +\infty$ in [Johansson 2000].

8. Asymptotics of Nonmixed Measures for Large n

As we have seen above, after mixing the study of our measures is not very difficult — we just need to look at the corresponding degenerations of the hypergeometric kernel. The picture before mixing is more subtle.

For $M_{k,l}^{(n)}$ and $M_{k,\infty}^{(n)}$ the asymptotics before and after mixing are different. In comparison to the mixed cases, there appear restrictions on the supports of the limit measures. These restrictions come from the trivial condition that the sum of lengths of rows of a Young diagram with n boxes is equal to n.

Consider the embedding of the set of Young diagrams with n boxes and length at most k into $\mathbb{R}_+^k$ defined by normalizing the lengths of rows of a Young diagram by n.

Proposition 3.2 and Proposition 3.4 lead to the following result.

PROPOSITION 8.1. *As $n \to \infty$, the images of the measures $M_{k,l}^{(n)}$ under the embeddings defined above converge to a measure concentrated on the set*

$$\{(x_1, \ldots, x_k) \in \mathbb{R}_+^k \mid x_1 \geq x_2 \geq \ldots \geq x_k, \sum_{i=1}^k x_i = 1\}.$$

The density of the limit measure with respect to the Lebesgue measure equals

$$\text{const} \cdot \prod_{1 \leq i < j \leq k} (x_i - x_j)^2 \prod_{i=1}^k x_i^a$$

(recall that $a = l - k \geq 0$).

As in Proposition 7.2, we have:

PROPOSITION 8.2. *As $n \to \infty$, the images of the measures $M_{k,\infty}^{(n)}$ under the embeddings defined above converge to the delta measure at the point $(1/k, \ldots, 1/k)$.*

Again, the fluctuations around the limit delta measure were determined by Johansson [1999].

Define an embedding of the set of Young diagrams with n boxes and length $\leq k$ into $\mathbb{R}^k$ setting the i-th coordinate of the image of $\lambda \in \mathbb{Y}_n$ equal to

$$\frac{\lambda_i - n/k}{\sqrt{2n/k}}$$

(compare Proposition 7.3).

PROPOSITION 8.3 [Johansson 1999]. *As $n \to \infty$, the images of the measures $M_{k,\infty}^{(n)}$ under the embeddings defined above converge to a measure concentrated on the set*

$$\left\{ (x_1, \ldots, x_k) \in \mathbb{R}^k \,\Big|\, x_1 \geq x_2 \geq \ldots \geq x_k,\ \sum_{i=1}^k x_i = 0 \right\}.$$

The density of the limit measure with respect to the Lebesgue measure equals

$$\text{const} \cdot \prod_{1 \leq i < j \leq k} (x_i - x_j)^2 \cdot e^{-x_1^2 - \cdots - x_k^2}.$$

For the values of $M_{k,\infty}^{(n)}$ on functions depending only on λ_1 the claim was proved by Tracy and Widom [1999].

In a sense, $M_{\infty,\infty}^{(n)}$ is the most pleasant measure. In this case the asymptotics of $M_{\infty,\infty}^{(n)}$ in the bulk of spectrum and at the edge of spectrum as $n \to \infty$ is exactly the same as the asymptotics of $M_{\infty,\infty,\theta}$ as $\theta \to \infty$. We can say that the asymptotics admits *depoissonization*; see [Borodin et al. 2000; Johansson 1999]. Let us explicitly state the analog of Proposition 7.4.

PROPOSITION 8.4 [Borodin et al. 2000; Johansson 1999]. *Let $\lambda \in \mathbb{Y}_n$ be distributed according to $M_{\infty,\infty}^{(n)}$. Then the random point configuration*

$$\left\{ \frac{\lambda_1 - 2\sqrt{n}}{n^{\frac{1}{6}}}, \frac{\lambda_2 - 2\sqrt{n}}{n^{\frac{1}{6}}}, \ldots \right\}$$

converges, as $n \to +\infty$, to the Airy ensemble.

Again, the convergence of distributions of first two points was proved in [Baik et al. 1999a; 1999b].

Depoissonization of the result in the bulk of spectrum requires different ideas from those used in the proof of Proposition 8.4. For the discussion of this case see [Borodin et al. 2000].

Proposition 7.5 also admits depoissonization.

PROPOSITION 8.5 [Johansson 1999]. *Let $\lambda \in \mathbb{Y}_n$ be distributed according to $M_{k,\infty}^{(n)}$. Then the random point configuration*

$$\left\{ \frac{\lambda_1 - n/k - 2\sqrt{n}}{(1+\sqrt{n}/k)^{\frac{2}{3}} n^{\frac{1}{6}}}, \frac{\lambda_2 - n/k - 2\sqrt{n}}{(1+\sqrt{n}/k)^{\frac{2}{3}} n^{\frac{1}{6}}}, \ldots \right\}$$

converges, as $k \to \infty$, $n \to \infty$ so that $(\ln n)^{\frac{1}{6}}/k \to 0$, to the Airy ensemble.

The structure of spectral z-measures $P_{z,z'}$ defined in Section 3 for general z and z' is fairly complicated. Note that $P_{z,z'}$ is the limit of the n-th level z-measures $M_{z,z'}^{(n)}$; see Proposition 3.4.

Every probability measure on Ω (definition in Section 3) can be viewed as a point process on $\mathbb{R}^*$, if we associate to every point $(\alpha, \beta) \in \Omega$ the point configuration $(\alpha_1, \alpha_2, \ldots, -\beta_1, -\beta_2, \ldots)$ (compare Proposition 6.4). The correlation functions of the process corresponding to $P_{z,z'}$ were all explicitly computed in [Borodin 1998a]. They do not have determinantal form and can be expressed through multivariate hypergeometric functions.

The situation after mixing is substantially simpler: the process associated to $\widetilde{P}_{z,z'}$ is the Whittaker ensemble (Proposition 6.4).

We refer to [Borodin 1998a; 1998b; 2000a; $\geq$ 2001; Borodin and Olshanski 1998a; 1998b; Olshanski 1998a; 1998b] for a detailed discussion of measures $P_{z,z'}$, $\widetilde{P}_{z,z'}$ and associated point processes.

9. Limit Transitions

The fact that numerous kernels and ensembles described above originated from the same hypergeometric kernel suggests a number of different limit transitions between them.

On the top of the hierarchy we have the hypergeometric kernel which degenerates to all ensembles described above. This corresponds to the fact that the hypergeometric function is on the top of the hierarchy of classical special functions in one variable. The kernel depends on three parameters z, z', ξ, and lives on the lattice $\mathbb{Z}'$.

The Meixner kernel is the specialization of the positive part of the hypergeometric kernel when one of the parameters z, z' is integral. To be concrete, we will assume below that $z \in \{1, 2, \ldots\}$.

The Charlier kernel and the Whittaker kernel are one step below — they both depend on two parameters, (z, θ) and (z, z'), respectively. The Charlier kernel is obtained from the Meixner kernel by taking the limit $z' - z \to +\infty$ with $\theta = zz'\xi$ fixed, the Whittaker kernel is obtained from the hypergeometric kernel via a scaling limit when $\xi \to 1$. The Charlier kernel lives on $\mathbb{Z}_+$, the Whittaker kernel lives on $\mathbb{R}^*$.

The Laguerre kernel is a particular case of the positive part of the Whittaker kernel when one of parameters (z, z') is integral. It can be also obtained from the Meixner kernel by taking the limit $\xi \to 1$ (Proposition 7.1). The Laguerre kernel depends on two parameters $(z, a = z' - z)$ and lives on $\mathbb{R}_+$.

The Plancherel kernel is one more step below — it lives on $\mathbb{Z}'$, depends on one parameter θ and can be obtained from the hypergeometric kernel via the limit $z, z' \to \infty$, $\xi \to 0$, $\theta = zz'\xi$ fixed. Its positive part can be obtained either from the Meixner kernel by letting $z, z' \to \infty$ with $\theta = zz'\xi$ fixed, or from the Charlier kernel by taking the limit $z \to \infty$. These two transitions are thoroughly discussed in [Johansson 1999].

The Hermite kernel also depending on one integral parameter z can be obtained from the Charlier kernel via the limit $\theta \to \infty$ (Proposition 7.3).

The Airy kernel is at the bottom — it has no parameters. It can be obtained in a number of different ways. For example, one can obtain the Airy kernel in the limit $\theta \to +\infty$ of the Plancherel kernel at the edge of spectrum (Proposition 7.4), or as the limit at the edge of spectrum of the Hermite kernel and the Laguerre kernel with parameter a fixed when the order z of these polynomial ensembles goes to infinity [Forrester 1993; Tracy and Widom 1994a]. It can also be obtained as a double limit of Charlier or Meixner kernels; see the end of Section 7, as well as [Johansson 2000; 1999].

Of course, this is not the end of the story. The discrete sine kernel and the conventional sine kernel can be obtained from the Plancherel kernel as $\theta \to \infty$; see Section 7. The so-called Bessel kernel can be extracted from the Laguerre kernel "at the hard edge of spectrum" [Forrester 1993; Nagao and Wadati 1993; Tracy and Widom 1994b]. The sine kernel can be obtained from the Laguerre and Hermite kernels in the bulk of spectrum; see [Nagao and Wadati 1991], for example. A number of new kernels can be obtained from the Whittaker kernel; see [Olshanski 1998b]. Presumably, all these kernels can also be obtained as double or triple limits of the hypergeometric kernel.

Thus, a variety of kernels known so far can be obtained from the hypergeometric kernel, often in several different ways. As we tried to demonstrate above, sometimes such degenerations also carry information about the asymptotic behavior of certain combinatorial objects.

Addendum

Recent developments in the harmonic analysis on the infinite *unitary* group led to a substantial expansion of the hierarchy of correlation kernels described above. See [Borodin and Olshanski $\geq$ 2001b].

References

[Alastuey and Forrester 1995] A. Alastuey and P. J. Forrester, "Correlations in two-component log-gas systems", *J. Statist. Phys.* **81**:3-4 (1995), 579–627.

[Baik and Rains 1999a] J. Baik and E. M. Rains, "Algebraic aspects of increasing subsequences", preprint, 1999. Available at http://arXiv.org/abs/math/9905083.

[Baik and Rains 1999b] J. Baik and E. M. Rains, "The asymptotics of monotone subsequences of involutions", preprint, 1999. Available at http://arXiv.org/abs/math/9905084.

[Baik et al. 1999a] J. Baik, P. Deift, and K. Johansson, "On the distribution of the length of the longest increasing subsequence of random permutations", *J. Amer. Math. Soc.* **12**:4 (1999), 1119–1178.

[Baik et al. 1999b] J. Baik, P. Deift, and K. Johansson, "On the distribution of the length of the second row of a Young diagram under Plancherel measure", preprint, 1999. Available at http://arXiv.org/abs/math/9901118.

[Borodin 1998a] A. Borodin, "Point processes and the infinite symmetric group. Part II: Higher correlation functions", preprint, 1998. Available at http://arXiv.org/abs/math/9804087.

[Borodin 1998b] A. Borodin, "Point processes and the infinite symmetric group. Part IV: Matrix Whittaker kernel", preprint, 1998. Available at http://arXiv.org/abs/math/9810013.

[Borodin 1999] A. Borodin, "Longest increasing subsequences of random colored permutations", *Electronic Journal of Combinatorics* **6** (1999), #R13.

[Borodin 2000a] A. Borodin, "Characters of symmetric groups and correlation functions of point processes", *Funktsional. Anal. i Prilozhen.* **34**:1 (2000), 12–28 (Russian); English translation in Funct. Anal. Appl. **34** (2000), no. 1.

[Borodin 2000b] A. Borodin, "Riemann–Hilbert problem and the discrete Bessel kernel", *Intern. Math. Research Notices* **2000**:9 (2000).

[Borodin ≥ 2001] A. Borodin, "Harmonic analysis on the infinite symmetric group and the Whittaker kernel", to appear in St. Petersburg Math. J.

[Borodin and Okounkov 2000] A. Borodin and A. Okounkov, "A Fredholm determinant formula for Toeplitz determinants", *Integral Equations Operator Theory* **37**:4 (2000), 386–396.

[Borodin and Olshanski 1998a] A. Borodin and G. Olshanski, "Point processes and the infinite symmetric group", *Math. Research Lett.* **5** (1998), 799–816. Preprint at http://arXiv.org/abs/math/9810015.

[Borodin and Olshanski 1998b] A. Borodin and G. Olshanski, "Point processes and the infinite symmetric group. Part III: Fermion point processes", preprint, 1998. Available at http://arXiv.org/abs/math/9804088.

[Borodin and Olshanski 2000a] A. Borodin and G. Olshanski, "Distributions on partitions, point processes, and the hypergeometric kernel", *Comm. Math. Phys.* **211**:2 (2000), 335–358. Preprint at http://arXiv.org/abs/math/9904010.

[Borodin and Olshanski 2000b] A. Borodin and G. Olshanski, "Harmonic functions on multiplicative graphs and interpolation polynomials", *Electr. J. Comb.* **7**:1, R.P. 38 (2000).

[Borodin and Olshanski ≥ 2001a] A. Borodin and G. Olshanski, in preparation.

[Borodin and Olshanski ≥ 2001b] A. Borodin and G. Olshanski, "Correlation kernels arising from the infinite unitary group and its representations", in preparation.

[Borodin et al. 2000] A. Borodin, A. Okounkov, and G. Olshanski, "Asymptotics of Plancherel measures for symmetric groups", *J. Amer. Math. Soc.* **13** (2000), 491–515. Preprint at http://arXiv.org/abs/math/9905032.

[Cornu and Jancovici 1987] F. Cornu and B. Jancovici, "On the two-dimensional Coulomb gas", *J. Statist. Phys.* **49**:1-2 (1987), 33–56.

[Cornu and Jancovici 1989] F. Cornu and B. Jancovici, "The electrical double layer: a solvable model", *Jour. Chem. Phys.* **90** (1989), 2444.

[Daley and Vere-Jones 1988] D. J. Daley and D. Vere-Jones, *An introduction to the theory of point processes*, Springer series in statistics, Springer, 1988.

[Dyson 1962] F. J. Dyson, "Statistical theory of the energy levels of complex systems I, II, III", *J. Math. Phys.* **3** (1962), 140–156, 157–165, 166–175.

[Erdélyi 1953a] A. Erdélyi (editor), *Higher transcendental functions*, vol. 1, Mc Graw–Hill, 1953.

[Erdélyi 1953b] A. Erdélyi (editor), *Higher transcendental functions*, vol. 2, Mc Graw–Hill, 1953.

[Forrester 1986] P. J. Forrester, "Positive and negative charged rods alternating along a line: exact results", *J. Statist. Phys.* **45**:1-2 (1986), 153–169.

[Forrester 1988] P. J. Forrester, "Solvable isotherms for a two-component system of charged rods on a line", *J. Statist. Phys.* **51**:3-4 (1988), 457–479.

[Forrester 1989] P. J. Forrester, "Exact results for correlations in a two-component log-gas", *J. Statist. Phys.* **54**:1-2 (1989), 57–79.

[Forrester 1993] P. J. Forrester, "The spectrum edge of random matrix ensembles", *Nucl. Phys. B* **402** (1993), 709–728.

[Fulton 1997] W. Fulton, *Young tableaux*, London Math. Soc. Student texts **35**, Cambridge University Press, Cambridge, 1997.

[Gaudin 1985] M. Gaudin, "L'isotherme critique d'un plasma sur réseau ($\beta = 2, d = 2, n = 2$)", *J. Physique* **46**:7 (1985), 1027–1042.

[Johansson 1999] K. Johansson, "Discrete orthogonal polynomial ensembles and the Plancherel measure", preprint, 1999. See http://arXiv.org/abs/math/9906120.

[Johansson 2000] K. Johansson, "Shape fluctuations and random matrices", *Commun. Math. Phys.* **209** (2000), 437–476. Preprint at http://arXiv.org/abs/math/9903134.

[Kerov 1993] S. V. Kerov, "Transition probabilities of continual Young diagrams and the Markov moment problem", *Funct. Anal. Appl.* **27** (1993), 104–117.

[Kerov 1994] S. V. Kerov, "The asymptotics of interlacing roots of orthogonal polynomials", *St. Petersburg Math. J.* **5** (1994), 925–941.

[Kerov 2000] S. V. Kerov, "Anisotropic Young diagrams and Jack symmetric functions", *Funktsion. Anal. i Prilozhen.* **34**:1 (2000), 51–64. In Russian; translation in *Funct. Anal. Appl.* **34**:1 (2000), 41–51 and at http://arXiv.org/abs/math/9712267.

[Kerov et al. 1993] S. Kerov, G. Olshanski, and A. Vershik, "Harmonic analysis on the infinite symmetric group: a deformation of the regular representation", *Comptes Rend. Acad. Sci. Paris, Sér. I* **316** (1993), 773–778.

[Kerov et al. 1998] S. Kerov, A. Okounkov, and G. Olshanski, "The boundary of Young graph with Jack edge multiplicities", *Intern. Math. Res. Notices* no. 4 (1998), 173–199.

[Knuth 1970] D. E. Knuth, "Permutations, matrices, and generalized Young tableaux", *Pacific J. Math.* **34** (1970), 709–727.

[Koekoek and Swarttouw 1998] R. Koekoek and R. F. Swarttouw, "The Askey–scheme of hypergeometric orthogonal polynomials and its q-analogue", preprint DUT-TWI-98-17, Technische Universiteit Delft, 1998. Available at http://aw.twi.tudelft.nl/~koekoek/askey.html.

[Macdonald 1995] I. G. Macdonald, *Symmetric functions and Hall polynomials*, 2nd ed., Oxford University Press, 1995.

[Mehta 1991] M. L. Mehta, *Random matrices*, 2nd ed., Academic Press, Boston, 1991.

[Nagao and Wadati 1991] T. Nagao and M. Wadati, "Correlation functions of random matrix ensembles related to classical orthogonal polynomials", *J. Phys. Soc. Japan* **60**:10 (1991), 3298–3322.

[Nagao and Wadati 1993] T. Nagao and M. Wadati, "Eigenvalue distribution of random matrices at the spectrum edge", *J. Phys. Soc. Japan* **62**:11 (1993), 3845–3856.

[Nikiforov et al. 1991] A. F. Nikiforov, S. K. Suslov, and V. B. Uvarov, *Classical orthogonal polynomials of a discrete variable*, Series in Computational Physics, Springer, Berlin, 1991.

[Okounkov 1999a] A. Okounkov, "Infinite wedge and measures on partitions", preprint, 1999. Available at http://arXiv.org/abs/math/9907127.

[Okounkov 1999b] A. Okounkov, "Random matrices and random permutations", preprint, 1999. Available at http://arXiv.org/abs/math/9903176.

[Okounkov 2001] A. Okounkov, "SL(2) and z-measures", in *Random matrices and their applications*, edited by P. Bleher and A. Its, Math. Sci. Res. Inst. Publications **40**, Cambridge Univ. Press, New York, 2001.

[Olshanski 1989] G. I. Olshanski, "Unitary representations of (G,K)-pairs connected with the infinite symmetric group $S(\infty)$", *Algebra i Analiz* **1**:4 (1989), 178–209. In Russian; translation in *Leningrad Math. J.*, **1**:4 (1990), 983–1014.

[Olshanski 1998a] G. Olshanski, "Point processes and the infinite symmetric group. Part I: The general formalism and the density function", preprint, 1998. Available at http://arXiv.org/abs/math/9804086.

[Olshanski 1998b] G. Olshanski, "Point processes and the infinite symmetric group. Part V: Analysis of the matrix Whittaker kernel", preprint, 1998. Available at http://arXiv.org/abs/math/9810014.

[Regev 1981] A. Regev, "Asymptotic values for degrees associated with strips of Young diagrams", *Adv. in Math.* **41** (1981), 115–136.

[Robinson 1938] G. Robinson, "On representations of the symmetric group", *Amer. J. Math.* **60** (1938), 745–760.

[Sagan 1991] B. E. Sagan, *The symmetric group: representations, combinatorial algorithms, and symmetric functions*, Wadsworth & Brooks/Cole, Pacific Grove, CA, 1991.

[Schensted 1961]   C. Schensted, "Longest increasing and decreasing subsequences", *Canad. J. Math.* **13** (1961), 179–191.

[Temme 1990]   N. M. Temme, "Asymptotic estimates for Laguerre polynomials", *J. Appl. Math. Physics (ZAMP)* **41** (1990), 114–126.

[Thoma 1964]   E. Thoma, "Die unzerlegbaren, positive–definiten Klassenfunktionen der abzählbar unendlichen, symmetrischen Gruppe", *Math. Zeitschr.* **85** (1964), 40–61.

[Thoma 1984]   E. Thoma, "Characters of infinite groups", pp. 211–216 in *Operator algebras and group representations* (Neptun, 1980), vol. II, Monographs Stud. Math. **18**, Pitman, Boston, 1984.

[Tracy and Widom 1994a]   C. A. Tracy and H. Widom, "Level-spacing distributions and the Airy kernel", *Comm. Math. Phys.* **159**:1 (1994), 151–174.

[Tracy and Widom 1994b]   C. A. Tracy and H. Widom, "Level spacing distributions and the Bessel kernel", *Comm. Math. Phys.* **161** (1994), 289–309.

[Tracy and Widom 1998]   C. A. Tracy and H. Widom, "Random unitary matrices, permutations and Painlevé", preprint, 1998. Available at http://arXiv.org/abs/math/9811154.

[Tracy and Widom 1999]   C. A. Tracy and H. Widom, "On the distributions of the lengths of the longest monotone subsequences in random words", preprint, 1999. Available at http://arXiv.org/abs/math/9904042.

[Vershik 1996]   A. M. Vershik, "Statistical mechanics of combinatorial partitions, and their limit shapes", *Funct. Anal. Appl.* **30** (1996), 90–105.

[Vershik and Kerov 1981a]   A. M. Vershik and S. V. Kerov, "Asymptotic theory of characters of the symmetric group", *Funct. Anal. Appl.* **15** (1981), 246–255.

[Vershik and Kerov 1981b]   A. M. Vershik and S. V. Kerov, "Character and factor representations of the infinite symmetric group", *Soviet Math. Doklady* **23**:2 (1981), 389–392.

[Wassermann 1981]   A. J. Wassermann, *Automorphic actions of compact groups on operator algebras*, thesis, University of Pennsylvania, 1981.

ALEXEI BORODIN
DEPARTMENT OF MATHEMATICS
THE UNIVERSITY OF PENNSYLVANIA
PHILADELPHIA, PA 19104-6395
UNITED STATES
borodine@math.upenn.edu

GRIGORI OLSHANSKI
DOBRUSHIN MATHEMATICS LABORATORY
INSTITUTE FOR PROBLEMS OF INFORMATION TRANSMISSION
BOLSHOY KARETNY 19
101447 MOSCOW GSP-4
RUSSIA
olsh@iitp.ru, olsh@online.ru

Phase Transitions and Random Matrices

GIOVANNI M. CICUTA

ABSTRACT. Phase transitions generically occur in random matrix models as the parameters in the joint probability distribution of the random variables are varied. They affect all main features of the theory and the interpretation of statistical models. In this paper a brief review of phase transitions in invariant ensembles is provided, with some comments to the singular values decomposition in complex non-hermitian ensembles.

1. Phase Transitions in Invariant Hermitian Ensembles

Random matrix ensembles have been extensively studied for several decades, since the early works of E. Wigner and F. Dyson, as effective mathematical reference models for the descriptions of statistical properties of the spectra of complex physical systems. In the past twenty years new applications spurned a large literature both in theoretical physics and among mathematicians. Several monographs review different sides of the physics literature of the past few decades, such as [7; 10; 18; 40; 41; 59; 92; 94]. Their combined bibliography, although very incomplete, exceeds a thousand papers. Sets of lecture notes are [102; 58; 5; 34; 85; 66]. The classic reference is Mehta's book [82].

For a long time studies and applications of random matrix theory in large part were limited to the choice of gaussian random variables for the independent entries of the random matrix. This was due both to the dominant role of the normal distribution in probability theory as well as to the nice analytic results which were obtained. Increasingly, in the past two decades, a wide variety of matrix ensembles were considered, where the joint probability distribution for the random entries depends on a number of parameters. As the latter are allowed to change, the generic occurrence of phase transitions emerged.

In this section, I begin by recalling the problem in the easiest case, the invariant ensemble of hermitian matrices.

Let $H = (H_{ij})_{i,j=1,\ldots,N}$ be hermitian random matrix with joint probability density for the independent entries

$$P(H_{11}, H_{12}, \ldots, H_{NN})\, dH = e^{-N\operatorname{Tr} V(H)}\, dH \Big/ \int e^{-N\operatorname{Tr} V(H)}\, dH,$$

$$V(H) = \frac{1}{2} a_2 H^2 + \frac{1}{4} a_4 H^4 + \cdots + \frac{1}{2p} a_{2p} H^{2p}, \quad a_{2p} > 0,$$

$$dH = \prod_{i<j} (d\operatorname{Re} H_{ij}\, d\operatorname{Im} H_{ij}) \prod_i dH_{ii}. \tag{1-1}$$

We are interested in the partition function $Z_N(a_2, a_4, \ldots, a_{2p})$, the free energy $F_N(a_2, a_4, \ldots, a_{2p}) = -\log Z_N(a_2, a_4, \ldots, a_{2p})$, the "one point resolvent" $G_N(z)$, the connected correlator $G_N^{(c)}(z_1, z_2)$

$$Z_N(a_2, a_4, \ldots, a_{2p}) = \int e^{-N\operatorname{Tr} V(H)}\, dH,$$

$$G_N(z) = \frac{1}{N} \operatorname{Tr} \left\langle \frac{1}{z-H} \right\rangle,$$

$$G_N^{(c)}(z_1, z_2) = \left\langle \operatorname{Tr} \frac{1}{z_1-H} \operatorname{Tr} \frac{1}{z_2-H} \right\rangle - \left\langle \operatorname{Tr} \frac{1}{z_1-H} \right\rangle \left\langle \operatorname{Tr} \frac{1}{z_2-H} \right\rangle. \tag{1-2}$$

The name invariant ensemble for the ensemble of these hermitian matrices reminds that since the density $P(H_{11}, H_{12}, \ldots, H_{NN})$ is invariant under a similarity transformation $H \to UHU^{-1}$ with arbitrary unitary matrix U, most of the interesting quantities, like those in (1–2), may be evaluated from the joint probability density of the eigenvalues.

Also important are the monic polynomials $P_n(z) = z^n + O(z^{n-1})$, orthogonal on the real line with the weight $e^{-NV(z)}$:

$$\int_{-\infty}^{\infty} P_n(z) P_m(z) e^{-NV(z)}\, dz = h_n \delta_{nm},$$

$$z P_n(z) = P_{n+1}(z) + R_n P_{n-1}(z),$$

$$R_n = \frac{h_n}{h_{n-1}} > 0,$$

$$Z_N(a_2, a_4, \ldots, a_{2p}) = N!\, (h_0)^N \prod_{n=1}^{N-1} (R_n)^{N-n}.$$

One obtains a non-linear recursion relation for the coefficients R_n. For instance, if

$$V(x) = \frac{a_2}{2} x^2 + \frac{a_4}{4} x^4 + \frac{a_6}{6} x^6,$$

one has the recurrence relation

$$\frac{n}{N} = R_n \Big(a_2 + a_4 (R_{n-1} + R_n + R_{n+1}) + a_6 (R_{n-1} + R_n + R_{n+1})^2$$

$$+ a_6 (R_{n-2} R_{n-1} - R_{n-1} R_{n+1} + R_{n+1} R_{n+2}) \Big)$$

(see [77; 63; 32]), occasionally called "pre-string equation" or the Freud equation. One also introduces the set of orthonormal functions $\psi_n(z)$ and the two point kernel $K_N(x,y)$, in terms of which all n-point correlation functions are expressible as

$$\psi_n(z) = \frac{1}{\sqrt{h_n}} e^{-NV(z)/2} P_n(z), \quad K_N(x,y) = \sum_{j=0}^{N-1} \psi_j(x)\psi_j(y).$$

If all the coefficients a_{2k} are positive, the statistics of the eigenvalues may be evaluated in the limit $N \to \infty$ (see [14]), $\lim_{N\to\infty} G_N(z) = G(z)$ is holomorphic in the complex z-plane, except for a segment $(-A, A)$ on the real axis. Furthermore

$$G(z) = \int_{-A}^{A} d\mu \, \frac{\rho(\mu)}{z-\mu}, \quad \rho(\mu) = \lim_{N\to\infty} \frac{1}{N} \text{Tr}\langle \delta(\mu - H)\rangle = \lim_{N\to\infty} K_N(\mu, \mu).$$

The limiting density of eigenvalues $\rho(\lambda)$, the one point correlation function, is the unique solution of the integral equation

$$V'(\lambda) = 2\mathcal{P} \int_{-A}^{A} d\mu \, \frac{\rho(\mu)}{\lambda - \mu}, \quad \int_{-A}^{A} \rho(\lambda) \, d\lambda = 1. \tag{1-3}$$

We have $\rho(\lambda) > 0$ on its support $(-A, A)$, and it vanishes as a square root at the boundary: $\rho(\lambda) \sim (A - |\lambda|)^{1/2}$. Furthermore the coefficients R_n approach a smooth limit $R(\frac{n}{N}) \sim R(x)$ and

$$F(a_2, a_4, \ldots, a_{2p})$$
$$= N^2 \left(\int_{-A}^{A} d\lambda \, \rho(\lambda) V(\lambda) - \int \int_{-A}^{A} d\lambda \, d\mu \, \rho(\lambda) \rho(\mu) \log|\lambda - \mu| + O(\frac{1}{N^2}) \right)$$
$$= N^2 \left(-\int_{0}^{1} dx \, (1-x) \log R(x) - \frac{1}{N} \log h_0 + O(\frac{1}{N^2}) \right).$$

The free energy $F(a_2, a_4, \ldots, a_{2p})$ is analytic in the couplings $(a_2, a_4, \ldots, a_{2p})$. The saddle point solution is equivalent to the resummation of the planar graphs, it is equivalent to the solution from the recurrence relations for R_n and to other techniques, such as the loop equations. The orthogonal polynomial technique and the loop equations are superior to evaluate in a systematic way the terms in the series in the parameter $(N^2)^{-k}$ corresponding to the resummation of the graphs which are embeddable on orientable surfaces with k handles [101].

It was also proved that the connected two point correlators exhibit two different forms of universality (they are independent of the set of coefficients $\{a_{2k}\}$ but depend only on the endpoints $\pm A$):

Global Universality. The limiting connected density-density correlator, after smoothing over a scale much larger than the level spacing Δ_N, is [4; 20]

$$\rho_c(\lambda, \lambda') = -\frac{1}{2\pi^2(\lambda-\lambda')^2} \frac{A^2-\lambda\lambda'}{\sqrt{A^2-\lambda^2}\sqrt{A^2-\lambda'^2}}, \quad \lambda \neq \lambda'.$$

Local Universality. The limiting two point kernel $K(x,y)$ has the sine law [90] for eigenvalues in the bulk of the spectrum, measured in units of Δ_N

$$K_{\text{bulk}}(s, s') = \frac{\sin(\pi(s-s'))}{\pi(s-s')}, \quad s = \lambda/\Delta_N, \; s' = \lambda'/\Delta_N,$$

or the Airy law, close to the tail of the spectrum (the soft edge) [15; 9]

$$K_{\text{soft}}(s, s') = \frac{\text{Ai}(s)\,\text{Ai}'(s') - \text{Ai}(s')\,\text{Ai}'(s)}{s-s'}, \quad s \sim N^{2/3}\left(\frac{\lambda}{A}-1\right).$$

A general derivation of spectral correlators is given in the recent paper by Kanzieper and Freilikher [70]. By following Shohat [97], the recurrence relation for the orthogonal polynomials $P_n(x)$ is turned into an exact second-order differential equation for the orthogonal functions $\psi_n(z)$

$$\frac{d^2\psi_n(\lambda)}{d\lambda^2} - \mathcal{F}_n(\lambda)\frac{d\psi_n(\lambda)}{d\lambda} + \mathcal{G}_n(\lambda)\psi_n(\lambda) = 0$$

Because of the smooth behavior of the coefficients R_n, in the case where the eigenvalues have support on a single segment, the complicated forms $\mathcal{F}_n(\lambda)$, $\mathcal{G}_n(\lambda)$ simplify at large order and the differential equation leads to the global and the local universality results mentioned above. (If a logarithmic singularity is present at the origin, the Bessel law is derived

$$K_{\text{origin}}(s, s') = \frac{\pi}{2}(ss')^{1/2}\frac{J_{\alpha+1/2}(\pi s)J_{\alpha-1/2}(\pi s') - J_{\alpha-1/2}(\pi s)J_{\alpha+1/2}(\pi s')}{s-s'},$$

where s and s' are scaled by the level spacing $\Delta_N(0)$ near the spectrum origin, $s = \lambda/\Delta_N(0)$, and $\alpha > -1/2$. For the simpler situations where the orthogonal polynomials are classical the Bessel law had been derived in [22; 87; 53; 88].)

All the quantities above are deeply affected by *phase transitions*.

The limiting eigenvalue density $\rho(\lambda; a_2, a_4, \ldots, a_{2p})$ continued to negative values for one or several coefficients a_{2k} (while keeping $a_{2p} > 0$) may not be positive definite. Then the integral equation (1–3) allows new solutions with eigenvalue density positive definite with support on two or more segments of the real axis. The correct solution minimizes the free energy. As one explores the space of the parameters, the free energy is evaluated on different saddle point solutions, which may coincide for certain critical values of the parameters. Then the free energy is usually a continuous but not analytic function of the parameters.

The recursion coefficients R_n no longer have a smooth "continuous" limit, which makes more difficult the orthogonal polynomials solution.

The lack of analyticity of the free energy at critical values of the parameters is analogous to phase transitions related to spontaneous symmetry breaking in classical statistical mechanics. For instance, in the simplest case of potential $V(x) = \frac{a_2}{2}x^2 + \frac{a_4}{4}x^4$, first analysed for negative values of a_2 in papers [96; 25], it is convenient to add a linear term, which explicitly breaks the Z_2 symmetry, $V(x) = a_1 x + \frac{a_2}{2}x^2 + \frac{a_4}{4}x^4$, next perform the "thermodynamic limit" $N \to \infty$, finally remove the symmetry breaking term $a_1 \to 0$, see [62], [26]. This allows the evaluation of the "order parameter" $\langle \text{Tr } H \rangle$:

$$\lim_{a_1 \to 0} \lim_{N \to \infty} \langle \text{Tr } H \rangle = \text{sign}(a_1)\theta(-a_2 - 2\sqrt{a_4}) f(a_2, a_4)$$

For the simplest case $V(x) = \frac{a_2}{2}x^2 + \frac{a_4}{4}x^4$, it was shown that the correct ansatz for the recursion coefficients R_n, if $a_2 < -2\sqrt{a_4}$ is that the even R_{2n} and odd R_{2n+1} approach two different smooth "continuous" functions [83]. This period-two ansatz requires a little generalization in the case which includes the infinitesimal symmetry breaking term [84; 23] because it leads to recurrence relations

$$zP_n(z) = P_{n+1}(z) + S_n P_n(z) + R_n P_{n-1}(z), \quad R_n = \frac{h_n}{h_{n-1}} > 0,$$

and it explains the origin of the period-two ansatz.

However in the next simplest case, like $V(x) = \frac{1}{2}a_2 x^2 + \frac{1}{4}a_4 x^4 + \frac{1}{6}a_6 x^6$, or higher order polynomials, corresponding to multiple well potentials, which may not be degenerate, the behaviour of the coefficients R_n is erratic and it is difficult to reproduce the results of the saddle point analysis by the orthogonal polynomials [69; 78; 79; 93; 95].

The lines of "phase transition" in the parameter space, related to the continuation to negative values of the coefficient a_{2p} of the monomial of highest order may be found in terms of previously discussed phase transitions by the addition of an infinitesimal monomial εx^{2p+2}; see for instance [27; 68].

Connected correlators, when the support of the eigenvalue density is two segments, were shown [6; 1] to have a different form of global universality, involving elliptic integrals. In the case of multicritical behaviour the local universality form has a modified Bessel law [3].

Important recent works seem to be so powerful and comprehensive to solve the above mentioned ambiguities. The Freud equation is expressed in a matrix Lax representation and the semiclassical asymptotics of the functions $\psi_n(z)$ is obtained in the whole complex z plane by solving a matrix Riemann-Hilbert problem, following earlier works [51; 52]. Very useful is the non-linear steepest descent method devised by Deift and Zhou [35; 36]. The works [12; 37; 38; 39] not only provide rigorous and more general solutions for methods and ansatzes previously used, but it seems to provide answers also for the models previously left unsolved (like the asymptotics of recurrence coefficients R_n in general cases). The recent and difficult developments will require some time to be exploited by physicists.

Finally, while almost all investigations related to the invariant ensemble of random matrices considered a probability distribution of the form of exponential of a polynomial like in (1–1), with possible addition of logarithmic terms, like the Penner model and the Kontsevich model, there exist probability distributions, still invariant under diagonalization by unitary matrices, which lead to different forms for the connected correlators [2]. Then the classification of universality classes perhaps is not yet complete, even in invariant one-matrix hermitian ensembles.

2. Further Matrix Ensembles and Singular Value Decomposition in Complex Non-Hermitian Random Matrices

Random matrix models more general than the hermitian one-matrix invariant ensemble are often more interesting because of the possibility to describe more interesting statistical models. The analytic solution of multi-matrix models both in the "perturbative phase" or in different phases is more complex. While an ensemble of hermitian random matrices describes triangulations of random orientable surfaces, multi-matrix ensembles are suitable to describe models of classical statistical mechanics on a random two-dimensional lattice. After the breakthrough of the Ising model [71; 13], it was possible to study random walks and loops [43], $O(N)$ model [55; 75; 76; 45; 46; 44], the Potts model [72; 33; 105], surfaces with holes [73; 30], a special case of 8-vertex model [74; 106], the chiral random matrix which simulate the spontaneously broken phase transition of QCD [103; 89; 98; 100; 65; 11]. Often it was possible to evaluate the critical exponents at phase transition. Multi-matrix models of hermitian matrices are also a good framework for combinatorial problems like the four-color theorem [28; 24; 47] or the enumeration of meanders [42; 81]. Most influential, for quantum field theorists were the phase transitions in models of random unitary matrices, describing one-plaquette of the lattice formulation of QCD [57; 17; 67], the saddle-point solution of the one-matrix ensemble [16], the Witten conjecture of a master field [104].

In this section I shall recall the singular value decomposition, which plays a role in the analysis of models with rectangular random matrices and square complex non-hermitian matrices.

Let

$$\phi = (\phi_{ij})_{\substack{i=1,\ldots,N \\ j=1,\ldots,M}}$$

be a rectangular random matrix with entries ϕ_{ij} real or complex numbers, and joint probability distribution $P(\phi)$ invariant under $\phi \to U\phi V$, with U unitary of order N and V unitary of order M:

$$P(\phi)\,d\phi = e^{-N\,\mathrm{Tr}\,V(\phi^\dagger \phi)}\,d\phi \bigg/ \int e^{-N\,\mathrm{Tr}\,V(\phi^\dagger \phi)}\,d\phi,$$

$$V(\phi^\dagger \phi) = \frac{1}{2}a_2(\phi^\dagger \phi) + \frac{1}{4}a_4(\phi^\dagger \phi)^2 + \cdots + \frac{1}{2p}a_{2p}(\phi^\dagger \phi)^p, \quad a_{2p} > 0,$$

$$d\phi = \begin{cases} \prod_{\substack{i=1,\ldots,N \\ j=1,\ldots,M}} d\phi_{ij} & \text{if } \phi \text{ is real,} \\ \prod_{\substack{i=1,\ldots,N \\ j=1,\ldots,M}} d\operatorname{Re}\phi_{ij}\, d\operatorname{Im}\phi_{ij} & \text{if } \phi \text{ is complex.} \end{cases} \quad (2\text{--}1)$$

The hermitian matrices $\phi^\dagger \phi$ and $\phi \phi^\dagger$ are positive semi-definite, have the same non-vanishing eigenvalues $t_i = \sigma_i^2$, where σ_i are the singular values of ϕ, themselves positive definite. It is straightforward to evaluate the statistics of the singular values in the limit $N \to \infty$, $M \to \infty$, while the ratio $N/M = L$ is fixed, by the ordinary saddle point analysis. Let us consider first $L \geq 1$. The probability distribution (2–1) is

$$e^{-N\sum_{i=1,\ldots,M} V(\sigma_i^2)} J(\sigma_i) \prod_{i=1,\ldots,M} d\sigma_i \Big/ \int e^{-N\sum_{i=1,\ldots,M} V(\sigma_i^2)} J(\sigma_i) \prod_{i=1,\ldots,M} d\sigma_i,$$

where the Jacobian $J(\sigma_i)$ may be evaluated with help from [60; 61]:

$$J(\sigma_i) = \begin{cases} \prod_{i=1,\ldots,M}(\sigma_i)^{N-M} \prod_{1\leq i<j\leq M} |\sigma_i^2 - \sigma_j^2| & \text{if } \phi \text{ is real,} \\ \prod_{i=1,\ldots,M}(\sigma_i)^{2N-2M+1} \prod_{1\leq i<j\leq M} |\sigma_i^2 - \sigma_j^2|^2 & \text{if } \phi \text{ is complex.} \end{cases}$$

For the simple case $V(\phi^\dagger \phi) = \frac{1}{2}a_2(\phi^\dagger \phi) + \frac{1}{4}a_4(\phi^\dagger \phi)^2$ one easily obtains [29] for complex rectangular matrix ϕ

$$G(z) = \lim_{\substack{N\to\infty \\ M\to\infty}} \frac{1}{M}\operatorname{Tr}_M \left\langle \frac{1}{z - \phi^\dagger \phi} \right\rangle = \int_A^B dt\, \frac{u(t)}{z - t}, \quad 0 \leq A \leq B$$

$$u(t) = \frac{1}{\pi}\sqrt{(B-t)(t-A)}\left(\frac{a_4}{4} + \frac{a_4(A+B)}{8t} + \frac{a_2}{2t}\right). \quad (2\text{--}2)$$

The extrema A, B are given by the usual pair of algebraic equations. Then for $a_4 = 0$ one finds the distribution of singular values for rectangular matrices, which is the generalization of Wigner "semicircle law"

$$u(\sigma) = \frac{a_2}{\pi}\frac{\sqrt{(B-\sigma^2)(\sigma^2 - A)}}{\sigma}, \quad \text{for } \sqrt{A} \leq \sigma \leq \sqrt{B},$$

with

$$A = \left(\frac{\sqrt{L}-1}{a_2}\right)^2, \quad B = \left(\frac{\sqrt{L}+1}{a_2}\right)^2.$$

Returning now to equation (2–2) for square complex matrices, $L = 1$, one finds a "perturbative" phase for $a_2 > -2\sqrt{a_4}$ with $A = 0$ where observables correspond to resummation of planar graphs, and a "non-perturbative" phase for $a_2 < -2\sqrt{a_4}$ where $A > 0$, quite similar to the random hermitian case. These results were rediscovered by several authors [8; 86; 48] who introduced the hermitian matrix

$$H = \begin{pmatrix} 0 & \phi \\ \phi^\dagger & 0 \end{pmatrix}$$

and the partition function

$$Z = \int DH \, e^{-\beta \, \text{Tr} \, V(H^\dagger H)}$$

$$\sim \int_{-\infty}^{\infty} \left(\prod_{i=1}^{M} dx_i \, e^{-2\beta V(x_i^2)} \right) \prod_{i=1}^{M} |x_i|^{2N-2M+1} \prod_{1 \le i < j \le M} (x_i^2 - x_j^2)^2.$$

The eigenvalues x_i of the "chiral" matrix H are in two to one correspondence with the singular values σ_i of A: $x_i = \pm \sigma_i$. The technique of using the auxiliary matrix H in the study of square complex non-hermitian matrix ϕ was developed by [49; 50; 21] into a powerful method to obtain the distribution $\rho(x, y)$ of complex eigenvalues λ_i, see also the similar and simultaneous paper [64]. Indeed,

$$\rho(x, y) = \frac{1}{N} \sum_{i=1}^{N} \langle \delta(x - \text{Re} \, \lambda_i) \delta(y - \text{Im} \, \lambda_i) \rangle$$

$$= \frac{1}{\pi} \frac{\partial}{\partial z} \frac{\partial}{\partial z^*} \frac{1}{N} \langle \text{Tr}_{(N)} \log(z - \phi)(z^* - \phi^\dagger) \rangle$$

$$= \frac{1}{\pi} \frac{\partial}{\partial z} \frac{\partial}{\partial z^*} \frac{1}{N} (\langle \text{Tr}_{(2N)} \log H \rangle - i\pi N^2),$$

where now

$$H = \begin{pmatrix} 0 & \phi - z \\ \phi^\dagger - z^* & 0 \end{pmatrix}.$$

Diagrammatic rules may then be used to evaluate the resolvent $\mathcal{G}(\eta; z, z^*)$

$$\mathcal{G}(\eta; z, z^*) = \frac{1}{2N} \left\langle \text{Tr}_{(2N)} \frac{1}{\eta - H} \right\rangle = \frac{\eta}{N} \left\langle \text{Tr}_N \frac{1}{\eta^2 - (z^* - \phi^\dagger)(z - \phi)} \right\rangle$$

Finally in terms of the integrated density of eigenvalues of H, $\Omega(\mu; z, z^*) = (2N)^{-1} \langle \text{Tr}_{(2N)} \, \theta(\mu - H) \rangle$, Feinberg and Zee [50] obtain

$$\rho(x, y) = -\frac{4}{\pi} \int_0^\infty d\mu \, \frac{\partial}{\partial z} \frac{\partial}{\partial z^*} \frac{\Omega(\mu; z, z^*)}{\mu}$$

As specific examples, Feinberg and Zee consider probability distributions $P(\phi, \phi^\dagger)$ for the complex matrix ϕ of the form (2–1), invariant under $\phi \to e^{i\alpha}\phi$, $\phi^\dagger \to e^{-i\alpha}\phi^\dagger$, where it is natural to expect that the distribution of complex eigenvalues $\rho(x, y)$ has rotational symmetry $\rho(x, y) = \rho(r)/(2\pi)$, as it indeed happens with the Ginibre gaussian ensemble [56]. For the simple model with only two coefficients a_2, a_4 they find that for $a_2 > -2\sqrt{a_4}$ the complex eigenvalues fill (non-uniformly) a disk centered at the origin, while for $a_2 < -2\sqrt{a_4}$ they fill a ring. This is expected from the analogous distribution of the singular values of the matrix ϕ. Next they prove the surprising "single ring theorem" asserting that for a generic polynomial potential of the form (2–1), even in the multiple-well cases where the distribution of the singular values of the matrix ϕ has support on several segments of the positive real axis, the distribution of eigenvalues of the matrix ϕ may only have one ring at most. It seems to me that in such cases the assumption of rotational symmetry should be checked. If it turns out, by using an explicitly symmetry breaking term to be removed after the thermodynamic limit $N \to \infty$, that one obtains different distributions, dependent on the direction of the symmetry breaking probe, it would be a remarkable example of spontaneous symmetry breaking of rotational symmetry.

3. Conclusion

Phase transitions generically occur in the study of ensembles of random matrices, as the parameters in the joint probability distribution of the random variables are varied. They are important for the physics interpretation of statistical models and they affect all the main features of random matrix theory. In invariant one-matrix ensembles recent progress of mathematicians seems to solve long standing problems related to multi-cut solutions with no symmetry. Phase transitions in ensembles of complex non-hermitian matrices were recently explored and it is likely that a richer variety of phase transitions will be discovered. Random matrix ensembles with a preferential basis, like band matrices were studied since the beginning of random matrix theory, see for instance [80; 31; 54; 99]. Even the simplest cases of tridiagonal matrices with random site or random hopping could be analytically solved only for a very limited choice of the probability distribution. Well known discrepancies between the moment method and numerical methods suggest the presence of phase transitions which seem more difficult to analyse than in case of invariant ensembles.

References

[1] G. Akemann, Higher genus correlators for the hermitian matrix model with multiple cuts, Nucl. Phys. **B482** (1996) 403; the generalization for singular values of complex matrices is G. Akemann, Universal correlators for multi-arc complex matrix models, Nucl. Phys. **B507** (1997) 475.

[2] G. Akemann, G. M. Cicuta, L. Molinari, G. Vernizzi, Compact support probability distributions in random matrix theory, Phys. Rev. **E59** (1999) 1489, and Non-universality of compact support probability distributions in random matrix theory, Phys. Rev. **E60** (1999) 5287–5292.

[3] G. Akemann, P. H. Damgaard, U. Magnea, S. M. Nishigaki, Multicritical microscopic spectral correlators of hermitian and complex matrices, Nucl. Phys. **B519** (1998) 682.

[4] J. Ambjorn, J. Jurkiewicz, Y. Makeenko, Multiloop correlators for two-dimensional quantum gravity, Phys. Lett. **B251** (1990) 517; the evaluation of correlators in $1/N^2$ expansion was also shown to be universal in J. Ambjorn, L. Chekhov, C. F.

Kristjansen, Y. Makeenko, Matrix model calculations beyond the spherical limit, Nucl. Phys. **B404** (1993) 127 and erratum in Nucl. Phys. **B449** (1995) 681; and generalized to supermatrices in [91].

[5] J. Ambjorn, Quantization of geometry, in 1994 Les Houches, ed. F. David, P. Ginsparg, J. Zinn-Justin, North-Holland 1996.

[6] J. Ambjorn, G. Akemann, New universal spectral correlators, J. Phys. **A29** (1996) L555.

[7] J. Ambjorn, B. Durhuus, T. Jonsson, Quantum Geometry, Cambridge Univ. Press 1997.

[8] A. Anderson, R. C. Myers, V. Periwal, Branched polymers from a double-scaling limit of matrix models, Nucl. Phys. **B360** (1991) 463–479.

[9] E. L. Basor, H. Widom, Determinants of Airy operators and applications to random matrices, J. Stat. Phys. **96** (1999) 1–20.

[10] C. W. J. Beenakker, Random-matrix theory of quantum transport, Rev. Mod. Phys. **69** (1997) 731.

[11] Berbenni-Bitsch, M. E. Meyer, S. Schafer, J. J. Verbaarschot, T. Wettig, Microscopic universality in the spectrum of the lattice Dirac operator, Phys. Rev. Lett. **80** (1998) 1146.

[12] P. Bleher, A. Its, Semiclassical asymptotics of orthogonal polynomials, Riemann–Hilbert problem, and universality in the matrix model, (1997) pag. 1–105.

[13] D. V. Boulatov, V. A. Kazakov, The Ising model on a random planar lattice: the structure of phase transition and the exact critical exponents, Phys. Lett. **B186** (1987) 379.

[14] A. Boutet de Monvel, L. Pastur, M. Shcherbina, On the statistical mechanics approach in the random matrix theory: integrated density of states, J. Stat. Phys. **79** (1995) 585.

[15] M. J. Bowick, E. Brezin, Universal scaling of the tail of the density of eigenvalues in random matrix models, Phys. Lett. **B268** (1991) 21.

[16] E. Brezin, C. Itzykson, G. Parisi, J. B. Zuber, Planar diagrams, Comm. Math. Phys. **59** (1978) 35.

[17] E. Brezin, D. J. Gross, The external field problem in the large N limit of QCD, Phys. Lett. **B97** (1980) 120.

[18] E. Brezin, S. R. Wadia, The large N expansion in quantum field theory and statistical physics, World scientific 1993.

[19] E. Brezin, A. Zee, Universality of the correlations between eigenvalues of large random matrices, Nucl. Phys. **B402** (1993) 613.

[20] E. Brezin, A. Zee, Correlation functions in disordered systems, Phys. Rev **E 49** (1994) 2588.

[21] E. Brezin, S. Hikami, A. Zee, Oscillating density of states near zero energy for matrices made of blocks with possible application to the random flux problem, Nucl. Phys. **B464** (1996) 411.

[22] B. V. Bronk, Exponential ensemble for random matrices, J. Math. Phys. **6** (1965) 228.

[23] R. C. Brower, N. Deo, S. Jain, C-I Tan, Symmetry breaking in the double-well hermitian matrix models, Nucl. Phys. **B405** (1993) 166–187.

[24] L. Ckekhov, C. Kristjansen, Hermitian matrix model with plaquette interaction, Nucl. Phys. **B479** (1996) 683.

[25] G. M. Cicuta, L. Molinari, E. Montaldi, Large N phase transitions in low dimensions, Mod. Phys. Lett. **1** (1986) 125–129.

[26] G. M. Cicuta, L. Molinari, E. Montaldi, Large-N spontaneous magnetization in zero dimension, J. Phys. **A 20** (1987) L67-L70.

[27] G. M. Cicuta, L. Molinari, E. Montaldi, Multicritical points in matrix models, J. Phys. **A 23** (1990) L421-L425.

[28] G. M. Cicuta, L. Molinari, E. Montaldi, Matrix models and graph colouring, Phys. Lett. **B306** (1993) 245.

[29] G. M. Cicuta, L. Molinari, E. Montaldi, R. Riva, Large rectangular random matrices, J. Math. Phys. **28** (1987) 1716. This work follows the previous investigations A. Barbieri, G. M. Cicuta, E. Montaldi, Nuovo Cimento **A84** (1984) 173 and C. M. Canali, G. M. Cicuta, L. Molinari, E. Montaldi, Nucl. Phys. **B265** (1986) 485.

[30] G. M. Cicuta, L. Molinari, E. Montaldi, S. Stramaglia, A matrix model for random surfaces with dynamical holes, J. Phys. **A29** (1996) 3769–3785.

[31] A. Crisanti, G. Paladin, A. Vulpiani, Products of random matrices in statistical physics, Springer series in solid-state sciences 104, Springer-Verlag 1993.

[32] C. Crnkovic, P. Ginsparg, G. Moore, The Ising model, the Yang-Lee edge singularity and the $2D$ quantum gravity, Phys. Lett. **237B** (1990) 196–201.

[33] J. M. Daul, Q-states Potts model on a random planar lattice, hep-th/9502014.

[34] F. David, Simplicial quantum gravity and random lattices, in Gravitation and Quantization, Les Houches 1992 Session LVII, B. Julia and J. Zinn-Justin eds. , North Holland 1995.

[35] P. Deift and X. Zhou, A steepest descent method for oscillatory Riemann–Hilbert problems. Asymptotics for the mKdV equation, Ann. of Math. **137** (1993) 295–370.

[36] P. Deift and X. Zhou, Asymptotics for the Painleve II equation, Comm. Pure and Appl. Math. **48** (1995) 277–337.

[37] P. Deift, A. R. Its, X. Zhou, A Riemann–Hilbert approach to asymptotic problems arising in the theory of random matrix models and also in the theory of integrable statistical mechanics, Ann. of Math. **146** (1997) 149–235.

[38] P. Deift, T. Kriecherbauer, K. T-R McLaughlin, S. Venakides, X. Zhou, Strong asymptotics of orthogonal polynomials with respect to exponential weights, preprint 1–71.

[39] P. Deift, T. Kriecherbauer, K. T-R McLaughlin, S. Venakides, X. Zhou, Uniform asymptotics for polynomials orthogonal with respect to varying exponential weights and applications to universality questions in random matrix theory, preprint 1–106.

[40] K. Demeterfi, Two-dimensional quantum gravity, matrix models and string theory, Int. J. Mod. Phys. **A8** (1993) 1185–1244.

[41] P. Di Francesco, P. Ginsparg, J. Zinn-Justin, $2D$ gravity and random matrices, Phys. Rep. **254** (1995) 1–133.

[42] P. Di Francesco, O. Golinelli, E. Guitter, Meander, folding and arch statistics, in Combinatorics and Physics, Mathematical and Computer Modelling 144 (1996).

[43] B. Duplantier, I. K. Kostov, Geometrical critical phenomena on a random surface of arbitrary genus, Nucl. Phys. **B340** (1990) 491.

[44] B. Durhuus, C. Kristjansen, Phase structure of the $O(n)$ model on a random lattice for $n > 2$, Nucl. Phys. **B483** (1997) 535–551.

[45] B. Eynard, C. Kristjansen, Exact solution of the $O(n)$ model on a random lattice, Nucl. Phys. **B455** (1995) 577–618.

[46] B. Eynard, C. Kristjansen, More on the exact solution of the $O(n)$ model on a random lattice and an investigation of the case $|n| > 2$, Nucl. Phys. **B466** (1996) 463.

[47] B. Eynard, C. Kristjansen, An iterative solution of the three-colour problem on a random lattice, Nucl. Phys. **B516** (1998) 529.

[48] J. Feinberg, A. Zee, Renormalizing rectangles and other topics in random matrix theory, cond-mat/9609190.

[49] J. Feinberg, A. Zee, Non-gaussian non-hermitian random matrix theory: phase transitions and addition formalism, Nucl. Phys. **B501** (1997) 643.

[50] J. Feinberg, A. Zee, Non-hermitian random matrix theory: method of hermitian reduction, Nucl. Phys. **B504** (1997) 579–608.

[51] A. S. Fokas, A. R. Its andA. V. Kitaev, The isomonodromy approach to matrix models in 2D quantum gravity, Comm. Math. Phys. **147** (1992) 395–430.

[52] A. S. Fokas, A. R. Its andA. V. Kitaev, Discrete Painleve equations and their appearance in quantum gravity, Comm. Math. Phys. **142** (1991) 313–344.

[53] P. J. Forrester, The spectrum of random matrix ensembles, Nucl. Phys. **B402** (1993) 709.

[54] Y. V. Fyodorov, A. D. Mirlin, Statistical properties of eigenfunctions of random quasi 1D one particle hamiltonians, Int. J. Mod. Phys. **B 8** (1994) 3795–3842.

[55] M. Gaudin, I. Kostov, $O(n)$ model on a fluctuating planar lattice: some exact results, Phys. Lett. **B220** (1989) 200.

[56] J. Ginibre, Statistical ensembles of complex, quaternion and real matrices, Jour. Math. Phys. **6** (1965) 440–449.

[57] D. J. Gross, E. Witten, Possible third-order phase transition in the large-N lattice gauge theory, Phys. Rev. **D21** (1980) 446–453.

[58] D. J. Gross, T. Piran, S. Weinberg, Two dimensional quantum gravity and random surfaces, World Scientific (1992)

[59] T. Guhr, A. Muller-Groeling, M. A. Weidenmuller, Random matrix theories in quantum physics: common concepts,Phys. Rep. **299** (1998) 190.

[60] S. Helgason, Differential geometry, Lie groups and symmetric spaces, Academic Press (1978).

[61] L. K. Hua, Harmonic analysis of functions of several complex variables in the classical domains, AMS Providence, Rhode Island (1963).

[62] E. M. Ilgenfritz, Yu. M. Makeenko, T. V. Shahbazyan, On the relation between many-color QCD and the free Nambu string on the lattice, Phys. Lett. **B172** (1986) 81–85.

[63] A. R. Its and A. V. Kitaev, Mathematical aspects of the non-perturbative $2D$ quantum gravity, Mod. Phys. Lett. A5 (1990) 2079.

[64] R. A. Janik, M. A. Nowak, G. Papp, I. Zahed, Nonhermitian random matrix models, Nucl. Phys. **B501** (1997) 603.

[65] R. A. Janik, M. A. Nowak, G. Papp, I. Zahed, The U(1) problem in chiral random matrix models, Nucl. Phys. **B498** (1997) 313–330.

[66] R. A. Janik, M. A. Nowak, G. Papp, I. Zahed, Various shades of Blue's functions, 1997 Zakopane lectures, hep-th/9710103.

[67] J. Jurkiewicz, K. Zalewski, Phase structure of $U(N \to \infty)$ gauge theory on a two-dimensional lattice for a broad class of variant actions, Nucl. Phys. **B220** (1983) 167–184.

[68] J. Jurkiewicz, Regularization of one-matrix models, Phys. Lett. **B245** (1990) 178–184.

[69] J. Jurkiewicz, Chaotic behaviour in one-matrix models, Phys. Lett. **B261** (1991) 260–268.

[70] E. Kanzieper, V. Freilikher, Spectra of large random matrices: a method of study, (1998) cond-mat/9809365.

[71] V. A. Kazakov, Ising model on a dynamical planar random lattice: exact solution, Phys. Lett. **A119** (1986) 140–144.

[72] V. A. Kazakov, Nucl. Phys. (Proc. Suppl.) **B4** (1988) 93.

[73] V. A. Kazakov, A simple solvable model of quantum field theory of open strings, Phys. Lett. **B237** (1990) 212.

[74] V. A. Kazakov, P. Zinn-Justin, Two matrix model with $ABAB$ interaction, Nucl. Phys. **B546** (1999) 647.

[75] I. Kostov, $O(N)$ vector model on a planar random lattice: spectrum of anomalous dimensions, Mod. Phys. Lett. **A4** (1989) 217–226.

[76] I. Kostov, M. Staudacher, Multicritical phases of the $O(n)$ model on a random lattice, Nucl. Phys. **B384** (1992) 459.

[77] O. Lechtenfeld, On eigenvalue tunneling in matrix models, Int. J. Mod. Phys. bf A7 (1992) 2335.

[78] O. Lechtenfeld, R. Ray, A. Ray, Phase diagram and orthogonal polynomials in multiple well matrix models, Int. J. Mod. Phys. **A6** (1991) 4491–4515.

[79] O. Lechtenfeld, Semiclassical approach to finite N matrix models, Int. J. Mod. Phys. **A7** (1992) 7097–7118.

[80] E. H. Lieb, D. C. Mattis, Mathematical physics in one dimension, Academic Press 1966.

[81] Y. Makeenko, H. Win Pe, Supersymmetric matrix models and the meander problem, (1996) hep-th/9601139.

[82] M. L. Mehta, Random Matrices, 2nd ed. , Academic Press, 1991.

[83] L. Molinari, Phase structure of matrix models through orthogonal polynomials, J. Phys. **A21** (1988) 1–6.

[84] L. Molinari, E. Montaldi, The large N magnetization in matrix models revisited, Nuovo Cimento **D15** (1993) 293.

[85] M. Mulase, Lectures on the asymptotic expansion of a hermitian matrix integral (1998) math-ph/9811023.

[86] R. C. Myers, V. Periwal, From polymers to quantum gravity: triple scaling in rectangular random matrix models, Nucl. Phys. **B390** (1993) 716–746.

[87] T. Nagao, K. Slevin, Nonuniversal correlations for random matrix ensembles, J. Math. Phys. **34** (1993) 2075; Laguerre ensembles of random matrices: nonuniversal correlation functions, J. Math. Phys. **34** (1993) 2317.

[88] T. Nagao, P. J. Forrester, Asymptotic correlations at the spectrum edge of random matrices, Nucl. Phys. **B435** (1995) 401.

[89] M. A. Nowak, J. J. Verbaarschot, I. Zahed, Chiral fermions in the instanton vacuum at finite temperature, Nucl. Phys. **B325** (1989) 581–592.

[90] L. A. Pastur, M. Shcherbina, Universality of the local eigenvalue statistics for a class of unitary invariant random matrix ensembles, J. Stat. Phys. **86** (1997) 109.

[91] J. C. Plefka, Iterative solution of the supereigenvalue model, Nucl. Phys. **B444** (1995) 333–352; The supereigenvalue model in the double-scaling limit, **B448** (1995) 355–372.

[92] P. Rossi, M. Campostrini, E. Vicari, The large-N expansion of unitary-matrix models, Phys. Rep. **302** (1998) 143–209.

[93] M. Sasaki, H. Suzuki, Matrix realization of random surfaces, Phys. Rev **D43** (1991) 4015–4028.

[94] G. W. Semenoff, R. J. Szabo, Fermionic matrix models, Int. J. Mod. Phys. **A12** (1997) 2135–2292.

[95] D. Senechal, Chaos in the hermitian one-matrix model, Int. J. Mod. Phys. **A7** (1992) 1491.

[96] Y. Shimamune, On the phase structure of large N matrix models and gauge models, Phys. Lett. **B108** (1982) 407. The authors of [25] and [26] were unaware of this letter.

[97] J. Shohat, A differential equation for orthogonal polynomials, Duke Math. J. **5** (1939) 401.

[98] E. Shuryak, J. J. Verbaarschot, Random matrix theory and spectral sum rules for the Dirac operator in QCD, Nucl. Phys. **A560** (1993) 306–320.

[99] P. G. Silvestrov, Summing graphs for random band matrices, Phys. Rev. **E 55** (1997) 6419–6432.

[100] Y. A. Simonov, Chiral-symmetry breaking in the disordered QCD vacuum, Phys. Rev. **D43** (1991) 3534–3540.

[101] The discovery by Gerard 't Hooft of the topological expansion, Nucl. Phys. **B72** (1974) 461 is probably the origin of the interest of quantum field theorists in random matrix theory.

[102] C. A. Tracy, H. Widom, Introduction to random matrices, in Springer lect. notes in physics 424, ed. G. F. Helminck, Geometric and quantum aspects of integrable systems (1993) 103.

[103] J. J. Verbaarschot, I. Zahed, Spectral density of the QCD Dirac operator near zero virtuality, Phys. Rev. Lett. **70** (1993) 3852; J. J. Verbaarschot, The spectrum of the QCD Dirac operator and chiral random matrix theory: the threefold way, Phys. Rev. Lett. **72** (1994) 2531; J. J. Verbaarschot, I. Zahed, Random matrix theory and QCD, Phys. Rev. Lett. **73** (1994) 2288.

[104] E. Witten, Baryons in the 1/N expansion, Nucl. Phys. **B160** (1979) 57–115.

[105] P. Zinn-Justin, The dilute Potts model on random surfaces, cond-mat/9903385.

[106] P. Zinn-Justin, The six-vertex model on random lattices, Europhys. Lett. **49** (2000) 15–21.

GIOVANNI M. CICUTA
DIPARTIMENTO DI FISICA
UNIVERSITÀ DI PARMA
VIALE DELLE SCIENZE
43100 PARMA
ITALY
cicuta@fis.unipr.it

Matrix Model Combinatorics: Applications to Folding and Coloring

PHILIPPE DI FRANCESCO

ABSTRACT. We present a detailed study of the combinatorial interpretation of matrix integrals, including the examples of tessellations of arbitrary genera, and loop models on random surfaces. After reviewing their methods of solution, we apply these to the study of various folding problems arising from physics, including: the meander (or polymer folding) problem "enumeration of all topologically inequivalent closed nonintersecting plane curves intersecting a line through a given number of points" and a fluid membrane folding problem reformulated as that of "enumerating all vertex-tricolored triangulations of arbitrary genus, with given numbers of vertices of either color".

Contents

1. Introduction	112
2. Matrix Models and Combinatorics	114
2.1. Gaussian Integrals and Wick's Theorem	114
2.2. One-Hermitian Matrix Model: Discrete Random Surfaces	116
2.3. Multi-Hermitian Matrix Case	122
2.4. A generating Function for Fatgraphs	125
3. Matrix Models: Solutions	127
3.1. Reduction to Eigenvalues	127
3.2. Orthogonal Polynomials	127
3.3. Large N asymptotics I: Orthogonal Polynomials	130
3.4. Large N asymptotics II: Saddle-Point Approximation	131
3.5. Critical Behavior and Asymptotic Enumeration	136
3.6. Gaussian Words	138
4. Folding Polymers: Meanders	139
4.1. Definitions and Generalities	139
4.2. A Simple Algorithm. Numerical Results	142
5. Algebraic Formulation: Temperley–Lieb Algebra	145
5.1. Definition	145
5.2. Meander Polynomials	146
5.3. Meander Determinants	147

6. Matrix Model for Meanders	147
6.1. The Black-And-White Model	148
6.2. Meander Polynomials and Gaussian Words	150
6.3. Exact Asymptotics for the Case of Arbitrary Many Rivers	153
6.4. Exact Meander Asymptotics from Fully-Packed Loop Models Coupled to Two-dimensional Quantum Gravity	154
7. Folding Triangulations	157
7.1. Folding the Triangular Lattice	158
7.2. Foldable Triangulations	159
8. Exact Solution	162
8.1. The Discrete Hirota Equation	162
8.2. Direct Expansion and Large N Asymptotics	165
9. Conclusion	167
Acknowledgements	167
References	167

1. Introduction

Our first aim of this article is to convince the reader that matrix integrals, exactly calculable or not, can always be interpreted in some sort of combinatorial way as generating functions for decorated graphs of given genus, with possibly specified vertex and/or face valencies. We show this by expressing pictorially the processes involved in computing Gaussian integrals over matrices, what physicists call generically Feynman rules. These matrix diagrammatic techniques have been first developed in the context of quantum chromodynamics in the limit of large number of colors (the size of the matrix) [1; 2], and more recently in the context of two-dimensional quantum gravity, namely the coupling of two-dimensional statistical models (matter theories) to the fluctuations of the two-dimensional space into surfaces of arbitrary topologies (gravity) [3]. These toy models for noncritical string theory are a nice testing ground for physical ideas, and have led to many confirmations of continuum field-theoretical results in quantum gravity. The purely combinatorial aspect of these models has often been treated as side-result, and we believe it deserves more attention, especially in view of some spectacular results. Indeed, the noncritical string machinery allows one to relate critical properties (such as singularities of thermodynamic quantities) of the flat space statistical models to those of the same models defined on random surfaces [4]. This has led for instance to recent progress in the study of random walks, by using the inverse relation to deduce flat space results from gravitational ones [5].

Once the connection is made between a combinatorial graph-related problem and some matrix model, we still have to compute the integral. Several powerful techniques have been developed, mainly in the context of various branches of physics, to compute those integrals. The original one is orthogonal polynomials [6], but it only applies to "simple" models. More general is the saddle-point technique [2], but it only allows for computing these integrals in the limit of large size of the matrices.

In this article, we wish to present applications of the combinatorics of matrix models to some specific questions arising in physics having to do with folding. Folding problems arise in biology and physics in particular when considering polymers or membranes. An ideal polymer is a chain of say n identical constituents represented by segments (chemical bonds) attached to one another by their ends (atoms), around which they can rotate freely. Membranes are two-dimensional generalizations of polymers, that is, reticulated networks made of vertices (atoms) linked by edges (chemical bonds), either in the form of a regular lattice (tethered membranes) or in the form of networks with arbitrary vertex valencies (fluid membranes). Ideally, imposing that all bonds be rigid, the only possibility for a polymer or membrane to change its spatial configuration is through folding, in which atoms (for polymers) or bonds (for membranes) serve as hinges. Quantities of interest for physicists are thermodynamic ones, characteristic of the systems when their size is large. Mathematically, these correspond to asymptotics of say the numbers of distinct configurations of folding of polymers or membranes of given length or area, when the latter tend to infinity. Both in the case of polymer and fluid membrane folding, we will present matrix models allowing for the calculation of such thermodynamic properties.

The polymer folding problem is better known in mathematics as the "meander problem", namely that of enumerating all the topologically inequivalent configurations of a closed nonselfintersecting plane curve intersecting a line through a given number of points. This problem apparently first emerged in some work by Poincaré in the beginning of the century, and reemerged in various areas of mathematics [7; 8; 9], from recreational mathematics to the 16th Hilbert problem to computer science to the theory of knots and links. In physics, the formulation as a folding problem and the relation to matrix integrals [10] have brought very different developments both algebraic [12; 13; 14; 15] and numerical [11; 16; 17]. As a highly nontrivial outcome of our study, we will present the exact meander asymptotics recently derived in [18].

The article is organized as follows. Section 2 gives a detailed presentation of the combinatorial interpretation of Hermitian matrix integrals as generating tools for fatgraphs, and Section 3 reviews their various methods of solution.

Three sections are then devoted to the application of these ideas to the problem of enumerating all distinct compact folding configurations of a closed or open polymer, namely the number of distinct ways to fold a (self-avoiding) chain of identical constituents onto itself: this is known as the meander or semimeander problem. After defining the problem and reviewing a few known results in Section 4, we present in Section 5 an algebraic formulation of the counting problem within the framework of the Temperley–Lieb algebra, omnipresent in the integrable statistical models, as well as the theory of knots and links. This connection produces remarkable results, like the exact expression for the meander determinant, a meander-related quantity. In Section 6, we connect the meander and related problems to a multimatrix integral, that allows for the exact determination of meandric configuration exponents, in particular through the abovementioned connection [4] between flat and curved space models.

The following two sections study a matrix model for generating foldable triangulations, a sort of two-dimensional generalization of the meander problem, but without the self-avoidance constraint. These triangulations form a simple model for fluid membranes encountered in physics and biology. After posing the problem in Section 7, we introduce a two-matrix integral generating these triangulations, and solve it in various ways in Section 8.

The last section gathers a few concluding remarks.

2. Matrix Models and Combinatorics

The aim of this and the next section is to familiarize the reader with the use of matrix integrals as tools for generating and enumerating graphs with various decorations. These correspond in turn to physical models of matter coupled to two-dimensional quantum gravity, in the form of fluctuating surfaces of arbitrary genus. In the following, we first present the combinatorial tools and give recipes to construct ad-hoc matrix integrals for various graph enumeration problems. We then expose various methods of computation of these matrix integrals, concentrating in particular on the planar graph limit.

2.1. Gaussian Integrals and Wick's Theorem. Consider the Gaussian average

$$\langle x^{2n} \rangle = \frac{1}{\sqrt{2\pi}} \int_{-\infty}^{\infty} e^{-x^2/2} x^{2n} \, dx = (2n-1)!! = \frac{(2n)!}{2^n n!}. \tag{2-1}$$

Among the many ways to compute this integral, let us choose the so-called source integral method, namely define the Gaussian source integral

$$\Sigma(s) = \langle e^{xs} \rangle = \frac{1}{\sqrt{2\pi}} \int_{-\infty}^{\infty} e^{-x^2/2 + sx} \, dx = e^{s^2/2}.$$

Then the average (2-1) is obtained by taking $2n$ derivatives of $\Sigma(s) = e^{s^2/2}$ with respect to s and by setting $s = 0$ in the end. It is then immediate to see that these derivatives must be taken by *pairs*, in which one derivative acts on the exponential and the other one on the prefactor s. Parallelly, we note that $(2n-1)!! = (2n-1)(2n-3) \times \cdots \times 3 \times 1$ is the total number of distinct associations of $2n$ objects into n pairs. We may therefore formulate pictorially the computation of (2-1) as follows.

We first draw a star-graph (see Figure 1), with one central vertex and $2n$ outcoming half-edges labeled 1 to $2n$ clockwise, one for each x in the integrand (this amounts to labeling the x's in x^{2n} from 1 to $2n$). Now the pairs of derivatives taken on the source integral are in one-to-one correspondence with pairs of edges

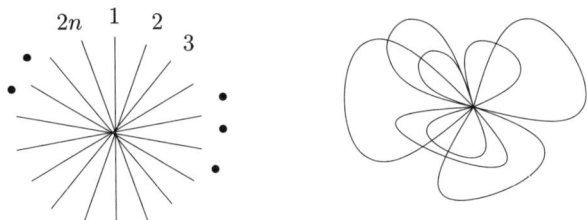

Figure 1. A star diagram with one vertex and $2n$ out-coming half-edges stands for the integrand x^{2n}. In the second diagram, we have represented one nonzero contribution to $\langle x^{2n} \rangle$ obtained by taking derivatives of $\Sigma(s)$ by pairs represented as the corresponding connections of half-edges (the x's and s's are dual to one another, hence derivatives with respect to s are in one to one correspondence with insertions of x in the integrand).

in the pictorial representation. Moreover, to get a nonzero contribution to $\langle x^{2n} \rangle$, we must saturate the set of $2n$ legs by taking n pairs of them. We represent each such saturation by connecting the corresponding edges as in Figure 1. We get exactly $(2n-1)!!$ distinct closed star-graphs with one vertex. We may therefore write the one-dimensional version of Wick's theorem

$$\langle x^{2n} \rangle = \sum_{\text{pairings}} \prod \langle x^2 \rangle, \qquad (2\text{--}2)$$

where the sum extends over all pairings saturating the $2n$ half-edges, and the weight is simply the product over all the edges formed of the corresponding averages $\langle x^2 \rangle = (d^2/ds^2)\Sigma(s)|_{s=0} = 1$. Each saturation forms a "Feynman diagram" of the Gaussian average. The edge pairings are called propagators (with value 1 here). The Feynman rules are simply the set of values of these propagators. This may appear like a complicated way of writing a rather trivial result, but it suits our purposes for generalization to matrix models and graphs. Note that the pictorial interpretation we have given for the computation of $\langle x^{2n} \rangle$ is not unique. For instance, we could have arbitrarily split the average into $\langle x^p x^q \rangle$ for some integers p, q with $p + q = 2n$. Then we would have rather represented *two* vertices with respectively p and q out-coming half-edges. Wick's theorem (2–2) states that we must now sum over (possibly disconnected) graphs obtained by saturating the $p + q$ half-edges by pairs. The result of course is still the same.

As a last remark, and to make the contact with graphs, we may consider for instance the formal series expansion

$$z(g_1, g_2, \ldots) = \langle e^{\sum_{i \geq 1} g_i x^i} \rangle \qquad (2\text{--}3)$$

in powers of $g_1, g_2, \ldots$, whose coefficients are computed using (2–1). Thanks to the previous pictorial interpretation, we may compute a typical term in the series expansion, say the coefficient $\langle \prod (x^i)^{V_i} \rangle$ of $\prod g_i^{V_i}/V_i!$, by drawing V_i star diagrams with i half-edges, $i = 1, 2, \ldots$, and saturating the $2E = \sum i V_i$ half-edges in all possible ways by forming E pairs. Hence computing (2–3) is reinterpreted as a generating function of graphs with specified numbers of vertices of given valencies.

2.2. One-Hermitian Matrix Model: Discrete Random Surfaces.

We now repeat the calculations of the previous section with the following Gaussian Hermitian matrix average of an arbitrary function f

$$\langle f(M) \rangle = \frac{1}{Z_0(N)} \int dM \, e^{-N \operatorname{Tr}(M^2/2)} f(M),$$

where the integral extends over Hermitian $N \times N$ matrices, with the standard Haar measure $dM = \prod_i dM_{ii} \prod_{i<j} dRe(M_{ij}) dIm(M_{ij})$, and the normalization factor $Z_0(N)$ is fixed by requiring that $\langle 1 \rangle = 1$ for $f = 1$. Typically, we may take for f a monomial of the form $f(M) = \prod_{(i,j) \in I} M_{ij}$, I a finite set of pairs of indices. Note the presence of the normalization factor N (=the size of the matrices) in the exponential. Note also the slight abuse of notation as we still denote averages with the same bracket sign as in previous section: we may simply include the case of the previous section as the particular case of integration over 1×1 Hermitian matrices (that is, real numbers) here.

As in the one-dimensional case of the previous section, for a given Hermitian $N \times N$ matrix S we introduce the source integral

$$\Sigma(S) = \langle e^{\operatorname{Tr}(SM)} \rangle = e^{\operatorname{Tr}(S^2)/(2N)},$$

easily obtained by completing the square $M^2 - N(SM + MS) = (M - NS)^2 - N^2 S^2$ and performing the change of variable $M' = M - NS$. We can use this equation to compute any average of the form

$$\langle M_{ij} M_{kl} \ldots \rangle = \frac{\partial}{\partial S_{ji}} \frac{\partial}{\partial S_{lk}} \cdots \Sigma(S)\Big|_{S=0}. \tag{2-4}$$

Note the interchange of the indices due to the trace $\operatorname{Tr}(MS) = \sum M_{ij} S_{ji}$. As before, derivatives with respect to elements of S must go by pairs, one of which acts on the exponential and the other one on the S element thus created. In particular, a fact also obvious from the parity of the Gaussian, (2–4) vanishes unless there are an even number of matrix elements of M in the average. In the simplest case of two matrix elements, we have

$$\langle M_{ij} M_{kl} \rangle = \frac{\partial}{\partial S_{lk}} \frac{1}{N} S_{ij} e^{\operatorname{Tr}(S^2)/(2N)}\Big|_{S=0} = \frac{1}{N} \delta_{il} \delta_{jk}. \tag{2-5}$$

Hence the pairs of derivatives must be taken with respect to S_{ij} and S_{ji} for some pair i, j of indices to yield a nonzero result. This leads naturally to the Matrix Wick's theorem:

$$\Big\langle \prod_{(i,j) \in I} M_{ij} \Big\rangle = \sum_{\text{pairings}} P \prod_{(ij),(kl) \in P} \langle M_{ij} M_{kl} \rangle, \tag{2-6}$$

where the sum extends over all pairings saturating the (pairs of) indices of M by pairs.

We see that in general, due to the restrictions (2–5) many terms in (2–6) will vanish. We now give a pictorial interpretation for the nonvanishing contributions in (2–6). We represent a matrix element M_{ij} as a half-edge (with a marked end) made of a double-line, each of which is oriented in an opposite direction. We decide that the line pointing from the mark carries the index i, while the other one, pointing to the mark, carries the index j. This reads

$$M_{ij} \longleftrightarrow \quad \substack{i \\ j}$$

The two-element result (2–5) becomes simply the construction of an edge (with both ends marked) out of two half-edges M_{ij} and M_{kl}, but is nonzero only if the indices i and j are conserved along the oriented lines. This gives pictorially

$$\langle M_{ij} M_{ji} \rangle \longleftrightarrow M_{ij} \substack{i \\ j} M_{ji} \qquad (2\text{–}7)$$

Similarly, an expression of the form $\text{Tr}(M^n)$ will be represented as a star diagram with one vertex connected to n double half-edges in such a way as to respect the identification of the various running indices, namely

$$\text{Tr}(M^n) = \sum_{i_1,i_2,\ldots,i_n} M_{i_1 i_2} M_{i_2 i_3} \ldots M_{i_n i_1} \longleftrightarrow \qquad (2\text{–}8)$$

As a first application of this diagrammatic interpretation of the Wick theorem (2–6), let us compute the large N asymptotics of $\langle \text{Tr}(M^n) \rangle$. To compute $\langle \text{Tr}(M^n) \rangle$, we must first draw a star diagram as in (2–8), then apply (2–6) to express the result as a sum over the saturations of the star with edges connecting its outcoming half-edges by pairs. To get a nonzero result, we must clearly have n even, say $n = 2p$. Note then that there are $(2p-1)!!$ such pairings, allowing us to recover the result of the previous section by setting $N = 1$, and simply replacing all oriented double lines by unoriented single ones. But if instead we take N to be large, we see that only a fraction of these $(2p-1)!!$ pairings will contribute at leading order. Indeed, assume first we restrict the set of pairings to "planar" ones (see Figure 2(a)), namely such that the saturated star diagrams have a petal structure in which edges only connect pairs of half-edges of the form (ij), $i < j$ and (kl), $k < l$ with either $j < k$ or $k < i < j < l$ or $i < k < l < j$ (in

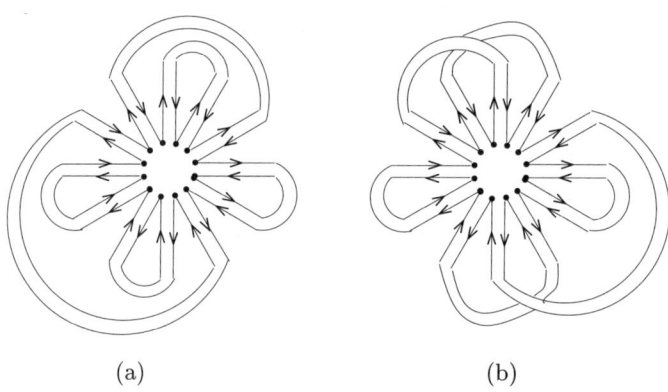

Figure 2. An example of planar (petal) diagram (a) and a nonplanar one (b). Both diagrams have $n = 2p = 12$ half-edges, connected with $p = 6$ edges. The diagram (a) has $p + 1 = 7$ faces bordered by oriented loops, whereas (b) only has 3 of them. The Euler characteristic reads $2 - 2h = F - E + 1$ ($V = 1$ in both cases), and gives the genus $h = 0$ for (a), and $h = 2$ for (b).

the labeling of half-edges, we have taken cyclic boundary conditions, namely the labels 1 and $n+1$ are identified). In other words, the petals are either juxtaposed or included into one-another. We may compute the genus of the petal diagrams by noting that they form a tessellation of the sphere (=plane plus point at infinity). This tessellation has $V = 1$ vertex (the star), $E = p$ edges, and F faces, including the "external" face containing the point at infinity. The planarity of the diagram simply expresses that its genus h vanishes, namely

$$2 - 2h = 2 = F - E + V = F + 1 - p \quad \Rightarrow \quad F = p + 1.$$

Such diagrams receive a total contribution $1/N^p$ from the propagators (weight $1/N$ per connecting edge), but we still have to sum over the remaining matrix indices $j_1, j_2, \ldots, j_{p+1}$ running over the $p + 1$ oriented loops we have created, which form the boundaries of the $F = p + 1$ faces. This gives an extra weight N per loop (or face) of the diagram, hence a total contribution of N^{p+1}. So all the petal diagrams contribute the same total factor N to $\langle \text{Tr}(M^n) \rangle$. Now any nonpetal (that is, nonplanar; see Figure 2(b)) diagram must have at least two less oriented loops, hence contributes at most $1/N$ to $\langle \text{Tr}(M^n) \rangle$. Indeed, its Euler characteristic is negative or zero, hence it has $F \leq E - V = p - 1$ and it contributes at most for $N^{F-p} \leq 1/N$. So, to leading order in N, only the genus zero (petal) diagrams contribute. We simply have to count them. This is a standard problem in combinatorics: one may for instance derive a recursion relation for the number c_p of petal diagrams with $2p$ half-edges, by fixing the left end of an edge (say at position 1), and summing over the positions of its right end (at positions $2j$, $j = 1, 2, \ldots, p$), and noting that the petal thus formed may contain c_{j-1} distinct petal diagrams and be next to c_{p-j} distinct ones. This gives the recursion relation

$$c_p = \sum_{j=1}^{p} c_{j-1} c_{p-j} \qquad c_0 = 1,$$

solved by the Catalan numbers

$$c_p = \frac{(2p)!}{(p+1)!\, p!}. \tag{2-9}$$

Finally, we get the one-matrix planar Gaussian average

$$\lim_{N \to \infty} \frac{1}{N} \langle \operatorname{Tr}(M^n) \rangle = \begin{cases} c_p & \text{if } n = 2p, \\ 0 & \text{otherwise.} \end{cases} \tag{2-10}$$

This exercise shows us what we have gained by considering $N \times N$ matrices rather than numbers: we have now a way of discriminating between the various genera of the graphs contributing to Gaussian averages. This fact will be fully exploited in the next example.

We would now like to present an application of (2–6) with important physical and mathematical consequences. We apply the matrix Wick theorem (2–6) to the following generating function $f(M) = \exp(N \sum_{i \geq 1} g_i \operatorname{Tr}(M^i)/i)$, to be understood as a formal power series of g_i, for $i = 1, 2, 3, 4, \ldots$:

$$\begin{aligned}
Z_N(g_1, g_2, \ldots) &= \langle e^{N \sum_{i \geq 1} g_i \operatorname{Tr}(M^i)/i} \rangle \\
&= \sum_{n_1, n_2, \ldots \geq 0} \prod_{i \geq 1} \frac{(N g_i)^{n_i}}{i^{n_i} n_i!} \left\langle \prod_{i \geq 1} \operatorname{Tr}(M^i)^{n_i} \right\rangle \\
&= \sum_{n_1, n_2, \ldots \geq 0} \prod_{i \geq 1} \frac{(N g_i)^{n_i}}{i^{n_i} n_i!} \sum N^{-E(\Gamma)} N^{F(\Gamma)}, \tag{2-11}
\end{aligned}$$

where the last sum is over all labeled fatgraphs Γ with n_i i-valent vertices. This comes from a direct application of (2–6).

gravi

In (2–11), we have first represented pictorially the integrand $\prod_i (\operatorname{Tr}(M^i))^{n_i}$ as a succession of n_i i-valent star diagrams like that of (2–8), $i = 1, 2, \ldots$. Then we have summed over all possible saturations of all the marked half-edges of all these stars, thus forming (not necessarily connected) ribbon or fatgraphs Γ with some labeling of their half-edges (see Figure 6–15 for an example of connected fatgraph). In (2–11), we have denoted by $E(\Gamma)$ the total number of edges of Γ, connecting half-edges by pairs, that is, the number of propagators needed (yielding a factor $1/N$ each, from (2–5)). The number $F(\Gamma)$ is the total number of faces of Γ. The faces of Γ are indeed well-defined because Γ is a fatgraph, that is, a graph with edges made of doubly oriented parallel lines carrying the corresponding matrix indices $i = 1, 2, \ldots, N$: the oriented loops we have created by the pairing process are interpreted as face boundaries, in one-to-one correspondence with faces of Γ. But the traces of the various powers

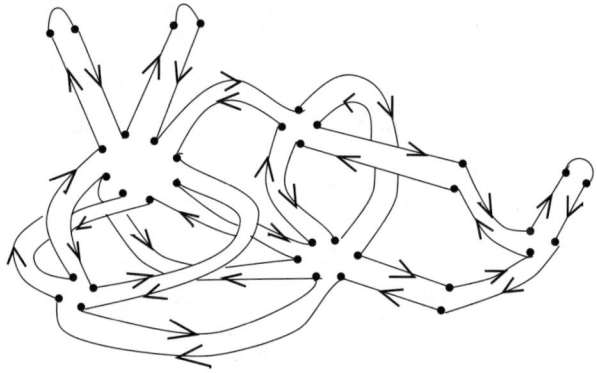

Figure 3. A typical connected fatgraph Γ, corresponding to the average $\langle \text{Tr}(M)^3 \text{Tr}(M^2)^2 \text{Tr}(M^3) \text{Tr}(M^4)^2 \text{Tr}(M^6) \text{Tr}(M^8) \rangle$. The graph was obtained by saturating the ten star diagrams corresponding to the ten trace terms, namely with $n_1 = 3$ univalent vertices, $n_2 = 2$ bi-valent ones, $n_3 = 1$ tri-valent one, $n_4 = 2$ four-valent ones, $n_6 = 1$ six-valent one and $n_8 = 1$ eight-valent one, hence a total of $V = 10$ vertices. This graph corresponds to one particular Wick pairing for which we have drawn the $E = 16$ connecting edges, giving rise to $F = 2$ oriented loops bordering the faces of Γ.

of M still have to be taken, which means all the indices running from 1 to N have to be summed over all these loops. This results in the factor N per face of Γ in (2–11). Finally, the sum extends over all (possibly disconnected) fatgraphs Γ with labeled half-edges. Each such labeled graph corresponds to exactly one Wick pairing of (2–6). Summing over all the possible labelings of a given unlabeled fatgraph Γ results in some partial cancellation of the symmetry prefactors $\prod_i 1/(i^{n_i} n_i!)$, which actually leaves us with the inverse of the order of the symmetry group of the unlabeled fatgraph Γ, denoted by $1/|\text{Aut}\,\Gamma|$. This gives the final form

$$Z_N(g_1, g_2, \ldots) = \sum_{\substack{\text{fatgraphs}\\ \Gamma}} \frac{N^{V(\Gamma) - E(\Gamma) + F(\Gamma)}}{|\text{Aut}\,\Gamma|} \prod_{i \geq 1} g_i^{n_i(\Gamma)}, \qquad (2\text{–}12)$$

where $n_i(\Gamma)$ denotes the total number of i-valent vertices of Γ and $V(\Gamma) = \sum_i n_i(\Gamma)$ is the total number of vertices of Γ. To restrict the sum in (2–12) to only connected graphs, we simply have to formally expand the logarithm of Z_N, resulting in the final identity

$$F_N(g_1, g_2, \ldots) = \text{Log}\, Z_N(g_1, g_2, \ldots) = \sum_{\substack{\text{connected}\\ \text{fatgraphs } \Gamma}} \frac{N^{2 - 2h(\Gamma)}}{|\text{Aut}\,\Gamma|} \prod_i g_i^{n_i(\Gamma)}, \qquad (2\text{–}13)$$

where we have identified the Euler characteristic $\chi(\Gamma) = F - E + V = 2 - 2h(\Gamma)$, where $h(\Gamma)$ is the genus of Γ (number of handles). Equation (2–13) gives a clear geometrical meaning to the Gaussian average of our choice of $f(M)$: it amounts to computing the generating function for fatgraphs of given genus and given vertex valencies. Such a fatgraph Γ is in turn dual to a tessellation Γ^* of a Riemann surface of same genus, by means of n_i i-valent polygonal tiles, $i = 1, 2, \ldots$.

The result (2–13) is therefore a statistical sum over discretized random surfaces (the tessellations), that can be interpreted in physical terms as the free energy of two-dimensional quantum gravity. The name free energy stands generically for the logarithm of the partition function Z_N. It simply identifies the Gaussian matrix integral with integrand $f(M)$ as a discrete sum over configurations of tessellated surfaces of arbitrary genera, weighted by some exponential factor. More precisely, imagine only $g_3 = g \neq 0$ while all other g_i's vanish. Then (2–13) becomes a sum over fatgraphs with cubic vertices (also called ϕ^3 fatgraphs), dual to triangulations T of Riemann surfaces of arbitrary genera. Assuming these triangles have all unit area, then $n_3(\Gamma) = A(T)$ is simply the total area of the triangulation T. Hence (2–13) becomes

$$F_N(g) = \sum_{\text{connected triangulations } T} \frac{g^{A(T)} N^{2-2h(T)}}{|\text{Aut } T|} \tag{2–14}$$

and the summand $g^A N^{2-2h} = e^{-S_E}$ is nothing but the exponential of the discrete version of Einstein's action for General Relativity in 2 dimensions, which reads, for any surface S

$$S_E(\Lambda, \mathcal{N}|S) = \Lambda \int_S d^2x \sqrt{|g|} + \frac{\mathcal{N}}{4\pi} \int_S d^2x \sqrt{|g|} R$$
$$= \Lambda A(S) + \mathcal{N}(2 - 2h(S)),$$

where g is the metric of S, R its scalar curvature, and the two multiplicative constants are respectively the cosmological constant Λ and the Newton constant $\mathcal{N}$. In the preceding equation, we have identified the two invariants of S: its area $A(S)$ and its Euler characteristic $\chi(S) = 2 - 2h(S)$ (using the Gauss–Bonnet formula). The contact with (2–14) is made by setting $g = e^{-\Lambda}$ and $N = e^{-\mathcal{N}}$. If we now include all g_i's in (2–13) we simply get a more elaborate discretized model, in which we can keep track of the valencies of vertices of Γ (or tiles of the dual Γ^*).

Going back to the purely mathematical interpretation of (2–13), we start to feel how simple matrix integrals can be used as tools for generating all sorts of graphs whose duals tessellate surfaces of arbitrary given topology. The size N of the matrix relates to the genus, whereas the details of the integrand relate to the structure of vertices. An important remark is also that the large N limit of (2–13) extracts the genus zero contribution, namely that coming from planar graphs. So as a by-product, it will be possible to extract from asymptotics of matrix integrals for large size N some results on planar graphs.

2.3. Multi-Hermitian Matrix Case.

The results of previous section can be easily generalized to multiple Gaussian integrals over several Hermitian matrices. More precisely, let $M_1, M_2, \ldots M_p$ denote p Hermitian matrices of same size $N \times N$, and $Q_{a,b}$, $a,b = 1, 2, \ldots, p$ the elements of a positive definite form Q. We consider the multiple Gaussian integrals of the form

$$\langle f(M_1, \ldots, M_p) \rangle = \frac{\int dM_1 \ldots dM_p \, e^{-(N/2) \sum_{a,b=1}^p \mathrm{Tr}(M_a Q_{ab} M_b)} f(M_1, \ldots, M_p)}{\int dM_1 \ldots dM_p \, e^{-(N/2) \sum_{a,b=1}^p \mathrm{Tr}(M_a Q_{ab} M_b)}}, \quad (2\text{-}15)$$

where by a slight abuse of notation we still denote averages by the same bracket sign as before: the one-Hermitian matrix case of the previous section corresponds simply to $p = 1$ and $Q_{1,1} = 1$. The averages (2-15) are computed by extending the source integral method of previous section: for some Hermitian matrices $S_1, \ldots, S_p$ of size $N \times N$, we define and compute the multisource integral

$$\Sigma(S_1, \ldots, S_p) = \langle e^{\sum_{a=1}^p \mathrm{Tr}(S_a M_a)} \rangle = e^{(1/2N) \sum_{a,b=1}^p \mathrm{Tr}(S_a (Q^{-1})_{a,b} S_b)}$$

and use multiple derivatives thereof to compute any expression of the form (2-15), before taking $S_a \to 0$. As before, derivatives with respect to elements of the S's must go by pairs to yield a nonzero result. For instance, in the case of two matrix elements of M's we find the propagators

$$\langle (M_a)_{ij} (M_b)_{kl} \rangle = \frac{1}{N} \delta_{il} \delta_{jk} (Q^{-1})_{a,b}. \quad (2\text{-}16)$$

In general we will apply the multimatrix Wick theorem

$$\left\langle \prod_{(a,i,j) \in J} (M_a)_{ij} \right\rangle = \sum_{\substack{\text{pairings} \\ P}} \prod_{\substack{\text{pairs} \\ (aij),(bkl) \in P}} \langle (M_a)_{ij} (M_b)_{kl} \rangle \quad (2\text{-}17)$$

expressing the multimatrix Gaussian average of any product of matrix elements of the M's as a sum over all pairings saturating the matrix half-edges, weighted by the corresponding value of the propagator (2-16). Note that half-edges must still be connected according to the rule (2-7), but that in addition, depending on the form of Q, some matrices may not be allowed to connect to one another (e.g. if $(Q^{-1})_{ab} = 0$ for some a and b, then $\langle M_a M_b \rangle = 0$, and there cannot be any edge connecting a matrix with index a to one with index b).

This gives us much freedom in "cooking up" multimatrix models to evaluate generating functions of graphs with specific decorations such as colorings, spin models, etc... This is expected to describe the coupling of matter systems (e.g. a spin model usually defined on a regular lattice) to two-dimensional quantum gravity (by letting the lattice fluctuate into tessellations of arbitrary genera).

An important example is the so-called gravitational $O(n)$ model [19], say on cubic fatgraphs. Its regular lattice version [24] is as spin model defined on the honeycomb (hexagonal) lattice, with only cubic (trivalent) vertices. A spin configuration of the model is simply a map from the set of vertices of the lattice to the n-dimensional unit sphere, hence the name inherited from the obvious symmetry of the target space. For $n = 1$, this is the Ising model, in which the spin variable may only take values ± 1. For $n \geq 2$ this gives an infinite continuum set of configurations, over which we have to perform a statistical sum, henceforth an integration over all the spin values. With a suitable choice of nearest neighbor interaction between adjacent spins, this statistical sum was shown to reduce to the following very simple loop model on the honeycomb lattice. The partition function $Z_{O(n)}$ of the regular lattice model is expressed as a sum over configurations of closed nonintersecting loops drawn on the honeycomb lattice, each weighed by the factor n, whereas each edge of each loop receives a Boltzmann factor K corresponding to the inverse temperature of the statistical mechanical model. The gravitational version thereof is very simple to figure out: we must replace the honeycomb lattice by its spatial fluctuations into arbitrary cubic fatgraphs. Then, on each fatgraph, we must consider arbitrary closed nonintersecting loop configurations, with the same weights as in the regular lattice case. The free energy of this model reads

$$F_N(g|K,n) = \sum_{\substack{\text{connected} \\ \text{cubic fatgraphs } \Gamma}} \frac{g^{A(\Gamma)} N^{2-2h(\Gamma)}}{|\text{Aut }\Gamma|} \sum_{\substack{\text{loop} \\ \text{configurations } \ell}} n^{C(\ell)} K^{E(\ell)}, \quad (2\text{–}18)$$

where as before we have inserted a weight g per cubic vertex ($A(\Gamma)$ is the total area of the triangulation dual to Γ, also the total number of vertices of Γ), and for each loop configuration ℓ on Γ, we have denoted by $C(\ell)$ and $E(\ell)$ respectively the total number of connected components of ℓ (number of loops) and the total number of edges occupied by the loops.

We now present a multimatrix model for (2–18). Loop models can be easily represented by using the so-called *replica trick*: as we wish to attach a weight n per loop, we may introduce a color variable $c = 1, 2, \ldots, n$ by which we may paint each connected component of loop independently. Summing over all possible colorings will yield the desired weight. Hence we are allowing the loops to be *replicated*, namely we may now choose among n different types of loops to build our configurations. To represent these configurations, we need $n + 1$ Hermitian matrices of same size $N \times N$, say $A_1, A_2, \ldots, A_n$ for the n colored loop half-edges, and B for the unoccupied half-edges. As loops are nonintersecting the only allowed vertices are of the form B^3 (unoccupied vertex) and $A_c^2 B$ (loop of color c going through a vertex, with exactly one unoccupied edge). The propagators must force loops to conserve the same color on each connected component, while unoccupied half-edges can only be connected to unoccupied ones, namely

$$\langle (A_c)_{ij}(A_d)_{kl} \rangle = \frac{K}{N}\delta_{cd}\delta_{il}\delta_{jk}, \quad \langle B_{ij}B_{kl} \rangle = \frac{1}{N}\delta_{il}\delta_{jk}, \quad \langle (A_c)_{ij}B_{kl} \rangle = 0,$$

where we also have added the weight K per edge of loop. This defines the inverse of the quadratic form Q we need for the corresponding matrix model, according to (2–16). This leads to the multimatrix integral

$$Z_N(g|K,n) = \frac{\int dB\, dA_1 \ldots dA_n\, e^{-N\,\mathrm{Tr}(V(B,A_1,\ldots,A_n))}}{\int dB\, dA_1 \ldots dA_n\, e^{-N\,\mathrm{Tr}(V_0(B,A_1,\ldots,A_n))}}$$

$$V(B,A_1,\ldots,A_n) = \frac{1}{2}B^2 + \frac{1}{2K}\sum_{c=1}^{n} A_c^2 - \frac{g}{3}B^3 - gB\sum_{c=1}^{n} A_c^2 \qquad (2\text{–}19)$$

$$V_0(B,A_1,\ldots,A_n) = \frac{1}{2}B^2 + \frac{1}{2K}\sum_{c=1}^{n} A_c^2.$$

To get the free energy of the gravitational $O(n)$ model (2–18), we simply have to take the logarithm of (2–19), to extract the contribution from connected fatgraphs only. As before, we may decorate the model with higher order terms of the form $\sum_i g_i B^i/i$ to allow for arbitrary unoccupied vertex configurations and keep track of the fine structure of the fatgraphs, without altering the loop configurations. Another possibility is to remove the B^3 term from (2–19), to restrict the loop configurations to compact ones, that is, covering all the vertices of Γ.

Another standard example, with a more complicated quadratic form, is the so-called gravitational q-states Potts model [20], defined say on arbitrary cubic diagrams Γ. Imagine that the diagram Γ is decorated by maps σ from its set of vertices $v(\Gamma)$ to $\mathbb{Z}_q$. We may alternatively view this as "spin" (or color) variables $\sigma \in \mathbb{Z}_q$ living on the vertices of Γ. The model is then defined by attaching Boltzmann factors corresponding to different energies of configurations of neighboring spins according to whether they are identical or distinct. More precisely, if v, v' are two vertices connected by an edge e in Γ, we define the weight

$$w_e(\sigma) = e^{K\delta_{\sigma(v),\sigma(v')}}$$

for some positive real parameter K (inverse temperature). The gravitational Potts model free energy is then defined as

$$F_N(g|K,q) = \sum_{\substack{\text{connected} \\ \text{cubic fatgraphs } \Gamma}} \frac{g^{A(\Gamma)} N^{2-2h(\Gamma)}}{|\mathrm{Aut}\,\Gamma|} \sum_{\substack{\text{maps} \\ \sigma: v(\Gamma)\to \mathbb{Z}_q}} \prod_{\substack{\text{edges} \\ e \text{ of } \Gamma}} w_e(\sigma). \qquad (2\text{–}20)$$

We now "cook up" a multimatrix model integral for (2–20). We wish to generate all connected cubic fatgraphs Γ: this would be easily done by considering the example of the previous section with all g's equal to zero except $g_3 = g$. To represent the spin configurations however, we need to be able to distinguish between q different types of vertices, according to the value of $\sigma(v) \in \mathbb{Z}_q$. This is done by introducing q Hermitian matrices $M_1, \ldots, M_q$ with the same size $N \times N$, with a Gaussian potential of the form $N\,\mathrm{Tr}(\sum M_a Q_{ab} M_b)/2$, such that

$$(Q^{-1})_{ab} = 1 + (e^K - 1)\delta_{ab} \qquad (2\text{–}21)$$

in order for the propagators (2–16) to receive the extra weight e^K when the "spins" a and b are identical, and 1 when they are distinct. Equation (2–21) is easily inverted into

$$Q_{ab} = \frac{1}{e^K - 1}\left(\delta_{ab} - \frac{1}{e^K - 1 + q}\right). \tag{2-22}$$

The free energy of the gravitational Potts model on cubic diagrams is therefore given by

$$F_N(g|K,q) = \text{Log}\, Z_N(g|K,q),$$

$$Z_N(g|K,q) = \frac{\int dM_1 \ldots dM_q\, e^{-N\,\text{Tr}\,V(M_1,\ldots,M_q)}}{\int dM_1 \ldots dM_q\, e^{-N\,\text{Tr}\,V_0(M_1,\ldots,M_q)}},$$

$$V(M_1,\ldots,M_q) = \frac{1}{2}\sum_{a,b=1}^{q} Q_{ab}M_a M_b - \frac{g}{3}\sum_{a=1}^{q} M_a^3, \tag{2-23}$$

$$V_0(M_1,\ldots,M_q) = \frac{1}{2}\sum_{a,b=1}^{q} Q_{ab}M_a M_b,$$

with Q as in (2–22). When $q = 2$ the model is nothing but the Ising model (also equivalent to the $O(n = 1)$ model, as mentioned above). To recover (2–20) from (2–23), we simply apply the multimatrix Wick theorem, and note that the overall contribution for each graph is $N^{V-E+F} = N^{2-2h}$ as before, as we get a weight N per vertex, $1/N$ per edge from propagators, and N per oriented loop of indices, irrespectively of the spin values.

2.4. A generating Function for Fatgraphs. In this section, we use the philosophy developed in the two previous ones to present a model of *dually weighted* fatgraphs, namely in which we wish to specify not only the structure of vertices but that of faces as well [21]. This uses the slightly more general notion of index-dependent propagators, obtained by considering one-Hermitian matrix models with Gaussian potentials of the form

$$V(M) = \frac{1}{2}\sum_{ijkl} M_{ij} Q_{ij;kl} M_{kl}, \tag{2-24}$$

where Q is some arbitrary tensor depending on the four matrix indices, but such that $\text{Tr}\,V(M)$ remains positive. Repeating the calculations of Section 2.3, we find propagators of the form

$$\langle M_{ij} M_{kl} \rangle = \frac{1}{2N}\left(\frac{1}{Q_{ij;kl}} + \frac{1}{Q_{kl;ij}}\right). \tag{2-25}$$

This gives even more freedom in tailoring specific models for combinatorial purposes. We now examine the choice

$$Q_{ij;kl} = \lambda_i \lambda_j \delta_{il} \delta_{jk}$$

for some $\lambda_1, \ldots, \lambda_N > 0$. We will also refer to the $N \times N$ diagonal matrix Λ with entries $\Lambda_{ij} = \lambda_i \delta_{ij}$. Then the propagator (2–25) becomes

$$\langle M_{ij} M_{kl} \rangle = \frac{1}{N \lambda_i \lambda_j} \delta_{il} \delta_{jk}. \qquad (2\text{–}26)$$

If we consider the Gaussian average with respect to (2–24) of the function $f(M) = \exp(-N \sum_{i \geq 1} g_i \, \mathrm{Tr}(M^i)/i)$, we find the following interpretation after taking the logarithm:

$$\begin{aligned} F_N(\Lambda; g_1, g_2, \ldots) &= \mathrm{Log} \langle e^{-N \sum_{i \geq 1} (g_i/i) \, \mathrm{Tr}(M^i)} \rangle \\ &= \sum_{\substack{\text{connected} \\ \text{fatgraphs } \Gamma}} \frac{N^{2-2h(\Gamma)}}{|\mathrm{Aut}\,\Gamma|} \prod_{i \geq 1} g_i^{n_i(\Gamma)} \prod_{\substack{\text{faces} \\ f \text{ of } \Gamma}} \sum_{i=1}^{N} \frac{1}{\lambda_i^{m(f)}}, \end{aligned} \qquad (2\text{–}27)$$

where, for each face f of Γ, we have denoted by $m(f)$ the perimeter of its boundary (number of adjacent edges, or valency). The latter term comes from the summation over all the indices running over the oriented loops of the graph. But this time, we do not use $\sum_{i=1}^{N} 1 = N$, but rather notice that each edge of a loop carrying the running index i comes with a factor $1/\lambda_i$ from the propagators (2–26). Hence each face comes with a total factor $1/\lambda_i^{m(f)}$, $m(f)$ the face valency, which has to be summed over from $i = 1$ to N, to yield (2–27). Introducing the variables

$$G_k = \mathrm{Tr}(\Lambda^{-k}) = \sum_{i=1}^{N} \frac{1}{\lambda_i^k}, \qquad (2\text{–}28)$$

we finally get the expansion

$$F_N(\Lambda; g_1, g_2, \ldots) = \sum_{\text{conn. fatgraphs } \Gamma} \frac{N^{2-2h(\Gamma)}}{|\mathrm{Aut}\,\Gamma|} \prod_{i \geq 1} g_i^{n_i(\Gamma)} G_i^{m_i(\Gamma)}, \qquad (2\text{–}29)$$

where $m_i(\Gamma)$ denotes the number of i-valent faces of Γ. The expression (2–29) looks almost symmetric in g_i and G_i: it would actually be the case if the G_i were an infinite set of independent variables like the g_i's, but they are not. Due to the definition (2–28), only the N first G's are generically independent variables, all the others are dependent. Here is how we should read (2–29). For any given N, let us expand $F_N(\Lambda; g_1, g_2, \ldots)$ as a sum over fatgraphs with $\leq N$ edges. Then the right-hand side only involves the numbers $G_1, \ldots, G_N$, as well as $g_1, \ldots, g_N$, and is manifestly symmetric in the g's and G's, as the two sets of variables are exchanged by going to the dual graphs. The coefficient of each monomial corresponds to the graphs with specified genus, numbers of i-valent vertices, and numbers of i-valent faces. To get information on larger graphs, we simply have to take N larger and larger.

3. Matrix Models: Solutions

We hope that this series of examples has convinced the reader of the relevance of matrix integrals to graph combinatorics. We will now see how to extract useful information from these matrix integrals. In this section, we will mainly cover the one-matrix integrals defined in Section 2.2. Multi-matrix techniques are very similar, and we will present them later, when we need them. More precisely, we will study the one-matrix integral

$$Z_N(V) = \frac{\int dM \, e^{-N \operatorname{Tr} V(M)}}{\int dM \, e^{-N \operatorname{Tr} V_0(M)}}, \qquad (3\text{--}1)$$

with an arbitrary polynomial potential $V(x) = \frac{x^2}{2} + \sum_{3 \leq i \leq k} \frac{g_i}{i} x^i$ and $V_0(x) = \frac{x^2}{2}$. This contains as a limiting case the partition function (2–11). We are not worrying at this point about convergence issues for these integrals, as they must be understood as formal tools allowing for computing well-defined coefficients in formal series expansions in the g's. (The reader may be more comfortable assuming $\operatorname{Re}(g_k) > 0$ to actually ensure convergence throughout this section.)

3.1. Reduction to Eigenvalues. The step zero in computing the integral (3–1) is the reduction to N one-dimensional integrals, namely over the real eigenvalues $m_1, \ldots, m_N$ of the Hermitian matrix M. This is done by performing the change of variables $M \to (m, U)$, where $m = \operatorname{diag}(m_1, \ldots, m_N)$, and U is a unitary diagonalization matrix such that $M = UmU^\dagger$, hence $U \in U(N)/U(1)^N$ as U may be multiplied by an arbitrary matrix of phases. The Jacobian of the transformation is readily found to be the squared Vandermonde determinant

$$J = \Delta(m)^2 = \prod_{1 \leq i < j \leq N} (m_i - m_j)^2.$$

Performing the change of variables in both the numerator and denominator of (3–1) we obtain

$$Z = \frac{\int_{\mathbb{R}^N} dm_1 \ldots dm_N \, \Delta(m)^2 e^{-N \sum_{i=1}^N V(m_i)}}{\int_{\mathbb{R}^N} dm_1 \ldots dm_N \, \Delta(m)^2 e^{-N \sum_{i=1}^N \frac{m_i^2}{2}}}. \qquad (3\text{--}2)$$

3.2. Orthogonal Polynomials. The standard technique of computation of (3–2) uses orthogonal polynomials. The idea is to disentangle the Vandermonde determinant squared interaction between the eigenvalues. The solution is based on the following simple lemma: If $p_m(x) = x^m + \sum_{j=0}^{m-1} p_{m,j} x^j$ are monic polynomials of degree m, for $m = 0, 1, \ldots, N-1$, then

$$\Delta(m) = \det(m_i^{j-1})_{1 \leq i,j \leq N} = \det(p_{j-1}(m_i))_{1 \leq i,j \leq N}. \qquad (3\text{--}3)$$

This is easily proved by performing suitable linear combinations of columns. We now introduce the unique set of monic polynomials p_m, of degree $m = $

$0, 1, \ldots, N-1$, that are orthogonal with respect to the one-dimensional measure $d\mu(x) = \exp(-NV(x))dx$, namely such that

$$(p_m, p_n) = \int_{\mathbb{R}} p_m(x) p_n(x)\, d\mu(x) = h_m \delta_{m,n}.$$

They allow us to rewrite the numerator of (3–2), using (3–3), as

$$\sum_{\sigma,\tau \in S_N} \varepsilon(\sigma\tau) \prod_{i=1}^{N} \int_{\mathbb{R}} d\mu(m_i)\, p_{\sigma(i)-1}(m_i) p_{\tau(i)-1}(m_i) e^{-NV(m_i)} = N! \prod_{j=0}^{N-1} h_j. \quad (3\text{--}4)$$

We may apply the same recipe to compute the denominator, with the result $N! \prod_{j=0}^{N-1} h_j^{(0)}$, where the $h_j^{(0)}$ are the squared norms of the orthogonal polynomials with respect to to the Gaussian measure $d\mu_0(x) = \exp(-Nx^2/2)\, dx$. Hence the h's determine $Z_N(V)$ entirely through

$$Z_N(V) = \prod_{i=0}^{N-1} \frac{h_i}{h_i^{(0)}}. \quad (3\text{--}5)$$

We may repeat this calculation in the presence of a "spectator" term of the form $\sum_i f(m_i)/N$ for some arbitrary function f. The term corresponding to $i=1$ in this sum simply reads, by obvious modification of (3–4),

$$\sum_{\sigma,\tau \in S_N} \varepsilon(\sigma\tau) \int_{\mathbb{R}} d\mu(m_1)\, \frac{f(m_1)}{N} p_{\sigma(1)-1}(m_1) p_{\tau(1)-1}(m_1) e^{-NV(m_1)}$$

$$\times \prod_{i=2}^{N} \int_{\mathbb{R}} d\mu(m_i)\, p_{\sigma(i)-1}(m_i) p_{\tau(i)-1}(m_i) e^{-NV(m_i)}$$

$$= N! \prod_{j=0}^{N-1} h_j \times \frac{1}{N} \int_{\mathbb{R}} dm\, f(m) \sum_{i=0}^{N-1} \frac{p_i(m)^2}{Nh_i} e^{-NV(m)},$$

which is independent of the choice of index i (which here equals 1). Therefore, the complete average with respect to the full matrix measure reads

$$\left\langle \frac{1}{N} \sum_{i=1}^{N} f(m_i) \right\rangle_V = \frac{\int_{\mathbb{R}^N} dm_1 \ldots dm_N\, \frac{1}{N} \sum_{i=1}^{N} f(m_i) \Delta(m)^2 e^{-N \sum V(m_i)}}{\int_{\mathbb{R}^N} dm_1 \ldots dm_N\, \Delta(m)^2 e^{-N \sum V(m_i)}}$$

$$= \int_{\mathbb{R}} dm\, f(m) \sum_{i=0}^{N-1} \frac{p_i(m)^2}{Nh_i} e^{-NV(m)}, \quad (3\text{--}6)$$

where we have added the subscript V to recall that the average is normalized with respect to the full measure (using V instead of V_0 in the denominator term). In the case where $f(m)$ is a polynomial, this can be immediately rewritten in terms of the r's and therefore the h's only, as we'll explain now.

To further compute the h's, we introduce operators Q and P, acting on the polynomials p_m:

$$Qp_m(x) = xp_m(x), \quad Pp_m(x) = \frac{d}{dx}p_m(x),$$

with the obvious commutation relation

$$[P, Q] = 1.$$

Using the self-adjointness of Q with respect to the scalar product $(f, g) = \int f(x)g(x)d\mu(x)$, it is easy to prove that

$$Qp_m(x) = xp_m(x) = p_{m+1}(x) + s_m p_m(x) + r_m p_{m-1}(x) \qquad (3\text{--}7)$$

for some constants r_m and s_m, and that $s_m = 0$ if the potential $V(x)$ is even. The same reasoning yields

$$r_m = \frac{h_m}{h_{m-1}} \quad \text{for } m = 1, 2, \ldots,$$

and we also set $r_0 = h_0$ for convenience.

Moreover, expressing both (Pp_m, p_m) and (Pp_m, p_{m-1}) in two ways, using integration by parts, we easily get the master equations

$$\frac{m}{N} = \frac{(V'(Q)p_m, p_{m-1})}{(p_{m-1}, p_{m-1})}, \quad 0 = (V'(Q)p_m, p_m), \qquad (3\text{--}8)$$

which amount to a recursive system for s_m and r_m. Note that the second line of (3–8) is automatically satisfied if V is even: it vanishes as the integral over $\mathbb{R}$ of an odd function. Indeed, this latter equation allows for computing the s's out of the r's, therefore becomes a tautology when all the s's vanish. Assuming for simplicity that V is even, the first equation of (3–8) gives a nonlinear recursion relation for the r's. The degree k of V actually determines the number of terms in the recursion, namely $k - 1$. So, we need to feed the $k - 2$ initial values of $r_0, r_1, r_2, \ldots, r_{k-3}$ into the recursion relation, and we obtain the exact value of $Z_N(V)$ by substituting $h_i = r_0 r_1 \ldots r_i$ in both the numerator and the denominator of (3–5). Note that for $V_0(x) = x^2/2$ the recursion (3–8) reduces simply to

$$\frac{m}{N} = \frac{(Qp_m^{(0)}, p_{m-1}^{(0)})}{(p_{m-1}^{(0)}, p_{m-1}^{(0)})} = r_m^{(0)},$$

and therefore $h_m^{(0)} = h_0^{(0)} m!/N^m = \sqrt{2\pi}m!/N^{m+1/2}$. The $p_m^{(0)}$ are simply the Hermite polynomials.

Finally, the free energy of the model (3–1) reads

$$F_N(V) = \text{Log } Z_N(V) = N \text{ Log } r_0 \sqrt{\frac{N}{2\pi}} + \sum_{i=1}^{N-1}(N - i) \text{ Log } \frac{Nr_i}{i} \qquad (3\text{--}9)$$

in terms of the r's. In the case when V is even and $f(m)$ a polynomial, the average (3–6) is clearly expressible in terms of the r's only, by use of the recursion relation (3–7) (with $s_m = 0$). Indeed, when $f(m) = m^n$, we get

$$\left\langle \frac{1}{N} \operatorname{Tr}(M^n) \right\rangle_V = \frac{1}{N} \sum_{i=0}^{N-1} \frac{(Q^n p_i, p_i)}{h_i}. \tag{3–10}$$

When $n = 2$ for instance, this simply reads

$$\left\langle \frac{1}{N} \operatorname{Tr}(M^2) \right\rangle_V = \frac{r_1}{N} + \sum_{i=1}^{N-1} \frac{r_{i+1} + r_i}{N}. \tag{3–11}$$

3.3. Large N asymptotics I: Orthogonal Polynomials.

As mentioned before, the large N limit of matrix integrals always has an interpretation as sum over planar (genus zero) fatgraphs. In particular, the free energy $F_N(V)$ of (3–9) can be expressed in an analogous way as (2–13) as a sum over connected fatgraphs, except that all but a finite number of g's vanish, namely $g_m = 0$ for $m \geq k+1$, and also for $m \leq 2$. The large N contribution is

$$F_N(V) \sim N^2 f_0(V) + O(1), \tag{3–12}$$

where $f_0(V)$ is the planar free energy, namely that obtained by restricting oneself to genus zero fatgraphs.

In view of the expression (3–9), it is straightforward to get asymptotics like (3–12), by first noting that as $h_0 \sim \sqrt{2\pi/N}$, the first term in (3–9) doesn't contribute to the leading order N^2 and then by approximating the sum by an integral of the form

$$f_0(V) = \lim_{N \to \infty} \frac{1}{N} \sum_{i=1}^{N-1} \left(1 - \frac{i}{N}\right) \operatorname{Log} \frac{r_i}{i/N} = \int_0^1 dz \, (1-z) \operatorname{Log} \frac{r(z)}{z}, \tag{3–13}$$

where we have assumed that the sequence r_i tends to a function $r_i \equiv r(i/N)$ of the variable $z = i/N$ when N becomes large. This can actually be proved rigorously. The limiting function $r(z)$ in (3–13) is determined by the equations (3–8), that become polynomial in this limit. In the case V even for instance, where $V(x) = x^2/2 + \sum_{i=2}^{p} g_{2i} x^{2i}/(2i)$ ($k = 2p$), we simply get

$$z = r(z) + \sum_{i=2}^{p} \binom{2i-1}{i} g_{2i} r(z)^i.$$

The function $r(z)$ is the unique root of this polynomial equation that tends to z for small z (it can be expressed using the Lagrange inversion method for instance, as a formal power series of the g's), and the free energy follows from (3–13). This allows also for computing the large N limit of averages of the form (3–6) or (3–10). For instance, the large N limit of (3–11) reads

$$\lim_{N \to \infty} \left\langle \frac{1}{N} \operatorname{Tr}(M^2) \right\rangle_V = 2 \int_0^1 dz \, r(z),$$

and more generally we have

$$\lim_{N\to\infty} \langle \frac{1}{N} \text{Tr}(M^n) \rangle_V = \begin{cases} \binom{2p}{p} \int_0^1 dz\, r(z)^p & \text{if } n = 2p, \\ 0 & \text{otherwise.} \end{cases} \quad (3\text{--}14)$$

In the case $V = V_0$ (and $r(z) = z$) of Gaussian averages, (3–14) reduces to the result 2–10.

3.4. Large N asymptotics II: Saddle-Point Approximation. We now present another solution, which does not rely on the orthogonal polynomial technique nor requires its applicability. We start from the N-dimensional integrals (3–2), that we rewrite

$$Z_N(V) = \frac{\int dm_1 \ldots dm_N e^{-N^2 S(m_1,\ldots,m_N)}}{\int dm_1 \ldots dm_N e^{-N^2 S_0(m_1,\ldots,m_N)}}, \quad (3\text{--}15)$$

where we have introduced the "actions"

$$S(m_1,\ldots,m_N) = \frac{1}{N}\sum_{i=1}^N V(m_i) - \frac{1}{N^2} \sum_{1\leq i\neq j\leq N} \text{Log}\,|m_i - m_j|,$$

$$S_0(m_1,\ldots,m_N) = \frac{1}{N}\sum_{i=1}^N V_0(m_i) - \frac{1}{N^2} \sum_{1\leq i\neq j\leq N} \text{Log}\,|m_i - m_j|.$$

For large N the numerator and denominator of (3–15) are dominated by the semiclassical (or saddle-point) minimum of S and S_0 respectively. For S, the saddle-point equations read

$$\frac{\partial S}{\partial m_j} = 0 \implies V'(m_j) = \frac{2}{N} \sum_{\substack{1\leq i\leq N \\ i\neq j}} \frac{1}{m_j - m_i} \quad (3\text{--}16)$$

for $j = 1, 2, \ldots, N$. Introducing the discrete resolvent

$$\omega_N(z) = \frac{1}{N} \sum_{i=1}^N \frac{1}{z - m_i}$$

and multiplying (3–16) by $1/(N(z - m_j))$ and summing over j, we easily get the equation

$$V'(z)\omega_N(z) + \frac{1}{N}\sum_{j=1}^N \frac{V'(m_j) - V'(z)}{z - m_j}$$

$$= \frac{1}{N^2} \sum_{1\leq i\neq j\leq N} \frac{1}{m_j - m_i}\left(\frac{1}{z - m_j} - \frac{1}{z - m_i}\right)$$

$$= \frac{1}{N^2} \sum_{1\leq i\neq j\leq N} \frac{1}{(z - m_i)(z - m_j)}$$

$$= \omega_N(z)^2 + \frac{1}{N}\omega'_N(z).$$

Assuming ω_N tends to a differentiable function $\omega(z)$ when $N \to \infty$ we may neglect the last derivative term, and we are left with the quadratic equation

$$\omega(z)^2 - V'(z)\omega(z) + P(z) = 0,$$

$$P(z) = \lim_{N \to \infty} \frac{1}{N} \sum_{j=1}^{N} \frac{V'(z) - V'(m_j)}{z - m_j}, \qquad (3\text{--}17)$$

where $P(z)$ is a polynomial of degree $k-2$. The existence of the limiting resolvent $\omega(z)$ boils down to that of the limiting density of distribution of eigenvalues

$$\rho(z) = \lim_{N \to \infty} \frac{1}{N} \sum_{j=1}^{N} \delta(z - m_j),$$

normalized by the condition

$$\int_{\mathbb{R}} \rho(z)\, dz = 1, \qquad (3\text{--}18)$$

since there are exactly N eigenvalues on the real axis. This density is related to the resolvent through

$$\omega(z) = \int \frac{\rho(x)}{z - x}\, dx = \sum_{m=1}^{\infty} \frac{1}{z^m} \int_{\mathbb{R}} x^{m-1} \rho(x)\, dx \qquad (3\text{--}19)$$

where the expansion holds in the large z limit, and the integral extends over the support of ρ, included in the real line. Conversely, the density is obtained from the resolvent by use of the discontinuity equation across its real support:

$$\rho(z) = \frac{1}{2i\pi} \lim_{\varepsilon \to 0} \omega(z + i\varepsilon) - \omega(z - i\varepsilon), \quad \text{for } z \in \text{supp}\,\rho. \qquad (3\text{--}20)$$

Solving the quadratic equation (3–17) as

$$\omega(z) = \frac{V'(z) - \sqrt{(V'(z))^2 - 4P(z)}}{2},$$

we must impose the large z behavior inherited from (3–18) and (3–19), namely that $\omega(z) \sim 1/z$ for large z. For $k \geq 2$, the polynomial in the square root has degree $2(k-1)$: expanding the square root for large z up to order $1/z$, all the terms cancel up to order 0 with $V'(z)$, and moreover the coefficient in front of $1/z$ must be 1 (this fixes the leading coefficient of P). The other coefficients of P are fixed by the higher moments of the measure $\rho(x)dx$. For instance, when $k = 2$ and $V = V_0$, we get

$$\omega_0(z) = \tfrac{1}{2}\bigl(z - \sqrt{z^2 - 4}\bigr).$$

It then follows from (3–20) that the density has the compact support $[-2, 2]$ and has the celebrated "Wigner's semicircle law" form

$$\rho_0(z) = \frac{1}{2\pi} \sqrt{4 - z^2}.$$

Viewing the resolvent ω_0 as the generating function for the moments of the measure whose density is ρ_0 (through the expansion (3–19)), we immediately get the values of the moments

$$\int_{\mathbb{R}} x^n \rho_0(x)\,dx = \begin{cases} c_p & \text{if } n = 2p, \\ 0 & \text{otherwise,} \end{cases} \tag{3–21}$$

with c_p as in (2–9). These are indeed immediately identified with the planar limit of the Gaussian Hermitian matrix averages (with potential $V_0(x) = x^2/2$) by using the following identity, valid for any V:

$$\lim_{N \to \infty} \langle \frac{1}{N} \operatorname{Tr} M^n \rangle_V = \int_{\mathbb{R}} x^n \rho(x)\,dx$$

The result (3–21) agrees with (2–10) and (3–14). Actually, comparing the preceding limit with (3–10), we deduce the following expression for the limiting density $\rho(x)$ in terms of the orthogonal polynomials of the previous solution:

$$\rho(x) = \lim_{N \to \infty} \frac{1}{N} \sum_{i=0}^{N-1} \frac{p_i(x)^2}{h_i} e^{-NV(x)}.$$

When $V = V_0$, we recover from this equation the Wigner's semicircle from standard asymptotics of the Hermite polynomials.

In the general case, the density reads

$$\rho(z) = \frac{1}{2\pi} \sqrt{4P(z) - (V'(z))^2}$$

and may have a disconnected support, made of a union of intervals. It is however interesting to restrict oneself to the case when the support of ρ is made of a single real interval $[a,b]$. It means that the polynomial $V'(z)^2 - 4P(z)$ has single roots at $z = a$ and $z = b$ and that all other roots have even multiplicities. In other words, we may write

$$\omega(z) = \frac{1}{2}(V'(z) - Q(z)\sqrt{(z-a)(z-b)}),$$

where $Q(z)$ is a polynomial of degree $k-2$, entirely fixed in terms of V by the asymptotics $\omega(z) \sim 1/z$ for large $|z|$. For instance, for an even quartic potential, we have

$$\begin{aligned} V(z) &= \frac{z^2}{2} - g\frac{z^4}{4} \\ \omega(z) &= \frac{1}{2}\left(z - gz^3 - \left(1 - g\frac{a^2}{2} - gz^2\right)\sqrt{z^2 - a^2}\right), \\ a^2 &= \frac{2}{3g}(1 - \sqrt{1-12g}), \\ \rho(z) &= \frac{1}{2\pi}(1 - \tfrac{1}{2}ga^2 - gz^2)\sqrt{a^2 - z^2}. \end{aligned} \tag{3–22}$$

The planar free energy (3–12) is finally obtained by substituting the limiting densities ρ, ρ_0 in the saddle point actions, namely

$$f_0(V) = S(\rho_0, V_0) - S(\rho, V),$$

$$S(\rho, V) = \int dx\, \rho(x) V(x) - \int dx\, dy\, \rho(x) \rho(y) \operatorname{Log} |x - y|.$$

This expression seems more involved than our previous result (3–13), but is equivalent to it. In the case of the quartic potential of (3–22), we find the genus zero free energy $f_0(g) \equiv f_0(V)$

$$f_0(g) = \frac{1}{2} \operatorname{Log} \frac{a^2}{4} + \frac{1}{384}(a^2 - 4)(a^2 - 36), \tag{3-23}$$

with a^2 as in (3–22).

In cases where the orthogonal polynomial technique does not apply (like in complicated multimatrix integrals), however, the saddle-point technique always gives access to the planar limit.

For completeness we quickly mention the corresponding results for the gravitational $O(n)$ model introduced in Section 2.3 (see [22] for details and a more general solution). We have to evaluate the large N asymptotics of the partition function

$$Z = \frac{\int dA_1 \ldots dA_n dB\, e^{-N \operatorname{Tr}(W(B, A_1, A_2, \ldots, A_n))}}{\int dA_1 \ldots dA_n dB\, e^{-N \operatorname{Tr}(W_0(B, A_1, A_2, \ldots, A_n))}},$$

$$W(B, A_1, A_2, \ldots, A_n) = V(B) + \frac{1}{2} \sum_{c=1}^{n} A_c^2 - gB \sum_{c=1}^{n} A_c^2, \tag{3-24}$$

$$W_0(B, A_1, A_2, \ldots, A_n) = \frac{1}{2} B^2 + \frac{1}{2} \sum_{c=1}^{n} A_c^2,$$

a simple generalization of (2–19) with some arbitrary potential $V(B)$. Note that W is simply quadratic in the A_c, so we can perform the Gaussian integrals overs the A's first. More precisely, the potential takes the form

$$\operatorname{Tr} W(B, A_1, A_2, \ldots, A_n) = \operatorname{Tr} V(B) + \frac{1}{2} \sum_{c=1}^{n} \sum_{ijkl=1}^{N} (A_c)_{ik} \mathbf{Q}_{ik;jl} (A_c)_{lj},$$

$$\mathbf{Q}_{ik;jl} = \delta_{ij}\delta_{kl} - g(\delta_{ij} B_{kl} + \delta_{kl} B_{ij}).$$

More compactly, the quadratic form reads

$$\mathbf{Q} = I \otimes I - g(I \otimes B + B \otimes I).$$

The Gaussian integral over the A's, normalized as in (3–24), gives $(\det \mathbf{Q})^{-1/2}$ for each integral; hence

$$Z = \frac{1}{\int dB\, e^{-N \operatorname{Tr} \frac{B^2}{2}}} \int dB\, e^{-N \operatorname{Tr} V(B)} \det(I \otimes I - g(I \otimes B + B \otimes I))^{-n/2}.$$

Thus the $O(n)$ model reduces to a one-matrix model, but with a complicated potential. We may now apply both the reduction to an eigenvalue integral and the large N technique sketched above. We get the eigenvalue integral

$$Z = \frac{1}{(2\pi)^{N/2}} \int db_1 \ldots db_N \, e^{-N^2 S(b_1,\ldots,b_N)},$$

$$S(b_1,\ldots,b_N) = \frac{1}{N}\sum_{i=1}^{N} V(b_i) - \frac{1}{N^2}\sum_{1\leq i\neq j\leq N} \mathrm{Log}\,|b_i - b_j|$$
$$+ \frac{n}{2N^2}\sum_{1\leq i,j\leq N} \mathrm{Log}(1 - g(b_i + b_j)). \quad (3\text{--}25)$$

In the last term we may impose the constraint $i \neq j$, as the terms $i = j$ only contribute for $O(\frac{1}{N})$ to S. The corresponding saddle-point equations $\partial_{b_i} S = 0$ read

$$V'(b_i) = \frac{2}{N}\sum_{j\neq i}\frac{1}{b_i - b_j} + \frac{gn}{N}\sum_{j\neq i}\frac{1}{1 - g(b_i + b_j)}$$

for $i = 1, 2, \ldots, N$. Setting $x_i = 1 - 2gb_i$, and $v'(x) = \frac{1}{2g}V'(b)$, we get

$$v'(x_i) = \frac{2}{N}\sum_{j\neq i}\frac{1}{x_i - x_j} + \frac{n}{N}\sum_{j\neq i}\frac{1}{x_i + x_j}. \quad (3\text{--}26)$$

Multiplying this by $\frac{1}{N}(\frac{1}{z-x_i} - \frac{1}{z+x_i})$ and summing over i, we arrive at the large N quadratic equation:

$$-Q(z) + v'(z)\omega(z) + v'(-z)\omega(-z) = \omega(z)^2 + \omega(-z)^2 + n\omega(z)\omega(-z),$$

$$\omega(z) = \lim_{N\to\infty} \frac{1}{N}\sum_{i=1}^{N}\frac{1}{z - x_i} \quad (3\text{--}27)$$

$$Q(z) = \lim_{N\to\infty} \frac{1}{N}\sum_{i=1}^{N}\frac{v'(z) - v'(x_i)}{z - x_i} - \frac{v'(-z) - v'(x_i)}{z + x_i},$$

while the saddle-point equation turns into a discontinuity equation across the support of the limiting eigenvalue distribution ρ:

$$\omega(z + i0) + \omega(z - i0) + n\omega(-z) = v'(z) \quad \text{for } z \in \mathrm{supp}\,\rho. \quad (3\text{--}28)$$

For generic n, $\omega(z)$ is well defined only in an infinite covering of the complex plane, the transition from one sheet to another being governed by (3–28). However, setting

$$n = 2\cos(\pi\nu),$$

we easily see that if $\nu = r/s$ is rational, this number becomes finite. Indeed, when expressed in terms of the shifted resolvent

$$w(z) = \frac{nv'(-z) - 2v'(z)}{n^2 - 4} + \omega(z), \quad (3\text{--}29)$$

the discontinuity and quadratic equations (3–28) and (3–27) become respectively

$$w(z+i0) + w(z-i0) + nw(-z) = 0,$$
$$w(z)^2 + w(-z)^2 + nw(z)w(-z) = P(z), \qquad (3\text{--}30)$$

where
$$P(z) = \frac{1}{n^2 - 4}(nv'(z)v'(-z) - v'(z)^2 - v'(-z)^2) - Q(z).$$

Introducing the two functions
$$w_\pm(z) = e^{\pm i\pi\frac{\nu}{2}} w(z) + e^{\mp i\pi\frac{\nu}{2}} w(-z) \qquad (3\text{--}31)$$

namely with $w_-(z) = w_+(-z)$, we immediately get from (3–30) that
$$w_+(z+i0) = -e^{i\pi\nu} w_-(z-i0),$$
$$w_-(z+i0) = -e^{-i\pi\nu} w_+(z-i0), \qquad (3\text{--}32)$$
$$w_+(z) w_-(z) = P(z).$$

For rational $\nu = r/s$, we see that the combination
$$w_+(z)^s + (-1)^{r+s} w_-(z)^s = 2S(z) \qquad (3\text{--}33)$$

is regular across the support of ρ. For instance, if $v(x)$ is polynomial or meromorphic, so are $P(z)$ and $S(z)$. We may now solve (3–32) and (3–33) for $w_\pm$ as
$$w_\pm(z) = \left(S(z) \pm \sqrt{S(z)^2 - P(z)^s}\right)^{\frac{1}{s}},$$

and finally get the resolvent using (3–31) and (3–29):
$$\omega(z) = \frac{2v'(z) - nv'(-z)}{n^2 - 4} - \frac{1}{2i\sin(\pi r/s)} \left(e^{i\pi r/(2s)} \left(S(z) + \sqrt{S(z)^2 - P(z)^s}\right)^{1/s} - e^{-i\pi r/(2s)} \left(S(z) - \sqrt{S(z)^2 - P(z)^s}\right)^{1/s} \right). \qquad (3\text{--}34)$$

Assuming that v is polynomial, we must further fix the coefficients of P and S by imposing the asymptotic behavior $\omega(z) \sim 1/z$ for large $|z|$. If $\deg(v) = k$, then $\deg(P) = 2(k-1)$ and $\deg(S) = s(k-1)$, and P and S are entirely fixed if we impose moreover that ρ has a support made of a single interval $[a, b]$.

3.5. Critical Behavior and Asymptotic Enumeration. We saw in Section 2 how to write the generating functions for various families of (decorated) fatgraphs in terms of Hermitian matrix integrals. It is a standard fact that the critical properties of these generating functions can be translated into asymptotics for these numbers of fatgraphs, say for large numbers of vertices. Indeed, writing such a generating function as $F(g) = \sum_{n \geq 0} F_n g^n$, where F_n denotes a number of connected (decorated) fatgraphs with n vertices, assume F has a critical singular part of the form

$$F(g)_{\text{sing}} \sim (g_c - g)^{2-\gamma} \qquad (3\text{--}35)$$

when g approaches some critical value g_c, then for large n the numbers F_n behave as

$$F_n \sim g_c^{-n} n^{-3+\gamma}.$$

We apply this to the example of (3-1) with the quartic potential $V(x) = \frac{1}{2}x^2 - \frac{1}{4}gx^4$. The generating function

$$F_N(g) = \mathrm{Log}\, Z_N(V) = \sum_{h \geq 0} N^{2-2h} f_h(g)$$

decomposes into the generating functions $h_h(g)$:

$$f_h(g) = \sum_{n \geq 0} g^n f_{h,n},$$

where $f_{h,n}$ denotes the number of fatgraphs of genus h with n vertices. Using the genus zero answer for $f_0(g)$ (3-23), we find that the critical singularity is attained when $g = g_c = \frac{1}{12}$, in which case $f_0(g)_{\mathrm{sing}} \sim (g_c - g)^{5/2}$ and $\gamma = -1/2$ in (3-35). This implies the following asymptotics for the numbers of connected genus zero fatgraphs

$$f_{0,n} \sim \frac{12^n}{n^{7/2}}.$$

More refined applications of the orthogonal polynomial techniques lead to the genus h result

$$f_{h,n} \sim \frac{12^n}{n^{1+(5/2)(1-h)}}$$

and to a score of interesting behaviors for the numbers of decorated fatgraphs for specific potentials V, as well as for the $O(n)$ model with also specific potentials, for which any positive rational value of $-\gamma$ in (3-35) may be reached. The exponent γ is called the string susceptibility exponent. More generally, it has been shown that the coupling of a critical matter theory to gravity results in a behavior (3-35) of the genus zero free energy with γ given by the formula [4]

$$\gamma = \frac{c - 1 - \sqrt{(25-c)(1-c)}}{12}, \qquad (3\text{-}36)$$

where c is the central charge of the conformal field theory underlying the flat space critical matter model [23]. The exponent $\gamma = -\frac{1}{2}$ is characteristic of the "pure gravity" models, in which there is no matter theory, in other words $c = 0$. In the case of the so-called dense critical phase of the $O(n)$ model, one has [24]

$$c(n) = 1 - 6\frac{e^2}{1-e}, \qquad n = 2\cos(\pi e), \qquad (3\text{-}37)$$

and therefore

$$\gamma(n) = -\frac{e}{1-e}$$

from (3-36). This is confirmed by the saddle-point results [22] mentioned above.

3.6. Gaussian Words.

In the case of multimatrix integrals, one may wonder how the beautifully simple result represented by (2–10) and (3–21) is generalized. More precisely, we are interested in the large N limit of the multi-Gaussian average of the trace of any word:

$$\eta_{n_1,n_2,\ldots,n_{mk}} = \langle\langle \mathrm{Tr}(M_1^{n_1} \ldots M_k^{n_k} M_1^{n_{k+1}} \ldots M_k^{n_{mk}}) \rangle\rangle$$

$$= \lim_{N\to\infty} \frac{\int dM_1 \ldots dM_k e^{-N\,\mathrm{Tr}\sum_{i=1}^k M_i^2} \mathrm{Tr}(M_1^{n_1} \ldots M_k^{n_k} M_1^{n_{k+1}} \ldots M_k^{n_{mk}})}{N \int dM_1 \ldots dM_k e^{-N\,\mathrm{Tr}\sum_{i=1}^k M_i^2}}. \quad (3\text{--}38)$$

In this setting, the result (2–10) corresponds to $k = m = 1$ and reads

$$\eta_n = \begin{cases} c_p & \text{if } n = 2p, \\ 0 & \text{otherwise,} \end{cases}$$

for all $n \geq 0$. The simplest way of computing (3–38) in a given situation is the blind application of Wick's theorem (2–17), by further restricting the Wick pairings to the *planar* ones only, as $N \to \infty$. For instance, for $k = 2$ and $m = 2$, we have

$$\eta_{n_1,n_2,n_3,n_4} = \langle\langle \mathrm{Tr}(M_1^{n_1} M_2^{n_2} M_1^{n_3} M_2^{n_4}) \rangle\rangle$$

$$= \eta_{n_1+n_3}\eta_{n_2}\eta_{n_4} + \eta_{n_2+n_4}\eta_{n_1}\eta_{n_3} - \eta_{n_1}\eta_{n_2}\eta_{n_3}\eta_{n_4},$$

where we have isolated respectively the pairings in which elements of the blocks $M_1^{n_1}$ and $M_1^{n_3}$ may be connected, then those in which elements of the blocks $M_2^{n_2}$ and $M_2^{n_4}$ may be connected, and we have subtracted the terms in which all pairings take place within the same powers of the M's to avoid counting them twice (once in each of the previous terms).

To compute the most general trace of a word in say k Gaussian Hermitian matrices, we simply have to apply the following quadratic recursion relation, for $\omega = e^{2i\pi/k}$

$$\eta_{n_1,\ldots,n_{mk}} = -\sum_{i=1}^{mk-1} \omega^i \eta_{n_1,\ldots,n_i} \eta_{n_{i+1},\ldots,n_{mk}}. \quad (3\text{--}39)$$

This relation is a compact rephrasing of the Wick theorem in the case of planar pairings, as the reader will check easily (the only relation to be used is

$$\sum_{0\leq i\leq k-1} \omega^i = 0;$$

see [10] for details). This gives a priori access to the average of any trace of word. The numbers η are natural multidimensional generalizations of the Catalan numbers (2–9).

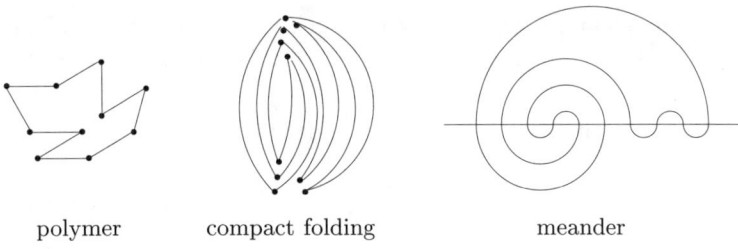

polymer   compact folding   meander

Figure 4. A typical polymer with $2n = 10$ segments is depicted, together with one of its compact folding configurations, in which the segments have been deformed and slightly pulled apart for clarity. The corresponding meander is obtained by drawing an horizontal line (river) intersecting the folding configuration (road) through $2n = 10$ points (bridges). The connected half segments of polymer have been replaced by semicircular portions of road for simplicity.

4. Folding Polymers: Meanders

We now present a first important application of matrix integrals to the fundamental combinatorial problem of meander enumeration, also equivalent to the compact folding problem of polymers. After defining meanders and reviewing various approaches to their enumeration, we will define in Section 6 a matrix model for meanders and discuss its solution in some particular cases. We finally present a general argument leading to the exact values of the meandric asymptotic configuration exponents.

4.1. Definitions and Generalities. A polymer is a chain of say m identical constituents, modelled by rigid segments attached by their ends, that serve as hinges in the folding process. A useful representation to bear in mind is that of a strip of stamps of size $1 \times m$, attached by their ends. Given a closed polymer chain with $m = 2n$ segments (one in which the chain is closed so as to form a loop; see Figure 4), we address the question of enumeration of all the ways of compactly folding the chain, in such a way that in the final (folded) state all the segments are piled up onto one another.

As illustrated in Figure 4, we may keep track of the folding by intersecting the final state with an oriented line (say east to west), and slightly pulling apart the segments of the folded polymer. Redrawing the result with a horizontal east-west oriented line, and representing any two connected half-segments by semicircles, we arrive at a meander, i.e., the configuration of a nonintersecting loop (road, made of semicircular bits) crossing the horizontal line (river, flowing east to west) through a given number $m = 2n$ of points (bridges), as depicted in Figure 4. The number of bridges will also be called the order of the meander. The number of distinct meanders of order $2n$ is denoted by M_{2n} [8].

We extend this definition to a set of k roads namely meanders with k possibly interlocking connected components and a total number of bridges $2n$ (corresponding to the simultaneous folding of k possibly interlocking polymers with a total number of $2n$ segments). The number of meanders of order $2n$ with k connected components is denoted by $M_{2n}^{(k)}$. Note that necessarily $1 \leq k \leq n$. These numbers are summarized in the meander polynomial [10]

$$m_{2n}(q) = \sum_{k=1}^{n} M_{2n}^{(k)} q^k. \tag{4-1}$$

Given a meander of order $2n$, the river cuts the plane into two parts. The upper part is called an arch configuration a of order $2n$: it is a set of n nonintersecting semicircles drawn in the upper half plane delimited by the river and joining the $2n$ equally spaced bridges by pairs. The lower part is clearly the reflection b^t of another arch configuration b of order $2n$. Let A_{2n} denote the set of arch configurations of order $2n$, and for any $a, b \in A_{2n}$, let $c(a, b)$ denote the number of connected components of road obtained by forming a (multicomponent) meander with a as upper half and b^t as lower half. Then

$$m_{2n}(q) = \sum_{a,b \in A_{2n}} q^{c(a,b)}.$$

The number of distinct arch configurations is nothing but the Catalan number

$$|A_{2n}| = c_n = \frac{(2n)!}{(n+1)!n!},$$

since A_{2n} can be easily mapped onto the planar Wick pairings of a single star with $2n$ branches leading to (2–10): indeed, we just have to pull all the bridges together so as to form a $2n$-legged vertex; this results in a planar petal diagram like that of Figure 2(a). As an immediate consequence we get $m_{2n}(1) = |A_{2n}|^2 = c_n^2 \sim 4^{2n}/(\pi n^3)$ by use of Stirling's formula for large n. In general, the meander polynomial is expected to behave for large n as

$$m_{2n}(q) \sim c(q) \frac{R(q)^{2n}}{n^{\alpha(q)}} \tag{4-2}$$

and we just proved that $R(1) = 4$, $\alpha(1) = 3$ and $c(1) = 1/\pi$. In Section 6.4 below, we will present an argument leading to the general determination of the exact value of the meander configuration exponent $\alpha(q)$ as a function of q. Note that

$$s(q) = \lim_{n \to \infty} \frac{1}{2n} \operatorname{Log} m_{2n}(q) = \operatorname{Log} R(q)$$

is nothing but the thermodynamic entropy of folding per segment of a multicomponent closed polymer.

We may also address the problem of compactly folding an open polymer chain of $n-1$ segments [10], attached to a wall by one of its ends (think of a strip of stamps, attached in a book by its left end). As illustrated in Figure 5, we may intersect the folded polymer with a circle that also crosses the book's end. Extending the polymer itself into a half-line, and deforming the circle accordingly

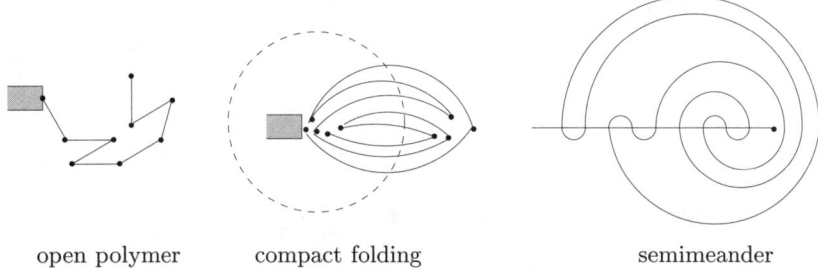

open polymer   compact folding   semimeander

Figure 5. An open polymer with 8 segments, together with one typical compactly folded configuration, and its semimeander version of order 9, obtained by intersecting the 8 segments of the configuration with a circle (represented in dashed line), that also intersects the support (wall) to which the open polymer is attached. The free end of the folded polymer is then stretched into a straight half-line, that becomes the semimeander's river, the free end becoming its source. The dashed circle is deformed into a winding road, crossing the river through 9 bridges. For simplicity, we have represented all the pieces of the road joining the various segments by means of semicircles. Note that here the road winds three times around the source of the river.

without creating new intersections, we arrive at a semimeander of order n, i.e., the configuration of a closed non-self-intersecting loop (road, the former circle) crossing a semi-infinite line (river with a source, the former polymer+book) through n points (bridges).

Note that, in a semimeander, the road may wind around the source of the river, as illustrated in Figure 5. We denote by $\bar{M}_n$ the number of topologically inequivalent semimeanders of order n, and by $\bar{M}_n^{(k)}$ the number of semimeanders with k connected components, $1 \leq k \leq n$. We also have the semimeander polynomial

$$\bar{m}_n(q) = \sum_{k=1}^{n} \bar{M}_n^{(k)} q^k.$$

A semimeander may be viewed as a particular kind of meander by opening the river as sketched in Figure 6 so as to double the number of bridges $n \to 2n$, and by connecting them as they were in the semimeander in the upper-half of the meander, and through the reflection of a rainbow arch configuration r_{2n}^t in the lower one, made of n concentric semicircles. The semimeander polynomial is easily rewritten as

$$\bar{m}_n(q) = \sum_{a \in A_{2n}} q^{c(a, r_{2n})} \tag{4-3}$$

and we have the value at $q = 1$: $\bar{m}_{2n}(1) = c_n \sim 4^n/(\sqrt{\pi} n^{3/2})$. We also expect the asymptotics

$$\bar{m}_{2n}(q) \sim \bar{c}(q) \frac{\bar{R}(q)^n}{n^{\bar{\alpha}(q)}} \tag{4-4}$$

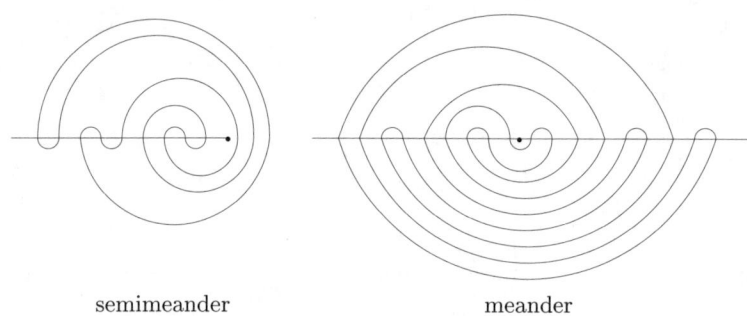

| semimeander | meander |

Figure 6. A semimeander of order n is opened into a meander of order $2n$. Think of the lower part of the river as pivoting around its source to form a straight line, while all bridge connections are deformed without any new intersections. The lower part of the meander is a rainbow, made of n concentric semicircles. The winding number is just the number of semicircular arches passing at the vertical of the center (former source).

and we have $\bar{R}(1) = 4$, $\bar{\alpha}(1) = \frac{3}{2}$ and $\bar{c}(q) = 1/\sqrt{\pi}$. Note again that the semimeander configuration exponent $\bar{\alpha}(q)$ will be determined as a function of q in Section 6.4 below.

Conversely, a meander may be viewed as a semimeander with no winding. It is therefore sufficient to solve the semimeander enumeration problem, provided we keep track of the winding numbers.

4.2. A Simple Algorithm. Numerical Results. Multi-component semimeanders of order n are in one-to-one correspondence with arch configurations of order $2n$ (4-3). Our algorithm [11] is based on a recursive construction of arch configurations that allows to keep track of both the numbers of connected components and windings. To build an arch configuration of order $2n + 2$ from one of order $2n$, we may do either of the following transformations, both involving the addition of two bridges along the river, respectively to the left and right of the previous ones:

(I) For each external arch (semicircle contained in no other) connecting say the bridges i and j, replace it by two semicircles, one connecting the new leftmost bridge to i, and one connecting j to the new rightmost bridge.

(II) Add a large external arch connecting the two added bridges: it circles the whole previous arch configuration.

The corresponding semimeanders are obtained by completing this arch with the reflection of the rainbow r_{2n+2} as lower part. It is easy to show that applying (I)–(II) to all of A_{2n} yields exactly A_{2n+2}. Moreover, (I) preserves the number of connected components, while (II) obviously increases it by 1 (the net result is to add a circle around the semimeander). Applying successions of $(I) - (II)$ on the "root" (semimeander of order 1), we may build the tree of semimeanders of Figure 7.

MATRIX MODEL COMBINATORICS, FOLDING AND COLORING 143

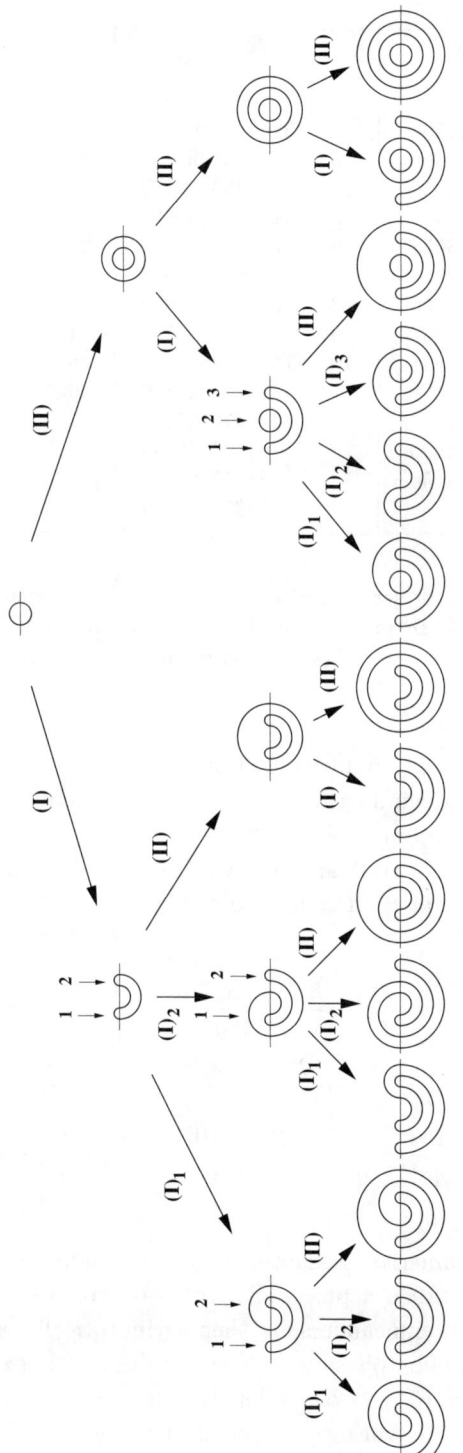

Figure 7. The tree of semimeanders down to order $n = 4$. This tree is constructed by repeated applications of the transformations (I) and (II) on the semimeander of order 1 (root). We have indicated by small vertical arrows the multiple choices for the process (I), each of which is indexed by its number. The number of connected components of a given semimeander is equal to the number of processes (II) in the path going from the root to it, plus one (that of the root).

n	$\bar{M}_n$	n	$\bar{M}_n$	k	$\bar{M}_{27}^{(k)}$	k	$\bar{M}_{27}^{(k)}$
1	1	16	1053874	1	369192702554	16	2376167414
2	1	17	3328188	2	2266436498400	17	628492938
3	2	18	10274466	3	6454265995454	18	153966062
4	4	19	32786630	4	11409453277272	19	34735627
5	10	20	102511418	5	14161346139866	20	7159268
6	24	21	329903058	6	13266154255196	21	1333214
7	66	22	1042277722	7	9870806627980	22	220892
8	174	23	3377919260	8	6074897248976	23	31851
9	504	24	10765024432	9	3199508682588	24	3866
10	1406	25	35095839848	10	1483533803900	25	374
11	4210	26	112670468128	11	619231827340	26	26
12	12198	27	369192702554	12	236416286832	27	1
13	37378	28	1192724674590	13	83407238044		
14	111278	29	3925446804750	14	27346198448		
15	346846			15	8352021621		

Table 1. The number $\bar{M}_n^{(k)}$ of semimeanders of order n with k connected components, obtained by exact enumeration on the computer. On the left, the number of one-component semimeander ($k = 1$) is given for $n \leq 29$; on the right, n is fixed to 27 and $1 \leq k \leq n$.

The preceding algorithm is easily implemented on a computer and yields all semimeander numbers with fixed winding and connected components up to some quite large orders. We give examples in Table 1.

These numbers allow in turn for probing the asymptotics (4–2) and (4–4), as functions of q. In particular, we have these large q asymptotics of the radii:

$$R(q) = 2\sqrt{q}\left(1 + \frac{1}{q} + \frac{3}{2q^2} - \frac{3}{2q^3} - \frac{29}{8q^4} - \frac{81}{8q^5} - \frac{89}{16q^6} + O\left(\frac{1}{q^7}\right)\right),$$

$$\bar{R}(q) = q + 1 + \frac{2}{q} + \frac{2}{q^2} + \frac{2}{q^3} - \frac{4}{q^5} - \frac{8}{q^6} - \frac{12}{q^7} - \frac{10}{q^8} - \frac{4}{q^9} + \frac{12}{q^{10}} + \frac{46}{q^{11}}$$
$$+ \frac{98}{q^{12}} + \frac{154}{q^{13}} + \frac{124}{q^{14}} + \frac{10}{q^{15}} - \frac{102}{q^{16}} + \frac{20}{q^{17}} - \frac{64}{q^{18}} + O\left(\frac{1}{q^{19}}\right).$$

At finite values of q, the numerical results displayed in Figure 8 reveal an interesting phase transition between a phase for $q < q_c$ of irrelevant winding: the numbers of meanders and semimeanders are then asymptotically equivalent $R(q) = \bar{R}(q)$), and a strong winding phase $q > q_c$, where the winding number is proportional to n, and therefore $\bar{R}(q) \gg R(q)$. The transition point is estimated as $q_c \simeq 2$ with poor precision. We will propose an exact value for q_c in Section 6.4 below.

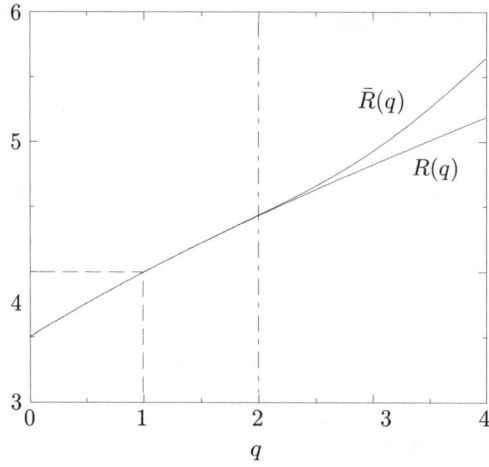

Figure 8. The functions $\bar{R}(q)$ and $R(q)$ for $0 \leq q \leq 4$ as results of large n extrapolations. The two curves coincide for $0 \leq q \leq 2$ and split for $q > 2$ with $\bar{R}(q) > R(q)$. Apart from the exact value $\bar{R}(1) = R(1) = 4$, we find the estimates $\bar{R}(0) = 3.50(1)$, $\bar{R}(2) = 4.44(1)$, $\bar{R}(3) = 4.93(1)$ and $\bar{R}(4) = 5.65(1)$.

5. Algebraic Formulation: Temperley–Lieb Algebra

5.1. Definition. The Temperley–Lieb algebra of order n and parameter q, denoted by $TL_n(q)$, is defined through its n generators $1, e_1, e_2, \ldots, e_{n-1}$ subject to the relations

(i) $\qquad e_i^2 = qe_i \quad$ for $i = 1, 2, \ldots, n-1$;

(ii) $\qquad [e_i, e_j] = 0 \quad$ if $|i - j| > 1$; $\hfill (5\text{–}1)$

(iii) $\qquad e_i e_{i\pm1} e_i = e_i \quad$ for $i = 1, 2, \ldots, n-1$.

This definition becomes clear in the "domino" pictorial representation, where the generators are represented as dominos as follows:

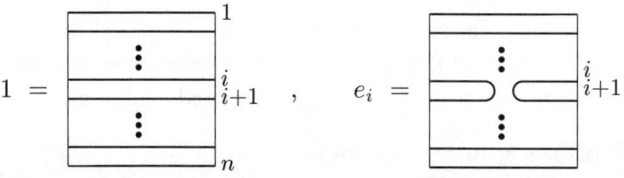

and a product of elements is represented by the concatenation of the corresponding dominos. Note that we have numbered the left and right "string ends" of the dominos from top to bottom, 1 to n. Relation (ii) expresses the locality of the e's, namely that the e's commute whenever they involve distant strings.

Relations (i) and (iii) read respectively

(i) $\quad e_i^2 = \boxed{\text{diagram}} = qe_i = q\boxed{\text{diagram}}$

(the loop has been erased, but affected the weight q) and

(iii) $\quad e_i e_{i+1} e_i = \boxed{\text{diagram}} = e_i = \boxed{\text{diagram}}$

(obtained by stretching the ($i+2$)-nd string).

Moreover the algebra is equipped with a trace, defined as follows. Given a domino d, we put it on a cylinder by identifying its left and right string ends. We then count the number $n(d)$ of connected strings in the resulting picture. The trace is then simply $\text{Tr}(d) = q^{n(d)}$. This definition is extended to any element of the algebra by linearity. This trace has the important Markov property which allows to compute it by induction:

$$\text{Tr}(d(e_1, e_2, \ldots, e_j)e_{j+1}) = \frac{1}{q} \text{Tr}(d(e_1, e_2, \ldots, e_j)) \qquad (5\text{-}2)$$

for any algebra element $d(e_1, \ldots, e_j)$ involving only the generators $e_1, e_2, \ldots e_j$.

The crucial remark actually concerns the left ideal $I_{2n}(q)$ of $TL_{2n}(q)$ generated by the element $e_1 e_3 \ldots e_{2n-1}$. The corresponding dominos have their right string ends paired by single arches linking the ends $2i-1$ and $2i$, $i = 1, 2, \ldots, n$. Therefore their $2n$ left string ends are also connected among themselves, thus forming an arch configuration of order $2n$ after eliminating the internal loops and stretching the strings (by use of (i)–(iii)) and giving them a semicircular shape. We therefore have a bijection between the set A_{2n} of arch configurations of order $2n$ and a basis of the ideal $I_{2n}(q) = TL_{2n}(q)e_1 e_3 \ldots e_{2n-1}$, made only of "reduced" dominos, namely with all their loops removed and their strings stretched. The set of reduced dominos of $I_{2n}(q)$ is denoted by D_{2n}. As a consequence we have $\dim(I_{2n}(q)) = c_n$. We denote by the same letter $a \in A_{2n}$ or D_{2n} the arch configuration or the corresponding reduced domino.

5.2. Meander Polynomials. Given a pair a, b of reduced dominos of $I_{2n}(q)$, we may form the scalar product

$$(a, b) = \frac{1}{q^n} \text{Tr}(b^t a),$$

where by b^t we mean the reflected domino with respect to one of its vertical edges. The middle of the concatenated domino $b^t a$ is nothing but the meander with a as upper-half and b^t as lower one (tilted by $90°$), while it also contains n small circles formed in the cylinder identification; hence

$$(a,b) = q^{c(a,b)}.$$

We deduce the following expressions for the meander and semimeander polynomials in purely algebraic terms [12]:

$$m_{2n}(q) = \sum_{a,b \in D_{2n}} (a,b), \quad \bar{m}_n(q) = \sum_{a \in D_{2n}} (a, r_{2n}).$$

Since the trace can be computed by induction, using (5–2), we now get an inductive way of computing the meander and semimeander polynomials. This is however far from explicit, and does not allow a priori for a good asymptotic study of these polynomials.

5.3. Meander Determinants. On the other hand, we may consider other meander-related quantities that can be explicitly calculated using Temperley–Lieb algebra representation theory. the most interesting of them is the "meander determinant" constructed as follows. We first define the Gram matrix for the reduced basis of $I_{2n}(q)$ as the $c_n \times c_n$ matrix $G_{2n}(q)$ with entries

$$G_{2n}(q)_{a,b} = (a,b) \quad \text{for all } a, b \in D_{2n}.$$

The meander determinant is then defined as the Gram determinant

$$\Delta_{2n}(q) = \det(G_{2n}(q)).$$

This determinant is computed by performing the explicit Gram–Schmidt orthonormalization of the scalar product $(\cdot,\cdot)$, and using along the way the representation theory of $TL_n(q)$. The result is remarkably simple [12; 13]:

$$\Delta_{2n}(q) = \prod_{m=1}^{n} U_m(q)^{a_{m,n}}, \quad a_{m,n} = \binom{2n}{n-m} - 2\binom{2n}{n-m-1} + \binom{2n}{n-m-2},$$

where the $U_m(q)$ are the Chebyshev polynomials of the second kind defined by $U_m(q) = \sin((m+1)\theta)/\sin(\theta)$ with $2\cos\theta = q$.

6. Matrix Model for Meanders

We have seen in Section 2 how to tailor matrix integrals to suit our combinatorial needs. The meanders are basically obtained by intersection of two types of curves: the roads and the rivers. In the following, we will define a matrix model that generates arbitrary configurations of any number of rivers (now viewed as closed curves, so in the case of just one river we should close it into a circle), crossed by any number of non-(self)-intersecting roads [10].

6.1. The Black-And-White Model.
We now construct a Hermitian matrix integral that generates the meander polynomials. The computation of such an integral must involve fatgraphs with double-line edges, which we will eventually interpret as the river(s) and the road(s). Let's paint in white the river edges, and in black the road edges. We therefore have a "black and white" graph made of black and white loops which intersect each other through simple intersections. To assign a weight say q per black loop (component of road) and p per white loop (component of river), the simplest way is to use the replica trick of Section 2.3: for positive integer values of p and q, introduce q "black" Hermitian matrices $B_1, B_2, \ldots, B_q$ and p "white" Hermitian matrices $W_1, W_2, \ldots, W_p$, all of size $N \times N$, with the only nonvanishing propagators

white edges: $\quad \langle (W_a)_{ij}(W_b)_{kl} \rangle = \frac{1}{N}\delta_{a,b}\delta_{il}\delta_{jk} =$ [diagram]

black edges: $\quad \langle (B_a)_{ij}(B_b)_{kl} \rangle = \frac{1}{N}\delta_{a,b}\delta_{il}\delta_{jk} =$ [diagram]

and only simple intersection vertices

$$\text{Tr}(W_a B_b W_a B_b) = \text{[diagram]}$$

for all $1 \leq x \leq p$ and $1 \leq y \leq q$.

The case of a unique river will then be recovered by taking the limit $p \to 0$. This suggests to introduce the "black and white" matrix integral

$$Z_{q,p}(N;x) = \frac{1}{Z_0} \int \prod_{a=1}^{q} dB_a \prod_{b=1}^{p} dW_b\, e^{-N\,\text{Tr}\,V(\{B_a\},\{W_b\})},$$

$$V(\{B_a\},\{W_b\}) = \frac{1}{2}\left(\sum_{a=1}^{q} B_a^2 + \sum_{b=1}^{p} W_b^2 - x\sum_{a=1}^{q}\sum_{b=1}^{p} B_a W_b B_a W_b\right),$$

(6-1)

where Z_0 is a normalization factor ensuring that $Z_{q,p}(N;0) = 1$. As before, the corresponding free energy may be formally expanded as a sum over all possible connected black and white graphs as

$$F_{q,p}(N;x) = \frac{1}{N^2} \log Z_{q,p}(N;x)$$
$$= \sum_{\text{black \& white } \Gamma} \frac{1}{|\text{Aut}\,\Gamma|} N^{-2g(\Gamma)} x^{v(\Gamma)} q^{L_b(\Gamma)} p^{L_w(\Gamma)},$$

(6-2)

where, as in Section 2.2, $g(\Gamma)$, $v(\Gamma)$ and $\text{Aut}(\Gamma)$ denote respectively the genus, number of vertices, and symmetry group of Γ, whereas $L_b(\Gamma)$ and $L_w(\Gamma)$ denote respectively the numbers of black and white loops and black and white edges in Γ. To get a generating function for meander polynomials from (6-2), we simply have to take $N \to \infty$ to only retain the planar graphs (genus zero), and then only compute the coefficient of w in the resulting expression as a power series of w, which leads to

$$F_q(x) \equiv \lim_{N \to \infty} \left. \frac{\partial F_{q,p}(N;x)}{\partial p} \right|_{p=0} = \sum_{n=1}^{\infty} \frac{x^{2n}}{4n} m_{2n}(q), \qquad (6\text{-}3)$$

where we simply have identified planar black and white fatgraphs with meanders (with the river closed into a loop), whose symmetry group is $\mathbb{Z}_{2n} \times \mathbb{Z}_2$ for the cyclic symmetry along the looplike river, and the symmetry between inside and outside of that loop. The meander polynomial is as in (4–1). The large n asymptotic behavior of the meander polynomial $m_{2n}(q)$ can be directly linked to the critical behavior of the generating function $F_q(x)$ (6–3). Indeed, (4–2) translates into a singular part

$$F_q(x)_{\text{sing}} \sim (x(q) - x)^{\mu(q)},$$

where

$$x(q) = \frac{1}{R(q)}, \qquad \mu(q) = \alpha(q).$$

So, if we can investigate the critical properties of $F_q(x)$, the meander asymptotics will follow.

The black-and-white model in the large N limit also allows for a simple representation of semimeanders as a correlation function. Indeed, considering the operator

$$\phi_1 = \lim_{N \to \infty} \frac{1}{N} \text{Tr}(W_1)$$

and computing the correlation function

$$\langle \phi_1 \phi_1 \rangle_{q,p} = \frac{1}{Z_0} \int \prod dW_b dB_a \phi_1 \phi_1 e^{-N \text{Tr} V(\{B_a\},\{W_b\})}$$

by use of the Wick theorem, we see that ϕ_1 creates a white line at a point, hence the fatgraphs contributing to the $p \to 0$ limit will simply have a white segment joining the two endpoints created by the two ϕ_1's, intersected by arbitrary configurations of road. Given such a planar graph, we may always send one of the two endpoints to infinity thus yielding a semi-infinite river, and the fatgraphs with n intersections are just the semimeanders of order n. Hence

$$\sum_{n \geq 1} \bar{m}_n(q) x^n = \langle \phi_1 \phi_1 \rangle_{q,0}.$$

Again, we will get the semimeander asymptotics from the singular behavior of this correlation function when $x \to x_c$.

Before taking the limit as $p \to 0$, we could have written

$$F_{q,p}(x) \equiv \lim_{N \to \infty} F_{q,p}(N;x) = \sum_{n=1}^{\infty} \frac{x^{2n}}{4n} m_{2n}(q,p), \qquad (6\text{-}4)$$

where we have defined a meander polynomial $m_{2n}(q,p)$ for multicomponent meanders with also multiple rivers (and a weight q per road and p per river). In particular $m_{2n}(q,p) \sim p m_{2n}(q)$ when $p \to 0$. Similarly,

$$\langle \phi_1 \phi_1 \rangle_{q,p} = \sum_{n \geq 1} \bar{m}_n(q,p) x^n \tag{6-5}$$

defines a semimeander polynomial with also multiple rivers, one of which is semi-infinite. In Section 6.3 below, we will derive the exact asymptotics of the polynomial $m_{2n}(q, p=1)$ from the black-and-white matrix model, while in Section 6.4, we will present an argument yielding all the critical configuration exponents for large n.

Note also that if we keep N finite, we get an all genus expansion

$$\left. \frac{\partial F_{q,p}(N;x)}{\partial p} \right|_{p=0} = \sum_{g \geq 0} N^{-2g} \sum_{n=1}^{\infty} \frac{x^n}{2n} m_n^{(g)}(q), \tag{6-6}$$

where we have defined the genus g meander polynomials $m_n^{(g)}(q)$ that counts the natural higher genus generalization of meanders of given genus and number of connected components of road, and with one river.

6.2. Meander Polynomials and Gaussian Words. The integral (6–1) can be considerably simplified by noting that it is just multi-Gaussian say in the white matrix vector $\vec{W} = (W_1, \ldots, W_p)$. Using the recipes of Section 3.4, we first identify the quadratic form involving each W_b in (6–1), namely

$$W_b^t \mathbf{Q} W_b = \sum_{i,j,k,l=1}^{N} (W_b)_{ji} Q_{ij,kl} (W_b)_{kl}$$

$$= \mathrm{Tr}\left(W_b^2 - x \sum_{a=1}^{q} W_b B_a W_b B_a \right);$$

hence

$$Q_{ij,kl} = \delta_{ik}\delta_{jl} - x \sum_{a=1}^{q} (B_a)_{ik} (B_a)_{lj},$$

or, more compactly,

$$\mathbf{Q} = I \otimes I - x \sum_{a=1}^{q} B_a \otimes B_a^t. \tag{6-7}$$

After integration over the W's in (6–1), we are left with

$$Z_{q,p}(N;x) = \frac{\int \prod_{a=1}^{q} dB_a \, e^{-\frac{N}{2} \mathrm{Tr} B_a^2} \det(\mathbf{Q})^{-p/2}}{\int \prod_{a=1}^{q} dB_a \, e^{-\frac{N}{2} \mathrm{Tr} B_a^2}}.$$

Extracting the $p \to 0$ limit is easy, with the result

$$\left. \frac{\partial F_{q,p}(N;x)}{\partial p} \right|_{p=0} = \frac{1}{N^2} \langle -\frac{1}{2} \mathrm{Tr} \, \mathrm{Log} \, \mathbf{Q} \rangle,$$

where the bracket stands as in Section 2.3 for the multi-Gaussian average with respect to the B's. Expanding Log $\mathbf{Q}$ of (6–7) as a formal power series of x, we get finally

$$\left.\frac{\partial F_{q,p}(N;x)}{\partial p}\right|_{p=0} = \sum_{m\geq 1} \frac{(-x)^m}{2m} \sum_{a_1,a_2,\ldots,a_m=1}^{q} \langle |\frac{1}{N}\mathrm{Tr}(B_{a_1}B_{a_2}\ldots B_{a_m})|^2\rangle.$$

Comparing this with (6–6), we identify the all-genus meander polynomial generating function

$$\sum_{g\geq 0} N^{-2g} m_n^{(g)}(q) = \sum_{a_1,a_2,\ldots,a_n=1}^{q} \langle |\frac{1}{N}\mathrm{Tr}(B_{a_1}B_{a_2}\ldots B_{a_n})|^2\rangle. \qquad (6\text{–}8)$$

Concentrating on the genus zero case, we may express the meander polynomial (4–1) as a multi-Gaussian integral

$$m_{2n}(q) = \sum_{a_1,a_2,\ldots,a_{2n}=1}^{q} \lim_{N\to\infty} \langle |\frac{1}{N}\mathrm{Tr}(B_{a_1}B_{a_2}\ldots B_{a_n})|^2\rangle, \qquad (6\text{–}9)$$

where we have noted that the right-hand side of (6–8) vanishes for large N if n is odd, by a simple parity argument. Moreover, the right-hand-side of (6–9) must be computed using the planar Wick theorem. It is clear however that no pairing can be made between B matrix elements pertaining to the two trace terms, as that would violate planarity. In other words, in the large N limit, we have $\langle \mathrm{Tr}\, f(B_j)\, \mathrm{Tr}\, g(B_j)\rangle = \langle \mathrm{Tr}\, f(B_j)\rangle \langle \mathrm{Tr}\, g(B_j)\rangle$ for any functions f and g of the B's. Hence, using the definition (3–38), we may finally express the meander polynomial in terms of planar multi-Gaussian averages of words

$$m_{2n}(q) = \sum_{a_1,a_2,\ldots,a_{2n}=1}^{q} |\gamma_{a_1,a_2,\ldots,a_{2n}}|^2, \qquad (6\text{–}10)$$

where, for $a_1,\ldots a_p \in \{1,2,\ldots,q\}$, we have set

$$\gamma_{a_1,a_2,\ldots,a_p} = \eta_{m_1,\ldots,m_{qr-1}} \quad \text{if } B_{a_1}B_{a_2}\ldots B_{a_p} = B_1^{m_1}\ldots B_q^{m_q} B_1^{m_{q+1}}\ldots B_q^{m_{qr-1}}.$$

Having proved (6–10) through some quite lengthy process, it is instructive to interpret (6–10) directly. Using the planar Wick theorem to compute the right-hand side of (6–10), we will represent for $\gamma_{a_1,\ldots,a_{2n}}$ the chain of $2n$ matrix elements as equally spaced points along a line, each with the corresponding color $a_i \in \{1,2,\ldots,q\}$ (see Figure 9 for an illustration). A given planar Wick pairing is nothing but an arch configuration linking these points by pairs, *provided they have the same color*. As we have to multiply two γ's, we get a pair of such colored arch configurations. The fact that the two γ's are conjugate of one another simply means that the two arch configurations may be superposed, and that their bridge colors match: indeed, $\gamma^*_{a_1,\ldots,a_{2n}} = \gamma_{a_{2n},\ldots,a_1}$, so we just have to represent its pairings head down and use the same line as for γ. So the net result is the production, for each (pair of) Wick pairing pertaining to the

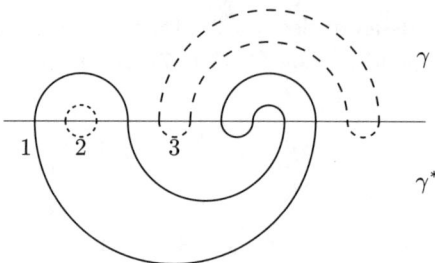

Figure 9. A typical pair of Wick pairings used to compute the quantity $|\gamma_{1,2,2,1,3,3,1,1,1,1,3,3}|^2$. Each pair is represented by a semicircle with the corresponding color (1 = solid, 2 = dashed, 3 = dotted). A pairing for γ is represented, as a colored arch configuration of order 12 here. A pairing for γ^* is also represented, but head-down, to match the bridge colors. The net result is a colored meander with 3 components of road.

two γ's and for each choice of colors $a_1, a_2, \ldots, a_{2n}$ of the bridges, of a *colored multicomponent meander*, namely with colored roads. As the colors are summed over, we get a final weight of q per connected component of road, which matches the definition of the meander polynomial.

This interpretation of the result (6-10) allows to immediately generalize it to the semimeander case:

$$\bar{m}_n(q) = \sum_{a_1, a_2, \ldots, a_n = 1}^{q} \gamma_{a_1, a_2, \ldots, a_n, a_n, a_{n-1}, \ldots, a_2, a_1}. \qquad (6\text{-}11)$$

Indeed, the planar Wick pairings for the γ form again colored arch configurations of order $2n$ with the constraint that they have symmetrically identical colors. This latter constraint can be represented by a lower rainbow diagram r_{2n} with the corresponding colors. The net result is a colored semimeander, and the result (6-11) follows exactly as before.

By analogy with (6-8), we may also define the genus g semimeander polynomials as

$$\sum_{g \geq 0} N^{-2g} \bar{m}_n^{(g)}(q) = \sum_{a_1, a_2, \ldots, a_n = 1}^{q} \left\langle \left| \frac{1}{N} \text{Tr}(B_{a_1} B_{a_2} \ldots B_{a_n} B_{a_n} \ldots B_{a_1}) \right| \right\rangle.$$

Having expressed all our quantities of interest in terms of γ's, we now have an alternative route for evaluating them. We have to use the recursion relations (3-39) to first compute the γ's and then to substitute them in the above expressions for (semi-)meander polynomials. This is still however a tedious work, and it doesn't really allow for any asymptotic study at large n. One advantage though of these expressions is that as the γ's are positive numbers, they allow for exact lower bounds on the meander polynomials.

A last remark is in order: in this section, the meander and semimeander polynomials have been computed only for positive integer values of q. Analytic continuation to any positive real q is however immediate as these are polynomials.

6.3. Exact Asymptotics for the Case of Arbitrary Many Rivers. In this section, following [25], we will use the black and white matrix model for meanders to evaluate the exact asymptotics of slightly different quantities, namely the "multimeander" polynomials $m_{2n}(q,1)$ counting the meanders with fixed numbers of connected components of roads and arbitrary numbers of rivers, closed into nonintersecting loops. However this side result already announces the flavor of what true meander asymptotics should look like.

We wish to compute the function $Z_{q,1}$ (6–1) by first integrating over all the B matrices, rather than the single W one. Using the form (6–7) with $B \to W$, we are then left with

$$Z_{q,1}(N;x) = \frac{\int dW e^{-N\operatorname{Tr}(W^2/2)} \det(1 \otimes 1 - xW \otimes W^t)^{-q/2}}{\int dW e^{-N\operatorname{Tr}(W^2/2)}}.$$

This is just a Gaussian average over one Hermitian matrix W. Applying the reduction method of Section 3.1, we arrive at the eigenvalue integral

$$Z_{q,1}(N;x) = \frac{1}{(2\pi N)^{N/2}} \int dw_1 \ldots dw_N \frac{\prod_{i<j}(w_i - w_j)^2}{\prod_{i,j}(1 - xw_i w_j)^{q/2}} e^{-N\sum_i w_i^2/2}.$$

This is very similar to the eigenvalue integral (3–25) we obtained in the case of the $O(n)$ model of Section 3.4. Actually, upon the change of variables

$$w = \frac{1}{\sqrt{x}} \frac{1-z}{1+z}$$

we have

$$Z_{q,1}(N;x) = \frac{2^{\frac{N^2}{2}(2-q)}}{(2\pi Nx)^{N/2}}$$
$$\times \int dz_1 \ldots dz_N \frac{\prod_{i<j}(z_i - z_j)^2}{\prod_{i,j}(z_i + z_j)^{q/2}} \prod_i (1+z_i)^{N(q-2)} e^{-N\sum_i \frac{1}{2x}\left(\frac{1-z_i}{1+z_i}\right)^2}.$$

We now evaluate the large N behavior of the integral using the saddle-point technique of Section 3.4. The action to be minimized reads now

$$S(z_1,\ldots,z_N) = \frac{1}{N}\sum_{i=1}^N v(z_i) - \frac{1}{N^2}\sum_{1\leq i\neq j\leq N}\operatorname{Log}|z_i - z_j| - \frac{q}{2N^2}\sum_{1\leq i,j\leq N}\operatorname{Log}(z_i + z_j)$$
$$v(z) = \frac{1}{2x}\left(\frac{1-z}{1+z}\right)^2 + (q-2)\operatorname{Log}(1+z)$$

(6–12)

Note that the sum in the last term of S may be restricted to $i \neq j$ up to a term of order $O(1/N)$ that does not contribute to the large N leading asymptotics. Expressing that $\partial_{z_i} S = 0$, we finally get

$$v'(z_i) = \frac{2}{N} \sum_{j \neq i} \frac{1}{z_i - z_j} + \frac{q}{N} \sum_{j \neq i} \frac{1}{z_i + z_j}.$$

This is exactly the saddle-point equation (3–26) for $n = q$ and the particular choice (6–12) of the potential $v(z)$. Note that v' is a meromorphic function of z with a third order pole at $z = -1$. As before, we assume that the limiting eigenvalue distribution has a support made of a single interval $[a, b]$, $0 < a < b$. This requirement turns out to fix entirely the meromorphic functions $S(z)$ and $P(z)$ in (3–34). The critical singularity of the genus zero free energy $f \sim (x(q) - x)^{2-\gamma}$ is found to lie at [25]

$$x = x(q) = \frac{f^2}{2\sin(\pi \frac{f}{2})},$$

where we have set $q = 2\cos(\pi f)$, and the corresponding critical exponent γ reads

$$\gamma = -\frac{f}{1-f}.$$

As mentioned before, these translate into multiriver meander asymptotics

$$m_{2n}(q, 1) \sim \frac{R(q, 1)^{2n}}{n^{\alpha(q,1)}}$$

as

$$R(q, 1) = \frac{1}{x(q)} = 2 \frac{\sin^2(\pi f/2)}{f^2}, \quad \alpha(q, 1) = \frac{2-f}{1-f}. \qquad (6\text{–}13)$$

In particular, we recover from these values the case of meanders with one river and arbitrary many connected components of road, with $q = 0$, $f = \frac{1}{2}$, and $R(1, 0) = R(0, 1) = 4$, $\alpha(1, 0) = \alpha(0, 1) = 3$. We list a few of these values for various fractions f in Table 2.

6.4. Exact Meander Asymptotics from Fully-Packed Loop Models coupled to Two-dimensional Quantum Gravity.

In a recent work [18] it was noticed that the black-and-white matrix model is the natural random surface version of some Fully Packed loop model on the square lattice. The latter is defined by assigning a color (B or W) to each edge of the square lattice, in such a way that two edges of each color meet at each vertex. These edges then form (Fully Packed) loops each of which is assigned a weight p or q for W and B loops respectively. The model is called the $FPL^2(p, q)$ model [26]. When defined on a random surface of genus zero, the model assigns colors to the edges of a random fatgraph with only vertices of the form $BWBW$ (crossing) or $BBWW$ (avoiding). The black-and-white model of Section 6.1 does not have the second kind of vertices. Therefore the original Fully Packed loop model has been further restricted.

q	f	$R(q,1)$	$\alpha(q,1)$	$R(q)$
0	$\frac{1}{2}$	4	3	3.50
1	$\frac{1}{3}$	$\frac{9}{2}$	$\frac{5}{2}$	4
$\sqrt{2}$	$\frac{1}{4}$	$16 - 8\sqrt{2} = 4.68\ldots$	$\frac{7}{3}$	$4.13\ldots$
$\sqrt{3}$	$\frac{1}{6}$	$36 - 18\sqrt{3} = 4.82\ldots$	$\frac{11}{5}$	$4.27\ldots$
2	0	$\pi^2/2 = 4.93\ldots$	2	$4.42\ldots$

Table 2. Multi-river meander asymptotics. We have listed a few values of $q = 2\cos\pi f$, together with the corresponding values of $R(q,1)$ and $\alpha(q,1)$, and the numerical values of $R(q)$ obtained from the direct enumeration results of Section 4.2 for comparison. Note that $R(q,1) \simeq R(q) + \frac{1}{2}$ with good precision.

The detailed study of the $FPL^2(p,q)$ model shows two remarkable facts: (i) it is critical for all values of $0 \leq p,q \leq 2$ (ii) it is represented in the continuum limit by a Conformal Theory with central charge

$$c_{FPL}(q,p) = 3 - 6\left(\frac{e^2}{1-e} + \frac{f^2}{1-f}\right),$$

where $p = 2\cos\pi e$ and $q = 2\cos\pi f$. This was proved by mapping the $FPL^2(p,q)$ model onto a three-dimensional height model, where the heights are defined in the center of each face, with an Ampère-like rule prescribing the transitions from one face to its neighbors. In the continuum limit, the height variable becomes a three-dimensional free field (conformal theory with central charge $c = 3$), and the corrective weights assigning the factors p and q per loop of each color account for the correction of c by electric charges e, f.

The restriction we impose here on the $FPL^2(p,q)$ model on a random surface amounts to restricting the height variable to only two dimensions instead of three. The correct formula for the flat space Fully Packed Loop theory is therefore

$$c(q,p) = 2 - 6\left(\frac{e^2}{1-e} + \frac{f^2}{1-f}\right), \quad p = 2\cos\pi e, \quad q = 2\cos\pi f, \qquad (6\text{--}14)$$

with $e, f \in [0, \frac{1}{2}]$ (so $0 \leq p, q \leq 2$). We therefore state that the black-and-white model of Section 6.1 is described in the planar (large N) limit by the gravitational version of a conformal theory with central charge (6–14), namely the same theory defined on fluctuating surfaces, that have to be summed over statistically. Note that the $O(n)$ model, whose gravitational version has been introduced in Section 2.3 is also described in the dense phase by a conformal theory of central charge $c = 1 - 6g^2/(1-g)$ where $n = 2\cos\pi g$. So we are now dealing with a sort of double $O(n)$ model coupled to gravity.

The coupling to gravity of a conformal theory with central charge $c \leq 1$ has been extensively studied within the context of noncritical string theory. The gravitational theory has a new parameter x, called the cosmological constant, coupled to the area of the surfaces we have to sum over. More precisely, the free energy for a conformal theory coupled to gravity in genus zero reads

$$F = \text{Log } Z = \sum_{A \geq 0} x^A \sum_{\substack{\text{connected surfaces } \Gamma \\ \text{of area } A}} Z_{CFT}(\Gamma), \qquad (6\text{--}15)$$

where $Z_{CFT}(\Gamma)$ denotes the partition function of the conformal theory on the genus zero connected surface Γ. Comparing Z with the black-and-white model partition function (6–1), we see that x plays the role of cosmological constant, as $n = A$ are the areas of the tessellations dual to the fatgraphs of the model. When the conformal theory has central charge c, the free energy (6–15) has been shown to have a singularity of the form (3–35), (3–36):

$$F \sim (x_c - x)^{2-\gamma} \qquad \gamma = \tfrac{1}{12}\left(c - 1 - \sqrt{(1-c)(25-c)}\right)$$

when x approaches some critical value x_c. This is easily translated into the large area asymptotics of the partition function of the model on surfaces of fixed area

$$F_A \sim \frac{x_c^{-A}}{A^{3-\gamma}}. \qquad (6\text{--}16)$$

Moreover, the operators of the conformal theory get "dressed" by gravity, and their correlation functions have singularities of the form

$$\langle \phi_{m_1} \ldots \phi_{m_k} \rangle \sim (x_c - x)^{\sum \Delta_{m_i} - \gamma + 2 - k},$$

where the "dressed dimensions" Δ_m are related to the conformal dimensions h_m of their undressed versions in the conformal theory through [4]

$$\Delta_m = \frac{\sqrt{1-c+24h_m} - \sqrt{1-c}}{\sqrt{25-c} - \sqrt{1-c}}. \qquad (6\text{--}17)$$

We now have all the necessary material to compute the configuration exponents of all the meandric numbers of interest. Applying the result (6–16) to the central charge (6–14), we find the configuration exponent of the multiriver meander polynomial (6–4)

$$m_{2n}(q,p) \sim \frac{R(q,p)^{2n}}{n^{\alpha(q,p)}},$$

$$\alpha(q,p) = 2 + \tfrac{1}{12}\sqrt{1-c(q,p)}\left(\sqrt{25-c(q,p)} + \sqrt{1-c(q,p)}\right),$$

as well as that of the multiriver semimeander polynomial (6–5)

$$\bar{m}_n(q,p) \sim \frac{R(q,p)^n}{n^{\bar{\alpha}(q,p)}}, \qquad \bar{\alpha}(q,p) = \alpha(q,p) - 1 + 2\Delta_1, \qquad (6\text{--}18)$$

where Δ_1 is the dressed dimension of the operator creating a white endpoint. In the conformal theory, this operator is known to have the dimension [27]

$$h_1 = \frac{1-e}{16} - \frac{e^2}{4(1-e)},$$

where $p = 2\cos\pi e$. We simply have to apply (6–17) to the preceding equation and substitute the value of Δ_1 back into (6–18).

For $p = 1$ ($e = \frac{1}{3}$) and q arbitrary, we find

$$\alpha(q,1) = \frac{2-f}{1-f}, \qquad \bar{\alpha}(q,1) = \frac{1}{1-f},$$

for all $q = 2\cos\pi f$. The first of these equations agrees with the saddle point result (6–13). The second is readily obtained by noticing that $\Delta_1 = 0$ when $e = \frac{1}{3}$, and therefore $\bar{\alpha}(q,1) = \alpha(q,1) - 1$.

For $p = q = 0$ ($e = f = \frac{1}{2}$), we get the exact values of the meander and semimeander configuration exponents:

$$\alpha(0,0) = 2 + \frac{1}{12}\sqrt{5}\bigl(\sqrt{5} + \sqrt{29}\bigr),$$

$$\bar{\alpha}(0,0) = 1 + \frac{1}{24}\sqrt{11}\bigl(\sqrt{5} + \sqrt{29}\bigr).$$

Note that the arguments of this section do not give any prediction for nonuniversal quantities such as $R(q,p)$ (which is expected to depend on q and p explicitly, not just on $c(q,p)$).

Finally, for $p = 0$ and q arbitrary, we find:

$$\alpha(q,0) = 2 + \frac{1}{12}\sqrt{1-c(q)}\bigl(\sqrt{25-c(q)} + \sqrt{1-c(q)}\bigr),$$

$$\bar{\alpha}(q,0) = 1 + \frac{1}{24}\sqrt{3-4c(q)}\bigl(\sqrt{25-c(q)} + \sqrt{1-c(q)}\bigr),$$

with $c(q)$ given by (3–37). Note that the second of these equations breaks down when $q = q_c$ corresponding to $c(q_c) = \frac{3}{4}$, namely

$$q_c = 2\cos\left(\pi\frac{\sqrt{97}-1}{48}\right).$$

We identify this as the critical value of q beyond which the winding becomes relevant in semimeanders, namely when circles (roads intersecting the river only once) dominate the semimeander configurations.

7. Folding Triangulations

We now turn to our second application of matrix integrals, having to do with the generation of foldable two-dimensional triangulations. These triangulations are a simple model for so-called tethered or fluid membranes, objects of physical and biological interest.

Although we now deal with the folding of two-dimensional objects (as opposed to the one-dimensional polymers of part B), we will rephrase the problem as that of enumerating vertex-tricolored triangulations. A suitable matrix model will be presented and solved using various techniques.

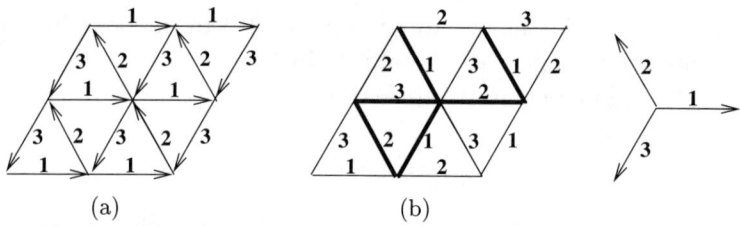

Figure 10. A choice (a) for the tangent vectors of the triangular lattice, together with the corresponding coloring of the edges by 1, 2, 3. This is the flat configuration of the membrane. A folding configuration (b) with the corresponding folded bonds (thick black lines) and edge coloring. The three colors correspond to the three unit vectors with vanishing sum represented above.

As mentioned above, one usually distinguishes between two distinct models for membranes:

Tethered Membranes are represented by reticulated networks, typically some domain of a regular lattice with physical vertices and rigid bonds, allowed only to change their spatial configuration through folding.

Fluid Membranes are represented also by now irregular reticulated networks in which the valency of vertices is no longer fixed, this disorder accounting for the fluidity. But the bonds are still rigid, and the membrane is still allowed to change its spatial configuration through folding.

In the following, we first briefly describe some results [28] on the triangular lattice folding (tethered membrane folding), before turning to the problem of folding triangulations [29] (fluid membranes). Here we only consider the so-called "phantom folding" of membranes; that is, we allow the network to interpenetrate itself, and are only interested in the statistics of the final folded states, independently of the actual feasibility of the folding process. Moreover, the folding of the triangular lattice will be two-dimensional, in that the folded configurations will be subsets of the original lattice.

7.1. Folding the Triangular Lattice. We consider the triangular lattice, or rather a rhombus-shaped portion of it with size $N \times P$. We view this as a reticulated network, with solid edges of unit length. Each such edge carries a unit tangent vector $\vec{t}$, with the constraint that

$$\sum_{\text{around faces}} \vec{t} = \vec{0}. \tag{7-1}$$

With the choice of tangent vectors indicated in Figure 10(a), a folding of the triangular lattice is a continuous map $\rho : S \to \mathbb{R}^2$, preserving the length of the tangent vectors and satisfying the condition (7–1) around each triangular face of S. Let $\vec{t}_1, \vec{t}_2, \vec{t}_3$ denote the unit tangent vectors to a given face of S. Their images $\rho(\vec{t}_i)$ are three unit vectors with vanishing sum, according to the face rule (7–1).

Fixing the image of one tangent vector of S to be a given unit vector $\vec{e_1}$, we see that the images of the tangent vectors of S may only take three values $\vec{e_1}, \vec{e_2}, \vec{e_3}$, where $\vec{e_1}, \vec{e_2}, \vec{e_3}$ are three unit vectors with vanishing sum, hence forming angles of $2\pi/3$. We associate colors numbered $1, 2, 3$ to these three possible images.

A folding map ρ of the triangular lattice is therefore a coloring of its edges, with the three colors $1, 2, 3$, such that the three colors of edges around each face are all distinct. An example of such a coloring is given in Figure 10(b) together with the corresponding folding configuration. The dual of this coloring model is the problem of tricoloring the edges of the hexagonal (honeycomb) lattice in such a way that the three edges adjacent to each vertex are painted with distinct colors $1, 2, 3$. It has been solved by Baxter [30], by use of the Bethe Ansatz. Baxter's results yield in particular the exact value for the thermodynamic entropy of folding per face of the triangular lattice $s = \lim_{N,P\to\infty} \frac{1}{NP} \text{Log} Z_{N,P}$, where $Z_{N,P}$ denotes the total number of distinct folded configurations of our portion of lattice. It reads

$$s = \text{Log}\left(\frac{\sqrt{3}}{2\pi}\Gamma(\tfrac{1}{3})^{3/2}\right). \tag{7-2}$$

This was originally proved by explicitly diagonalizing a large (transfer) matrix, indexed by the coloring configurations of rows of N edges in the honeycomb lattice, and describing the "row-to-row transfer", that is, the allowed coloring configurations for two such neighboring rows. The thermodynamic entropy (7–2) is then the logarithm of the largest (Perron–Frobenius) eigenvalue of this matrix. The diagonalization is performed using a particular ansatz for the form of the eigenvectors, the Bethe Ansatz. The proof of (7–2) being highly technical, we will not reproduce it here, but refer the interested reader to the original papers [30]. We simply mention that this model is part of the class of Two-dimensional Integrable Lattice Models, for which a Bethe Ansatz solution exists.

A score of other lattice folding problems have been studied [31]. Actually, one can classify all the (compactly) foldable lattices [32], and define higher-dimensional generalizations of their folding problems. Remarkably, all of them lead to some equivalent coloring problems, also rephrased into Fully Packed loop enumeration problems. For instance, in the case of folding the triangular lattice, we have seen that the model is equivalent to that of tricoloring the edges of the lattice. Concentrating say on the colors 1 and 2, we see that edges of alternating colors 1212... form loops on the dual (honeycomb) lattice, and moreover each vertex of the dual is visited by exactly one of these loops: the loops 1212... are therefore fully packed, as well as the 2323... and 1313... Remarkably, the same type of Fully Packed loop models have emerged in our study of meanders (see Section 6.4).

7.2. Foldable Triangulations. Fluid membranes are modelled by irregular networks of vertices linked by edges, in which the valencies of the vertices are arbitrary, as well as the genus of the underlying surface, which might have an arbitrary topology. As advocated in Section 2, matrix models provide us with a means of generating such graphs.

We restrict ourselves to networks with only triangular faces, namely triangulations. The initial question one can ask is: are all triangulations foldable? It is indeed desirable to consider only triangulations with a large number of folded configurations, otherwise the effect of folding might be wiped out in the limit of large size. We therefore demand that our triangulations be *compactly* foldable, namely that one can fold them completely onto just one of their (equilateral triangular) faces. With this constraint, it is clear that not all triangulations turn out to be "foldable". Indeed, let us paint by three distinct colors $1, 2, 3$ the three vertices of the image triangle, and paint accordingly the vertices of the preimages under the folding map. This results in the tricoloring of the vertices of the initial triangulation, in such a way that the three colors around each triangular face are distinct. So only the vertex-tricolorable triangulations will be compactly foldable. Another way of viewing this restriction is to recall that we first need to attach tangent vectors to the edges of the triangulation, in such a way that (7-1) is satisfied. It is straightforward to see that this is possible only if the vertices of the triangulation are all even, as around such a vertex, we must have an alternance of tangent vectors pointing to and from it. This condition turns out to be sufficient in genus zero to grant the tricolorability of the triangulation. The situation in higher genus is unclear [33].

We now introduce a generating function $Z(x_1, x_2, x_3; t; N)$ for possibly disconnected vertex-tricolored triangulations of arbitrary genus, such that

$$F(x_1, x_2, x_3; t; N) = \text{Log } Z(x_1, x_2, x_3; t; N)$$
$$= \sum_{\substack{\text{vertex-tricolored} \\ \text{connected triangulations } T}} x_1^{n_1(T)} x_2^{n_2(T)} x_3^{n_3(T)} \frac{t^{A(T)/2} N^{2-2h(T)}}{|\text{Aut } T|},$$

where $n_i(T)$ denote the total numbers of vertices of color i, $A(T)$ the total number of faces, $h(T)$ the genus and $|\text{Aut } T|$ the order of the symmetry group of the tricolored triangulation T.

The construction of a matrix model to represent $Z(x_1, x_2, x_3; t; N)$ is based on the following simple remark: in a given tricolored triangulation T if we remove say all the vertices of color 3 and all edges connected to them, we end up with a bicolored graph, with unconstrained vertex valencies. Such bicolored graphs are easily built out of the Feynman graphs of a two Hermitian matrix model, say M_1 and M_2, the index 1 and 2 standing for the color, with colored vertices

$$\text{Tr}(M_1^n) = \quad \longleftrightarrow N x_1, \qquad \text{Tr}(M_2^n) = \quad \longleftrightarrow N x_2,$$

connected through propagators imposing the alternance of colors, namely

$$\langle (M_a)_{ij}(M_b)_{kl}\rangle = (1-\delta_{ab})\delta_{jk}\delta_{il}\frac{t}{N} = \;\substack{\longrightarrow \\ \dashleftarrow}\;,$$

where $a,b = 1,2$. We introduce the corresponding matrix integral, but keep N fixed while the matrices are taken of size $n \times n$, n possibly different from N. This gives the partition function

$$Z_n(x_1,x_2;t;N) = \frac{1}{\varphi_n(t,N)}\int dM_1\,dM_2\,e^{-N\,\mathrm{Tr}\,V(M_1,M_2;x_1,x_2,t)}, \qquad (7\text{--}3)$$

$$V(M_1,M_2;x_1,x_2,t) = x_1\,\mathrm{Log}(1-M_1) + x_2\,\mathrm{Log}(1-M_2) + \frac{1}{t}M_1M_2,$$

where the normalization factor $\varphi_n(t,N)$ ensures that $Z_n(0,0;t;N) = 1$. Clearly we must take the integral (7–3) only at the level of formal series, by expanding it in power series of x_1 and x_2, and computing the Gaussian integrals with measure $dM_1 dM_2 e^{-(N/t)M_1M_2}$ by the Feynman procedure. In particular, this gives the following rule for two-dimensional integrals

$$\langle x^\alpha y^\beta\rangle = \frac{N}{t}\int dx\,dy\,e^{-(N/t)xy}x^\alpha y^\beta$$
$$= \delta_{\alpha\beta}\Gamma(\alpha+1)\left(\frac{t}{N}\right)^\alpha. \qquad (7\text{--}4)$$

The integral could be made rigorous and convergent by considering instead a normal matrix $M_1 \to M$, $M_2 \to M^*$, but we will content ourselves with formal power series anyway.

We now compute the Feynman graph expansion of the free energy

$$F_n(x_1,x_2;t;N) = \mathrm{Log}\,Z_n(x_1,x_2;t;N)$$
$$= \sum_{\substack{\text{bicolored connected} \\ \text{graphs }\Gamma}} \frac{1}{|\mathrm{Aut}\,\Gamma|} x_1^{n_1(\Gamma)} x_2^{n_2(\Gamma)} t^{E(\Gamma)} N^{(V(\Gamma)-E(\Gamma))} n^{F(\Gamma)},$$

where we have denoted by $n_i(\Gamma)$ the number of vertices of color i, $V(\Gamma) = n_1(\Gamma)+n_2(\Gamma)$, $E(\Gamma)$ the number of edges, $F(\Gamma)$ the number of faces (the boundary of which are loops of running indices, accounting for a factor n each), and $|\mathrm{Aut}\,\Gamma|$ the order of the symmetry group of the bicolored fatgraph Γ. Adding a central vertex of color 3 in the middle of each face of Γ, and connecting it to all the vertices around the face with edges will result in a vertex-tricolored triangulation T. The number of such added vertices is nothing but $n_3(T) = F(\Gamma)$. Introducing

$$x_3 = \frac{n}{N},$$

we may rewrite

$$F_n(x_1,x_2;t;N) = F(x_1,x_2,x_3;t;N) \qquad (7\text{--}5)$$

by using the Euler relation $2 - 2h(\gamma) = 2 - 2h(T) = V(\Gamma) - E(\Gamma) + F(\Gamma)$ and the fact that $A(T) = 2E(\Gamma)$, as each edge of Γ gives rise to two triangles of T, one in each of the two faces adjacent to the edge. It is also a simple exercise to show that $|\operatorname{Aut} T| = |\operatorname{Aut} \Gamma|$. Considering (7-5) as a formal power series of t with polynomial coefficients in x_1, x_2, x_3, we see that the knowledge of $F_n(x_1, x_2; t; N)$ for integer values of n determines completely the polynomial dependence on $x_3 = n/N$ at each order in t. Hence computing $F_n(x_1, x_2; t; N)$ through the integral formulation (7-3) will yield the generating function for compactly foldable triangulations.

8. Exact Solution

We now present the exact computation of the generating function for compactly foldable triangulations, $F(x_1, x_2, x_3; t; N)$. Although it can be obtained through orthogonal polynomial techniques generalizing that of Section 3.2, we choose to present an alternative and powerful approach, using the Discrete Hirota equation, in the form of a recursion relation for the quantities $F_n(x_1, x_2; t; N)$, in terms of n, Nx_1 and Nx_2. We will then write some direct formal series expansion for the solution, and use saddle-point techniques to extract the large N behavior of the free energy. This will give a nice formula for the generating function of vertex-tricolored triangulations of genus zero.

8.1. The Discrete Hirota Equation. The step zero in computing the integral (7-3) is like in Section 3.1 the reduction to eigenvalue integrals. Diagonalizing both matrices as $M_i = U_i m_i U_i^\dagger$, with m_i diagonal and U_i unitary, the change of variables $(M_1, M_2) \to (U_1, m_1; U_2, m_2)$ has the Jacobian $J = \Delta(m_1)^2 \Delta(m_2)^2$. We encounter a problem as the potential $V(M_1, M_2; x_1, x_2, t)$ is not invariant under unitary conjugation, namely the term $\operatorname{Tr}(M_1 M_2) = \operatorname{Tr}(\Omega m_1 \Omega^\dagger m_2)$, with the unitary matrix $\Omega = U_2^\dagger U_1$. So the integral over U_i is no longer trivial, and yields a factor

$$\int d\Omega e^{-(N/t)\operatorname{Tr}(\Omega m_1 \Omega^\dagger m_2))} \propto \frac{\det\left[e^{-(N/t)m_{1,i}m_{2,j}}\right]_{1\le i,j\le n}}{\Delta(m_1)\Delta(m_2)} \qquad (8\text{-}1)$$

up to some normalization factor depending on n and N only. This is the celebrated Itzykson–Zuber formula of integration over the unitary group [35], itself a particular case of the Duistermaat–Heckmann localization formula [36], since the determinant in (8-1) may be viewed as a sum over the critical (saddle-) points of the integrand, at $\Omega = P_\sigma$ the permutations of the eigenvalues.

The integral (7-3) then reduces to

$$Z_n(x_1, x_2; t; N) = \frac{1}{\psi_n(t, N)} \int dm_1 dm_2 \Delta(m_1) \Delta(m_2) e^{-N \operatorname{Tr}(V(m_1, m_2; x_1, x_2, t))},$$

where the normalization factor $\psi_n(t, n)$ ensures that $Z_n(0, 0; t; N) = 1$. It is easily derived by expanding the two determinants as sums over permutations, with the result

$$\psi_n(t,n) = \int dm_1 dm_2 \Delta(m_1)\Delta(m_2) e^{-Nu\,\mathrm{Tr}(m_1 m_2)}$$

$$= \sum_{\sigma,\tau \in S_n} \mathrm{sgn}(\sigma\tau) \prod_{i=1}^{n} \int dm_{1,i} dm_{2,i} m_{1,i}^{\sigma(i)-1} m_{2,i}^{\tau(i)-1} e^{-(N/t)m_{1,i}m_{2,i}}$$

$$= n! \prod_{i=1}^{n} \frac{(i-1)!}{(N/t)^i}, \qquad (8\text{--}2)$$

by use of (7–4).

Writing

$$Z_n(x_1, x_2; t; N) = \frac{1}{\psi_n(t,N)} \int_n \Delta(m_1)\Delta(m_2)$$
$$\times \prod_{i=1}^{n} e^{-(N/t)m_{1,i}m_{2,i}} (1-m_{1,i})^{-a}(1-m_{2,i})^{-b} dm_{1,i} dm_{2,i}$$

with $a = Nx_1$, $b = Nx_2$, and using the basic definition of determinants

$$\prod_{i=1}^{n} (1-m_{k,i})^{-a_k} \Delta(1-m_k) = \det\left[(1-m_{k,i})^{j-a_k-1}\right]_{1 \le i,j \le n}$$
$$= \sum_{\sigma \in S_n} \mathrm{sgn}(\sigma) \prod_{i=1}^{n} (1-m_{k,i})^{\sigma(i)-a_k-1}$$

for $k = 1, 2$, and the shorthand notation

$$Z_n(a,b) = Z_n(x_1, x_2; t; N) \quad \text{for } a = Nx_1, \ b = Nx_2,$$

we finally get

$$Z_n(a,b) = \frac{1}{\psi_n(t,N)} \sum_{\sigma,\tau \in S_n} \mathrm{sgn}(\sigma\tau) \prod_{i=1}^{n} \int dm_{1,i}\,dm_{2,i}$$
$$\times (1-m_{1,i})^{\sigma(i)-a-1}(1-m_{2,i})^{\tau(i)-b-1} e^{-(N/t)m_{1,i}m_{2,i}}$$
$$= \frac{n!}{\psi_n(t,N)} \sum_{\nu \in S_n} \mathrm{sgn}(\nu) \prod_{i=1}^{n} \int dx\,dy\,(1-x)^{i-a-1}(1-y)^{\nu(i)-b-1} e^{-(N/t)xy},$$

where we have set $\nu = \tau\sigma^{-1}$, with the same signature as $\sigma\tau$, and explicitly factored out the sum over σ. Moreover, the dummy integration variables have been rebaptized x and y, and the integral can be computed by expanding the integrand as a power series of x,y and then using term by term the prescription (7–4). The partition function takes therefore the form

$$Z_n(a,b) = \frac{n!}{\psi_n(t,N)} D_n(a,b), \qquad (8\text{--}3)$$

where $D_n(a,b)$ is the $n \times n$ determinant

$$D_n(a,b) = \det\left[\int dx\,dy\,(1-x)^{i-a-1}(1-y)^{j-b-1}e^{-Nuxy}\right]_{1\le i,j\le n} \quad (8\text{-}4)$$

and $\psi_n(t,N)/n! = D_n(0,0) = \prod_{1\le i\le n}(i-1)!/(N/t)^i$.

The Hirota equation [34] is simply the rephrasing in terms of $D_n(a,b)$ of the following identity satisfied by any $(n+1)\times(n+1)$ determinant D and the minors $D_{i,j}$ obtained by erasing the i-th row and j-th column, as well as the minors $D_{i_1,i_2;j_1,j_2}$ obtained by erasing the rows i_1,i_2 and columns j_1,j_2 in D:

$$DD_{1,n+1;1,n+1} = D_{n+1,n+1}D_{1,1} - D_{1,n+1}D_{n+1,1}.$$

When expressed in terms of $D = D_{n+1}(a+1,b+1)$, this implies the quadratic equation

$$D_{n+1}(a+1,b+1)D_{n-1}(a,b)$$
$$= D_n(a+1,b+1)D_n(a,b) - D_n(a,b+1)D_n(a+1,b). \quad (8\text{-}5)$$

Finally, using $\psi_{n+1}(t,N)\psi_{n-1}(t,N)/\psi_n(t,N)^2 = n/(Nu) = nt/N$ from (8-2), we get the Hirota Bilinear equation for $Z_n(a,b)$ by substituting (8-3) into (8-5):

$$n\frac{t}{N}Z_{n+1}(a+1,b+1)Z_{n-1}(a,b)$$
$$= Z_n(a+1,b+1)Z_n(a,b) - Z_n(a,b+1)Z_n(a+1,b). \quad (8\text{-}6)$$

This recursion relation determines the partition function $Z_n(a,b)$ entirely, once we apply the following initial conditions. Let

$$F_n(a,b) = \text{Log}\,Z_n(a,b) = \sum_{m\ge 1}(t/N)^m \omega_m(a,b,n)$$

be the generating function for connected tricolored triangulations, with $a = Nx_1, b = Nx_2$ and $n = Nx_3$. Then, introducing the shorthand notation $\delta_x f(x) = f(x+1) - f(x)$ for finite differences, the Hirota equation (8-6) turns into a finite difference equation for $F_n(a,b)$:

$$\delta_a \delta_b F_n(a,b) = -\text{Log}\left(1 - n\frac{t}{N}e^{\delta_n F_n(a+1,b+1) - \delta_n F_{n-1}(a,b)}\right). \quad (8\text{-}7)$$

This turns into a nonlinear recursion relation for the coefficients $\omega_m(a,b,n)$. But thanks to their interpretation in terms of tricolored triangulation counting, namely $\sum \omega_m(Nx_1,Nx_2,Nx_3)(t/N)^m = \sum N^{2-2h}x_1^{n_1}x_2^{n_2}x_3^{n_3}\#(\text{tricol.triang.})$, all of these are polynomials of a,b,n, and all have at least a term abn in factor, as there is always at least one vertex of each color $1,2,3$ in such a triangulation. This allows for performing the discrete integration step involved in the recursion, and yields the complete solution for $F_n(a,b)$ as a formal power series of t. The first few terms read:

$$\omega_1 = nab,$$
$$\omega_2 = \frac{nab}{2}(n+a+b),$$
$$\omega_3 = \frac{nab}{3}(n^2 + 3(a+b)n + a^2 + 3ab + b^2 + 1),$$
$$\omega_4 = \frac{nab}{4}(n^3 + 6(a+b)n^2 + (6a^2+17ab+6b^2+5)n + (a+b)(a^2+5ab+b^2+5)).$$

Note the symmetry in a,b,n, now manifest. In the limit of large N, the genus zero free energy

$$f_0(x_1, x_2, x_3; t) = \lim_{N \to \infty} \frac{1}{N^2} F_{Nx_3}(Nx_1, Nx_2) \qquad (8\text{--}8)$$

satisfies the differential equation

$$\partial_{x_1}\partial_{x_2} f_0 = -\text{Log}\left(1 - tx_3 e^{\partial_{x_3}(\partial_{x_1} + \partial_{x_2} + \partial_{x_3})f_0}\right). \qquad (8\text{--}9)$$

as a consequence of (8–7).

8.2. Direct Expansion and Large N Asymptotics. We will explicitly compute the partition function $Z_n(a,b)$ (8–3) by expressing the determinant (8–4) as a formal series expansion in t. In the integrand of (8–4) we expand all terms of the form

$$(1-u)^{-c} = \sum_{k \geq 0} u^k \frac{\Gamma(k+c)}{\Gamma(c)k!},$$

for $(u,c) = (x, a+1-i)$ and $(y, b+1-j)$ respectively, and then compute the defining integral by use of the prescription (7–4). After a little algebra, we arrive at

$$D_n(a,b) = \det\left[\sum_{k \geq 0} \frac{1}{k!} \frac{\Gamma(k+a-i+1)}{\Gamma(1+a-i)} \frac{\Gamma(k+b-j+1)}{\Gamma(1+b-j)} \left(\frac{t}{N}\right)^{k+1}\right]_{1 \leq i,j \leq n}$$

$$= \sum_{k_1,\ldots,k_n \geq 0} \prod_{i=1}^n \frac{(t/N)^{k_i+1}\Gamma(k_i+a-i+1)}{k_i!\Gamma(1+a-i)\Gamma(1+b-i)} \det\left[\Gamma(k_i+b-j+1)\right]_{1 \leq i,j \leq n}.$$

Factoring $\Gamma(k_i+b-n+1)$ out of each line of the remaining determinant, we are left with the computation of the determinant

$$\det\left[(k_i+b-j)(k_i+b-j-1)\ldots(k_i+b-n+1)\right] = \det\left[q_{n-j}(k_i)\right],$$

where the polynomials q_m are monic of degree m, and as such satisfy

$$\det\left[q_{n-j}(k_i)\right] = \det\left[k_i^{j-1}\right] = \Delta(k)$$

from (3–3), which gives the Vandermonde determinant of the k's. Repeating the same trick with the columns of the determinant, we finally get the symmetric expression

$$Z_n(a,b) = \sum_{k_1,\ldots,k_n \geq 0} \Delta(k)^2 \prod_{i=1}^{n} \frac{(t/N)^{k_i+1-i}}{i!\, k_i!} \frac{\Gamma(k_i + a - n + 1)}{\Gamma(1 + a - i)} \frac{\Gamma(k_i + b - n + 1)}{\Gamma(1 + b - i)}. \tag{8-10}$$

By construction, this formal series of t is a solution to the Hirota equation (8–6). Note the remarkable similarity between the expansion (8–10) and the reduction to eigenvalues of the one-matrix integral (3–2), except that the integration over eigenvalues is replaced by a sum over nonnegative integers. We indeed have the same two types of terms: (i) the squared Vandermonde determinant, that is, the repulsion term, and (ii) the product of ratios of gamma functions, that can be exponentiated into a potential term. The expansion (8–10) allows us for a computation of the large N limit of the free energy, by applying the saddle-point techniques. Indeed, writing $k_i = N\alpha_i$, we may estimate the expansion by a real integral over the α's when N is large, of the form $\int d\alpha_1 \ldots d\alpha_n \exp(-N^2 S(\{\alpha_i\}))$, where the index i itself ranging from 1 to $n = Nx_3$ may be written as $i = Ns$, $s \in [0,1]$. Expressing that this integral is dominated by the minimum of S, it is a straightforward, though tedious calculation to get the large N asymptotics of Z. We refer the reader to [29] for details, and simply give the beautifully simple result here. The genus zero free energy of the model (8–8) satisfies the following

$$\begin{aligned} t(t\partial_t)^2 f_0(x_1, x_2, x_3; t) &= F_1 F_2 F_3, \\ F_1(1 - F_2 - F_3) &= tx_1, \\ F_2(1 - F_3 - F_1) &= tx_2, \\ F_3(1 - F_1 - F_2) &= tx_3, \end{aligned} \tag{8-11}$$

where $F_i \equiv F_i(x_1, x_2, x_3; t) = tx_i + O(t^2)$ are formal series of t with polynomial coefficients of the x's. Actually it is easy to see that F_i is the generating function for tricolored rooted trees, whose root has color i. This result still awaits a good combinatorial interpretation. One way of proving (8–11) *a posteriori* is to check that it satisfies the differential equation (8–9). As an independent check, in the particular case $x_1 = x_2 = x_3 = z$ we find $F_1 = F_2 = F_3 = F$ and $F(1-2F) = tz$, in agreement with a former result of Tutte [37].

Analyzing the critical properties of f_0 as a function of t, we find a singular behavior of the form $f_0(t) \sim (t_c - t)^{5/2}$ for generic values of the x's ($x_i > 0$ and $x_i \neq x_j$ for $i \neq j$). This leads to a number of tricolored triangulations behaving as $T(x_1, x_2, x_3)^A / A^{7/2}$ in terms of their fixed area A. We recover therefore the value $\gamma = -1/2$ (3–36) of the pure gravity, meaning that the tricoloring constraint does not affect the configuration exponent of triangulations, but does affect the leading behavior through the function $T(x_1, x_2, x_3) \neq 12$.

9. Conclusion

The combinatorial applications of matrix integrals are various and many, and it would be impossible to even enumerate them all here. In this article we have chosen to concentrate on folding problems with simple physical interpretations, but this is mainly a matter of taste.

In the study of meanders, we have shown that even an almost one-dimensional problem had to be first viewed as a random graph (or surface) problem, and then matrix and quantum gravity techniques have allowed for obtaining many interesting results.

In the case of foldable triangulation enumeration, the matrix model has allowed for the derivation of a very simple formula for the genus zero counting function, that still has to be interpreted combinatorially.

The main lesson we would like to draw is that the matrix models give a maybe less intuitive but definitely different angle on graph-related combinatorial problems.

Acknowledgements

I would like to thank A. Its and P. Bleher, organizers of the semester "Random Matrices and Applications" at MSRI (Spring 1999) for their hospitality and their suggestion to write this article. This work was partially supported by the NSF grant PHY-9722060.

References

[1] G. 't Hooft, *Nucl. Phys.* **B72** (1974), 461.

[2] E. Brézin, C. Itzykson, G. Parisi and J.-B. Zuber, "Planar diagrams", *Comm. Math. Phys.* **59** (1978), 35–51.

[3] See for instance the review by P. Di Francesco, P. Ginsparg and J. Zinn-Justin, "2D gravity and random matrices", *Physics Reports* **254** (1995), 1–131, and references therein.

[4] V. G. Knizhnik, A. M. Polyakov and A. B. Zamolodchikov, "Fractal structure of 2D-quantum gravity", *Mod. Phys. Lett.* **A3** (1988), 819–826; F. David, "Conformal field theories coupled to 2-D gravity in the conformal gauge", *Mod. Phys. Lett.* **A3** (1988), 1651–1656; J. Distler and H. Kawai, "Conformal field theory and 2D quantum gravity", *Nucl. Phys.* **B321** (1989), 509–527.

[5] B. Duplantier, "Two-dimensional copolymers and exact conformal multifractality", *Phys. Rev. Lett.* **82** (1999), 880–883; "Exact multifractal exponents for two-dimensional percolation", *Phys. Rev. Lett.* **82** (1999), 3940–3943; "Conformally invariant fractals and potential theory", *Phys. Rev. Lett.* **84** (2000), 1363–1367.

[6] D. Bessis, "A new method in the combinatorics of the topological expansion", *Comm. Math. Phys.* **69** (1979), 147–163; D. Bessis, C. Itzykson and J.-B. Zuber, "Quantum field theory techniques in graphical enumeration", *Adv. Appl. Math.* **1** (1980), 109–157.

[7] A. Sainte-Laguë, *Avec des nombres et des lignes (Récréations Mathématiques)*, Vuibert, Paris, 1937; J. Touchard, "Contributions à l'étude du problème des timbres poste", Canad. J. Math. **2** (1950), 385–398; W. Lunnon, "A map-folding problem", Math. of Computation **22** (1968), 193–199; DiK. Hoffman, K. Mehlhorn, P. Rosenstiehl and R. Tarjan, "Sorting Jordan sequences in linear time using level-linked search trees", Information and Control **68** (1986), 170–184; V. Arnold, "The branched covering of $CP_2 \to S_4$, hyperbolicity and projective topology", Siberian Math. Jour. **29** (1988), 717–726; K. H. Ko and L. Smolinsky, "A combinatorial matrix in 3-manifold theory", Pacific. J. Math **149** (1991), 319–336.

[8] S. Lando and A. Zvonkin, "Plane and projective meanders", *Theor. Comp. Science* **117** (1993), 227–241; "Meanders", *Selecta Math. Sov.* **11** (1992), 117–144.

[9] R. Bacher, "Meander algebras", prépublication de l'Institut Fourier n° 478, 1999.

[10] P. Di Francesco, O. Golinelli and E. Guitter, "Meander, folding and arch statistics", *Math. Comput. Modelling* **26** (1997), 97–147.

[11] P. Francesco, O. Golinelli and E. Guitter, "Meanders: a direct enumeration approach", *Nuc. Phys.* **B482** [FS] (1996), 497–535.

[12] P. Di Francesco, O. Golinelli and E. Guitter, "Meanders and the Temperley-Lieb algebra", *Commun. Math. Phys.* **186** (1997), 1–59.

[13] P. Di Francesco, "SU(N) Meander determinants", *J. Math. Phys.* **38** (1997), 5905–5943; "Meander determinants", *Commun. Math. Phys.* **191** (1998), 543–583.

[14] P. Di Francesco, "Truncated Meanders", preprint UNC-CH-MATH-98/3, to appear in *Proc. Amer. Math. Soc.*

[15] Y. Makeenko, "Strings, matrix models and meanders", *Nuclear Phys. B Proc. Suppl.* **49** (1995), 226–237; Y. Makeenko and H. Win Pe, "Supersymmetric matrix models and the meander problem", preprint ITEP-TH-13/95 (1996); G. Semenoff and R. Szabo, "Fermionic Matrix Models", *Internat. J. Modern Phys.* **A12**, (1997), 2135–2291.

[16] O. Golinelli, "A Monte-Carlo study of meanders", Eur. Phys. J. **B14** (2000), 145–155.

[17] I. Jensen, "Enumerations of plane meanders", preprint cond-mat/9910313. InstDi

[18] P. Di Francesco, O. Golinelli and E. Guitter, "Meanders: exact asymptotics", *Nucl. Phys.* **B570** (2000), 699–712.

[19] I. Kostov, Mod. Phys. Lett. **A4** (1989) 217; M. Gaudin and I. Kostov, "$O(n)$ model on a fluctuating planar lattice: Some exact results", *Phys. Lett.* **B220** (1989), 200–206; I. Kostov and M. Staudacher, "Multicritical phases of the $O(n)$ model on a random lattice", *Nucl. Phys.* **B384** (1992), 459–483.

[20] V. Kazakov, Nucl. Phys. **B4** (Proc. Suppl.) (1998), 93; J.-M. Daul, "Q-states Potts model on a random planar lattice", preprint hep-th/9502014; P. Zinn-Justin, "The dilute Potts model on random surfaces", preprint cond-mat/9903385; B. Eynard and G. Bonnet, "The Potts-q random matrix model: loop equations, critical exponents, and rational case", Phys. Lett. **B463** (1999), 273–279.

[21] P. Francesco and C. Itzykson, "A generating function for fatgraphs", *Ann. . Henri Poincaré*, **59**:2 (1993), Nienhuis117–139; V. Kazakov, M. Staudacher and T. Wynter, "Character expansion methods and for matrix models of dually weighted graphs",

Comm. Math. Phys. **177**, (1996) 451–468, "Almost flat planar diagrams", *Comm. Math. Phys.* **179** (1996), 235–256; Exact solution of discrete two-dimensional R^2 gravity, *Nucl. Phys.* **B471** (1996), 309–333.

[22] B. Eynard and J. Zinn-Justin, "The $O(n)$ model on a random surface: critical points and large-order behaviour", *Nucl. Phys.* **B386** (1992), 558; B. Eynard and C. Kristjansen, "Exact solution of the $O(n)$ model on a random lattice", *Nucl. Phys.* **B455** (1995), 577–618; "More on the exact solution of the $O(n)$ model on a random lattice and an investigation of the case $|n| > 2$", *Nucl. Phys.* **B466** (1996), 463–487.

[23] P. Di Francesco, P. Mathieu and D. Sénéchal, *Conformal field theory*, Graduate Texts in Contemporary Physics, Springer, New York, 1996.

[24] B. in *Phase Transitions and Critical Phenomena*, Eynardvol. **11**, edited by C. Domb and J. L. Lebowitz, Academic Press, 1987.

[25] L. Chekhov and C. Kristjansen, "Hermitian matrix model with plaquette interaction", *Nucl. Phys.* **B479** (1996), 683–696.

[26] J. Jacobsen and J. Kondev, "Field theory of compact polymers on the square lattice", *Nucl. Phys.* **B 532** [FS], (1998), 635–688; "Transition from the compact to the dense phase of two-dimensional polymers", *J. Stat. Phys.* **96**, (1999), 21–48.

[27] F. David and B. Duplantier, "Exact partition functions and correlation functions of multiple Hamiltonian walks on the Manhattan lattice", *J. Stat. Phys.* **51**, (1988), 327–434.

[28] P. Di Francesco and E. Guitter, *Europhys. Lett.* **26** (1994), 455.

[29] P. Di Francesco, B. and E. Guitter, "Coloring random triangulations", *Nucl. Phys.* **B516** [FS] (1998), 543–587.

[30] R. J. Baxter, "Colorings of a hexagonal lattice", *J. Math. Phys.* **11** (1970), 784–789; "q colourings of the triangular lattice", *J. Phys.* **A19** (1986), 2821–2839.

[31] P. Di Francesco, "Folding the square-diagonal lattice", *Nucl. Phys.* **B525** [FS] (1998), 507–548; "Folding transitions of the square-diagonal lattice", *Nucl. Phys.* **B528** [FS] (1998), 453–465; P. Di Francesco and E. Guitter, "Folding transition of the triangular lattice", *Phys. Rev.* **E50** (1994), 4418; M. Bowick, P. Di Francesco, O. Golinelli and E. Guitter, "3D folding of the triangular lattice", *Nucl. Phys.* **B450** [FS] (1995), 463–494.

[32] P. Di Francesco, "Folding and coloring problems in mathematics and physics", *Bull. Amer. Math. Soc.* **37**:3 (2000), 251–307.

[33] *Mathematical Intelligencer* **19**:4 (1997), 48; **20**:3 (1998), 29.

[34] I. Krichever, O. Lipan, P. Wiegmann and A. Zabrodin, "Quantum integrable systems and elliptic solutions of classical discrete nonlinear equations", *Comm. Math. Phys.* **188** (1997), 267–304.

[35] C. Itzykson and J.-B. Zuber, "The planar approximation II", *J. Math. Phys.* **21** (1980), 411–421.

[36] Harish-Chandra, "Differential operators on a semi-simple Lie algebra", *Amer. Jour. of Math.* **79** (1957), 87–120; J. Duistermaat and G. Heckman, "On the variation of cohomology of the symplectic form of the reduced phase space", *Inv. Math.* **69** (1982), 259–268.

[37] W. Tutte, "A census of planar maps", *Canad. Jour. of Math.* **15** (1963) 249–271.

PHILIPPE DI FRANCESCO
CEA-SACLAY
SERVICE DE PHYSIQUE THÉORIQUE
F-91191 GIF SUR YVETTE CEDEX
FRANCE
   philippe@spht.saclay.cea.fr

Interrelationships Between Orthogonal, Unitary and Symplectic Matrix Ensembles

PETER J. FORRESTER AND ERIC M. RAINS

ABSTRACT. We consider the following problem: When do alternate eigenvalues taken from a matrix ensemble themselves form a matrix ensemble? More precisely, we classify all weight functions for which alternate eigenvalues from the corresponding orthogonal ensemble form a symplectic ensemble, and similarly classify those weights for which alternate eigenvalues from a union of two orthogonal ensembles forms a unitary ensemble. Also considered are the k-point distributions for the decimated orthogonal ensembles.

1. Introduction

Given a probability measure on a space of matrices, the eigenvalue PDF (probability density function) follows by a change of variables. For example, consider the space of $n \times n$ real symmetric matrices $A = [a_{j,k}]_{0 \leq j,k < n}$ with probability measure proportional to

$$e^{-\operatorname{Tr}(A^2)/2}(dA), \qquad (dA) := \prod_{j \leq k} a_{jk}. \tag{1-1}$$

The eigenvalues $x_0 < \cdots < x_{n-1}$ are introduced via the spectral decomposition $A = RLR^T$ where R is a real orthogonal matrix with columns given by the eigenvectors of A and $L = \operatorname{diag}(x_0, \ldots, x_{n-1})$. Since $e^{-\operatorname{Tr}(A^2)/2} = e^{-\sum_{j=0}^{n-1} x_j^2/2}$ the change of variables is immediate for the weight function; however the change of variables in (dA) cannot be carried out with such expedience.

The essential point of the latter task is to compute the Jacobian for the change of variables from the independent elements of A to the eigenvalues and the independent variables associated with the eigenvectors. Also, because only the eigenvalue PDF is being computed, one must integrate out the eigenvector dependence. In fact the dependence in the Jacobian on the eigenvalues separates from the dependence on the eigenvectors, so the task of performing the integration does not become an issue. Explictly, one finds (see e.g. [21])

$$(dA) = \Delta(x) \prod_{j=0}^{n-1} dx_j (R^T \, dR)$$

where $\Delta(x) := \Delta(x_0, \ldots, x_{n-1}) := \prod_{0 \le j < k < n}(x_k - x_j)$, and thus the eigenvalue PDF corresponding to (1–1) is proportional to

$$\prod_{j=0}^{n-1} g(x_j) |\Delta(x)|, \tag{1-2}$$

with $g(x) = e^{-x^2/2}$ (taking the absolute value of $\Delta(x)$ allows the ordering restriction on the eigenvalues to be dropped).

More generally, the above working shows that a space of $n \times n$ real symmetric matrices with probability measure proportional to

$$\exp\Big(\sum_{j=1}^{\infty} \alpha_j \mathrm{Tr}(A^j)\Big)(dA) \tag{1-3}$$

will have eigenvalue PDF (1–2) with $g(x) = \exp(\sum_{j=1}^{\infty} \alpha_j x^j)$. Because (1–3) is unchanged by similarity transformations $A \mapsto RAR^T$ with R real orthogonal, for general g (1–2) is said to be the eigenvalue PDF of an orthogonal ensemble, and denoted $\mathrm{OE}_n(g)$.

Real symmetric matrices are Hermitian matrices with all elements constrained to be real. If one considers Hermitian matrices without this constraint, so the off diagonal elements can now be complex, the eigenvalue PDF corresponding to the ensemble (1–3) is again given by (1–2) but with $|\Delta(x)|$ replaced by $(\Delta(x))^2$. Because (1–3) is then unchanged by $A \mapsto UAU^\dagger$ for U unitary, (1–2) so modified is referred to as a unitary ensemble and denoted $\mathrm{UE}_n(g)$. The third and final possibility [7] is to consider $n \times n$ Hermitian matrices in which each element is itself a 2×2 matrix of the form

$$\begin{bmatrix} z & w \\ -\bar{w} & \bar{z} \end{bmatrix}.$$

This class of 2×2 matrices form the real quaternion number field $\mathbb{H}$. The spectrum of such matrices, regarded as $2n \times 2n$ matrices with complex entries, is doubly degenerate. The ensemble of matrices (1–3) is now invariant under the transformations $A \mapsto BAB^\dagger$ for B symplectic unitary, and so referred to as a symplectic ensemble. The eigenvalue PDF of the distinct eigenvalues is given by (1–2) with $|\Delta(x)|$ replaced by $(\Delta(x))^4$, and this is denoted $\mathrm{SE}_n(g)$.

The matrix ensembles corresponding to the eigenvalue PDFs

$$\mathrm{OE}_n(e^{-x^2/2}), \quad \mathrm{UE}_n(e^{-x^2}), \quad \mathrm{SE}_n(e^{-x^2})$$

are given the special labels GOE_n, GUE_n and GSE_n respectively (the G standing for Gaussian). As seen from (1–1) they can be realized by an appropriate Gaussian weight function in the probability space. Because for A real symmetric

$$e^{-\text{Tr}(A^2)/2} = \prod_{j=0}^{n-1} e^{-a_{jj}^2/2} \prod_{j<k}^{n-1} e^{-a_{jk}^2},$$

and similarly for A Hermitian with complex or real quaternion elements, independent elements of the Gaussian ensemble are independently distributed Gaussian random variables.

There are also a number of other known random matrix ensembles with this latter property, and which have eigenvalue PDF of the form $\text{OE}_n(g)$, $\text{UE}_n(g)$ or $\text{SE}_n(g)$ for some g. Seven such ensembles result by taking the Hermitian part of the matrix Lie algebras related to Cartan's ten families of infinite symmetric spaces [30]. We specify five of them:

$\text{Mat}(p,q;\mathbb{R})$ $p \times q$ matrices over $\mathbb{R}$ $(p \geq q)$,

$\text{Mat}(p,q;\mathbb{C})$ $p \times q$ matrices over $\mathbb{C}$ $(p \geq q)$,

$\text{Mat}(p,q;\mathbb{H})$ $p \times q$ matrices over $\mathbb{H}$ $(p \geq q)$,

$\text{Symm}(n;\mathbb{C})$ $n \times n$ symmetric complex matrices,

$\text{Anti}(n;\mathbb{C})$ $n \times n$ antisymmetric complex matrices.

The quantities of interest are the square of the non-zero singular values, or equivalently the eigenvalues of $A^\dagger A$ for A a member of the ensemble, in each case. The first two of these ensembles were studied long ago in mathematical statistics [29; 14]; these two together with the third have occurred in recent physical applications (see [3] and references therein), while the final two (in a different guise) have also arisen in a physical context [3]. The distribution of the eigenvalues of $A^\dagger A$ can be computed in a number of ways; one approach is to make use of the correspondence [30] to a symmetric space (of types BDI, $AIII$, CII, CI and $DIII$ respectively), which allows the tables in [16] to be utilized. Abusing notation, we have

$$\text{Mat}(p,q;\mathbb{R}) = \text{OE}_q(x^{(p-q-1)/2}e^{-x/2}),$$
$$\text{Mat}(p,q;\mathbb{C}) = \text{UE}_q(x^{p-q}e^{-x}),$$
$$\text{Mat}(p,q;\mathbb{H}) = \text{SE}_q(x^{2(p-q)+1}e^{-x}),$$
$$\text{Symm}(n;\mathbb{C}) = \text{OE}_n(e^{-x/2}),$$
$$\text{Anti}(2n;\mathbb{C}) = \text{SE}_n(e^{-x}),$$
$$\text{Anti}(2n+1;\mathbb{C}) = \text{SE}_n(x^2 e^{-x}). \tag{1-4}$$

Up to the scale of x, all the above weight functions are of the Laguerre form $x^\alpha e^{-x}$ and so by definition are examples of Laguerre matrix ensembles.

Another class of matrix ensembles in which the entries of the underlying matrices are independently distributed Gaussian random variables are known in mathematical statistics [21]. With $a \in \mathrm{Mat}(p_1, q; \mathbb{F})$ (where $\mathbb{F} = \mathbb{R}, \mathbb{C}$ or $\mathbb{H}$), $b \in \mathrm{Mat}(p_2, q, \mathbb{F})$, and $A = a^\dagger a$, $B = b^\dagger b$, these distributions are described by

$$\mathrm{Beta}(p_1, p_2, q; \mathbb{F}) \quad q \times q \text{ matrices } A(A+B)^{-1}.$$

They have corresponding eigenvalue PDF (abusing notation as in (1–4))

$$\mathrm{Beta}(p_1, p_2, q; \mathbb{R}) = \mathrm{OE}_q(x^{(p_1-q-1)/2}(1-x)^{(p_2-q-1)/2}),$$
$$\mathrm{Beta}(p_1, p_2, q; \mathbb{C}) = \mathrm{UE}_q(x^{p_1-q}(1-x)^{p_2-q}),$$
$$\mathrm{Beta}(p_1, p_2, q; \mathbb{H}) = \mathrm{SE}_q(x^{2(p_1-q)+1}(1-x)^{2(p_2-q)+1}),$$

where $0 < x < 1$, and thus involve weight functions of the Jacobi type.

The revision above demonstrates that it is possible to realize, in terms of matrices with entries which are independently distributed Gaussian random variables, the distributions $\mathrm{OE}_n(g)$, $\mathrm{UE}_n(g)$ and $\mathrm{SE}_n(g)$ for g one of the forms

$$e^{-x^2}, \quad x^\alpha e^{-x}, \quad x^a(1-x)^b. \tag{1-5}$$

These same weight functions occur in the theory of orthogonal polynomials [26] — they are associated with the three families of classical orthogonal polynomials Hermite, Laguerre and Jacobi respectively, and are themselves referred to as classical weight functions. The classical polynomials share many special properties not enjoyed by orthogonal polynomials associated with other weight functions. In the present study of matrix ensembles, we will see that the distributions $\mathrm{OE}_n(g)$, $\mathrm{UE}_n(g)$ and $\mathrm{SE}_n(g)$ also have special features for g a classical weight function (1–5).

Our interest is in the properties of alternate eigenvalues in matrix ensembles. In particular we seek to determine the weights g for which alternate eigenvalues taken from a random union of two orthogonal ensembles form a unitary ensemble. Similarly we seek the weights g for which alternate eigenvalues from an orthogonal ensemble form a symplectic ensemble. The motivation for this study comes from recent work of Baik and Rains [4]. Consider the distribution $\mathrm{OE}_n(e^{-x})$, n even, and order the eigenvalues $x_0 < x_1 < \cdots < x_{n-1}$. In [4] it was proved that after integrating out every second eigenvalue x_{n-1}, x_{n-3}, etc., the remaining eigenvalues have the distribution $\mathrm{SE}_{n/2}(e^{-x})$. The proof of Baik and Rains is particular to the $a = 0$ case of the Laguerre ensemble. However other considerations lead these authors [5] to conjecture that in an appropriate scaled limit the distribution of the largest eigenvalue in the GSE corresponds to that of the second largest eigenvalue in the GOE. From this it is remarked that presumably the joint distribution of every second eigenvalue in the GOE coincides with the joint distribution of all the eigenvalues in the GSE, with an appropriate number of eigenvalues.

Baik and Rains [5] were also led to consider two GOE_n spectra, superimposing them at random, and integrating out every second eigenvalue of the resulting sequence. Results were presented which suggest that in the scaled $n \to \infty$ limit at the soft edge the distribution becomes that of GUE_∞, appropriately scaled.

Such interrelationships between ensembles first occurred in the work of Dyson [8] on the circular ensembles of random unitary matrices. These ensembles have eigenvalue PDF proportional to

$$\prod_{0 \le j < k < n} |e^{i\theta_k} - e^{i\theta_j}|^\beta, \quad 0 \le \theta_j < 2\pi, \tag{1-6}$$

for $\beta = 1, 2,$ and 4 (COE_n, CUE_n and CSE_n) respectively. Dyson conjectured that

$$\text{alt}(COE_n \cup COE_n) = CUE_n \tag{1-7}$$

which means that if two spectra from the COE_n distribution are superimposed at random with every second eigenvalue integrated out, the CUE_n distribution results. This was subsequently proved by Gunson [15]. Also, Mehta and Dyson [20] proved that integrating out every second eigenvalue from the distribution COE_n with n even gives the distribution $CSE_{n/2}$, or symbolically

$$\text{alt}(COE_n) = CSE_{n/2}. \tag{1-8}$$

The circular ensembles can be analyzed in the course of the present study of ensembles with real valued eigenvalues by making the stereographic projection

$$e^{i\theta_j} = \frac{1 - ix_j}{1 + ix_j}.$$

The PDF (1-6) then maps to

$$\prod_{j=0}^{n-1} \frac{1}{(1+x_j^2)^{\beta(n-1)/2+1}} \prod_{0 \le j < k < n} |x_k - x_j|^\beta$$

which is of the general type under consideration. Here the weight function is of the form

$$\frac{1}{(1+x^2)^\alpha}, \quad \alpha > 1. \tag{1-9}$$

This only has a finite number of well defined moments and thus in this respect differs from the classical weight functions (1-5). On the other hand the corresponding orthogonal polynomials are $\{P_n^{(-\alpha,-\alpha)}(ix)\}_{n<\alpha-1/2}$ [25], with $P_n^{(\alpha,\beta)}$ denoting the Jacobi polynomial, thus implying (1-9) can be viewed as a fourth classical weight function.

2. Pseudo-Ensembles

We begin with the orthogonal ensemble eigenvalue PDF (1–2), taking away the modulus sign, replacing n by l (to avoid overuse of the former) and rewriting the product as a determinant using the Vandermonde formula to obtain

$$\Delta(x) \prod_i g(x_i) = \det(x_i^j)_{0 \leq i,j < l} \prod_i g(x_i) = \det\bigl(g(x_i) x_i^j\bigr)_{0 \leq i,j < l}.$$

In particular, we note that each row corresponds to a variable, while each column corresponds to a function. Given a collection of n functions $F_i : \mathbb{R} \to \mathbb{R}$, we thus define the associated "orthogonal pseudo-ensemble" by the following "density":

$$\det\bigl(F_j(x_i)\bigr)_{0 \leq i,j < l}.$$

Thus any orthogonal ensemble is also an orthogonal pseudo-ensemble, but certainly not vice versa. Indeed, one has:

THEOREM 2.1. *Fix an integer $l > 0$, and let $G : \mathbb{R} \to \mathbb{R}$ be a function supported on at least n points. Then for a collection of n functions $F_0, F_1, \ldots, F_{l-1}$, we have*

$$\det\bigl(F_j(x_i)\bigr)_{0 \leq i,j < l} \propto \prod_i G(x_i) \Delta(x) \tag{2-1}$$

if and only if there exist l linearly independent polynomials p_i of degree at most $l-1$ such that $F_i(x) = p_i(x) G(x)$ for all i and x.

PROOF. The "if" portion is easy enough:

$$\det\bigl(G(x_i) p_j(x_i)\bigr)_{0 \leq i,j < l} = \prod_i G(x_i) \det\bigl(p_j(x_i)\bigr)_{0 \leq i,j < l} \propto \prod_i G(x_i) \Delta(x),$$

since the polynomials are assumed linearly independent.

Now, suppose (2–1) holds. It will turn out to be convenient to restate the equation in terms of exterior products. Define a vector-valued function $V_F(x)$ by

$$V_F(x)_i = F_i(x).$$

Then we can write

$$\det\bigl(F_j(x_i)\bigr)_{0 \leq i,j < l} = \Bigl\langle V_F(x_0), \Bigl(\bigwedge_{1 \leq i < l} V_F(x_i)\Bigr) \Bigr\rangle,$$

where $\langle \cdot, \cdot \rangle$ stands for the standard duality between 1-forms and $l-1$-forms. Consider this as a function of x_0 as the other variables range over the support of G; we have:

$$V_F(x) \cdot \Bigl(\bigwedge_{1 \leq i < l} V_F(x_i)\Bigr) \propto G(x) \prod_{1 \leq i < l} (x - x_i).$$

Now, since G has at least l elements in its support, these functions span an l-dimensional space (this follows, for instance, from Lagrange's interpolation formula). On the other hand, the functions must clearly be linear combinations of the F_i. Since there only l functions F_i, it follows that we can write the F_i as linear combinations of the functions $G(x)x^i$, $0 \le i < l$. But this is precisely what we wanted to prove. $\square$

Similarly, the density function of a symplectic ensemble can also be written as a determinant, namely
$$\Delta(x)^4 \prod_i g(x_i)^2 = \det\bigl(g(x_i)x_i^j, jg(x_i)x_i^{j-1}\bigr)_{0 \le i < l, 0 \le j < 2l};$$
this follows by differentiating the Vandermonde determinant. When $\log(g)$ is differentiable, we can perform column transformations to put this determinant in the form
$$\det\bigl(F_j(x_i), F_j'(x_i)\bigr)_{0 \le i < l, 0 \le j < 2l}; \qquad (2\text{--}2)$$
simply take $F_j(x) = g(x)x^j$, and observe that
$$F_j'(x) - \frac{g'(x)}{g(x)} F_j(x) = jg(x)x^{j-1}.$$

In fact, we can often define (2–2) even when the functions F_j are not differentiable, by expressing it in terms of the 2-form-valued function
$$V_F^{(2)}(x) = \lim_{y \to x} \frac{1}{(x-y)} (V_F(x) \wedge V_F(y)).$$
For $F_j(x) = g(x)x^j$, we find that this is defined wherever g is continuous.

THEOREM 2.2. *Fix an integer $l > 1$, let O be a nonempty open subset of $\mathbb{R}$, and let $G : O \to \mathbb{R}$ be a continuous function supported on O. Then for a collection of continuous functions $F_j : O \to \mathbb{R}$ such that (2–2) is well-defined,*
$$\det(F_j(x_i), F_j'(x_i))_{0 \le i < l, 0 \le j < 2l} = \Delta(x)^4 \prod_i G(x_i)^2$$
on O^{2l} if and only if there exist linearly independent polynomials p_j of degree at most $2l - 1$ with $F_j(x) = G(x)p_j(x)$.

PROOF. Again the "if" case is straightforward. In the other direction, we can clearly divide each F_j by G, and thus may assume without loss of generality that $G = 1$ on O.

We first consider the case $l = 2$, for which
$$\det\bigl(F_j(x)\ F_j'(x)\ F_j(y)\ F_j'(y)\bigr)_{0 \le j < 4} = \bigl\langle V_F^{(2)}(x), V_F^{(2)}(y) \bigr\rangle = (x-y)^4.$$
As y varies over O, this spans a 5-dimensional function space; it follows that as y varies, $V_F^{(2)}(y)$ spans a 5-dimensional space (the dimension must be either 5 or 6; 6 clearly leads to a contradiction). In other words, there must be a linear dependence between the coefficients of $V_F^{(2)}(y)$. By replacing the F_i with an

orthogonal linear combination, we find that this dependence is without loss of generality of the form
$$V_F^{(2)}(y)_{01} = CV_F^{(2)}(y)_{23},$$
for some constant C. Now, if C were 0, then we would have
$$F_0(x)F_1'(x) = F_1(x)F_0'(x).$$
Now, let $I \subset O$ be an open interval in O. If either F_0 or F_1 were identically 0 on I, our determinant would be identically 0 on I^2 (contradiction); it follows that we may choose I so that both F_0 and F_1 are nonzero. Then we can divide both sides of the preceding equation by $F_0(x)F_1(x)$ and integrate; we find that $F_0 \propto F_1$ on I. But this again makes the determinant 0. We conclude that the linear dependence satisfied by $V_F^{(2)}$ must take the form
$$V_F^{(2)}(y)_{01} = CV_F^{(2)}(y)_{23}$$
with $C \neq 0$.

In particular, we find that the 2-form V' orthogonal to $V_F^{(2)}(y)$ is not itself in the span of $V_F^{(2)}(y)$. In particular, any 2-form can be written as a linear combination of V' and some of the $V_F^{(2)}(y)$. Taking the inner product with $V_F^{(2)}(x)$, we conclude that for $0 \leq i < j \leq 3$, we have
$$V_F^{(2)}(x)_{ij} = p_{ij}(x)$$
for some polynomial p of degree at most 4. Now, since $V_F(x) \wedge V_F^{(2)}(x) = 0$, we find:
$$F_0(x)p_{12}(x) - F_1(x)p_{02}(x) + F_2(x)p_{01}(x) = 0$$
for all i, j, k. Similarly, since
$$\frac{d}{dx} V_F^{(2)}(x) = V_F(x) \wedge V_F''(x),$$
we have
$$F_0(x)p_{12}'(x) - F_1(x)p_{02}'(x) + F_2(x)p_{01}'(x) = 0$$
Now, since $V_F^{(2)}(x)_{ki}$ is linearly independent of $V_F^{(2)}(x)_{ij}$, we can solve these two equations for F_1 and F_2 as rational multiples of F_0, substitute into the equation $V_F^{(2)}(x)_{12} = p_{12}(x)$, then solve for F_0. We find
$$F_0 = \frac{p_{01}p_{02}' - p_{02}p_{01}'}{\sqrt{D}}, \quad F_1 = \frac{p_{01}p_{12}' - p_{12}p_{01}'}{\sqrt{D}}, \quad F_2 = \frac{p_{02}p_{12}' - p_{12}p_{02}'}{\sqrt{D}},$$
where
$$D = \det \begin{pmatrix} p_{01} & p_{12} & p_{20} \\ p_{01}' & p_{12}' & p_{20}' \\ p_{01}'' & p_{12}'' & p_{20}'' \end{pmatrix}.$$
We observe that each numerator has degree at most 6, as does the polynomial D. In particular, if we exclude any given F, we can express the squares of the other F as rational functions with common denominator of degree at most 6. It

follows that the functions F^2 have at most 8 poles between them, and thus that we can write

$$F_0(x) = p_0(x)p(x)^{-1/2}, \quad F_1(x) = p_1(x)p(x)^{-1/2},$$
$$F_2(x) = p_2(x)p(x)^{-1/2}, \quad F_3(x) = p_3(x)p(x)^{-1/2},$$

where p_0, p_1, p_2, and p_3 are polynomials of degree at most 7 and p is a polynomial of degree at most 8.

We now need to show that, in fact, each F_i is a polynomial of degree at most 3. By the usual factorization, we find:

$$\det\bigl(p_j(x) \ p'_j(x) \ p_j(y) \ p'_j(y)\bigr)_{0 \le j < 4} \propto p(x)p(y)(x-y)^4, \tag{2-3}$$

valid on $\mathbb{R}$. Without loss of generality, we may assume that the constant of proportionality is 1, and that $p(0) = 1$. Dividing both sides by $(x-y)^4$ and taking the limit as $x, y \to 0$, we find:

$$\det \begin{pmatrix} p_0(0) & p_1(0) & p_2(0) & p_3(0) \\ p'_0(0) & p'_1(0) & p'_2(0) & p'_3(0) \\ p''_0(0) & p''_1(0) & p''_2(0) & p''_3(0) \\ p'''_0(0) & p'''_1(0) & p'''_2(0) & p'''_3(0) \end{pmatrix} = 1$$

Applying a suitable linear transformation to the polynomials p_i, we have, without loss of generality,

$$p_0(x) = 1 + p_{04}x^4 + p_{05}x^5 + p_{06}x^6 + p_{07}x^7,$$
$$p_1(x) = x + p_{14}x^4 + p_{15}x^5 + p_{16}x^6 + p_{17}x^7,$$
$$p_2(x) = x^2 + p_{24}x^4 + p_{25}x^5 + p_{26}x^6 + p_{27}x^7,$$
$$p_3(x) = x^3 + p_{34}x^4 + p_{35}x^5 + p_{36}x^6 + p_{37}x^7.$$

We can then solve for $p(x)$ by taking $y = 0$ above; we find

$$p(x) = x^{-4}(p_2(x)p'_3(x) - p_3(x)p'_2(x)).$$

At this point, we can compare coefficients on both sides of (2–3), obtaining a number of polynomial equations relating the coefficients p_{ij}, where $0 \le i \le 3$ and $4 \le j \le 7$. The resulting ideal can be verified (using Magma, for instance) to contain the polynomials $(p_{25} + p_{34}p_{35} - p_{36})^2$, $(p_{26} + p_{34}p_{36} - p_{37})^2$, and $(p_{27} + p_{34}p_{37})^2$; passing to the radical, we can then solve for p_{ij}, $0 \le i \le 2$, $4 \le j \le 7$. Substituting in, we find that $p(x)$ is now a square, and that each $p_i(x)$ is a multiple of $\sqrt{p(x)}$. In other words, each $F_i(x)$ is a polynomial of degree at most 3, and we are done with the case $l = 2$.

It remains only to show that we can reduce the cases $l > 2$ to cases of lower dimension. Choose a particular element $x_0 \in O$. By replacing the F_i with appropriate linear transformations, we may assume

$$V_F^{(2)}(x_0) = Ce_1 \wedge e_2,$$

for some nonzero constant C. In particular, we find that

$$\det\bigl(F_j(x_i)\bigr)_{0\leq i<l,\ 0\leq j\leq 2l}$$
$$= C \det\bigl(F_j(x_i)\bigr)_{1\leq i<l,\ 2\leq j\leq 2l} \propto G(x_0)\Delta(x_1,x_2,\ldots,x_{l-1})^4 \prod_{1\leq i<l} x_i^4 G(x_i)^2.$$

By induction, it follows that for $2 \leq i < l$, there exist polynomials $p_i(x)$ of degree at most $2l - 1$ such that $F_i(x) = (x-x_0)^2 G(x) p_i(x)$ on O. Undoing our linear transformations, we find that for every polynomial $p(x)$ of degree at most $2l - 1$ vanishing to second order at x_0, we can write $G(x)p(x)$ as a linear combination of the $F_i(x)$. But this was independent of our choice of x_0. In particular, taking x_0' to be any other element of O, we have

$$1 = \frac{3(x_0-x_0')(x-x_0)^2 + 2(x-x_0)^3 + 3(x_0-x_1')(x-x_0')^2 - 2(x-x_0')^3}{(x_0-x_0')^3}$$

and

$$(x-x_0) = \frac{-2(x_0-x_0')(x-x_0)^2 - (x-x_0)^3 - (x_0-x_0')(x-x_0')^2 + (x-x_0')^3}{(x_0-x_0')^2}.$$

It follows that for any polynomial $p(x)$ of degree at most $2l - 1$, $G(x)p(x)$ is a linear combination of the $F_i(x)$. By dimensionality, it follows that each $F_i(x)$ is itself of the form $G(x)p(x)$, and we are done. □

3. Linear Fractional Transformations

It will be convenient in the sequel to determine how matrix ensembles behave under a linear fractional transformation (LFT) or change of variables. To be precise, let f be a weight function, and consider what the density of one of its associated matrix ensemble is in terms of the variables y_i defined by $x_i = (\alpha y_i + \beta)/(\gamma y_i + \delta)$. Clearly, we need only determine how $\Delta(x)$ and $\prod_i dx_i$ transform.

We readily compute:

$$dx = \frac{\alpha\delta - \beta\gamma}{(\gamma y + \delta)^2},$$

thus answering that question. As for Δ:

LEMMA 3.1. *Let $y_0, y_1, \ldots, y_{l-1}$ be a collection of l real numbers. Then for any α, β, γ and δ such that $\gamma y_i + \delta$ is never 0,*

$$\Delta\left(\frac{\alpha y_i + \beta}{\gamma y_i + \delta}\right) = (\alpha\delta - \beta\gamma)^{l(l-1)/2} \prod_i (\gamma y_i + \delta)^{1-l} \Delta(y_i).$$

PROOF. For each $i < j$, we have

$$\frac{\alpha y_j + \beta}{\gamma y_j + \delta} - \frac{\alpha y_i + \beta}{\gamma y_i + \delta} = \frac{\alpha\delta - \beta\gamma}{(\gamma y_i + \delta)(\gamma y_j + \delta)}(y_j - y_i).$$

Multiplying over $i < j$, we are done. □

We thus obtain the following transformation rules:

THEOREM 3.2. *Let f be any weight function. Under the change of variables $x_i = (\alpha y_i + \beta)/(\gamma y_i + \delta)$, we have*

$$\text{OE}_l(f(x)) \to \text{OE}_l\big(|\alpha\delta - \beta\gamma|^{(l+1)/2}(\gamma y + \delta)^{-1-l}\tilde{f}(y)\big),$$
$$\text{UE}_l(f(x)) \to \text{UE}_l\big(|\alpha\delta - \beta\gamma|^l(\gamma y + \delta)^{-2l}\tilde{f}(y)\big),$$
$$\text{SE}_l(f(x)) \to \text{SE}_l\big(|\alpha\delta - \beta\gamma|^{2l-1}(\gamma y + \delta)^{2-4l}\tilde{f}(y)\big),$$

where $\tilde{f}(y) = f\big((\alpha y + \beta)/(\gamma y + \delta)\big)$, the normalization constants are the same on both sides, and for OE_{2l}, $\gamma y + \delta$ must be positive over the support of $\tilde{f}$.

PROOF. When $\alpha\delta - \beta\gamma < 0$, the LFT reverses the order of integration, thus justifying the extra factor of $(-1)^l$ introduced for UE_l and SE_l. For OE_l, there is a more subtle difficulty, namely that the relative order of the eigenvalues is significant, and can change. If we simply reverse the order, this is not a problem (the total effect is $(-1)^{l(l+1)/2}$, thus cancelling out the sign of $\alpha\delta - \beta\gamma$). So we can restrict to the case $\alpha\delta - \beta\gamma > 0$. The effect of the LFT is then to cyclically shift the ordering, taking the eigenvalues with $x > \alpha/\gamma$ and making them smallest. If there are k such eigenvalues, the sign of the Vandermonde matrix is changed by $(-1)^{k(l-1)}$; thus if l is odd, there is no problem. On the other hand, if l is odd, we have a problem unless the eigenvalues are restricted to only one side of α/γ, or equivalently that $\gamma y + \delta$ has constant sign over the support of $\tilde{(f)}$. Since

$$\frac{\alpha y + \beta}{\gamma y + \delta} = \frac{-\alpha y - \beta}{-\gamma y - \delta},$$

we may take this sign to be positive. □

REMARK. For algebraic purposes, we can often ignore the constraint $\gamma y + \delta > 0$, since the transform still has the correct form to be a matrix ensemble density, despite not being nonnegative.

The upshot of this is that we can use this freedom to send a suitably chosen point to ∞, thus simplifying our analysis below.

4. The Main Results

For a matrix ensemble M, we define $\text{even}(M)$ to be the ensemble obtained by taking the second largest, fourth largest, etc. eigenvalues of M, and similarly for $\text{odd}(M)$.

When considering $\text{even}(M)$ or $\text{odd}(M)$ in the cases $M = \text{OE}_n \cup \text{OE}_n$ and $M = \text{OE}_n \cup \text{OE}_{n+1}$, the following lemma is crucial:

LEMMA 4.1. *For any integer $n > 0$,*

$$\sum_{\substack{S \subset \{0,1,\ldots,2n-1\} \\ |S|=n}} \Delta(x_S)\Delta(x_{\{0,1,\ldots,2n-1\}-S}) = 2^n \Delta(x_{\{0,2,\ldots,2n-2\}})\Delta(x_{\{1,3,\ldots,2n-1\}})$$

and
$$\sum_{\substack{S\subset\{0,1,\ldots,2n\}\\|S|=n+1}} \Delta(x_S)\Delta(x_{\{0,1,\ldots,2n\}-S}) = 2^n \Delta(x_{\{0,2,\ldots,2n\}})\Delta(x_{\{1,3,\ldots,2n-1\}})$$

PROOF. Consider what happens when we exchange x_i and x_{i+2} in a term of either equation. If $i, i+2 \in S$ or $i, i+2 \notin S$, then
$$\Delta(x_S)\Delta(x_{\{0,1,\ldots,l-1\}-S}) \to -\Delta(x_S)\Delta(x_{\{0,1,\ldots,l-1\}-S}),$$
since Δ is alternating. Otherwise, we see that every factor $x_j - x_k$ with $j > k$ is taken to another such factor, *except* for the factor $x_{i+1} - x_i$ or $x_{i+2} - x_{i+1}$, whichever is present. So each term in our sum is taken to the negative of a term from our sum; it follows that the sum is alternating under parity-preserving permutations. It follows that it must be a multiple of
$$\Delta(x_{\{0,2,\ldots,2\lceil l/2\rceil-2\}})\Delta(x_{\{1,3,\ldots,2\lfloor l/2\rfloor-1\}}).$$

By degree considerations, it remains only to verify the constant, which we can do by considering the coefficient of largest degree in x_0, and applying induction. $\square$

REMARK. The even case of this lemma is implicit in [15], where it was used to analyze $\text{even}(\text{OE}_n \cup \text{OE}_n)$ with respect to the weight function 1 on the unit circle.

From the lemma, it follows that the density of $\text{OE}_n(f) \cup \text{OE}_n(f)$, expressed in terms of ordered variables, is proportional to
$$\prod_{0\leq i\leq 2n} f(x_i)\Delta(x_{\{0,2,\ldots,2n-2\}})\Delta(x_{\{1,3,\ldots,2n-1\}});$$
similarly, the density of $\text{OE}_n(f) \cup \text{OE}_{n+1}(f)$ is proportional to
$$\prod_{0\leq i\leq 2n} f(x_i)\Delta(x_{\{0,2,\ldots,2n\}})\Delta(x_{\{1,3,\ldots,2n-1\}}).$$

For some weight functions f, if we integrate over the odd/even variables, the resulting density is the density of a unitary ensemble; we wish to determine precisely when that is. We first consider the case $\text{even}(\text{OE}_2(f) \cup \text{OE}_3(f))$.

THEOREM 4.2. *Let $f : \mathbb{R} \to \mathbb{R}$ be a function which is differentiable on a possibly unbounded open interval $I \subset \mathbb{R}$ and 0 elsewhere. Suppose $\text{even}(\text{OE}_2(f) \cup \text{OE}_3(f)) = \text{UE}_2(g)$ for some function g. Then up to a linear transformation of variables, f must have one of the following forms. On the interval $(0,1)$:*
$$f(x) \propto x^\alpha(1-x)^\beta(1-rx)^{-4-\alpha-\beta}, \quad \alpha,\beta > -1,\ r < 1$$
$$f(x) \propto x^{-4-\alpha}e^{-1/x}(1-x)^\alpha, \quad \alpha > -1.$$

On the interval $(0, \infty)$:
$$f(x) \propto x^\alpha (1-rx)^\beta, \quad \alpha > -1, \, \alpha+\beta < -3, \, r < 0,$$
$$f(x) \propto x^{-4-\alpha} e^{-1/x}, \quad \alpha > -1,$$
$$f(x) \propto x^\alpha e^{-x}, \quad \alpha > -1.$$

Finally, on the entire real line:
$$f(x) \propto (1+x^2)^\alpha, \quad \alpha < -\tfrac{3}{2}$$
$$f(x) \propto e^{-x^2/2}.$$

PROOF. We need to integrate this over the variables x_{2i}, and thus need to evaluate the determinant
$$\det \left(\int_{[x_{2i-1}, x_{2i+1}]} f(x) x^j \, dx \right)_{0 \le i,j \le 2},$$
where we take $x_{-1} = a$ to be the left endpoint of I, and $x_5 = b$ to be the right endpoint of I. In particular, we need to determine when there exists a function $g(x)$ with
$$\det \left(\int_{[x_{2i-1}, x_{2i+1}]} f(x) x^j \, dx \right)_{0 \le i,j \le 2} \propto g(x_1) g(x_3) (x_3 - x_1).$$

As in [19, Section 10.6], we may use row operations to transform this to
$$\det \left(F_j(x_{2i+1}) \right)_{0 \le i,j \le 2},$$
where we define
$$F_j(y) = \int_{[a,y]} f(x) x^j \, dx.$$

We cannot quite apply Theorem 2.1, however, since the last column of our determinant is constant. However, we clearly have $F_0(b) > 0$, so we can eliminate that column, obtaining
$$F_0(b) \det \left(F_j(x_{2i+1}) - \frac{F_j(b)}{F_0(b)} F_0(x_{2i+1}) \right)_{0 \le i,j \le 1}.$$

This, then, satisfies the hypotheses of Theorem 2.1; there thus exist linear polynomials p_j such that
$$p_2(x)(F_1(x) - C_1 F_0(x)) = p_1(x)(F_2(x) - C_j F_0(x)), \tag{4-1}$$
where we have set
$$C_i = \frac{F_i(b)}{F_0(b)}.$$

Differentiating twice and using the definition of F_i, we find
$$\left(p_2(x)(x - C_1) - p_1(x)(x^2 - C_2) \right) f'(x)$$
$$= \left(-2(x - C_1) p_2'(x) + 2(x^2 - C_2) p_1'(x) - p_2(x) + 2x p_1(x) \right) f(x).$$

We can thus solve this for $f'(x)/f(x)$; we find that $f'(x)/f(x)$ has the form $p(x)/q(x)$ with $\deg(p) \leq 2$, $\deg(q) \leq 3$, and $\deg(xp + 4q) \leq 2$. We observe that these conditions are, naturally, preserved by linear fractional transformations. In particular, by applying a suitable linear fractional transformation, we may insist that q be strictly cubic, and that both endpoints of I be finite (possibly equal). (The result may very well no longer be a matrix ensemble, but as we noted above, this does not affect any algebraic conclusions.)

Now, consider how $f(x)$ and $q(x)$ must behave at 0 and 1. Differentiating (4-1) once and taking a limit $x \to x_{-1}$ we find, since each $F_i(x_{-1}) = 0$,

$$\lim_{x \to x_{-1}} (p_2(x)(x - C_1) + p_1(x)(C_2 - x))f(x) = 0.$$

But this is just $\lim_{x \to x_{-1}} q(x)f(x)$. If $q(x_{-1}) \neq 0$, we must have $\lim_{x \to x_{-1}} f(x) = 0$. Then

$$\lim_{x \to x_{-1}} \frac{f'(x)}{f(x)} = \infty.$$

The only way this can happen is if $q(x_{-1}) = 0$ after all. Similarly, we have $q(x_{2n+1}) = 0$.

Suppose first that $a \neq b$. Then up to LFT, we may insist that $a = 0$ and $b = 1$, and thus $q(0) = q(1) = 0$. We thus have two possibilities. The first is that $q(x)$ has an additional zero, neither 0 nor 1. In this case, integrating f'/f and taking into account the constraints on $p(x)$, we obtain

$$f(x) = x^\alpha (1-x)^\beta (1-rx)^{-4-\alpha-\beta}$$

Now, for $\int f(x)$ not to diverge at 0, we must have $\alpha > -1$, and similarly $\beta > -1$. But then $-4 - \alpha - \beta < -2$; it follows that $(1 - rx)$ must be nonzero on $(0, 1)$; in particular, $r < 1$. The other possibility is that $q(x)$ has a double root, without loss of generality, at 0. Upon integrating f'/f, we obtain

$$f(x) = x^{-4-\alpha} e^{-\beta/x} (1-x)^\alpha,$$

and find $\alpha > -1$, $\beta > 0$. The possibilities for $(0, \infty)$ then follow by LFT.

The other case we must consider is $I = \mathbb{R}$, and thus $\deg(q) = 2$, $\deg(p) = 1$. If q had a simple root in $\mathbb{R}$, it would have two (say 0 and 1, without loss of generality), and thus f would have the form $x^\alpha(1-x)^\beta$ with $\alpha, \beta > -1$. But then the integral for F_2 would diverge. Similarly, if q had a double root at 0, f would have the form $x^\alpha e^{-\beta/x}$, which would diverge on one side of 0. Thus either q has a pair of complex roots, or $\deg(q) = 0$. In the first case, a linear transformation takes the roots to $\pm i$, and thus

$$f(x) = (1+x^2)^\alpha.$$

For F_2 to be well-defined, we must have $\alpha < -3/2$. The other possibility gives $\log(f(x)) = ax^2 + bx + c$; thus a linear transformation gives

$$f(x) = e^{-x^2/2}. \qquad \square$$

We can now extend this to $n \geq 2$.

THEOREM 4.3. *Fix an integer $n \geq 2$. Let $f : \mathbb{R} \to \mathbb{R}$ be a function which is differentiable on a possibly unbounded open interval $I \subset \mathbb{R}$ and 0 elsewhere. Suppose $\text{even}(\text{OE}_n(f) \cup \text{OE}_{n+1}(f)) = \text{UE}_n(g)$ for some function g. Then up to a linear transformation of variables, f and g can have precisely the following forms. On the interval $(0,1)$:*

$$f(x) \propto x^\alpha (1-x)^\beta (1-rx)^{-n-2-\alpha-\beta}, \quad \alpha, \beta > -1, \; r < 1,$$
$$g(x) \propto x^{2\alpha+1}(1-x)^{2\beta+1}(1-rx)^{-2n-3-2\alpha-2\beta},$$
$$f(x) \propto x^{-n-2-\alpha} e^{-1/2x}(1-x)^\alpha, \quad \alpha > -1,$$
$$g(x) \propto x^{-2n-2-2\alpha} e^{-1/x}(1-x)^{2\alpha+1}.$$

On the interval $(0, \infty)$:

$$f(x) \propto x^\alpha (1-rx)^\beta, \quad \alpha > -1, \; \alpha + \beta < -n-1, \; r < 0,$$
$$g(x) \propto x^{2\alpha+1}(1-rx)^{2\beta+1},$$
$$f(x) \propto x^{-n-2-\alpha} e^{-1/2x}, \quad \alpha > -1,$$
$$g(x) \propto x^{-2n-2-2\alpha} e^{-1/x},$$
$$f(x) \propto x^\alpha e^{-x/2}, \quad \alpha > -1,$$
$$g(x) \propto x^{2\alpha+1} e^{-x}.$$

Finally, on the entire real line:

$$f(x) \propto (1+x^2)^\alpha, \quad \alpha < -(n+1)/2,$$
$$g(x) \propto (1+x^2)^{2\alpha+1},$$
$$f(x) \propto e^{-x^2/2},$$
$$g(x) \propto e^{-x^2}.$$

PROOF. As for $n = 2$, the issue is when

$$\det\bigl(F_j(x_{2i+1})\bigr)_{0 \leq i,j \leq n}$$

takes the form of an orthogonal ensemble. Applying an LFT as necessary, we may assume that $a = -\infty$. Now differentiate with respect to x_1, divide by $x_1^{n-1} f(x_1)$, and take a limit as $x_1 \to -\infty$. On the one hand, this operation takes orthogonal ensembles to orthogonal ensembles. On the other hand, we can then expand along the first column, finding that

$$\det\bigl(F_{j-1}(x_{2i+1})\bigr)_{1 \leq i,j \leq n}$$

must take the form of an orthogonal ensemble. By induction, we find that $f(x)$ must satisfy the constraints valid for $n - 1$. Upon undoing the LFT, we obtain the desired "only if" result.

It remains to show that each of the above weight functions actually do work. We need only consider the following possibilities:

$$f(x) = x^\alpha(1-x)^\beta \quad \text{on } (0,1),$$
$$f(x) = x^\alpha e^{-x} \quad \text{on } (0,\infty),$$
$$f(x) = (1+x^2)^\alpha \quad \text{on } \mathbb{R},$$
$$f(x) = e^{-x^2/2} \quad \text{on } \mathbb{R},$$

since the others are all images of these under LFTs.

For $f(x) = x^\alpha(1-x)^\beta$, observe that

$$(\alpha+\beta+j+2)F_{j+1}(x) - (\alpha+j+1)F_j(x) = -x^j(x^{\alpha+1}(1-x)^{\beta+1});$$

this is true for $x = 0$, and both sides have the same derivative. In particular, for each j, we have a polynomial $p_j(x)$ of degree $j-1$ and a constant C_j with

$$F_j(x) = C_j F_0(x) + p_j(x)(x^{\alpha+1}(1-x)^{\beta+1}).$$

In particular, this must be true for $x = 1$, and thus $C_j = F_j(1)/F_0(1)$ as required. We thus find that we obtain a unitary ensemble with weight function proportional to $x^{2\alpha+1}(1-x)^{2\beta+1}$. Similarly, for $f(x) = x^\alpha e^{-x}$, we have

$$F_{j+1}(x) - (\alpha+j+1)F_j(x) = -x^{\alpha+j+1}e^{-x},$$

so $g(x) \propto x^{2\alpha+1}e^{-2x}$.

For $f(x) = (1+x^2)^\alpha$, we find

$$(2\alpha+j+2)F_{j+1}(x) + jF_{j-1} = x^j(1+x^2)^{\alpha+1}$$

This allows us to solve for each F_j except F_0; we obtain $g(x) \propto (1+x^2)^{2\alpha+1}$. Finally, for $f(x) = e^{-x^2/2}$, we have:

$$F_{j+1} - jF_{j-1} = -x^j e^{-x^2/2},$$

and $g(x) \propto e^{-x^2}$. □

REMARK. We observe that in each case $g(x) \propto f(x)^2 q(x)$.

For even($\text{OE}_n \cup \text{OE}_n$), the calculations are analogous, and we have:

THEOREM 4.4. *Fix an integer $n \geq 2$. Let $f : \mathbb{R} \to \mathbb{R}$ be a function which is differentiable on a possibly unbounded open interval $I \subset \mathbb{R}$ and 0 elsewhere. Suppose* even($\text{OE}_n(f) \cup \text{OE}_n(f)$) = $\text{UE}_n(g)$ *for some function g. Then up to an order-preserving linear transformation of variables, f and g must have one of the following forms. On an interval with right endpoint 0:*

$$f(x) \propto (-x)^\alpha(1-rx)^{-n-1-\alpha}, \quad \alpha > -1,\ 1 \notin rI,$$
$$g(x) \propto (-x)^{2\alpha+1}(1-rx)^{-2n-1-2\alpha},$$
$$f(x) \propto (-x)^{-n-1}e^{1/x},$$
$$g(x) \propto (-x)^{-2n}e^{2/x}.$$

On an interval of the form (a, ∞), $a > -\infty$:

$$f(x) \propto (1 - rx)^\alpha, \quad \alpha < -n, \; r < 0,$$
$$g(x) \propto (1 - rx)^{2\alpha+1},$$
$$f(x) \propto e^{-x},$$
$$g(x) \propto e^{-2x}$$

On the entire real line, no possibilities exist.

REMARK. The relation between g and f is here slightly modified, by removing the factor of q corresponding to the left endpoint. For odd($OE_n \cup OE_n$), we remove the factor corresponding to the right endpoint, and for odd($OE_{n-1} \cup OE_n$), we remove both factors.

For odd($OE_n \cup OE_n$), we simply reverse the ordering. For odd($OE_n \cup OE_{n+1}$), we have:

THEOREM 4.5. *Fix an integer $n \geq 2$. Let $f : \mathbb{R} \to \mathbb{R}$ be a function which is differentiable on a possibly unbounded open interval $I \subset \mathbb{R}$ and 0 elsewhere. Suppose $\mathrm{odd}(OE_{n-1}(f) \cup OE_n(f)) = UE_n(g)$ for some function g. Then f and g have the form*

$$f(x) \propto (1 - rx)^{-n},$$
$$g(x) \propto (1 - rx)^{-2n+1}$$

for some r (possibly ∞) with $1/r \notin I$.

PROOF. The only tricky aspect of this case is that the determinant we must analyze is no longer of the form to which Theorem 2.1 applies; to be precise, we need

$$\det \left(F_j(x_{2i+2}) - F_j(x_{2i}) \right)_{0 \leq i < n}$$

to have orthogonal ensemble form. But this determinant is clearly equal to the determinant of the block matrix

$$\begin{pmatrix} (1) & (0)_{0 \leq i < n} \\ (F_j(x_0))_{0 \leq j < n} & (F_j(x_{2i+2}) - F_j(x_{2i}))_{0 \leq i, j < n} \end{pmatrix}.$$

Adding the first column to the other columns, we can then apply Theorem 2.1, and argue as above. □

We finally consider a fifth possibility for decimation. Recall that for the circular ensemble results cited above, while there was a *local* notion of order, there was no notion of largest. This suggests that we consider the ensemble derived by choosing randomly between $\mathrm{odd}(M)$ and $\mathrm{even}(M)$. More precisely, for an ensemble with an even number of variables, we define $\mathrm{alt}(M)$ to be $\mathrm{even}(M)$ with probability $\frac{1}{2}$ and $\mathrm{odd}(M)$ with probability $\frac{1}{2}$.

THEOREM 4.6. *Fix an integer $n \geq 2$. Let $f : \mathbb{R} \to \mathbb{R}$ be a function which is differentiable on a possibly unbounded open interval $I \subset \mathbb{R}$ and 0 elsewhere. Suppose* $\mathrm{alt}(\mathrm{OE}_n(f) \cup \mathrm{OE}_n(f)) = \mathrm{UE}_n(g)$ *for some function g. Then up to a linear transformation of variables, f and g have the form*

$$f = (1+x^2)^{-(n+1)/2},$$
$$g = (1+x^2)^{-n}.$$

PROOF. Consider the determinants associated to $\mathrm{even}(\mathrm{OE}_n(f) \cup \mathrm{OE}_n(f))$ and $\mathrm{odd}(\mathrm{OE}_n(f) \cup \mathrm{OE}_n(f))$. Up to cyclic shift, only one column differs between the two determinants, thus allowing us to express their sum as a determinant. When n is even, the 'special' column takes the form

$$(F_j(x_0) + F_j(x_{2n-1}) - F_j(I))_{0 \leq j < n};$$

here $F_j(I) = \int_{x \in I} x^j f(x)$. Taking appropriate linear combinations, we obtain the determinant

$$\det\bigl(F_j(x_i) - F_j(I)/2\bigr)_{0 \leq i,j < n}.$$

When n is odd, the special column takes the form

$$(F_j(x_{2n-1}) - F_j(x_0) - F_j(I))_{0 \leq j < n};$$

this leads (up to sign) to the $n+1 \times n+1$ block determinant

$$\det \begin{pmatrix} 0 & (F_j(I))_{0 \leq j < n} \\ 1 & (F_j(x_i))_{0 \leq i,j < n} \end{pmatrix}.$$

We first analyze the case n odd. In this case, the usual theory tells us that there exist polynomials $p_j(x)$ and $q(x)$ of degree at most $n-1$ with

$$F_j(x) - C_j F_0(x) = p_j(x)/q(x)$$

for all j, with C_j as above. Now, evaluating this at an endpoint of I, we find that the polynomials p_j must have a common root (possibly ∞). In particular, it follows that f must satisfy the conditions of Theorem 4.3. On the other hand, we find that

$$x^j f(x) - C_j f(x) = \frac{d}{dx} \frac{p_j(x)}{q(x)}$$

for each j; in particular, $f(x)$ must be a rational function. We therefore have the following possibilities to consider:

$$f(x) = x^\alpha (1-x)^\beta \quad \text{for } \alpha, \beta \in \mathbb{N},$$
$$f(x) = (1+x^2)^{-\alpha} \quad \text{for } \alpha \in \mathbb{N}, \alpha > n/2,$$

on $(0,1)$ and $\mathbb{R}$ respectively. In the first case, we find that each

$$F_j(x) - C_j F_0(x)$$

is a polynomial of degree $j+2$. In particular, $F_{n-1}(x) - C_{n-1} F_0(x)$ is a polynomial of degree $n+1$, contradicting the bound on $\deg(p_j(x))$.

In the second case, we observe that

$$F_j(x) - C_j F_0(x) = r_j(x)(1+x^2)^{1-\alpha}$$

for polynomials $r_j(x)$ of degree $j-1$. In particular, we find that $p_1(x)/q(x) \propto (1+x^2)^{1-\alpha}$, implying, since $\alpha > n/2 > 1$, that

$$q(x) \propto (1+x^2)^{\alpha-1}$$

Since $\deg(q) \leq n-1$, we have $n/2 < \alpha \leq (n+1)/2$, the only integral solution of which is $\alpha = (n+1)/2$. In this case, the relevant degree bounds all hold, and thus the determinant is indeed of the correct form, giving $g(x) \propto (1+x^2)^{1-2\alpha} = (1+x^2)^{-n}$ as required.

For n even, we must have polynomials p_i of degree at most $n-1$ with

$$p_i(x)(F_j(x) - C'_j) = p_j(x)(F_i(x) - C'_i),$$

where we write C'_j for $F_j(I)/2$. We can rewrite this as:

$$p_0(x)(F_j(x) - C_j F_0(x)) = (p_j(x) - C_j p_0(x))(F_0(x) - C'_0),$$

using the fact that $C_j = C'_j/C'_0$. For $n > 2$, we conclude that $f(x)$ must satisfy the conditions of Theorem 4.3. Now, if the endpoints of I are different, then we find, since $F_0(x) - C'_0 = \pm C'_0$ at both endpoints, that each $p_j(x) - C_j p_0(x) = 0$ at both endpoints. But this causes the polynomials to be linearly dependent, a contradiction. On the other hand, in the other cases, we know that $C_1 = 0$ and $p_1 \propto 1$. In both cases, we obtain from the identity for $F_1(x) - C_1 F_0(x)$ a differential equation for $p_0(x)$. For $e^{-x^2/2}$, no polynomial solution to the equation exists. For $(1+x^2)^{-\alpha}$, we can find an explicit power series solution to the equation, and find that a polynomial solution exists only when α is half-integral, when the solution has degree $2\alpha - 2$. As above, this leaves only one possibility for α, namely $\alpha = (n+1)/2$, as required.

It remains to consider $n = 2$. Here we can twice differentiate the equation

$$p_0(x)(F_1(x) - C'_1) = p_1(x)(F_0(x) - C'_0)$$

(with p_0 and p_1 linear) to deduce that

$$f'(x)/f(x) = p(x)/q(x)$$

with $\deg(p) \leq 1$, $\deg(q) \leq 2$, and $\deg(xp + 3q) \leq 1$. So up to LFT, $f(x)$ must have one of the forms

$$\begin{aligned} f(x) &= x^\alpha && \text{for } \alpha \geq -\tfrac{3}{2}, \\ f(x) &= e^{-x}, \\ f(x) &= (1+x^2)^{-3/2}. \end{aligned}$$

(note that if we exchange 0 and ∞ in the first case, we replace α with $-3 - \alpha$, justifying our restriction on α.) In the first case, if $\alpha \neq -1$, the following:

$$\frac{x^{\alpha+2} - D_1}{x^{\alpha+1} - D_0}$$

must be a linear polynomial for suitable constants D_1 and D_0 respectively proportional to C'_1 and C'_0 (and thus $D_0 \neq 0$). We deduce therefore that $\alpha = 0$. But then we readily determine that only the empty interval satisfies the requirements. For x^{-1} and e^{-x}, there are not even appropriate choices for D_0 and D_1 (since both $\log(x)$ and e^{-x} are transcendental functions). Finally, for the third choice, we readily verify that decimation indeed works as required. □

We now turn our attention to decimations of single orthogonal ensembles. We have, quite simply:

THEOREM 4.7. *Fix an integer $n \geq 2$. For any functions f and g with f differentiable on a possibly unbounded open interval $I \subset \mathbb{R}$ and 0 elsewhere, the following equivalences hold (where we write $\mathrm{OE}_{2n} f = \mathrm{OE}_{2n}(f)$ and so on):*

$$\mathrm{even}(\mathrm{OE}_{2n} f \cup \mathrm{OE}_{2n+1} f) = \mathrm{UE}_{2n} g \iff \mathrm{even}(\mathrm{OE}_{2n+1} f) = \mathrm{SE}_n((g/f)^2), \quad (4\text{-}2)$$
$$\mathrm{even}(\mathrm{OE}_{2n} f \cup \mathrm{OE}_{2n} f) = \mathrm{UE}_{2n} g \iff \mathrm{even}(\mathrm{OE}_{2n} f) = \mathrm{SE}_n((g/f)^2), \quad (4\text{-}3)$$
$$\mathrm{odd}(\mathrm{OE}_{2n} f \cup \mathrm{OE}_{2n} f) = \mathrm{UE}_{2n} g \iff \mathrm{odd}(\mathrm{OE}_{2n} f) = \mathrm{SE}_n((g/f)^2), \quad (4\text{-}4)$$
$$\mathrm{odd}(\mathrm{OE}_{2n-1} f \cup \mathrm{OE}_{2n} f) = \mathrm{UE}_{2n} g \iff \mathrm{odd}(\mathrm{OE}_{2n-1} f) = \mathrm{SE}_n((g/f)^2), \quad (4\text{-}5)$$
$$\mathrm{alt}(\mathrm{OE}_{2n} f \cup \mathrm{OE}_{2n} f) = \mathrm{UE}_{2n} g \iff \mathrm{alt}(\mathrm{OE}_{2n} f) = \mathrm{SE}_n((g/f)^2). \quad (4\text{-}6)$$

PROOF. Consider, for instance, $\mathrm{even}(\mathrm{OE}_{2n+1}(f))$. Once we integrate along the largest, third largest, etc., variables and do some simplification, the resulting matrix has columns $(F_j(x_{2i+1}))_{0 \leq j \leq 2n}$ and $(x_{2i+1}^j f(x_{2i+1}))_{0 \leq j \leq 2n}$, with the last column given by $(F_j(b))_{0 \leq j \leq 2n}$. In particular, we note that aside from the last, constant, column, the columns come in pairs, one the derivative of the other. Thus the determinant is essentially of the form considered in Theorem 2.2. In particular, it has a symplectic ensemble form if and only if the determinant

$$\det\bigl(F_j(x_{i+1})\bigr)_{0 \leq i,j \leq 2n}$$

(of which our determinant is a derivative) has an orthogonal ensemble form. But by the proof of Theorems 4.2 and 4.3, this precisely what we needed to show. (The statement about the resulting weight functions is straightforward.) Similar arguments apply for the remaining equivalences. □

5. Random matrix applications

In random matrix applications f, g and $(g/f)^2$ must be (up to the scale of x) one of the four classical forms (1–5) and (1–9). So specializing Theorems 4.3–4.5 we can read off for which of the classical forms the statement of Theorem 4.7 is valid.

THEOREM 5.1. *Restricting attention to the classical weights (1–5) and (1–9), statement (4–2) holds for*

$$(f,g) = \begin{cases} (e^{-x^2/2}, e^{-x^2}), \\ (x^{(a-1)/2}e^{-x/2}, x^a e^{-x}) & \text{with } x > 0, \\ (x^{(a-1)/2}(1-x)^{(b-1)/2}, x^a(1-x)^b) & \text{with } 0 < x < 1, \\ ((1+x^2)^{-(\alpha+1)/2}, (1+x^2)^{-\alpha}); \end{cases} \quad (5\text{–}1)$$

the statement (4–3) holds for

$$(f,g) = \begin{cases} (e^{-x/2}, e^{-x}) & \text{with } x > 0, \\ ((1-x)^{(a-1)/2}, (1-x)^a) & \text{with } 0 < x < 1; \end{cases}$$

(4–4) is valid for the particular pair of Jacobi weights

$$(f,g) = (x^{(a-1)/2}, x^a) \quad \text{with } 0 < x < 1;$$

(4–5) is valid for the particular pair of Jacobi weights

$$(f,g) = (1,1) \quad \text{with } 0 < x < 1;$$

and (4–6) is valid for the weights

$$((1+x^2)^{-(n+1)/2}, (1+x^2)^{-n}).$$

Because the weights in Theorem 5.1 occur in the matrix ensembles listed in the Introduction, the theorems of Section 4 imply interrelationships between the different ensembles.

THEOREM 5.2. *The following relations hold between the above matrix ensembles under decimation, for all $n > 0$:*

$$\text{even}(\text{GOE}_{2n+1}) = \text{GSE}_n,$$
$$\text{even}(\text{Symm}(n;\mathbb{C})) = \text{Anti}(n;\mathbb{C}),$$
$$\text{even}(\text{Mat}(2p+1, 2q+1;\mathbb{R})) = \text{Mat}(p, q;\mathbb{H}),$$
$$\text{even}(\text{Beta}(2p_1+1, 2p_2+1, 2q+1;\mathbb{R})) = \text{Beta}(p_1, p_2, q;\mathbb{H}),$$
$$\text{even}(\text{GOE}_n \cup \text{GOE}_{n+1}) = \text{GUE}_n,$$
$$\text{even}(\text{Symm}(n;\mathbb{C}) \cup \text{Symm}(n;\mathbb{C})) = \text{Mat}(n, n;\mathbb{C}),$$
$$\text{even}(\text{Symm}(n;\mathbb{C}) \cup \text{Symm}(n+1;\mathbb{C})) = \text{Mat}(n+1, n;\mathbb{C}),$$
$$\text{even}(\text{Mat}(p, q;\mathbb{R}) \cup \text{Mat}(p+1, q+1;\mathbb{R})) = \text{Mat}(p, q;\mathbb{C}),$$
$$\text{even}(\text{Beta}(p_1, p_2, q;\mathbb{R}) \cup \text{Beta}(p_1+1, p_2+1, q+1;\mathbb{R})) = \text{Beta}(p_1, p_2, q;\mathbb{C}).$$

REMARK 1. *It would be very nice to have a direct, matrix-theoretic, proof of any of these relations.*

REMARK 2. There are actually a few more relations, all of which follow from the preceding ones together with the relation

$$\mathrm{Mat}(n+1,n;\mathbb{R}) = \mathrm{Symm}(n;\mathbb{C}).$$

Again, a matrix-theoretic proof of this would be nice.

We now turn our attention to the implications of Theorem 5.1 with respect to gap probabilities. In circular ensemble theory the results (1–7) and (1–8) were shown [8] to imply interrelationships between the probability of an eigenvalue free region amongst the various symmetry classes. With $E^{(\beta)}(p;J;n)$ denoting the probability that, for the ensembles COE_n ($\beta=1$), CUE_n ($\beta=2$) and CSE_n ($\beta=4$), there are exactly p eigenvalues in the interval J, the interrelationships are

$$E^{(2)}(0;(0,s);n) = E^{(1)}(0;(0,s);n)\big(E^{(1)}(0;(0,s);n) + E^{(1)}(1;(0,s);n)\big),$$
$$E^{(4)}(p;(0,s);n) = E^{(1)}(2p;(0,s);2n)$$
$$+ \tfrac{1}{2}E^{(1)}(2p-1;(0,s);2n) + \tfrac{1}{2}E^{(1)}(2p+1;(0,s);2n), \quad (5\text{--}2)$$

where

$$E^{(\beta)}(p;J;n) := 0, \quad \text{for} \quad p<0. \tag{5--3}$$

Similar interrelationships between gap probabilities, but now with the eigenvalue free interval including an endpoint of the support of the interval, can be deduced from the pairs of statements of Theorem 4.7.

THEOREM 5.3. *Let $E^{(\beta)}(p;J;g;n)$ denote the probability that, for the ensembles $\mathrm{OE}_n(g)$ ($\beta=1$), $\mathrm{UE}_n(g)$ ($\beta=2$) and $\mathrm{SE}_n(g)$ ($\beta=4$) the interval J contains exactly n eigenvalues.*

(i) *The statements (4–2) imply*

$$E^{(2)}(0;J;g;2n) = E^{(1)}(0;J;f;2n)E^{(1)}(0;J;f;2n+1)$$
$$+ E^{(1)}(0;J;f;2n)E^{(1)}(1;J;f;2n+1) + E^{(1)}(0;J;f;2n+1)E^{(1)}(1;J;f;2n),$$
$$E^{(4)}(p;J;(g/f)^2;n) = E^{(1)}(2p;J;f;2n+1) + E^{(1)}(2p+1;J;f;2n+1), \quad (5\text{--}4)$$

for $J=(-\infty,-s)$ or (s,∞).

(ii) *The statements (4–3) imply*

$$E^{(2)}(0;(s,\infty);g;2n) = \big(E^{(1)}(0;(s,\infty);f;2n)\big)^2$$
$$+ 2E^{(1)}(0;(s,\infty);f;2n)E^{(1)}(1;(s,\infty);f;2n),$$
$$E^{(4)}(p;(s,\infty);(g/f)^2;n) = E^{(1)}(2p;(s,\infty);f;2n)+E^{(1)}(2p+1;(s,\infty);f;2n) \quad (5\text{--}5)$$

and

$$E^{(2)}(0;(-\infty,-s);g;2n) = \big(E^{(1)}(0;(-\infty,-s);f;2n)\big)^2,$$
$$E^{(4)}(p;(-\infty,-s);(g/f)^2;n) = E^{(1)}(2p;(-\infty,-s);f;2n)$$
$$+ E^{(1)}(2p-1;(-\infty,-s);f;2n). \quad (5\text{--}6)$$

(iii) *The statements* (4–4) *imply*

$$E^{(2)}(0;(s,\infty);g;2n) = \left(E^{(1)}(0;(s,\infty);f;2n)\right)^2,$$
$$E^{(4)}(p;(s,\infty);(g/f)^2;n) = E^{(1)}(2p;(s,\infty);f;2n) + E^{(1)}(2p-1;(s,\infty);f;2n) \quad (5\text{-}7)$$

and

$$E^{(2)}(0;(-\infty,-s);g;2n) = \left(E^{(1)}(0;(-\infty,-s);f;2n)\right)^2$$
$$+ 2E^{(1)}(0;(-\infty,-s);f;2n)E^{(1)}(1;(-\infty,-s);f;2n),$$
$$E^{(4)}(p;(-\infty,-s);(g/f)^2;n) = E^{(1)}(2p;(-\infty,-s);f;2n)$$
$$+ E^{(1)}(2p+1;(-\infty,-s);f;2n). \quad (5\text{-}8)$$

(iv) *The statements* (4–5) *imply*

$$E^{(2)}(0;J;g;2n) = E^{(1)}(0;J;f;2n)E^{(1)}(0;J;f;2n-1),$$
$$E^{(4)}(p;J;(g/f)^2;n) = E^{(1)}(2p;J;f;2n-1) + E^{(1)}(2p-1;J;f;2n-1), \quad (5\text{-}9)$$

for $J = (-\infty,-s)$ *or* (s,∞).

(v) *The statements* (4–6) *imply the relations* (5–2) *with n replaced by $2n$ in the first equation and $(0,s)$ replaced throughout by J, $J = (-\infty,-s)$ or (s,∞).*

PROOF. We will consider only the deductions from (4–2), as the other cases are similar. Let J be a single interval which includes an endpoint of the support of f and g. From the first statement in (4–2) we see that the event of a sequence of eigenvalues from $\text{UE}_{2n}(g)$ not being contained in J occurs in three ways relative to the ensemble $\text{OE}_{2n}(f) \cup \text{OE}_{2n+1}(f)$: (i) the eigenvalues from $\text{OE}_{2n}(f)$ and those from $\text{OE}_{2n+1}(f)$ are not contained in J; or (ii) one eigenvalue from $\text{OE}_{2n+1}(f)$ is contained in J and no eigenvalue from $\text{OE}_{2n}(f)$ is contained in J (note that the one eigenvalue must be either the largest (smallest) eigenvalue when J contains the right (left) hand end point); or (iii) one eigenvalue from $\text{OE}_{2n}(f)$ is contained in J and no eigenvalue from $\text{OE}_{2n+1}(f)$ is contained in J. This gives the first equation in (5–4). From the second statement in (4–2) we see that the event of a sequence of eigenvalues from $\text{SE}_n((g/f)^2)$ containing p eigenvalues in J can occur in two ways relative to $\text{OE}_{2n+1}(f)$: (i) there are $2p$ eigenvalues from $\text{OE}_{2n+1}(f)$ in J; or (ii) there are $2p+1$ eigenvalues from $\text{OE}_{2n+1}(f)$ in J (of which $p+1$ are integrated out in forming even($\text{OE}_{2n+1}(f)$)). This implies the second equation in (5–4). □

Recalling (5–3) we see that in the case $p = 0$ the equations (5–6), (5–7) and (5–9) give particularly simple interrelationships between the $E^{(\beta)}(0;\ldots)$. In fact referring back to Theorem 5.1 for the permissible pairs (f,g) in these cases it is a simple exercise in changing variables to compute the $E^{(\beta)}(0;\ldots)$ in terms of elementary functions. Recalling that

$$E^{(\beta)}(0;J;w;n) := \frac{1}{C} \int_J dx_0 \cdots \int_J dx_{n-1} \prod_{l=0}^{n-1} w(x_l) \prod_{0 \leq j < k \leq n-1} |x_k - x_j|^\beta,$$

where $\bar{J} = (-\infty, \infty) - J$ and C is such that $E^{(\beta)}(0; \emptyset; w; n) = 1$ we find

$$E^{(1)}(0; (0,s); e^{-x/2}; n) = e^{-sn/2},$$
$$E^{(2)}(0; (0,s); e^{-x}; n) = E^{(4)}(0; (0,s); e^{-x}; n) = e^{-sn},$$
$$E^{(1)}(0; (0,s); (1-x)^{(a-1)/2}; n) = E^{(1)}(0; (1-s,1); x^{(a-1)/2}; n) = (1-s)^{n(n+a)/2},$$
$$E^{(2)}(0; (0,s); (1-x)^a; n) = E^{(2)}(0; (1-s,1); x^a; n) = (1-s)^{n(a+n)},$$
$$E^{(4)}(0; (0,s); (1-x)^{a+1}; n) = E^{(4)}(0; (1-s,1); x^{a+1}; n) = (1-s)^{2n^2+na}$$

(in the first two cases the weight functions are restricted to $x > 0$, while in the remaining cases $0 < x < 1$). The equations (5–6), (5–7) and (5–9) for $p = 0$ can be checked immediately.

The pairs of equations (5–4), (5–5) and (5–8) contain $E^{(1)}(1; \ldots)$ as well as $E^{(\beta)}(0; \ldots)$. In the equations (5–5) and (5–8) the dependence on $E^{(1)}(1; \ldots)$ can be eliminated. Noting from Theorem 5.1 the allowed pairs (f, g) for the validity of equations (5–5) and (5–8) the following result is obtained.

PROPOSITION 5.4. *Assume that* $(f, g) = (e^{-x/2}, e^{-x})$ *with* $x > 0$ *and that* $J = (s, \infty)$; *or that* $(f, g) = ((1-x)^{(a-1)/2}, (1-x)^a)$ *with* $0 < x < 1$ *and that* $J = (1-s, 1)$. *Then*

$$E^{(4)}(0; J; (g/f)^2; n) = \frac{1}{2}\left(E^{(1)}(0; J; f; 2n) + \frac{E^{(2)}(0; J; g; 2n)}{E^{(1)}(0; J; f; 2n)}\right). \quad (5\text{–}10)$$

In the scaled $n \to \infty$ limit, as appropriate for the particular choice of weight function in (5–1), the pair of equations (4–2) also imply an equation of the form (5–10). First consider the Gaussian ensembles with the scaling [11]

$$x \mapsto (2n)^{1/2} + \frac{x}{2^{1/2}n^{1/6}},$$

which corresponds to studying the distribution of the eigenvalues at the (soft) edge of the leading order support of the spectrum. Defining

$$E^{(1)}_{\text{soft}}(p; (s, \infty)) := \lim_{n \to \infty} E^{(1)}\left(p; \left((n)^{1/2} + \frac{s}{2^{1/2}n^{1/6}}, \infty\right) e^{-x^2/2}; n\right)$$

$$E^{(2)}_{\text{soft}}(p; (s, \infty)) := \lim_{n \to \infty} E^{(2)}\left(p; \left((n)^{1/2} + \frac{s}{2^{1/2}n^{1/6}}, \infty\right) e^{-x^2}; n\right)$$

$$E^{(4)}_{\text{soft}}(p; (s, \infty)) := \lim_{n \to \infty} E^{(4)}\left(p; \left((n)^{1/2} + \frac{s}{2^{1/2}n^{1/6}}, \infty\right) e^{-x^2}; n/2\right),$$

(the existence of these limits is known from explicit calculation [27]; see below). The equations (5–4) imply:

PROPOSITION 5.5. *For the scaled infinite Gaussian ensembles at the soft edge*

$$E^{(4)}_{\text{soft}}(0; (s, \infty)) = \frac{1}{2}\left(E^{(1)}_{\text{soft}}(0; (s, \infty)) + \frac{E^{(2)}_{\text{soft}}(0; (s, \infty))}{E^{(1)}_{\text{soft}}(0; (s, \infty))}\right), \quad (5\text{–}11)$$

$$E^{(1)}_{\text{soft}}(1;(s,\infty)) = E^{(4)}_{\text{soft}}(0;(s,\infty)) - E^{(1)}_{\text{soft}}(0;(s,\infty)). \tag{5-12}$$

As alluded to above, the $E^{(\beta)}_{\text{soft}}(0;(s,\infty))$ are known exactly from the work of Tracy and Widom [27]. To present these results, let $q(s)$ denote the solution of the particular Painlevé II equation

$$q'' = sq + 2q^3,$$

which satisfies the boundary condition $q(s) \sim \text{Ai}(s)$ as $s \to \infty$. Then we have

$$E^{(2)}_{\text{soft}}(0;(s,\infty)) = \exp\left(-\int_s^\infty (t-s)q^2(t)\,dt\right),$$

$$\left(E^{(1)}_{\text{soft}}(0;(s,\infty))\right)^2 = E^{(2)}_{\text{soft}}(0;(s,\infty)) \exp\left(-\int_s^\infty q(t)\,dt\right), \tag{5-13}$$

$$\left(E^{(4)}_{\text{soft}}(0;(s,\infty))\right)^2 = E^{(2)}_{\text{soft}}(0;(s,\infty)) \cosh^2\left(\frac{1}{2}\int_s^\infty q(t)\,dt\right)$$

(in [27] $E^{(4)}_{\text{soft}}$ is defined with $s \mapsto s/2^{1/2}$ relative to our definition). Equation (5–11) is immediately seen to be satisfied, while the second equation gives

$$\left(E^{(1)}_{\text{soft}}(1;(s,\infty))\right)^2 = E^{(2)}_{\text{soft}}(0;(s,\infty)) \sinh^2\left(\frac{1}{2}\int_s^\infty q(t)\,dt\right). \tag{5-14}$$

Next consider the scaled limit at an edge for which the weight function is strictly zero on one side. For the classical ensembles this occurs in the Laguerre and Jacobi case; for definiteness consider the Laguerre case. The appropriate scaling is [11]

$$x \mapsto \frac{x}{4n},$$

and we define

$$E^{(1)}_{\text{hard}}(p;(0,s);(a-1)/2) := \lim_{n\to\infty} E^{(1)}(p;(0,s/4n);x^{(a-1)/2}e^{-x};n),$$

$$E^{(2)}_{\text{hard}}(p;(0,s);a) := \lim_{n\to\infty} E^{(2)}(p;(0,s/4n);x^a e^{-x};n),$$

$$E^{(4)}_{\text{hard}}(p;(0,s);a+1) := \lim_{n\to\infty} E^{(4)}(p;(0,s/4n);x^{a+1}e^{-x};n/2)$$

(the existence of these limits for general $a > -1$ can be deduced from the existence of the k-point distributions in the same scaled limits [22]). Use of (4–2) then gives the analogue of Proposition 5.5 for the hard edge.

PROPOSITION 5.6. *For the scaled infinite Laguerre ensembles at the hard edge, we have*

$$E^{(4)}_{\text{hard}}(0;(0,s);a+1) = \frac{1}{2}\left(E^{(1)}_{\text{hard}}(0;(0,s);\tfrac{1}{2}(a-1)) + \frac{E^{(2)}_{\text{hard}}(0;(0,s);a)}{E^{(1)}_{\text{hard}}(0;(0,s);(a-1)/2)}\right) \tag{5-15}$$

and

$$E^{(1)}_{\text{hard}}(1;(0,s);\tfrac{1}{2}(a-1)) = E^{(4)}_{\text{hard}}(0;(0,s);a+1) - E^{(1)}_{\text{hard}}(0;(0,s);(a-1)/2).$$

Exact Pfaffian formulas are known for

$$E^{(1)}_{\text{hard}}(0;(0,s);(a-1)/2) \quad \text{and} \quad E^{(4)}_{\text{hard}}(0;(0,s);a+1)$$

in the case of a an odd positive integer [23], while $E^{(2)}_{\text{hard}}(0;(0,s);a)$ can then be expressed as a determinant [13] (the dimension of the Pfaffians and the determinants are proportional to a), although (5–15) is not a natural consequence of these formulas. There are also multiple integral expressions for the same expression [10], but again they do not naturally satisfy (5–15).

Note Added. An analogous formula to that in (5–13) for $E^{(1)}_{\text{soft}}(0;(s,\infty))$, but now involving a certain Painlevé V transcendent, has recently been derived [12] for

$$E^{(1)}_{\text{hard}}(0;(0,s);\tfrac{1}{2}(a-1)),$$

and formula (5–15) then used to compute $E^{(4)}_{\text{hard}}(0;(0,s);a+1)$.

6. Distribution Functions for Superimposed Spectra

In general, for a symmetric PDF $p(x_0,\ldots,x_{n-1})$ the k-point distribution function ρ_k is defined by

$$\rho_k(x_0,\ldots,x_{k-1}) \\ := n(n-1)\cdots(n-k+1) \int_{(-\infty,\infty)^{n-k}} p(x_0,\ldots,x_{n-1}) dx_k \cdots dx_{n-1}. \quad (6\text{--}1)$$

In this section we take up the task of computing ρ_k for even(M), odd(M), alt(M) with $M = \text{OE}_n(f) \cup \text{OE}_n(f)$ and even(M), odd(M) with $M = \text{OE}_n(f) \cup \text{OE}_{n+1}(f)$.

For $M = \text{OE}_n(f) \cup \text{OE}_n(f)$, write

$$D^{\text{even}(M)}(x_0,\ldots,x_{n-1}) := \det\bigl(F_j(x_i) - F_j(I)\bigr)_{0\leq i,j<n},$$

$$D^{\text{odd}(M)}(x_0,\ldots,x_{n-1}) := \det\bigl(F_j(x_i)\bigr)_{0\leq i,j<n},$$

$$D^{\text{alt}(M)}(x_0,\ldots,x_{n-1}) := \det\bigl(F_j(x_i) - \tfrac{1}{2}F_j(I)\bigr)_{0\leq i,j<n} \quad (n \text{ even}), \qquad (6\text{--}2)$$

$$D^{\text{alt}(M)}(x_0,\ldots,x_{n-1}) := \det\begin{pmatrix} 0 & (F_j(I))_{0\leq j<n}, \\ (1)_{1\leq i<n-1} & (F_j(x_i))_{0\leq i,j<n} \end{pmatrix} \quad (n \text{ odd}),$$

and for $M = \text{OE}_n(f) \cup \text{OE}_{n+1}(f)$ let

$$D^{\text{even}(M)}(x_0,\ldots,x_{n-1}) := \det\begin{pmatrix} (F_j(x_i))_{0\leq i<n,\,0\leq j<n+1} \\ (F_j(I))_{0\leq j<n+1} \end{pmatrix},$$

$$D^{\text{odd}(M)}(x_0,\ldots,x_{n-1}) := \det\begin{pmatrix} (1)_{0\leq j<n+1} \\ (F_j(x_i))_{0\leq i<n,\,0\leq j<n+1} \end{pmatrix}. \qquad (6\text{--}3)$$

In each case, workings contained in the proofs of Theorems 4.2 and 4.6 show (after relabelling the coordinates) that the PDF is proportional to

$$\prod_{i=0}^{n-1} f(x_i) \Delta(x_0, \ldots, x_{n-1}) D(x_0, \ldots, x_{n-1}) \tag{6-4}$$

for D as specified. Now introduce a set of functions $\{\eta_j(x)\}_{0 \le j < n}$ such that

$$D(x_0, \ldots, x_{n-1}) \propto \det(\eta_j(x_i))_{0 \le i,j < n}$$

and a set of monic polynomials $\{q_j(x)\}_{0 \le j < n}$, q_j of degree j, such that the biorthogonality property

$$\int_{-\infty}^{\infty} f(x) q_i(x) \eta_j(x) \, dx = \delta_{i,j}$$

holds (assuming such biorthogonal families exist). The k-point distribution can be expressed in terms of these functions.

LEMMA 6.1. *For the PDF* (6-4) *and* $\{\eta_j(x)\}_{0 \le j < n}$, $\{q_j(x)\}_{0 \le j < n}$ *specified as above, we have*

$$\rho_k(x_0, \ldots, x_{k-1}) = \prod_{j=0}^{k-1} f(x_i) \det\left(\sum_{l=0}^{n-1} q_l(x_i) \eta_l(x_j)\right)_{0 \le i,j < k}. \tag{6-5}$$

PROOF. From the definitions of $\{\eta_j(x)\}_{0 \le j < n}$ and $\{q_j(x)\}_{0 \le j < n}$ we see that (6-4) is proportional to

$$\prod_{i=0}^{n-1} f(x_i) \det(q_j(x_i))_{0 \le i,j < n} \det(\eta_j(x_i))_{0 \le i,j < n}.$$

The biorthogonal property allows the integrations required by the definition (6-1) to be computed to give (6-5). □

REMARK. Suppose for some $\{\xi_j(x)\}_{j=0,\ldots,n-1}$ we can write

$$D(x_0, \ldots, x_{n-1}) \propto \det(\xi_j(x_i))_{0 \le i,j < n-1}.$$

It is easy to show [18; 6] that sufficient conditions for the existence of the biorthogonal sets is that

$$\det\left(\int_{-\infty}^{\infty} f(x) x^i \xi_j(x) \, dx\right)_{0 \le i,j < p} \ne 0$$

for $p = 0, \ldots, n-1$.

For f a classical weight and so of the form (1–5) or (1–9), the biorthogonal functions can be computed explicitly. This is possible because of the following special property of the classical weights and their corresponding orthogonality [1].

LEMMA 6.2. *Consider the pairs (f,g) of classical weight functions (5–1). Let $\{p_j(x)\}_{j=0,1,\ldots}$ be the set of monic orthogonal polynomials, $p_j(x)$ of degree j corresponding to the weight function g, let $(p_k, p_k)_2$ denote their normalization with respect to integration over the measure $g(x)dx$, and define γ_k so that $\gamma_k(p_k, p_k)_2$ equals respectively 1, $\frac{1}{2}$, $\frac{1}{2}(2k+2+a+b)$, or $\alpha-k-1$ in the four cases. With*

$$\boldsymbol{n} := \frac{1}{f(x)} \frac{d}{dx} \frac{g(x)}{f(x)}, \qquad c_k := \gamma_k (p_k, p_k)_2 (p_{k+1}, p_{k+1})_2$$

we have

$$\boldsymbol{n}\, p_k(x) = -\frac{c_k}{(p_{k+1}, p_{k+1})_2} p_{k+1}(x) + \frac{c_{k-1}}{(p_{k-1}, p_{k-1})_2} p_{k-1}(x).$$

PROOF. This is a simple consequence of the following property [2]:

$$(\phi, \boldsymbol{n}\,\psi)_2 = -(\boldsymbol{n}\,\phi, \psi)_2. \qquad \square$$

As a consequence, the determinant formulas (6–2) and (6–3) for $D(x_0, \ldots, x_{n-1})$ can be simplified. Let

$$(u(x))_j := \sum_{l=j}^{\infty} \frac{(u(x), p_l(x))_2}{(p_l, p_l)_2} p_l(x) = u(x) - \sum_{l=0}^{j-1} \frac{(u(x), p_l(x))_2}{(p_l, p_l)_2} p_l(x),$$

and define

$$r_j^{(1)}(x) = \left(\frac{f(x)}{g(x)} \int_x^\infty f(t)\,dt\right)_j,$$

$$r_j^{(2)}(x) = \left(\frac{f(x)}{g(x)} \int_{-\infty}^x f(t)\,dt\right)_j,$$

$$r_j^{(3)}(x) = \left(\frac{f(x)}{g(x)} \left(\int_{-\infty}^x f(t)\,dt - \frac{1}{2}\int_{-\infty}^\infty f(t)\,dt\right)\right)_j,$$

$$r_j^{(4)}(x) = \left(\frac{f(x)}{g(x)}\right)_j,$$

$$r_j^{(5)}(x) = r_{j-1}^{(4)}(x) - \frac{(r_{j-1}^{(4)}, r_{j-1}^{(4)})_2}{(r_{j-1}^{(4)}, r_{j-1}^{(2)})_2} r_{j-1}^{(2)}(x).$$

Then, by adding appropriate linear combinations of the columns in the determinant formulas (6–2) and (6–3) and making use of Lemma 6.2, we readily find for

$M = \mathrm{OE}_n(f) \cup \mathrm{OE}_n(f)$

$$D^{\mathrm{even}(M)}(x_0, \ldots, x_{n-1}) \propto \prod_{i=1}^{n-1} \frac{g(x_i)}{f(x_i)} \det\left((p_j(x_i))_{\substack{0 \leq i < n \\ 0 \leq j < n-1}} \ (r^{(1)}_{n-1}(x_i))_{0 \leq i < n} \right),$$

$$D^{\mathrm{odd}(M)}(x_0, \ldots, x_{n-1}) \propto \prod_{i=1}^{n-1} \frac{g(x_i)}{f(x_i)} \det\left((p_j(x_i))_{\substack{0 \leq i < n \\ 0 \leq j < n-1}} \ (r^{(2)}_{n-1}(x_i))_{0 \leq i < n} \right),$$

$$D^{\mathrm{alt}(M)}(x_0, \ldots, x_{n-1}) \propto \prod_{i=1}^{n-1} \frac{g(x_i)}{f(x_i)} \det\left((p_j(x_i))_{\substack{0 \leq i < n \\ 0 \leq j < n-1}} \ (r^{(3)}_{n-1}(x_i))_{0 \leq i < n} \right)$$
$$(n \text{ even}),$$

$$D^{\mathrm{alt}(M)}(x_0, \ldots, x_{n-1}) \propto \prod_{i=1}^{n-1} \frac{g(x_i)}{f(x_i)} \det\left((p_j(x_i))_{\substack{0 \leq i < n \\ 0 \leq j < n-2}} \ (r^{(4)}_{n-1}(x_i))_{0 \leq i < n} \right)$$
$$(n \text{ odd}),$$

while for $M = \mathrm{OE}_n(f) \cup \mathrm{OE}_{n+1}(f)$

$$D^{\mathrm{even}(M)}(x_0, \ldots, x_{n-1}) \propto \prod_{i=1}^{n-1} \frac{g(x_i)}{f(x_i)} \det\left(p_j(x_i) \right)_{0 \leq i,j < n},$$

$$D^{\mathrm{odd}(M)}(x_0, \ldots, x_{n-1})$$
$$\propto \prod_{i=1}^{n-1} \frac{g(x_i)}{f(x_i)} \det\left((p_j(x_i))_{\substack{0 \leq i < n \\ 0 \leq j < n-2}} \ (r^{(4)}_{n-2}(x_i))_{0 \leq i < n} \ (r^{(5)}_{n-1})_{0 \leq i < n} \right).$$

In each case, setting $\eta_j(x) / \int_{-\infty}^{\infty} f(x) p_i(x) \eta_j(x) \, dx$ equal to $g(x)/f(x)$ times the function in column j and $q_i(x) = p_i(x)$ we have

$$\int_{-\infty}^{\infty} f(x) q_i(x) \eta_j(x) \, dx = \delta_{i,j} \quad (i, j = 0, \ldots, n-1),$$

which is the desired biorthogonality property. Hence substitution of these values into (6–5) gives the k-point distribution in each case.

In particular with $M = \mathrm{OE}_n(f) \cup \mathrm{OE}_{n+1}(f)$ and f one of the classical weights in (5–1) we read off that

$$\rho_k^{\mathrm{even}(M)}(x_0, \ldots, x_{k-1}) = \det\left((g(x_i) g(x_j))^{1/2} \sum_{l=0}^{n-1} \frac{p_l(x_i) p_l(x_j)}{(p_l, p_l)_2} \right)_{0 \leq i,j < k}.$$

This is the well known expression for ρ_k in $\mathrm{UE}(g)$, and thus is in keeping with the result of Theorem 4.3, giving $\mathrm{even}(\mathrm{OE}_n(f) \cup \mathrm{OE}_{n+1}(f)) = \mathrm{UE}_n(g)$ for each of the pairs (f, g) in (5–1). Furthermore, the Christoffel–Darboux formula evaluates the sum as

$$S_2(x, y) := (g(x) g(y))^{1/2} \sum_{l=0}^{n-1} \frac{p_l(x) p_l(y)}{(p_l, p_l)_2}$$
$$= \frac{(g(x) g(y))^{1/2}}{(p_{n-1}, p_{n-1})_2} \frac{p_n(x) p_{n-1}(y) - p_{n-1}(x) p_n(y)}{x - y}. \quad (6\text{–}6)$$

7. Distribution Functions for Alternate Eigenvalues in a Single OE_n

The k-point distribution function for the alternate eigenvalues in a single OE_n has a different structure to ρ_k for the superimposed OE_n spectra. The cases n even and n odd must be treated separately.

Case of n even. Consider first $\text{even}(OE_n(f))$ with n even. From the manipulations sketched in the proof of Theorem 4.7 we have that the PDF of this ensemble is given by

$$\frac{1}{C} \prod_{l=0}^{n/2-1} f(x_{2l}) \det \begin{pmatrix} x_{2i}^j \\ \int_{x_{2i}}^\infty t^j f(t)\, dt \end{pmatrix}_{\substack{0 \le i < n \\ 0 \le j < 2n}}. \tag{7-1}$$

To perform the integration required by (6-1) we introduce the skew inner product

$$\langle u|v\rangle_1 := \frac{1}{2} \int_{-\infty}^\infty dx\, f(x) \left(u(x) \int_x^\infty dy\, f(y) v(y) - v(x) \int_x^\infty dy\, f(y) u(y) \right)$$

$$= \frac{1}{2} \int_{-\infty}^\infty dx\, f(x) u(x) \int_{-\infty}^\infty dy\, v(y)\, \text{sgn}(y-x), \tag{7-2}$$

together with a corresponding family of monic skew orthogonal polynomials $\{R_i(x)\}_{i=0,1,\ldots}$ which are defined so that

$$\langle R_{2i}|R_{2j+1}\rangle_1 = -\langle R_{2j+1}|R_{2i}\rangle_1 = r_j \delta_{i,j},$$
$$\langle R_{2i}|R_{2j}\rangle_1 = \langle R_{2i+1}|R_{2j+1}\rangle_1 = 0. \tag{7-3}$$

Note that the skew orthogonality property still holds if we make the replacement

$$R_{2i+1}(x) \mapsto R_{2i+1}(x) + \gamma_{2i} R_{2i}(x)$$

for arbitrary γ_{2i}. However, a construction of Gram–Schmidt type shows that $\{R_i(x)\}_{i=0,1,\ldots}$ is unique up to this transformation.

We will start by expressing (7-1) as a quaternion determinant involving $\{R_i(x)\}_{i=0,1,\ldots}$ and then show how the property (7-3) can be used to perform the integrations. This requires the definition of a quaternion determinant. We regard a quaternion as a 2×2 matrix, and a quaternion matrix as a matrix with quaternion elements. With n even and

$$Z_n := \mathbf{1}_{n/2} \otimes \begin{pmatrix} 0 & -1 \\ 1 & 0 \end{pmatrix},$$

a $n/2 \times n/2$ quaternion matrix Q is said to be self dual if

$$Q^D := Z_n Q^T Z_n^{-1} = Q.$$

In terms of its 2×2 sub-blocks this means that the quaternion element in position (kj) is related to the element in position (jk), $j<k$ by

$$q_{kj} = \begin{pmatrix} d & -b \\ -c & a \end{pmatrix} \quad \text{if} \quad q_{jk} = \begin{pmatrix} a & b \\ c & d \end{pmatrix}.$$

Now for a self dual quaternion matrix the determinant, to be denoted qdet, is defined by [9]

$$\operatorname{qdet} Q = \sum_{P \in S_{n/2}} (-1)^{n/2-l} \prod_l (q_{ab}q_{bc}\cdots q_{da})^{(0)} \tag{7-4}$$

where the superscript (0) denotes the operation $\tfrac{1}{2}\operatorname{Tr}$, P is any permutation of the indices $(1,\ldots,n/2)$ consisting of l exclusive cycles of the form $(a\to b\to c\to \cdots d\to a)$ and $(-1)^{n/2-l}$ is the parity of P. Furthermore, $\operatorname{qdet} Q$ is related to the Pfaffian via the formula [9]

$$\operatorname{qdet} Q = \operatorname{Pf} QZ_n^{-1},$$

which since $(\operatorname{Pf} QZ_n^{-1})^2 = \det Q$ (where Q is regarded as an ordinary $n\times n$ matrix) implies [17]

$$\det Q = \operatorname{qdet}(QQ^D), \tag{7-5}$$

assuming $\det Q$ is positive.

PROPOSITION 7.1. *With $p(x_0,x_2,\ldots,x_{n-2})$ denoting the probability density function (7-1), $\{R_i(x)\}_{i=0,1,\ldots}$ the monic orthogonal polynomials with respect to (7-2) and $\{r_i\}_{i=0,1,\ldots}$ the corresponding normalizations we can write*

$$p(x_0,x_2,\ldots,x_{n-2}) = \frac{1}{C}\prod_{k=0}^{n/2-1}(2r_k)\operatorname{qdet}\bigl(T(x_{2j},x_{2k})\bigr)_{0\le j,k<n/2}, \tag{7-6}$$

where

$$T(x,y) := \sum_{k=0}^{n/2-1}\frac{1}{2r_k}\bigl(\chi_k(y)\chi_k^D(x)\bigr)^T = \begin{pmatrix} S(x,y) & I(x,y) \\ D(x,y) & S(y,x) \end{pmatrix},$$

$$\chi_k(x) := \begin{pmatrix} f(x)R_{2k}(x) & f(x)R_{2k+1}(x) \\ \int_x^\infty f(t)R_{2k}(t)\,dt & \int_x^\infty f(t)R_{2k+1}(t)\,dt \end{pmatrix},$$

$$S(x,y) = \sum_{k=0}^{N/2-1}\frac{f(y)}{2r_k}\left(R_{2k}(y)\int_x^\infty f(t)R_{2k+1}(t)\,dt - R_{2k+1}(y)\int_x^\infty f(t)R_{2k}(t)\,dt\right),$$

$$I(x,y) = -\int_x^y S(x,y')\,dy',$$

$$D(x,y) = \frac{\partial}{\partial y}S(x,y). \tag{7-7}$$

PROOF. Because the polynomials $\{R_k(x)\}_{k=0,1,\ldots}$ are monic we can add multiples of columns in (7–1) to obtain

$$p(x_0, x_2, \ldots, x_{n-2}) = \frac{1}{C} \det \begin{pmatrix} f(x_{2j}) R_{k-1}(x_{2j}) \\ \int_{x_{2j}}^{\infty} f(t) R_k(t)\, dt \end{pmatrix}_{\substack{0 \le j < n/2 \\ 0 \le k < n}} = \frac{1}{C} \prod_{k=0}^{n/2-1} (2r_k)$$

$$\times \det \begin{pmatrix} f(x_{2j})(2r_k)^{-1/2} R_{2k}(x_{2j}) & f(x_{2j})(2r_k)^{-1/2} R_{2k+1}(x_{2j}) \\ (2r_k)^{-1/2} \int_{x_{2j}}^{\infty} f(t) R_{2k}(t)\, dt & (2r_k)^{-1/2} \int_{x_{2j}}^{\infty} f(t) R_{2k+1}(t)\, dt \end{pmatrix}_{0 \le j,k < n/2} \quad (7\text{–}8)$$

Application of (7–5) and the formula $\text{qdet}\, A = \text{qdet}\, A^T$ gives the formula (7–6) with $S(x,y)$ as specified and formulas for $I(x,y)$ and $D(x,y)$ which are easily seen to be expressible in terms of $S(x,y)$ as stated. $\square$

A special feature of $T(x,y)$, which follows from its definition in (7–7) in terms of $\chi_k(y) \chi_k^D(x)$ and the skew orthogonality of $\{R_k(x)\}_{k=0,1,\ldots}$ with respect to (7–2), is the integration formulas

$$\int_{-\infty}^{\infty} T(x,x)\, dx = N/2,$$

$$\int_{-\infty}^{\infty} T(x,y) T(y,z)\, dy = T(x,z). \quad (7\text{–}9)$$

As a consequence of (7–9) and the quaternion formula (7–6), the integrations required to compute (6–1) can be carried out. Thus with (7–9) holding it is generally true that [17]

$$\int_{-\infty}^{\infty} dx_{2m}\, \text{qdet}\, \bigl(T(x_{2i}, x_{2j})\bigr)_{0 \le i,j \le m} = (n/2 - (m-1))\, \text{qdet}\, \bigl(T(x_{2i}, x_{2j})\bigr)_{0 \le i,j \le m-1}. \quad (7\text{–}10)$$

Consequently we see from (7–6) that

$$\rho_k(x_0, \ldots, x_{2k-2}) = \text{qdet}\, \bigl(T(x_{2i}, x_{2j})\bigr)_{0 \le i,j < k}. \quad (7\text{–}11)$$

If instead of considering even($\text{OE}_n(f)$) we consider odd($\text{OE}_n(f)$), the reasoning is essentially unchanged. Thus (7–6) and (7–7) hold with the replacements

$$\int_x^{\infty} \mapsto \int_{-\infty}^x \quad \text{and} \quad \{x_0, x_2, \ldots, x_{n-2}\} \mapsto \{x_1, x_3, \ldots, x_{n-1}\}, \quad (7\text{–}12)$$

and with this modification of $T(x,y)$ the formula (7–11) for ρ_k holds with the replacements

$$\{x_0, x_2, \ldots, x_{2k-2}\} \mapsto \{x_1, \ldots, x_{2k-1}\}. \quad (7\text{–}13)$$

The structure of (7–11) with $T(x,y)$ given by (7–7) is very similar to the general expression for ρ_k as computed for the ensemble $\text{SE}_n((g/f)^2)$. First it is necessary to introduce monic skew orthogonal polynomials $\{Q_k(x)\}_{k=0,1,\ldots}$ and corresponding normalizations $\{q_k\}_{k=0,1,\ldots}$ with respect to the skew inner product

$$\langle u|v\rangle_4 := \int_{-\infty}^{\infty} dx\, (g(x)/f(x))^2 \bigl(u(x)v'(x) - u(x)v'(x)\bigr).$$

We then have [24] (see also [28])

$$\rho_k(x_0,\ldots,x_{k-1}) = \mathrm{qdet}\,\bigl(T_4(x_i,x_j)\bigr)_{0\leq i,j\leq k} \qquad (7\text{-}14)$$

where

$$T_4(x,y) := \begin{pmatrix} S_4(x,y) & I_4(x,y) \\ D_4(x,y) & S_4(y,x) \end{pmatrix},$$

$$S_4(x,y) = \sum_{k=0}^{N-1} \frac{f(y)}{2q_k}\left(Q_{2k}(y)\frac{d}{dx}\bigl(f(x)Q_{2k+1}(x)\bigr) - Q_{2k+1}(x)\frac{d}{dx}\bigl(f(x)Q_{2k}(x)\bigr)\right),$$

$$I_4(x,y) = -\int_x^y S_4(x,y')\,dy',$$

$$D_4(x,y) = \frac{\partial}{\partial y} S_4(x,y). \qquad (7\text{-}15)$$

Case of n odd. The PDF for the distribution $\mathrm{even}(\mathrm{OE}_n(f))$ with n odd is

$$\frac{1}{C} \prod_{l=0}^{(n-3)/2} f(x_{2l}) \det \begin{pmatrix} \left(x_{2i}^{2j} \atop \int_{x_{2j}}^{\infty} f(t)t^k\,dt \right)_{0\leq i<(n-1)/2 \atop 0\leq j<n} \\ \left(\int_{-\infty}^{\infty} w_1(t)t^j\,dt\right)_{0\leq j<n} \end{pmatrix}.$$

As in (7–8), we can introduce the monic polynomials $\{R_j(x)\}_{j=0,1,\ldots}$ to rewrite this as

$$\frac{1}{C}\det\begin{pmatrix} \left(f(x_{2i})R_j(x_{2i}) \atop \int_{x_{2i}}^{\infty} f(t)R_j(t)\,dt \right)_{0\leq i<(n-1)/2 \atop 0\leq j<n} \\ \left(\int_{-\infty}^{\infty} f(t)R_j(t)\,dt\right)_{j=0,\ldots,n-2} \end{pmatrix}.$$

Subtracting appropriate multiples of the last column from the columns $0, 1, \ldots, n-2$ so as to eliminate the element of the column in the final row then gives

$$\frac{1}{C}\left(\int_{-\infty}^{\infty} f(t)R_{n-1}(t)\,dt\right) \det\begin{pmatrix} f(x_{2i})\hat{R}_j(x_{2i}) \\ \int_{x_{2i}}^{\infty} f(t)\hat{R}_j(t)\,dt \end{pmatrix}_{0\leq i<(n-1)/2 \atop 0\leq j<n}, \qquad (7\text{-}16)$$

where

$$\hat{R}_j(x) := R_j(x) - \left(\frac{\int_{-\infty}^{\infty} f(t)R_j(t)\,dt}{\int_{-\infty}^{\infty} f(t)R_{n-1}(t)\,dt}\right) R_{n-1}(x). \qquad (7\text{-}17)$$

The determinant in (7–16) is formally the same as that in (7–8). Thus in the case n odd $p(x_0, x_2, \ldots, x_{n-3})$ can be written as in (7–6) but with

$$n \mapsto n-1, \quad R_i \mapsto \hat{R}_i \qquad (7\text{-}18)$$

and $C \mapsto C'$ for some normalization C'.

Now we can check from the definition (7–17) that for $j = 1, \ldots, n-1$ the polynomials $\hat{R}_{j-1}$ satisfy the skew orthogonality property (7–3). This means that the integration formula (7–10) again applies in this modified setting and consequently the k-point distribution is given by

$$\rho_k(x_0, \ldots, x_{2k-2}) = \operatorname{qdet}\left(T^{\text{odd}}(x_{2i}, x_{2j})\right)_{0 \le i,j < k}, \qquad (7\text{--}19)$$

where T^{odd} is defined as $T(x,y)$ in (7–7) but with the replacements (7–18). In the case of $\operatorname{odd}(\mathrm{OE}_n(f))$ the replacements (7–12) and (7–18) must be made in (7–11) and (7–7), and the replacement (7–13) made in (7–19).

7.1. Summation Formulas.

It has already been remarked that ρ_k for $\mathrm{SE}_n((g/f)^2)$ has the quaternion determinant form (7–14) and (7–15). Furthermore it is known [1] that with f one of the classical weights in (5–1), the quantity S_4 in (7–15) can be summed to give an expression independent of the skew orthogonal polynomials associated with g, and dependent only on the monic orthogonal polynomials $\{p_i(x)\}_{i=0,1,\ldots}$ associated with the weight function $g(x)$. Explicitly,

$$2S_4(x,y) = \left(\frac{g(x)}{g(y)}\right)^{1/2} \frac{f(y)}{f(x)} S_2(x,y)\bigg|_{n \mapsto 2n} - \gamma_{2n-1} f(y) p_{2n}(y) \int_x^\infty f(t) p_{2n-1}(t)\, dt, \qquad (7\text{--}20)$$

where S_2 is specified by (6–6) and γ_{2n-1} by Lemma 6.2. Here we will use results from [1] to obtain an analogous summation for the quantity $S(x,y)$ in (7–7).

Suppose n is even and write

$$\Phi_j(x) := \frac{1}{2} \int_{-\infty}^\infty f(t)\operatorname{sgn}(x-t) R_j(t)\, dt.$$

Then straightforward manipulation of the definition of $S(x,y)$ allows it to be rewritten

$$S(x,y) = \frac{1}{2}\Big(S_1(x,y) - S_1(\infty, y)\Big) \qquad (7\text{--}21)$$

where

$$S_1(x,y) = \sum_{k=0}^{n/2-1} \frac{f(y)}{r_k}\Big(\Phi_{2k}(x) R_{2k+1}(y) - \Phi_{2k+1}(x) R_{2k}(y)\Big)$$

The quantity $S_1(x,y)$ occurs in the quaternion determinant formula for k-point distribution of $\mathrm{OE}_n(f)$. With f one of the classical forms (5–1) it can be summed to give [1]

$$S_1(x,y) = \left(\frac{g(x)}{g(y)}\right)^{1/2} \frac{f(y)}{f(x)} S_2(x,y)\bigg|_{n \mapsto n-1}$$
$$+ \gamma_{n-2} f(y) p_{n-1}(y) \frac{1}{2} \int_{-\infty}^\infty \operatorname{sgn}(x-t) f(t) p_{n-2}(t)\, dt. \qquad (7\text{--}22)$$

From this it follows that
$$S_1(\infty, y) = \gamma_{n-2} f(y) p_{n-1}(y) \frac{1}{2} \int_{-\infty}^{\infty} f(t) p_{n-2}(t)\, dt,$$
and so by (7–21) we can evaluate $S(x, y)$.

PROPOSITION 7.2. *For (f, g) a classical pair (5–1), $\{p_j(x)\}_{j=0,1,\ldots}$ monic orthogonal polynomials with respect to the weight function $g(x)$, and n even the quantity $S(x, y)$ in (7–7) has the evaluation*
$$2S(x, y) = \left(\frac{g(x)}{g(y)}\right)^{1/2} \frac{f(y)}{f(x)} S_2(x, y)\bigg|_{n \mapsto n-1} - \gamma_{n-2} f(y) p_{n-1}(y) \int_x^{\infty} f(t) p_{n-2}(t)\, dt \tag{7-23}$$
(compare (7–20)).

This summation fully determines even($OE_n(f)$) with n even. For odd($OE_n(f)$), n even, the prescription (7–12) says the replacement $\int_x^{\infty} \mapsto \int_{-\infty}^x$ should be made in (7–23).

It remains to consider the case n odd. Consider first even($OE_n(f)$). In fact the formulas in [1] giving the analogous formula to (7–22) for n odd allows us to deduce that the summation (7–23) remains valid for n odd. For n odd comparison of (7–23) and (7–20) shows that
$$S(x, y) = S_4(x, y)\bigg|_{n \mapsto (n-1)/2},$$
which because of the formulas (7–7) (with $n \mapsto n-1$), (7–11), (7–14), and (7–15) implies
$$\rho_k^{\text{even}(OE_{2n+1}(f))}(x_0, x_2, \ldots, x_{2k-2}) = \rho_k^{SE_n((g/f)^2)}(x_0, x_2, \ldots, x_{2k-2}).$$
This is equivalent to the second statement of (4–2), which we already know from Theorem 5.1 is valid for the pairs (f, g) in (5–1). In the case of odd($OE_n(f)$) with n odd, again the prescription (7–12) says we simply make the replacement $\int_x^{\infty} \mapsto \int_{-\infty}^x$ in (7–23).

Acknowledgements

Forrester thanks J. Baik for drawing his attention to the conjecture noted in the paragraph above the paragraph containing (1–6), and acknowledges the Australian Research Council for financial support.

References

[1] M. Adler, P. J. Forrester, T. Nagao, and P. van Moerbeke, "Classical skew orthogonal polynomials and random matrices", *J. Stat. Phys.* **99** (2000), 141–170.

[2] M. Adler and P. van Moerbeke, "Matrix integrals, Toda symmetries, Virasoro constraints and orthogonal polynomials", *Duke Math. Journal* **80** (1995), 863–911.

[3] A. Altland and M. R. Zirnbauer, "Nonstandard symmetry classes in mesoscopic normal-superconducting hybrid compounds", *Phys. Rev. B* **55** (1997), 1142.

[4] J. Baik and E. M. Rains, "Algebraic aspects of increasing subsequences", preprint, math.CO/9905083, 1999.

[5] J. Baik and E. M. Rains, "The asymptotics of monotone subsequences of involutions", Preprint, math.CO/9905084, 1999.

[6] A. Borodin, "Biorthogonal ensembles", *Nucl. Phys. B* **536** (1999), 704–732.

[7] F. J. Dyson, "Statistical theory of energy levels of complex systems I", *J. Math. Phys.* **3** (1962), 140–156.

[8] F. J. Dyson, "The three fold way. Algebraic structure of symmetry groups and ensembles in quantum mechanics", *J. Math. Phys.* **3** (1962), 1199–1215.

[9] F. J. Dyson, "Correlations between the eigenvalues of a random matrix", *Commun. Math. Phys.* **19** (1970), 235–250.

[10] P. J. Forrester, "Exact results and universal asymptotics in the Laguerre random matrix ensemble", *J. Math. Phys.* **35** (1993), 2539–2551.

[11] P. J. Forrester, "The spectrum edge of random matrix ensembles", *Nucl. Phys. B* **402** (1993), 709–728.

[12] P. J. Forrester, "Painlevé transcendent evaluation of the scaled distribution of the smallest eigenvalue in the Laguerre orthogonal and symplectic ensembles", preprint, nlin.SI/0005064, 2000.

[13] P. J. Forrester and T. D. Hughes, "Complex Wishart matrices and conductance in mesoscopic systems: exact results. *J. Math. Phys.* **35** (1994), 6736–6747.

[14] N. R. Goodman, "Statistical analysis based on a certain multivariate complex Gaussian distribution (an introduction)", *Ann. Math. Stat.* **34** (1963), 152–176.

[15] J. Gunson, "Proof of a conjecture of Dyson in the statistical theory of energy levels", *J. Math. Phys.* **4** (1962), 752–753.

[16] S. Helgason, *Differential geometry and symmetric spaces*, Academic Press, New York, 1962.

[17] G. Mahoux and M. L. Mehta, "A method of integration over matrix variables IV", *J. Physique I (France)* **1** (1991), 1093–1108.

[18] G. Mahoux, M. L. Mehta, and J.-M. Normand, "Matrices coupled in a chain, II: Spacing functions", *J. Phys. A* **31** (1999), 4457–4464.

[19] M. L. Mehta, *Random Matrices*, 2nd ed., Academic Press, New York, 1991.

[20] M. L. Mehta and F. J. Dyson, "Statistical theory of the energy levels of complex systems, V", *J. Math. Phys.* **4** (1963), 713–719.

[21] R. J. Muirhead, *Aspects of multivariable statistical theory*. Wiley, New York, 1982.

[22] T. Nagao and P. J. Forrester, "Asymptotic correlations at the spectrum edge of random matrices", *Nucl. Phys. B* **435** (1995), 401–420.

[23] T. Nagao and P. J. Forrester, "The smallest eigenvalue at the spectrum edge of random matrices", *Nucl. Phys. B* **509** (1998), 561–598.

[24] T. Nagao and M. Wadati, "Correlation functions of random matrix ensembles related to classical orthogonal polynomials II, III", *J. Phys. Soc. Japan*, **61** (1992), 78–88, 1910–1918.

[25] V. Romanovsky, "Sur quelques classes nouvelles de polynômes orthogonaux", *C. R. Acad. Sci. Paris* **188** (1929), 1023–1025.

[26] G. Szegö, *Orthogonal polynomials*, 4th ed., Amer. Math. Soc., Providence, RI, 1975.

[27] C. A. Tracy and H. Widom, "On orthogonal and symplectic matrix ensembles", *Commun. Math. Phys.* **177** (1996), 727–754.

[28] C. A. Tracy and H. Widom, "Correlation functions, cluster functions and spacing distributions in random matrices", *J. Stat. Phys.* **92** (1998), 809–835.

[29] J. Wishart, "The generalized product moment distribution in samples from a normal multivariate population", *Biometrika* **20A** (1928), 32–43.

[30] M. R. Zirnbauer, "Riemannian symmetric superspaces and their origin", *J. Math. Phys.* **10** (1997), 4986–5018.

PETER J. FORRESTER
DEPARTMENT OF MATHEMATICS AND STATISTICS
UNIVERSITY OF MELBOURNE
PARKVILLE, VICTORIA 3052
AUSTRALIA
p.forrester@ms.unimelb.edu.au

ERIC M. RAINS
AT&T RESEARCH
FLORHAM PARK, NJ 07932
UNITED STATES
rains@research.att.com

Dual Isomonodromic Tau Functions and Determinants of Integrable Fredholm Operators

J. HARNAD

ABSTRACT. The Hamiltonian approach to dual isomonodromic deformations in the setting of rational R-matrix structures on loop algebras is reviewed. The construction of a particular class of solutions to the deformation equations, for which the isomonodromic τ-functions are given by the Fredholm determinants of a special class of integrable integral operators, is shown to follow from the matrix Riemann–Hilbert approach of Its, Izergin, Korepin and Slavnov. This leads to an interpretation of the notion of duality in terms of the data defining the Riemann–Hilbert problem, and Laplace–Fourier transforms of the corresponding Fredholm integral operators.

1. Introduction

1a. Isomonodromic Deformation Equations. We consider rational covariant derivative operators on the punctured Riemann sphere, having the form

$$\mathcal{D}_\lambda = \frac{\partial}{\partial \lambda} - \mathcal{N}(\lambda),$$

$$\mathcal{N}(\lambda) := B + \sum_{i=1}^{n} \frac{N_i}{\lambda - \alpha_j}, \tag{1-1}$$

where

$$B = \mathrm{diag}(\beta_1, \ldots, \beta_r), \quad N_j \in \mathfrak{gl}(r, \mathbb{C}).$$

They have regular singular points at $\{\lambda = \alpha_i\}_{i=1,\ldots,n}$ and an irregular singularity at $\lambda = \infty$ with Poincare index 1. If the residue matrices $\{N_i\}_{i=1,\ldots,n}$ are deformed differentiably with respect to the parameters $\{\alpha_i\}_{i=1,\ldots,n}$ and $\{\beta_a\}_{a=1,\ldots,r}$, the monodromy (including Stokes parameters and connection matrices) of the operator $\mathcal{D}_\lambda$ will be invariant under such deformations, as was shown in [Jimbo et al. 1980; Jimbo et al. 1981], provided the differential equations implied by the commutativity conditions

$$[\mathcal{D}_\lambda, \mathcal{D}_{\alpha_i}] = 0, \quad i = 1, \ldots, n, \tag{1-2}$$

$$[\mathcal{D}_\lambda, \mathcal{D}_{\beta_a}] = 0, \quad a = 1, \ldots, r, \tag{1-3}$$

are satisfied, where the differential operators $\mathcal{D}_{\alpha_j}$, $\mathcal{D}_{\beta_a}$ are defined by

$$\mathcal{D}_{\alpha_i} := \frac{\partial}{\partial \alpha_i} - U_i,$$

$$\mathcal{D}_{\beta_a} := \frac{\partial}{\partial \beta_a} - V_a,$$

$$U_i := -\frac{N_i}{\lambda - \alpha_j}, \quad i = 1, \ldots, n,$$

$$V_a := \lambda E_a + \sum_{\substack{b=1 \\ b \neq a}}^r \frac{E_a \left(\sum_{j=1}^n N_j\right) E_b + E_b \left(\sum_{j=1}^n N_j\right) E_a}{\beta_a - \beta_b}, \quad a = 1, \ldots, r,$$

and E_a is the elementary $r \times r$ matrix with elements

$$(E_a)_{bc} := \delta_{ab}\delta_{ac}.$$

These also imply the commutativity conditions

$$[\mathcal{D}_{\alpha_i}, \mathcal{D}_{\alpha_j}] = [\mathcal{D}_{\alpha_i}, \mathcal{D}_{\beta_a}] = [\mathcal{D}_{\beta_a}, \mathcal{D}_{\beta_b}] = 0, \quad \text{for } i, j = 1, \ldots, n \text{ and } a, b = 1, \ldots, r. \tag{1-4}$$

Equations (1–2), (1–3), and (1–4) define a Frobenius integrable system of PDE's for the residue matrices $\{N_i\}_{i=1,\ldots,n}$. They are "zero curvature" equations, implying the consistency of the overdetermined system

$$\mathcal{D}_\lambda \Psi = 0, \quad \mathcal{D}_{\alpha_i} \Psi = 0, \quad \mathcal{D}_{\beta_a} \Psi = 0. \tag{1-5}$$

They may also be interpreted as Hamiltonian equations [Harnad 1994] with respect to the Lie Poisson structure on the space $(\mathfrak{gl}(r))^{*n} = \{N_1, \ldots, N_n\}$ of residue matrices in $\mathcal{N}(\lambda)$, defined by

$$\{(N_i)_{ab}, (N_j)_{cd}\} = \delta_{ij}[(N_i)_{ad}\delta_{bc} - (N_i)_{bc}\delta_{ad}] \tag{1-6}$$

(where we identify $\mathfrak{gl}(r)$ and its dual space $(\mathfrak{gl}(r))^*$ through the trace pairing $(X, Y) = \operatorname{tr}(XY)$). The Hamiltonians $\{H_i, K_a\}_{j=1,\ldots,n,\ a=1,\ldots,r}$ generating them are given by

$$H_i := \tfrac{1}{2} \operatorname{res}_{\lambda=\alpha_i} \operatorname{tr}(\mathcal{N}(\lambda)^2) = \operatorname{tr}(BN_i) + \sum_{\substack{j=1 \\ j \neq i}}^n \frac{\operatorname{tr}(N_i N_j)}{\alpha_i - \alpha_j},$$

$$K_a := \sum_{j=1}^n \alpha_j (N_j)_{aa} + \sum_{\substack{b=1 \\ b \neq a}}^r \frac{\left(\sum_{j=1}^n N_j\right)_{ab} \left(\sum_{k=1}^n N_k\right)_{ba}}{\beta_a - \beta_b}. \tag{1-7}$$

It follows from the Poisson bracket relations (1–6) that

$$\{\mathcal{N}(\lambda), H_i\} = [U_i, \mathcal{N}(\lambda)],$$
$$\{\mathcal{N}(\lambda), K_a\} = [V_a, \mathcal{N}(\lambda)],$$

which imply, together with the identities

$$\mathcal{N}(\lambda)_{\alpha_i} = \frac{\partial U_i}{\partial \lambda} = \frac{N_i}{(\lambda - \alpha_i)^2}, \qquad (1\text{–}8)$$

$$\mathcal{N}(\lambda)_{\beta_a} = \frac{\partial V_a}{\partial \lambda} = E_a \qquad (1\text{–}9)$$

(where the subscripts in $\mathcal{N}(\lambda)_{\alpha_i}$ and $\mathcal{N}(\lambda)_{\beta_a}$ denote derivation only with respect to the explicit dependence on the parameters appearing in the definition of $\mathcal{N}(\lambda)$), that the equations obtained by equating the residues at the poles $\{\lambda = \alpha_j\}$ in (1–2)–(1–3) and the leading terms at $\lambda = \infty$ are the nonautonomous Hamiltonian equations generated by the H_i's and K_a's when the α_i's and β_a's are identified with the respective "time" parameters.

The compatibility of these equations may be seen as a consequence of the fact that all the Hamiltonians Poisson commute:

$$\{H_i, H_j\} = 0, \quad \{H_i, K_a\} = 0, \quad \{K_a, K_b\} = 0,$$

for $i, j = 1, \ldots, n$ and $a, b = 1, \ldots, r$. This further implies [Jimbo et al. 1980; Jimbo et al. 1981] that the differential 1-form

$$\theta := \sum_{i=1}^{n} H_i \, d\alpha_i + \sum_{a=1}^{b} K_a \, d\beta_a, \qquad (1\text{–}10)$$

on the parameter space, taken along any solution to this system of equations, is closed, and hence locally exact, implying the existence of the isomonodromic τ-function $\tau(\alpha_1, \ldots, \alpha_n, \beta_1, \ldots, \beta_r)$, defined up to a multiplicative constant by:

$$d \ln(\tau) = \theta. \qquad (1\text{–}11)$$

The Hamiltonian structure of the above equations may be seen to follow from a more general setting, involving commuting Hamiltonians flows on loop algebras generated by spectral invariant functions, with respect to a rational R-matrix structure, adapted to the case of nonautonomous Hamiltonians [Harnad 1994]. For our purposes, it is sufficient to consider the Poisson space $\widetilde{\mathfrak{gl}}_{\mathrm{rat}}(r)$ consisting of $r \times r$ matrix-valued rational functions $X(\lambda)$ of the complex parameter λ, and to split $\widetilde{\mathfrak{gl}}_{\mathrm{rat}}(r)$ into the direct sum

$$\widetilde{\mathfrak{gl}}_{\mathrm{rat}}(r) := \mathfrak{g}_+ + \mathfrak{g}_-$$

of the subspaces $\mathfrak{g}_+$ consisting of polynomial $X(\lambda)$'s, and $\mathfrak{g}_-$ consisting of $X(\lambda)$'s satisfying $X(\infty) = 0$. Viewing each $X(\lambda)$ as an endomorphism of $\mathbb{C}^r$, we may concisely represent the Poisson bracket structure on the space $\widetilde{\mathfrak{gl}}_{\mathrm{rat}}(r)$ by simultaneously taking tensor products on the space of endomorphisms and giving an equation in the space $\mathrm{End}(\mathbb{C}^r \otimes \mathbb{C}^r)$ that determines the Poisson brackets of all the matrix elements of the $X(\lambda)$'s:

$$\{X(\lambda) \overset{\otimes}{,} X(\mu)\} = [r(\lambda-\mu), X(\lambda) \otimes \boldsymbol{I} + \boldsymbol{I} \otimes X(\mu)], \tag{1-12}$$

where

$$r(\lambda - \mu) := \frac{P_{12}}{\lambda - \mu} \in \mathrm{End}(\mathbb{C}^r \otimes \mathbb{C}^r), \qquad P_{12}(u \otimes v) := v \otimes u,$$

is the rational classical R-matrix. We may view $\mathcal{N}(\lambda)$ as the image of a map

$$\mathcal{N}_B^A : (\mathfrak{gl}(r))^{*n} \to \widetilde{\mathfrak{gl}}_{\mathrm{rat}}(r)$$

$$\mathcal{N}_B^A : \{N_1, \ldots, N_n\} \mapsto \mathcal{N}(\lambda) = B + \sum_{j=1}^{n} \frac{N_j}{\lambda - \alpha_j}. \tag{1-13}$$

It is easily verified that this defines a Poisson embedding of $(\mathfrak{gl}(r))^{*n}$ as an affine subspace in $\widetilde{\mathfrak{gl}}_{\mathrm{rat}}(r)$. (If we take the union over all $r \times r$ matrices B, this becomes a linear Poisson subspace, but since the coefficients of the matrices B are in the centre of the Poisson algebra on this space, B may as well be chosen to have fixed constant values.)

Now let $\mathfrak{I}$ denote the ring of polynomial functions of the coefficients of elements of $\widetilde{\mathfrak{gl}}_{\mathrm{rat}}(r)$ that are invariant under conjugation by λ-dependent invertible matrices, restricted to a finite dimensional Poisson submanifold such as, for example, the image of the map $\mathcal{N}_B^A$. This is just the ring of spectral invariants, generated by the coefficients of the characteristic polynomial

$$\det(X(\lambda) - z\boldsymbol{I}) := \mathcal{P}(\lambda, z).$$

The classical R-matrix theorem, adapted to the case of explicit time dependence in the Hamiltonians and in the elements $X(\lambda) \in \widetilde{\mathfrak{gl}}_{\mathrm{rat}}(r)$, then tells us that the elements of $\mathfrak{I}$ Poisson commute, and the Hamiltonian equations generated by any $\phi \in \mathfrak{I}$ may be expressed as

$$\frac{dX(\lambda)}{dt} = \pm[(d\phi)_\pm, X(\lambda)] + X(\lambda)_t, \tag{1-14}$$

where $X(\lambda)_t$ denotes the explicit time derivative, the differential $d\phi$ is identified as an element of the same space $\widetilde{\mathfrak{gl}}_{\mathrm{rat}}(r)$ through the dual pairing $\langle X, Y \rangle := \mathrm{res}_{\lambda=\infty} \mathrm{tr}((X(\lambda)Y\lambda))$ and $(d\phi)_\pm$ denotes projection to the subspaces $\mathfrak{g}_\pm$.

If it happens also that the term $X(\lambda)_t$ equals the λ derivative of either $d\phi_+$ or $-d\phi_-$

$$X(\lambda)_t = \pm \frac{\partial d(\phi_\pm)}{\partial \lambda}, \tag{1-15}$$

then equation (1–14) becomes a commutativity condition

$$\left[\frac{\partial}{\partial \lambda} - X(\lambda), \frac{\partial}{\partial t} \mp (d\phi)_\pm \right] = 0. \tag{1-16}$$

(More generally, we could replace $+d\phi_-$ and $-d\phi_+$ by any element along the line $(1+c)d\phi_+ + cd\phi_-$ through them.) In particular, this is the case if we choose ϕ as any of the Hamiltonians $\{H_i\}_{i=1,\ldots,n}$

$$H_i = \frac{1}{2}\operatorname{res}_{\lambda=\alpha_i} \operatorname{tr}(X^2(\lambda)),$$

which clearly are elements of the spectral ring $\mathcal{J}$ which, when evaluated on $X(\lambda) = \mathcal{N}(\lambda)$ of the form (1-1), give

$$-(dH_i)_- = U_i = -\frac{N_i}{\lambda - \alpha_i}.$$

Condition (1-15) is satisfied on this subspace if we identify the time parameter as $t = \alpha_i$, since this reduces to the identity (1-8), while (1-16) gives the equations (1-2).

To obtain a similar interpretation of the equations (1-3), we note that the Hamiltonians $\{K_a\}_{a=1,\ldots,r}$ can be expressed as follows:

$$K_a = \tfrac{1}{2}\operatorname{res}_{\lambda=\infty}\left(\operatorname{res}_{z=\beta_a} \lambda[(B-z\boldsymbol{I})^{-1}\mathcal{N}(\lambda)]^2 - 2\operatorname{tr}[(B-z\boldsymbol{I})^{-1}\mathcal{N}(\lambda)]\right),$$

which shows that they also belong to the spectral ring $\mathcal{J}$. Taking the $\mathfrak{g}_+$ projections of the differentials dK_a evaluated at $X(\lambda) = \mathcal{N}(\lambda)$ gives

$$(dK_a)_+ = V_a = \lambda E_a + \sum_{\substack{b=1\\b\neq a}}^r \frac{E_a\left(\sum_{j=1}^n N_j\right)E_b + E_b\left(\sum_{j=1}^n N_j\right)E_a}{\beta_a - \beta_b}.$$

Identifying the time parameter as $t = \beta_a$, condition (1-15) reduces to (1-9), and (1-16) gives the equations (1-3).

1b. Symplectic Lift and Duality. The Hamiltonian structure of the isomonodromic deformation equations presented above involves a degenerate Poisson structure. (The center of the Poisson algebra consists of the elements of $\mathcal{J}$ obtained by localizing the spectral invariants at the points $\{\alpha_1,\ldots,\alpha_n,\infty\}$.) It is possible, however, to view this space as a quotient of a symplectic space $\mathcal{M}$ under a suitable Hamiltonian group action. Moreover, doing so shows that the rôles of the deformation parameters $\{\alpha_1,\ldots,\alpha_n\}$ and $\{\beta_1,\ldots,\beta_r\}$ are in some sense interchangeable, and there exists another "dual" isomonodromic deformation system, obtained also as a Hamiltonian quotient of the system on $\mathcal{M}$, in which the parameters $\{\beta_1,\ldots,\beta_r\}$ appear as the locations of the regular singular points, while $\{\alpha_1,\ldots,\alpha_n\}$ become the eigenvalues at ∞ of the rational matrix defining this dual system.

To see this, suppose that the rank of the residue matrix N_i is k_i. We may express N_i in a factored form as the product of two maximal rank matrices of dimensions $r \times k_i$ and $k_i \times r$:

$$N_i = -G_i^T F_i, \qquad F_i, G_i \in \operatorname{Mat}^{k_i \times r}. \tag{1-17}$$

Of course, this factorization is arbitrary up to the following action of the group $\mathrm{GL}(k_i, \mathbb{C})$ on the space of such pairs (F_i, G_i):

$$g_i : (F_i, G_i) \mapsto (g_i F_i, (g_i^T)^{-1}), \quad g_i \in \mathrm{GL}(k_i, \mathbb{C}).$$

Making a similar factorization of all the residue matrices, we let

$$N := \sum_{i=1}^{n} k_i$$

and define the space $\mathcal{M}$ to consist of the set of pairs (F, G) of $N \times r$ matrices formed from n vertical blocks of $k_i \times r$ matrices of maximal rank:

$$F := \begin{pmatrix} F_1 \\ \cdot \\ F_i \\ \cdot \\ F_n \end{pmatrix}, \quad G := \begin{pmatrix} G_1 \\ \cdot \\ G_i \\ \cdot \\ G_n \end{pmatrix}. \tag{1-18}$$

Let $A \in \mathfrak{gl}(N, \mathbb{C})$ be the diagonal matrix with eigenvalues $(\alpha_1, \ldots, \alpha_n)$ appearing with respective multiplicities $(k_1, \ldots, k_n)$.

$$A = \mathrm{diag}(\alpha_1, \ldots, \alpha_i, \ldots, \alpha_n).$$

Using the resolvent matrix $(A - \lambda \boldsymbol{I})^{-1}$, we can express the rational matrix $\mathcal{N}(\lambda)$ as follows:

$$\mathcal{N}(\lambda) = B + G^T (A - \lambda \boldsymbol{I})^{-1} F,$$

where the different possible choices for the pairs (F, G) form an orbit under the block diagonal subgroup $G_A := \mathrm{GL}(k_1, \mathbb{C}) \times \cdots \times \mathrm{GL}(k_n, \mathbb{C}) \subset \mathrm{GL}(N, \mathbb{C})$, under the action $G_A \times \mathcal{M} \to \mathcal{M}$ defined by

$$(g, (F, G)) \mapsto (gF, (g^T)^{-1} G), \text{ for } g \in G_A. \tag{1-19}$$

This subgroup $G_A \subset \mathrm{GL}(N, \mathbb{C})$ is just the stabilizer of $A \in \mathfrak{gl}(N, \mathbb{C})$ under the adjoint (conjugation) action. Choosing the canonical symplectic structure

$$\omega := \mathrm{tr}(dF^T \wedge dG) \tag{1-20}$$

on $\mathcal{M}$, the G_A action is a free Hamiltonian group action generated by the equivariant moment map

$$J_k^N : (F, G) \to (F_1 G_1^T, \ldots, F_i G_i^T, \ldots, F_n G_n^T)$$
$$\in \mathfrak{gl}^*(k_1, \mathbb{C}) \oplus \cdots \oplus \mathfrak{gl}^*(k_i, \mathbb{C}) \oplus \cdots \oplus \mathfrak{gl}^*(k_n, \mathbb{C}),$$

where the dual space $\mathfrak{gl}^*(k_i, \mathbb{C})$ is identified with the space $\mathfrak{gl}(k_i, \mathbb{C})$ through the trace pairing. The Poisson subspace of $(\mathfrak{gl}(r))^{*n}$ consisting of matrices $\{N_1, \ldots, N_n\}$ having respective ranks $\{k_1, \ldots, k_n\}$ may thus be viewed as a quotient $\mathcal{M}/G_A$ by the Hamiltonian group action (1-19). Composing the projection map $\pi_A^{\mathcal{M}} : \mathcal{M} \to \mathcal{M}/G_A$ with the Poisson embedding map $\mathcal{N}_B^A$ defined in (1–13), we obtain a Poisson map $J_B^A : \mathcal{M} \to \widetilde{\mathfrak{gl}}_{\text{rat}}(r)$ given by

$$J_B^A : (F, G) \mapsto \mathcal{N}(\lambda) := B + G^T(A - \lambda \mathbf{I})^{-1} F, \qquad (1\text{–}21)$$

whose fibres are the orbits under the free G_A-action (1–19), allowing us to identify the image J_B^A both as a quotient space $\mathcal{M}/G_A$ and a Poisson subspace of $\widetilde{\mathfrak{gl}}_{\text{rat}}(r)$ (i.e., the space consisting of those $\mathcal{N}(\lambda)$'s for which the residue matrices $\{N_1, \ldots, N_n\}$ have ranks $\{k_1, \ldots, k_n\}$). Moreover, the ring $\mathcal{I}$ of spectral invariants, restricted to this subspace, may be pulled back to $\mathcal{M}$ to define a Poisson commuting ring

$$\mathcal{I}_B^A := J_B^{A*}(\mathcal{I})$$

of G_A-invariant functions on $\mathcal{M}$. The Hamiltonian vector fields generated by the elements of $\mathcal{I}_B^A$ project to the corresponding vector fields on the quotient, as do their integral curves. In particular, if we identify the parameters $\{\alpha_1, \ldots, \alpha_n\}$ and $\{\beta_1, \ldots, \beta_r\}$ with multi-time variables associated to the pullbacks of the Hamiltonians $\{H_1, \ldots, H_n\}$ and $\{K_1, \ldots, K_r\}$, the corresponding nonautonomous Hamiltonian systems on $\mathcal{M}$ are given by

$$\begin{aligned}\frac{\partial F}{\partial \alpha_i} &= \{F_i,\ J_B^{A*} H_i\}, & \frac{\partial G}{\partial \alpha_i} &= \{G_i,\ J_B^{A*} H_i\}, \\ \frac{\partial G}{\partial \beta_a} &= \{G_i,\ J_B^{A*} K_a\}, & \frac{\partial G}{\partial \beta_a} &= \{G_i,\ J_B^{A*} K_a\},\end{aligned} \qquad (1\text{–}22)$$

where $\{\cdot, \cdot\}$ denotes the Poisson brackets on $\mathcal{M}$ determined by the symplectic form (1–20).

There is also a Hamiltonian action of the subgroup $G_B \subset \text{GL}(r, \mathbb{C})$ stabilizing the matrix B under conjugation, which commutes with the G_A action (1–19), namely the action $G_B \times \mathcal{M} \to \mathcal{M}$ defined by

$$(g, (F, G)) \mapsto (Fg^{-1}, Gg^T), \qquad g \in G_B. \qquad (1\text{–}23)$$

If the eigenvalues $\{\beta_1, \ldots, \beta_r\}$ are required to be distinct, the group G_B just consists of the invertible diagonal matrices in $\text{GL}(r, \mathbb{C})$. (More generally, like $G_A \subset \text{GL}(N, \mathbb{C})$, the group G_B is identified with the block diagonal subgroup $G_B := \text{GL}(l_1, \mathbb{C}) \times \cdots \times \text{GL}(l_p, \mathbb{C}) \subset \text{GL}(r, \mathbb{C})$, where $\{l_1, \ldots, l_p\}$ are the multiplicities of the eigenvalues of B.) We make this restriction henceforth, and also assume $N \geq r$. From its definition, the quotient map J_B^A intertwines the G_B action on $\mathcal{M}$ with the conjugation action of $G_B \subset \text{GL}(r, \mathbb{C})$ on $\widetilde{\mathfrak{gl}}_{\text{rat}}(r)$. Since the elements of the ring $\mathcal{I}$ are invariant under this action, they project to a Poisson commuting ring on the double quotient $G_B \backslash \mathcal{M} / G_A$.

It is natural to now ask what happens if we interchange the rôles of the matrices A and B and the corresponding groups G_A and G_B. We may consider the space $\widetilde{\mathfrak{gl}}_{\text{rat}}(N)$ consisting of $N \times N$ matrices $Y(z)$ depending rationally on an auxiliary complex variable z, with the rational R-matrix structure (1–12) (with the replacements $r \to N$, $\lambda \to z$, $X \to Y$). Restricting analogously to the Poisson subspace consisting of elements of the form

$$\mathcal{M}(z) = A + \sum_{a=1}^{r} \frac{M_a}{z - \beta_a}, \qquad M_a \in \mathfrak{gl}(N, \mathbb{C}), \tag{1-24}$$

we must, consistently with our assumption that the matrix B has a simple spectrum, require the residue matrices $\{M_1, \ldots, M_r\}$ to all have rank one. We may now repeat the entire Hamiltonian quotienting process as above by defining a Poisson map $J_A^B : \mathcal{M} \to \widetilde{\mathfrak{gl}}_{\text{rat}}(N)$ as in (1-21):

$$J_A^B : (F, G) \mapsto \mathcal{N}(\lambda) = A + F(B - z\mathbf{I})^{-1} G^T. \tag{1-25}$$

The fibres of this map are the orbits of the free Hamiltonian G_B-action (1-23), so we may identify the quotient space (which we express as a left quotient) $G_B \backslash \mathcal{M}$ simultaneously as the image of the map J_A^B and as the Poisson subspace of $\widetilde{\mathfrak{gl}}_{\text{rat}}(N)$ consisting of elements of the form (1-24) with rank one residue matrices at the poles $z = \beta_a$. Again, the Poisson map J_A^B intertwines the G_A action on $\mathcal{M}$ with the conjugation action of $G_A \subset \text{GL}(N, \mathbb{C})$ on $\widetilde{\mathfrak{gl}}_{\text{rat}}(N)$. We may define the ring $\tilde{\mathcal{I}}$ to consist of the spectral invariant polynomial functions formed from the $\mathcal{M}(z)$'s (i.e., generated by their characteristic polynomials), and obtain the Poisson commutative ring $\mathcal{I}_A^B := J_A^{B*}(\tilde{\mathcal{I}})$ by pulling back the elements of $\tilde{\mathcal{I}}$ under the map J_A^B. Defining the projections $\pi_A : \widetilde{\mathfrak{gl}}_{\text{rat}}(r) \to \widetilde{\mathfrak{gl}}_{\text{rat}}(r)/G_A$ and $\pi_B : \widetilde{\mathfrak{gl}}_{\text{rat}}(N) \to G_B \backslash \widetilde{\mathfrak{gl}}_{\text{rat}}(N)$ to the quotient space under the respective conjugation actions, we see that the composite maps $\pi_A \circ J_B^A$ and $\pi_B \circ J_A^B$ coincide, defining the projection from $\mathcal{M}$ to the double quotient $G_B \backslash \mathcal{M} / G_A$.

We can now consider the analog of the overdetermined system (1-5)

$$\widetilde{\mathcal{D}}_\lambda \widetilde{\Psi} = 0, \quad \widetilde{\mathcal{D}}_{\alpha_i} \widetilde{\Psi} = 0, \quad \widetilde{\mathcal{D}}_{\beta_a} \widetilde{\Psi} = 0 \tag{1-26}$$

with respect to the operators $\widetilde{\mathcal{D}}_\lambda$, $\widetilde{\mathcal{D}}_{\alpha_i}$, and $\widetilde{\mathcal{D}}_{\beta_a}$ defined by

$$\left. \begin{aligned} \widetilde{\mathcal{D}}_z &:= \frac{\partial}{\partial z} - \mathcal{M}(z), \\ \widetilde{\mathcal{D}}_{\beta_a} &:= \frac{\partial}{\partial \alpha_i} - \widetilde{V}_a, \\ \widetilde{\mathcal{D}}_{\alpha_i} &:= \frac{\partial}{\partial \beta_a} - \widetilde{U}_i, \\ \widetilde{V}_a &:= -\frac{M_a}{z - \beta_a}, \quad a = 1, \ldots, r, \\ \widetilde{U}_i &:= zE_i + \sum_{\substack{j=1 \\ j \neq i}}^{r} \frac{E_i \left(\sum_{b=1}^{r} M_b\right) E_j + E_j \left(\sum_{b=1}^{n} M_b\right) E_i}{\alpha_i - \alpha_j}, \quad i = 1, \ldots, n. \end{aligned} \right\} \tag{1-27}$$

The Frobenius integrable set of compatibility conditions for this system are again given by the commutativity of the operators $\widetilde{\mathcal{D}}_\lambda$, $\widetilde{\mathcal{D}}_{\alpha_i}$, and $\widetilde{\mathcal{D}}_{\beta_a}$, and these may again be viewed as nonautonomous Hamiltonian systems either in the quotient space $G_B\backslash\mathcal{M}$ or lifted to $\mathcal{M}$. The Hamiltonians in the ring $\widetilde{\mathcal{J}}$ corresponding to the α_i deformations and the β_a deformations are given respectively by

$$\widetilde{H}_i = \tfrac{1}{2}\operatorname{res}_{z=\infty}\left(\operatorname{res}_{\lambda=\alpha_i} z[(A-\lambda I)^{-1}\mathcal{M}(z)]^2 - 2\operatorname{tr}[(A-\lambda I)^{-1}\mathcal{M}(z)]\right),$$
$$\widetilde{K}_a = \tfrac{1}{2}\operatorname{res}_{z=\beta_a}\operatorname{tr}(\mathcal{M}^2(z)).$$

(1–28)

The main result relating these systems to the ones introduced in the previous subsection is contained in the following theorem.

THEOREM 1.1 [Harnad 1994]. *The two Poisson commuting rings $\mathcal{J}_B^A$ and $\mathcal{J}_A^B$ coincide, and, in particular, we have the equalities*

$$J_B^{A*}(H_i) = J_A^{B*}(\widetilde{H}_i), \qquad J_B^{A*}(K_a) = J_A^{B*}(\widetilde{K}_a), \qquad i=1,\ldots,n,\ a=1,\ldots,r.$$

Therefore, the lifted systems in $\mathcal{M}$ coincide, as do the projected systems in $G_B\backslash\mathcal{M}/G_A$.

2. The Riemann–Hilbert Problem and Integrable Fredholm Operators

In this section, we show how a class of solutions to the isomonodromic deformation equations considered here result from the solution of a particular type of matrix Riemann–Hilbert problem, and how the corresponding τ-function may be identified as the Fredholm determinant of a special class of "integrable" integral operators. In this, we follow the general approach developed in [Its et al. 1990]. The results presented here are based on joint work with Alexander Its, and are presented in greater detail in [Harnad and Its 1997], where further developments may also be found.

2a. Riemann–Hilbert Problem. Let Γ be an oriented curve in the complex λ-plane passing sequentially through the points $\{\alpha_1,\ldots,\alpha_n\}$. In the following, we take $n=2m$ to be even, (although we may also let it be odd by considering ∞ as the last point). Denote by Γ_j be the segment of Γ from α_{2j-1} to α_{2j}. Now choose a set of constant pairs of maximal rank matrices $\{(f_j, g_j)\}_{j=1,\ldots,m}$ of dimensions $\{r\times k_j\}_{j=1,\ldots,m}$, where $k_j \leq r$, satisfying the orthogonality conditions:

$$g_j^T f_j = 0, \quad j=1,\ldots,m. \tag{2-1}$$

Let $\theta_j(\lambda)$ denote the characteristic function for the segment G_j, viewed as a function defined along Γ, and define the piecewise constant functions

$$f_0(\lambda) := \sum_{j=1}^m f_j \theta_j(\lambda), \quad g_0(\lambda) := \sum_{j=1}^m g_j \theta_j(\lambda), \tag{2-2}$$

supported on $\cup_{j=1}^{m}\Gamma_j$. Now let $\Psi_0(\lambda)$ denote the exponential "vacuum" isomonodromic solution

$$\Psi_0(\lambda) := e^{\lambda B}$$

satisfying

$$\frac{\partial \Psi_0}{\partial \lambda} = B\Psi_0, \quad \frac{\partial \Psi_0}{\partial \alpha_i} = 0, \quad \frac{\partial \Psi_0}{\partial \beta_a} = \lambda E_a \Psi_0,$$

and let

$$f(\lambda) := \Psi_0(\lambda) f_0(\lambda), \qquad g(\lambda) := (\Psi_0^T(\lambda))^{-1} g_0(\lambda). \tag{2-3}$$

Define a $GL(r,\mathbb{C})$-valued piecewise continuous exponential function along the curve Γ by

$$H(\lambda) := \Psi_0(\lambda) H_0(\lambda) \Psi_0^{-1}(\lambda) = \boldsymbol{I} + 2\pi f(\lambda) g^T(\lambda),$$

where $H_0(\lambda)$ is the piecewise constant $GL(r,\mathbb{C})$-valued function

$$H_0(\lambda) := \boldsymbol{I} + 2\pi f_0(\lambda) g_0^T(\lambda).$$

In terms of these quantities, we pose the following Riemann–Hilbert problem: Find a $GL(r,\mathbb{C})$-valued function $X(\lambda)$ that is holomorphic on the complement $\mathbb{C}\setminus\Gamma$ of the curve Γ and at $\lambda = \infty$, with the following asymptotic form near $\lambda = \infty$,

$$X(\lambda) = \boldsymbol{I} + O(\lambda^{-1}),$$

and such that the limits $X_\pm(\lambda)$ of the values of $X(\lambda)$ when approaching the curve Γ from the left (+) and right (−) are related by

$$X_+(\lambda) = X_-(\lambda) H(\lambda).$$

Moreover, we require that the local behaviour of the singularity in $X(\lambda)$ in a neighborhood of any of the points $\{\alpha_1, \ldots, \alpha_n\}$ should be just logarithmic. (Its uniqueness follows from the analyticity conditions imposed.) Define

$$\Psi(\lambda) := X(\lambda) \Psi_0(\lambda). \tag{2-4}$$

Because of the orthogonality conditions (2–1), the limits

$$\begin{aligned} F_i &:= \lim_{\lambda \to \alpha_i} (X(\lambda) f(\lambda))^T, \\ G_i &:= (-1)^j \lim_{\lambda \to \alpha_i} (X(\lambda) g(\lambda))^T \end{aligned} \tag{2-5}$$

(taken within the curve segments Γ_j) exist. Now define the pair (F,G) entering in (1–21) as in (1–18), and the residue matrices in $\mathcal{N}(\lambda)$ by (1–17). Then:

THEOREM 2.1 [Harnad and Its 1997]. *The matrix-valued function* $\Psi(\lambda, a_1, \ldots, n, \beta_1, \ldots, \beta_r)$ *defined in* (2–4) *satisfies the linear equations* (1–5) *with the matrix* $\mathcal{N}(\lambda)$ *and its residues* N_i *determined through* (1–17) *in terms of the matrices* $\{F_i, G_i\}_{i=1,\ldots,n}$ *defined in* (2–5). *The corresponding pairs* $(F,G) \in \mathcal{M}$ *satisfy*

Hamilton's equations (1–22). The local behaviour of the function $\Psi(\lambda)$ in a neighborhood of the curve segment Γ_j is of the form

$$\Psi(\lambda) \sim \Psi_{\mathrm{an}}^j(\lambda) \left(\frac{\lambda - \alpha_{2j-1}}{\lambda - \alpha_{2j}} \right)^{f_j g_j^T}, \tag{2-6}$$

where $\Psi_{\mathrm{an}}^j(\lambda)$ is analytic in this neighborhood. Therefore the monodromy representation is generated by the following matrices $\{M_i\}_{i=1,\ldots,2m}$, corresponding to simple positively oriented loops from an arbitrary base point λ_0 going once around the singular points $\{a_i\}_{i=1,\ldots,2m}$:

$$M_{2j-1} = \exp(2\pi i f_j g_j^T) = \boldsymbol{I} + 2\pi i f_j g_j^T,$$
$$M_{2j} = \exp(-2\pi i f_j g_j^T) = \boldsymbol{I} - 2\pi i f_j g_j^T, \qquad j = 1,\ldots,m.$$

(*There is no monodromy at* $\lambda = \infty$, *and the Stokes matrices are just the identity element.*)

The proof of this result is elementary; the local behaviour (2–6) follows from the conditions of the associated Riemann–Hilbert problem, and the differential equations (1–5) follow from explicit differentiation to obtain the local pole structure in $(\partial \Psi/\partial \lambda)\Psi^{-1}$, and application of Liouville's theorem. (Of course, the actual solution of the Riemann–Hilbert problem is highly nontrivial.)

2b. The Fredholm Determinant. We now choose all the ranks $\{k_i\}_{i=1,\ldots,n}$ equal to k, and define a matrix Fredholm integral operator $\boldsymbol{K} : L^2(\Gamma, \mathbb{C}^k) \to L^2(\Gamma, \mathbb{C}^k)$ by

$$\boldsymbol{K}(v)(\lambda) = \int_\Gamma K(\lambda, \mu) v(\mu) d\mu, \quad v \in L^2(\Gamma, \mathbb{C}^k),$$

where the kernel is chosen to have the special form

$$K(\lambda, \mu) = \frac{f^T(\lambda) g(\mu)}{\lambda - \mu},$$

with $f(\lambda)$ and $g(\lambda)$ defined in (2–3). (For the case $r = 2$, $k = 1$, and $\beta_1 = -\beta_2$, this is just the sine kernel occurring in the computation of spectral distributions for random matrices in the GUE [Tracy and Widom 1993; 1994]. Because of the orthogonality conditions (2–1), we have

$$f^T(\lambda) g(\lambda) = 0,$$

and hence the kernel is nonsingular, with diagonal values

$$K(\lambda, \lambda) = f'^T(\lambda) g(\lambda) = -f^T(\lambda) g'(\lambda).$$

The main result relating the Riemann–Hilbert problem discussed above with this Fredholm operator (see [Its et al. 1990]) is that its solution is equivalent to the determination of the resolvent operator

$$\boldsymbol{R} := (\boldsymbol{I} - \boldsymbol{K})^{-1} \boldsymbol{K}.$$

Specifically:

LEMMA 2.2. *The resolvent operator also has the special form*
$$\boldsymbol{R}(\boldsymbol{v})(\lambda) = \int_\Gamma R(\lambda,\mu)\boldsymbol{v}(\mu)d\mu$$
where the kernel is
$$R(\lambda,\mu) := \frac{F^T(\lambda)G(\mu)}{\lambda-\mu}.$$
with $F(\lambda)$, $G(\lambda)$ given by
$$F(\lambda) = X(\lambda)f(\lambda), \quad G(\lambda) = (X^T)^{-1}(\lambda)g(\lambda).$$
Conversely, the matrix $X(\lambda)$ solving the above Riemann–Hilbert problem is given by the integral formula
$$X(\lambda) = \boldsymbol{I}_r + \int_\Gamma \frac{F(\mu)g^T(\mu)}{\lambda-\mu}d\mu.$$

This result follows directly from the Cauchy integral representation for $X(\lambda)$, given its specified analytic properties [Its et al. 1990; Harnad and Its 1997]. Using it, we find a remarkable deformation formula for the Fredholm determinant $\det(\boldsymbol{I}-\boldsymbol{K})$:

THEOREM 2.3.
$$d\ln\det(\boldsymbol{I}-\boldsymbol{K}) = \sum_{k=1}^n H_k\,d\alpha_k + \sum_{a=1}^r K_a\,d\beta_a,$$
where the differential is understood as taken with respect to the deformation parameters $\{\alpha_1,\ldots,\alpha_n\}$ and $\{\beta_1,\ldots,\beta_r\}$, and the coefficients $\{H_i\}_{i=1,\ldots,n}$ and $\{K_a\}_{a=1,\ldots,r}$ are given by the formulae (1–7) defining the Hamiltonians generating the isomonodromic deformations, with the residue matrices $\{N_i\}_{i=1,\ldots,n}$ given in terms of the matrices $\{F_i,G_i\}_{i=1,\ldots,n}$ defined in (2–5) by (1–17).

The proof of this result is given in [Harnad and Its 1997]. It implies that the Fredholm determinant $\det(\boldsymbol{I}-\boldsymbol{K})$ may be identified with the isomonodromic τ-function defined by formulae (1–10) and (1–11).

2c. The Dual Riemann–Hilbert Problem. The results of the previous subsection can of course be repeated with the rôles of the matrices A and B interchanged. For this, we assume that $r = 2s$ is even, the multiplicities k_i of the eigenvalues of A are all equal to 1, so $n = N$, and we choose the eigenvalues $\{\beta_a\}_{a=1,\ldots,2s}$ of the matrix B to all have the same multiplicity $l \leq n$ (the analog of k above), so the matrix B is now of dimension $2ls \times 2ls$. We also choose a set of $2s$ pairs $\{\tilde{f}_a,\tilde{g}_a\}_{a=1\ldots,s}$ of fixed maximal rank matrices of dimension $n \times l$ satisfying the orthogonality conditions
$$\tilde{f}_a^T \tilde{g}_b = 0, \quad a,b = 1,\ldots,s.$$

As before, we choose an oriented simple curve $\widetilde{\Gamma}$ in the complex z-plane passing sequentially through the points $\{\beta_1, \ldots, \beta_{2s}\}$, and denote the segments from β_{2a-1} to β_{2a} by $\widetilde{\Gamma}_a$. Now let

$$\tilde{f}(z) = e^{zA} \sum_{a=1}^{s} \tilde{f}_a \theta_a(z), \quad \tilde{g}(z) = e^{-zA} \sum_{a=1}^{s} \tilde{g}_a \theta_a(z), \tag{2-7}$$

where θ_a is now the characteristic function of the curve segment $\widetilde{\Gamma}_a$. As above, we associate a Riemann–Hilbert problem to this data, consisting of finding an $n \times n$ matrix valued function $\widetilde{X}(z)$ that is nonsingular and holomorphic on the complement of $\widetilde{\Gamma}$, with asymptotic form near $z = \infty$

$$\widetilde{X}(z) = \boldsymbol{I} + O(z^{-1}),$$

and discontinuities along $\widetilde{\Gamma}$ supported on the segments $\widetilde{\Gamma}_a$ defined by

$$\widetilde{X}_-(z) = \widetilde{X}_+(z) \widetilde{H}(z), \quad z \in \widetilde{\Gamma},$$

where

$$\widetilde{H}(z) = \boldsymbol{I} + 2\pi i \tilde{f}(z) \tilde{g}^T(z) = \exp 2\pi i \tilde{f}(z) \tilde{g}^T(z).$$

Once again, the solution of this Riemann–Hilbert problem allows us to define a matrix valued function

$$\widetilde{\Psi}(z) := \widetilde{X}(z) e^{zA}$$

that satisfies the linear system (1–26), where the operators (1–27) are determined in terms of the residue matrices M_a of the matrix $\mathcal{M}(z)$ at the points β_a by:

$$M_a := -\widetilde{F}_a \widetilde{G}_a^T,$$

with

$$\widetilde{F}_a := \lim_{z \to \beta_a} (\widetilde{X}(z) \tilde{f}(z)), \quad \widetilde{G}_a := (-1)^a \lim_{z \to \beta_a} (\widetilde{X}(z) \tilde{g}(z)).$$

As above, we may define a Fredholm integral operator along the curve acting on $\mathbb{C}^q$-valued functions $\tilde{v}$ on $\widetilde{\Gamma}$ by

$$\boldsymbol{\widetilde{K}}(\tilde{v})(z) = \int_{\widetilde{\Gamma}} \widetilde{K}(z, w) \tilde{v}(w) \, dw,$$

where

$$\widetilde{\boldsymbol{K}}(z, w) = \frac{\tilde{f}^T(z) \tilde{g}(w)}{z - w}.$$

The corresponding resolvent operator

$$\widetilde{\boldsymbol{R}} = (\boldsymbol{I} - \widetilde{\boldsymbol{K}})^{-1} \widetilde{\boldsymbol{K}}$$

again has the form

$$\widetilde{\boldsymbol{R}}(\tilde{v})(z) = \int_{\widetilde{\Gamma}} \widetilde{R}(z, w) \tilde{v}(w),$$

where the kernel

$$\widetilde{R}(z, w) = \frac{\widetilde{F}^T(z) \widetilde{G}(w)}{z - w}$$

is given by
$$\widetilde{F}^T(z) = \widetilde{X}(z)\tilde{f}(z), \quad \widetilde{G}(z) = (\widetilde{X}^T(z))^{-1}\tilde{g}(z).$$

As before, the deformation formula for the Fredholm determinant gives
$$d\ln\det(1 - \widetilde{\boldsymbol{K}}) = d\ln\widetilde{\tau} = \sum_{j=1}^{n} \widetilde{H}_j\, d\alpha_j + \sum_{a=1}^{r} \widetilde{K}_a\, d\beta_a,$$

where the coefficients are given by the formulae (1–28).

Defining the pair of $n \times rl$ matrices $(\widetilde{F}, \widetilde{G})$ formed from the blocks $\{\widetilde{F}_a, \widetilde{G}_a\}$
$$\begin{aligned}\widetilde{F} &:= (\widetilde{F}_1 \quad \cdots \quad \widetilde{F}_a \quad \cdots \quad \widetilde{F}_r), \\ \widetilde{G} &:= (\widetilde{G}_1 \quad \cdots \quad \widetilde{G}_a \quad \cdots \quad \widetilde{G}_r),\end{aligned} \tag{2-8}$$

we may express the matrix $\mathcal{M}(z)$ as
$$\mathcal{M}(z) :== A + \widetilde{F}(B - z\boldsymbol{I})^{-1}\widetilde{G}^T.$$

Returning to the special case $l = 1$, we may ask whether there are choices of the matrices $\{(f_i, g_i)\}_{i=1,\ldots,m}$ and $\{(\tilde{f}_a, \tilde{g}_a)\}_{a=1,\ldots,s}$ defining the respective Riemann–Hilbert problems for which the $n \times r$ pairs of matrices (F, G) and $\widetilde{F}, \widetilde{G})$ defined by (1–18) and (2–8) coincide, defining the same solution to the respective dual Hamiltonian systems and dual isomonodromic deformation equations.

The following provides a particular answer to this question. Choosing $k = q = 1$, the (f_i, g_i)'s and $(\tilde{f}_a, \tilde{g}_a)$'s become pairs of r-component and n-component column vectors, respectively. We pick a fixed $m \times s$ matrix with elements $\{c_{ja}\}_{j=1,\ldots,m,\ a=1,\ldots,s}$ and choose the components of these vectors to be
$$(f_i)_{2a} = (f_i)_{2a-1} = (\tilde{f}_a)_{2i} = (\tilde{f}_a)_{2i-1} = 1,$$
$$(\tilde{g}_i)_{2a} = -(\tilde{g}_i)_{2a-1} = (\tilde{g}_a)_{2j} = -(\tilde{g}_a)_{2j-1} =: c_{ij},$$

for $i = 1,\ldots,m$ and $a = 1,\ldots,s$. Now define, on the product $\Gamma \times \widetilde{\Gamma}$, the locally constant function
$$\hat{K}(\lambda, z) := \sum_{j=1}^{m}\sum_{a=1}^{s} c_{ja}\theta_j(\lambda)\tilde{\theta}_a(z).$$

Taking the Fourier–Laplace transform with respect to the variables z and λ along the curves $\widetilde{\Gamma}$ and Γ, respectively, gives the two Fredholm kernels
$$K(\lambda, \mu) = \int_{\widetilde{\Gamma}} \hat{K}(\mu, z)e^{z(\lambda-\mu)}dz = \frac{f^T(\lambda)g(\mu)}{\lambda - \mu},$$
$$\widetilde{K}(w, z) = \int_{\Gamma} \hat{K}(\mu, z)e^{\mu(w-z)}d\mu = \frac{\tilde{f}^T(w)\tilde{g}(z)}{w - z},$$

where $f(\lambda)$, $g(\lambda)$, $\tilde{f}(z)$, $\tilde{g}(z)$ define the Riemann–Hilbert data for these choices as in (2-2), (2-3), and (2-7). Then:

THEOREM 2.4. *The Fredholm determinants of these two operators are equal and so are the matrix pairs* (F, G) *and* $(\widetilde{F}, \widetilde{G})$ *constructed from the associated Riemann–Hilbert data.*

The proof is based on a straightforward application of the Neumann expansion for the resolvent and may be found in [Harnad and Its 1997]. In that work, further results are presented extending the above analysis to more general classes of isomonodromic deformation problems, corresponding to polynomial asymptotic terms in the matrix $\mathcal{N}(\lambda)$, as well as symplectic reductions by discrete symmetries. (Further cases corresponding to higher order poles in $\mathcal{N}(\lambda)$ and applications may be found in [Harnad and Routhier 1995; Harnad and Wisse 1996; Harnad et al. 1993].) The τ-functions associated with the special data discussed above in relation to the dual Fredholm operators $\boldsymbol{K}$ and $\widetilde{\boldsymbol{K}}$ are also shown in [Harnad and Its 1997], to be interpretable in a manner similar to multi-component KP τ-functions, as determinants of projection operators over suitably defined infinite dimensional Grassmannians. The interchange of data underlying the duality is then seen as an interchange of the rôles of the data determining the initial point W in the Grassmannian and the abelian group elements determining the flow.

Acknowledgements

This research was supported in part by the Natural Sciences and Engineering Research Council of Canada, the Fonds FCAR du Québec and the National Science Foundation, grant DMS-9501559. The author would like to thank the organizers of the Jan-June 1999 MSRI program on Random Matrices and their Applications, P. Bleher, A. Edelman, A. Its, C. Tracy, and H. Widom for the kind invitation to spend this period at the MSRI and to take part in this very stimulating program, as well MSRI's director D. Eisenbud, and all the staff there, for providing such a welcoming and agreeable environment.

References

[Harnad 1994] J. Harnad, "Dual isomonodromic deformations and moment maps to loop algebras", *Comm. Math. Phys.* **166**:2 (1994), 337–365.

[Harnad and Its 1997] J. Harnad and A. R. Its, "Integrable Fredholm operators and Dual Isomonodromic Deformations", preprint CRM-2477, Centre de recherches mathématiques, Montreal, 1997. See http://www.arxiv.org/abs/solv-int/9706002.

[Harnad and Routhier 1995] J. Harnad and M. Routhier, "*R*-matrix construction of electromagnetic models for the Painlevé transcendents", *J. Math. Phys.* **36**:9 (1995), 4863–4881.

[Harnad and Wisse 1996] J. Harnad and M. A. Wisse, "Loop algebra moment maps and Hamiltonian models for the Painlevé transcendants", pp. 155–169 in *Mechanics day* (Waterloo, ON, 1992), edited by P. S. Krishnaprasad et al., Fields Inst. Commun. **7**, Amer. Math. Soc., Providence, RI, 1996.

[Harnad et al. 1993] J. Harnad, C. A. Tracy, and H. Widom, "Hamiltonian structure of equations appearing in random matrices", pp. 231–245 in *Low-dimensional topology and quantum field theory* (Cambridge, 1992), edited by H. Osborn, Plenum, New York, 1993.

[Its et al. 1990] A. R. Its, A. G. Izergin, V. E. Korepin, and N. A. Slavnov, "Differential equations for quantum correlation functions", *Internat. J. Modern Phys. B* **4**:5 (1990), 1003–1037.

[Jimbo et al. 1980] M. Jimbo, T. Miwa, Y. Môri, and M. Sato, "Density matrix of an impenetrable Bose gas and the fifth Painlevé transcendent", *Phys. D* **1**:1 (1980), 80–158.

[Jimbo et al. 1981] M. Jimbo, T. Miwa, and K. Ueno, "Monodromy preserving deformation of linear ordinary differential equations with rational coefficients. I. General theory and τ-function", *Phys. D* **2**:2 (1981), 306–352.

[Tracy and Widom 1993] C. A. Tracy and H. Widom, "Introduction to random matrices", pp. 103–130 in *Geometric and quantum aspects of integrable systems* (Scheveningen, 1992), edited by G. F. Helminck, Lecture Notes in Physics **424**, Springer, Berlin, 1993.

[Tracy and Widom 1994] C. A. Tracy and H. Widom, "Fredholm determinants, differential equations and matrix models", *Comm. Math. Phys.* **163**:1 (1994), 33–72.

J. HARNAD

DEPARTMENT OF MATHEMATICS AND STATISTICS
CONCORDIA UNIVERSITY
7141 SHERBROOKE W.
MONTRÉAL, QUÉBEC H4B 1R6
CANADA

CENTRE DE RECHERCHES MATHÉMATIQUES
UNIVERSITÉ DE MONTRÉAL
C. P. 6128, SUCC. CENTRE VILLE
MONTRÉAL, QUÉBEC, H3C 3J7
CANADA
harnad@crm.umontreal.ca

Functional Equations and Electrostatic Models for Orthogonal Polynomials

MOURAD E. H. ISMAIL

To my dear friend Walter K. Hayman on the occasion of his 75th birthday.

ABSTRACT. This article deals with connections between orthogonal polynomials, functional equations they satisfy, and some extremal problems. We state Stieltjes electrostatic models and Dyson's Coulomb fluid method. We also mention the evaluation of the discriminant of Jacobi polynomials by Stieltjes and Hilbert. We show how these problems can be extended to general orthogonal polynomials with absolutely continuous measures or having purely discrete orthogonality measures whose masses are located at at most two sequences of geometric progressions.

1. Introduction

This is a survey article dealing with connections between orthogonal polynomials, functional equations they satisfy, and some extremal problems. Although the results surveyed are not new we believe that we are putting together results from different sources which appear together for the first time, many of them are of recent vintage.

One question in the theory of orthogonal polynomials is how the zeros of a parameter dependent sequence of orthogonal polynomials change with the parameters involved. Stieltjes [1885a; 1885b] proved that the zeros of Jacobi polynomials $P_N^{(\alpha,\beta)}(x)$ increase with β and decrease with α for $a > -1$ and $\beta > -1$. The Jacobi polynomials satisfy the following orthogonality relation [Szegő 1975, (4.3.3)]:

1991 *Mathematics Subject Classification.* Primary 42C05; Secondary 33C45.

Key words and phrases. Discriminants, orthogonal polynomials, zeros, interacting particles.

Research partially supported by NSF grant DMS-99-70865.

$$\int_{-1}^{1}(1-x)^\alpha(1+x)^\beta P_m^{(\alpha,\beta)}(x)P_n^{(\alpha,\beta)}(x)\,dx$$
$$= \frac{2^{\alpha+\beta+1}}{2n+\alpha+\beta+1}\frac{\Gamma(n+\alpha+1)\Gamma(n+\beta+1)}{\Gamma(n+1)\Gamma(n+\alpha+\beta+1)}\delta_{m,n}. \quad (1\text{-}1)$$

In Section 2 we state Stieltjes's results and describe the circle of ideas around them. Section 3 surveys the Coulomb Fluid method of Freeman Dyson and its potential theoretic set-up.

We shall use the shifted factorial notion [Andrews et al. 1999]

$$(\lambda)_n = \prod_{k=1}^{n}(\lambda+k-1),$$

and the hypergeometric notation [Andrews et al. 1999]

$$_rF_s\left(\begin{matrix}a_1,\ldots,a_r\\b_1,\ldots,b_s\end{matrix}\bigg|z\right) = {}_rF_s(a_1,\ldots,a_r;b_1,\ldots,b_s;z)$$
$$= \sum_{n=0}^{\infty}\frac{(a_1)_n\ldots(a_r)_n}{(b_1)_n\ldots(b_s)_n}\frac{z^n}{n!}.$$

The Jacobi polynomials have the explicit form

$$P_n^{(\alpha,\beta)}(x) = \frac{(\alpha+1)_n}{n!}{}_2F_1\left(\begin{matrix}-n,\,n+\alpha+\beta+1\\\alpha+1\end{matrix}\bigg|\frac{1-x}{2}\right). \quad (1\text{-}2)$$

The discriminant of a polynomial g_n,

$$g_n(x) := \gamma_n x^n + \text{ lower order terms}, \quad \gamma_n \ne 0, \quad (1\text{-}3)$$

is defined by

$$D(g_n) := \gamma_n^{2n-2}\prod_{1\le j<k\le n}(x_j-x_k)^2, \quad (1\text{-}4)$$

where $x_1, x_2, \ldots, x_n$ are the zeros of $g_n(x)$; see [Dickson 1939]. The discriminant has the alternate representation

$$D(g_n) = (-1)^{n(n-1)/2}\gamma_n^{n-2}\prod_{j=1}^{n}g_n'(x_j). \quad (1\text{-}5)$$

See, for example, Dickson [1939, § 100].

2. Stieltjes

Stieltjes [1885a; 1885b] considered the following electrostatic model. Fix two charges $(\alpha+1)/2$ and $(\beta+1)/2$ at $x=1$ and $x=-1$, respectively, then put N movable unit charges at distinct points in $(-1,1)$. The potential here is a logarithmic potential so the potential energy of a system of two charges e_1 and

e_2 located at x and y is $-2e_1 e_2 \ln|x-y|$. Let the position of the unit charges be at $x_1, x_2, \ldots, x_N$. Thus the energy of this system is

$$E_N(\boldsymbol{x}) = -(\alpha+1)\sum_{k=1}^{N} \ln|1-x_k| - (\beta+1)\sum_{k=1}^{N} \ln|1+x_k|$$
$$-2\sum_{1\leq j<k\leq N} \ln|x_j - x_k|,$$

where

$$\boldsymbol{x} = (x_1, x_2, \ldots, x_N). \qquad (2\text{--}1)$$

For convenience we consider the function

$$T_N(\boldsymbol{x}) := \exp(-E_N(\boldsymbol{x})),$$

that is,

$$T_N(\boldsymbol{x}) = \prod_{k=1}^{N}(1-x_k)^{\alpha+1}(1+x_k)^{\beta+1} \prod_{1\leq i<j\leq N}[x_i - x_j]^2. \qquad (2\text{--}2)$$

The equilibrium position of the system occurs at the points which minimize $E_N(\boldsymbol{x})$ or equivalently maximize $T_N(\boldsymbol{x})$.

THEOREM 2.1 (STIELTJES). *The maximum of the function $T_N(\boldsymbol{x})$ taken over $\boldsymbol{x} \in \mathbb{R}^N$ is attained when $\boldsymbol{x}$ is formed by the zeros of $P_N^{(\alpha,\beta)}(x)$. In other words the equilibrium position of the movable charges in the electrostatic model described above is attained at the zeros of the Jacobi polynomial $P_N^{(\alpha,\beta)}(x)$.*

In Section 4, we will give a proof of a generalization of Theorem 2.1.

Stieltjes found a closed form expression for the maximum value of $T_N(\boldsymbol{x})$ in Theorem 2.1. He observed that (1–2) implies

$$P_N^{(\alpha,\beta)}(x) = (N+\alpha+\beta+1)_N \frac{2^{-N}}{N!} x^N + \text{lower order terms},$$

so that if $x_{1,N} > x_{2,N} > \cdots > x_{N,N}$ are the zeros of $P_N^{(\alpha,\beta)}(x)$ then

$$\prod_{k=1}^{N}(\pm x - x_{k,N}) = \frac{2^N N!}{(N+\alpha+\beta+1)_N} P_N^{(\alpha,\beta)}(\pm x),$$

so that at equilibrium the first product in (2–2) can be found from

$$\prod_{k=1}^{N}(1-x_{j,N}) = \frac{2^N N!}{(N+\alpha+\beta+1)_N} P_N^{(\alpha,\beta)}(1)$$

$$\prod_{k=1}^{N}(1+x_{j,N}) = \frac{(-1)^N 2^N N!}{(N+\alpha+\beta+1)_N} P_N^{(\alpha,\beta)}(-1).$$

Clearly (1–2) implies

$$P_N^{(\alpha,\beta)}(1) = (\alpha+1)_N/N!.$$

Since the weight function and the right-hand side in (1–1) are symmetric under the exchange $(x,\alpha,\beta) \to (-x,\beta,\alpha)$ then

$$P_N^{(\alpha,\beta)}(-1) = (-1)^N P_N^{(\beta,\alpha)}(1) = (-1)^N \frac{(\beta+1)_N}{N!}.$$

The second product in (2–2) at equilibrium is the discriminant of Jacobi polynomials. Stieltjes then found explicit formulas for the discriminants of the Jacobi polynomials. His formula in our notation is

$$D(P_N^{(\beta,\alpha)}) = 2^{-N(N-1)} \prod_{k=1}^{N} k^{k+2-2N}(k+\alpha)^{k-1}(k+\beta)^{k-1}(N+k+\alpha+\beta)^{N-k}. \tag{2-3}$$

Since the Hermite and Laguerre polynomials are limiting cases of Jacobi polynomials, (2–3) yields explicit evaluations for the discriminants of the Hermite and Laguerre polynomials. Shortly after Stieltjes work appeared, Hilbert [1888] gave another proof of (2–3). Schur [1931] also gave a very elegant proof. We reproduce the proof here to make the paper as self-contained as possible and also because Schur's method proved to be central in the generalizations of Stieltjes results to general orthogonal polynomials.

LEMMA 2.2 [Schur 1931]. *Assume that $\{\rho_n(x)\}$ is a sequence orthogonal polynomials satisfying a three-term recurrence relation of the form*

$$\rho_{n+1}(x) = (\xi_{n+1}x + \eta_{n+1})\rho_n(x) - \zeta_{n+1}\rho_{n-1}(x), \tag{2-4}$$

and the initial conditions

$$\rho_0(x) = 1, \quad \rho_1(x) = \xi_1 x + \eta_1. \tag{2-5}$$

If

$$x_{1,n} > x_{2,n} > \cdots > x_{n,n} \tag{2-6}$$

are the zeros of $\rho_n(x)$ then

$$\prod_{k=1}^{n} \rho_{n-1}(x_{k,n}) = (-1)^{n(n-1)/2} \prod_{k=1}^{n} \xi_k^{n-2k+1} \zeta_k^{k-1}, \tag{2-7}$$

with $\zeta_1 := 1$.

PROOF. Let Δ_n denote the left-hand side of (2–7). The coefficient of x^n in $\rho_n(x)$ is $\xi_1 \xi_2 \ldots \xi_n$. Thus by expressing ρ_n in terms and ρ_{n+1} of their zeros we find

$$\Delta_{n+1} = (\xi_1 \xi_2 \ldots \xi_n)^{n+1} \prod_{k=1}^{n+1} \prod_{j=1}^{n} (x_{k,n+1} - x_{j,n})$$

$$= (-1)^{n(n+1)} (\xi_1 \xi_2 \ldots \xi_n)^{n+1} \prod_{j=1}^{n} \prod_{k=1}^{n+1} (x_{j,n} - x_{k,n+1})$$

$$= \frac{(\xi_1 \xi_2 \ldots \xi_n)^{n+1}}{(\xi_1 \xi_2 \ldots \xi_{n+1})^n} \prod_{j=1}^{n} \rho_{n+1}(x_{j,n}).$$

On the other hand, the three-term recurrence relation (2–4) simplifies the extreme right-hand side in the above equation and we get

$$\Delta_{n+1} = \xi_1 \xi_2 \ldots \xi_n \, \xi_{n+1}^{-n} (-\zeta_{n+1})^n \Delta_n.$$

By iterating this relation we establish (2–7).

The relevance of Schur's lemma to the evaluation of discriminants of Jacobi polynomials is the fact that the Jacobi polynomials satisfy a lowering (annihilation) relation of the type

$$\frac{d}{dx} P_N^{(\alpha,\beta)}(x) = A_N(x) P_{N-1}^{(\alpha,\beta)}(x) - B_N(x) P_N^{(\alpha,\beta)}(x),$$

with [Rainville 1960, (7), § 136]

$$A_N(x) = \frac{2(N+\alpha)(N+\beta)}{(\alpha+\beta+2N)(1-x^2)}, \quad B_N(x) = \frac{N(\beta-\alpha+(\alpha+\beta+2N)x)}{(\alpha+\beta+2N)(1-x^2)}.$$

Thus

$$\frac{d}{dx} P_N^{(\alpha,\beta)}(x_{k,N}) = A_N(x_{k,N}) P_{N-1}^{(\alpha,\beta)}(x_{k,N})$$

and (1–5) and Schur's lemma lead to the evaluation of the discriminant. □

We now formulate this procedure as a general property of polynomials satisfying three term recurrence relations and possessing a lowering operator.

THEOREM 2.3. *Let a system of polynomials $\{p_n(x)\}$ be generated by (2–4) and (2–5) and assume that the zeros of $p_n(x)$ be arranged as in (2–6). If $\{p_n(x)\}$ satisfies the differential recurrence relation*

$$\frac{d}{dx} p_n(x) = A_n(x) p_{n-1}(x) - B_n(x) p_n(x), \tag{2–8}$$

then the discriminant of $p_n(x)$ is given by

$$D(p_n) = \left(\prod_{j=1}^{n} A_n(x_{j,n}) \right) \prod_{k=1}^{n} \xi_k^{2n-2k-1} \zeta_k^{k-1}.$$

PROOF. The γ_n in (1–3) is $\xi_1\ldots\xi_n$. Thus (1–5) and Lemma 2.2 give

$$D(p_n) = (-1)^{n(n-1)/2}(\xi_1\ldots\xi_n)^{n-2}(-1)^{n(n-1)/2}\prod_{k=1}^{n}A_n(x_{k,n})\xi_k^{n-2k+1}\zeta_k^{k-1},$$

which establishes the theorem. □

It is clear that Theorem 2.3 implies the evaluation (2–3) of the discriminant of the Jacobi polynomials once we know the three-term recurrence relation satisfied by Jacobi polynomials [Rainville 1960, (1) §137]. From [Bauldry 1990; Bonan and Clark 1990; Chen and Ismail 1997], we now know that every polynomial sequence orthogonal with respect to a weight function, satisfying certain smoothness conditions (see §4), satisfies a differential recurrence relation of the type (2–8). Hence, Theorem 2.3 holds for orthogonal polynomials in this generality, a result from [Ismail 1998].

Selberg [1944] proved

$$\int_{[0,1]^n}\left(\prod_{j=1}^{n}t_j^{x-1}(1-t_j)^{y-1}\right)\prod_{1\leq i<k\leq n}|t_i-t_k|^{2z}\,dt_1\ldots dt_n$$
$$=\prod_{j=1}^{n}\frac{\Gamma(x+(n-j)z)\Gamma(y+(n-j)z)\Gamma(jz+1)}{\Gamma(x+y+(2n-j-1)z)\Gamma(z+1)},\quad (2\text{--}9)$$

for $\operatorname{Re} x > 0$, $\operatorname{Re} y > 0$, and $\operatorname{Re} z > -\min\{1/n, \operatorname{Re} x/n - 1, \operatorname{Re} y/(n-1)\}$. Here $[0,1]^n$ is the unit cube in $\mathbb{R}^n$. The integral evaluation (2–9) is the multivariate generalization of the beta integral and is now called the Selberg integral [Andrews et al. 1999]. It is important to note that if we normalize the Jacobi polynomials to be orthogonal on $[0,1]$ then the Stieltjes–Hilbert results provide the L_∞ norm of

$$\left(\prod_{j=1}^{n}t_j^{\alpha}(1-t_j)^{\beta}\right)\prod_{1\leq i<k\leq n}|t_i-t_k|^2,\quad t_j\in(-1,1)\text{ for }1\leq j\leq n.\quad (2\text{--}10)$$

On the other hand, the Selberg integral (2–9) essentially gives the L_p norm of the expression in (2–10). One is then led to view the Stieltjes–Hilbert results as a limiting case of the Selberg integral.

3. Dyson and Potential Theory

As a generalization of the Stieltjes electrostatic problem we consider a system of N logarithmically repelling particles obeying Boltzmann statistics subject to a common external potential $v(x)$ in one dimension. Let $\Phi(\boldsymbol{x})$ denote the total energy of the system, that is

$$\Phi(\boldsymbol{x}) := \sum_{1\leq j\leq N}v(x_j) - 2\sum_{1\leq j<k\leq N}\ln|x_j-x_k|,$$

and x is as in (2–1). Here the particles are assumed to be confined to a real interval K, finite, semi-infinite, or infinite. Dyson's idea [1962a; 1962b; 1962c] was that for large N, one would expect that this collection of particle can be approximated by a continuous fluid where techniques of macroscopic physics such as thermodynamics and electrostatics can be applied. The Coulomb fluid approximation is described by an equilibrium density $\sigma(x)$, supported on a set $J \subset K$, which is obtained by minimizing the free energy functional, $F[\sigma]$

$$F[\sigma] = \int_J v(x)\sigma(x)\,dx - \int_J\int_J \sigma(x)\ln|x-y|\,\sigma(y)\,dy\,dx, \qquad (3\text{--}1)$$

subject to the side condition

$$\int_J \sigma(x)\,dx = N. \qquad (3\text{--}2)$$

From the Euler–Lagrange equations for the system (3–1) it follows that the density $\sigma(x)$ satisfies the singular integral equation

$$v'(x) = 2P\int_J \frac{\sigma(y)}{x-y}\,dy, \quad x \in J. \qquad (3\text{--}3)$$

Therefore

$$A = v(x) - 2\int_J \sigma(y)\ln|x-y|\,dy, \quad x \in J, \qquad (3\text{--}4)$$

where A is the Lagrange multiplier for the constraint (3–2) and is recognized as the chemical potential for the fluid. Determining J is part of the solution of the variational problem.

Let $v(x)$ is convex for $x \in \mathbb{R}$, so that $v''(x) \geq 0$ almost everywhere. We shall assume $v'(x) > 0$ on a set of positive measures. At this stage, some physical considerations are invoked. With the condition $v(x) > 0$ it follows that J is a single interval denoted by (a,b). Intuitively, this physical principle can be understood by using an analogy from elasticity theory [Muskhelishvili 1953], where the fluid density $\sigma(x)$ is identified with the pressure under a stamp pressing vertically downwards against an elastic half-plane. If the applied force is moderate, the end points of the interval, a and b, are the points for which the elastic material come into contact with the rigid stamp. On the other hand if the force applied to the stamp is too great the end points will be fixed as the end points of the boundary of the stamp.

We seek a solution of (3–3) which is nonnegative on (a,b). If imposing the boundary conditions $\sigma(a) = \sigma(b) = 0$ lead to a σ satisfying $\sigma(x) \geq 0$ on (a,b) then, according to the standard theory of singular integral equations [Gakhov 1990], the solution of (3–3) is

$$\sigma(x) = \frac{\sqrt{(b-x)(x-a)}}{2\pi^2}P\int_a^b \frac{v'(x)}{(y-x)\sqrt{(b-y)(y-a)}}\,dy, \qquad (3\text{--}5)$$

and a and b must satisfy the constraint (3–2) as well as the supplementary condition,

$$0 = \int_a^b \frac{v'(x)}{\sqrt{(b-x)(x-a)}}\, dx. \tag{3-6}$$

Using (3–6) the normalization condition becomes

$$N = \frac{1}{2\pi} \int_a^b \frac{x v'(x)}{\sqrt{(b-x)(x-a)}}\, dx. \tag{3-7}$$

The end points of the support of the density, a and b, that are solutions of (3–6) and (3–7) are denoted by $a(N)$ and $b(N)$.

Sometimes the boundary conditions that $\sigma(x)$ vanishes at the endpoints of J do not lead to a solution $\sigma(x)$ which is nonnegative on J. In this case other forms of solutions of (3–4) can be used. A good application of this physical approach is to the Freud weights

$$w(x,\alpha) = \exp(-|x|^\alpha), \quad \alpha > 0, \tag{3-8}$$

so that $v(x) = |x|^\alpha$. Since w is even then $a(N) = -b(N)$. From (3–5) and (3–6) we find that the density, denoted by $\sigma(x,\alpha)$, to be

$$\sigma(x,\alpha) = \frac{\alpha}{\pi} \frac{2^{1-\alpha}\Gamma(\alpha)}{(\Gamma(\alpha/2))^2} (b(N))^{\alpha-2} \sqrt{(b(N))^2 - x^2}\, {}_2F_1\left(1-\alpha/2, 1; \tfrac{3}{2}; 1-(x/b(N))^2\right).$$

The form (3–7) of side condition (3–2) gives

$$N = \frac{2^{1-\alpha} 2^{1-\alpha} \Gamma(\alpha)}{\Gamma^2(\alpha/2)} (b(N))^\alpha.$$

Since the Coulomb fluid approximation works for large N then we find the asymptotic result

$$b(N) \asymp \left(\frac{\Gamma^2(\alpha/2) 2^{\alpha-1} N}{\Gamma(\alpha)}\right)^{1/\alpha}.$$

This gives the large N behavior of the position of the charges on the extreme right ($= b(N)$) and extreme left ($= -b(N)$). Furthermore, motivated by the Stieltjes result, this was believed to be the positions of the largest and smallest zeros of the polynomials orthogonal with respect to $\exp(-|x|^\alpha)$.

What is outlined so far was started by Dyson [1962a; 1962b; 1962c] for the circular ensemble of random matrix theory. It was further developed and applied to orthogonal polynomials by many physicists, too many to be cited in this short article. Having said this, perhaps it is not too impertinent to mention a sample of the work of Yang Chen and his collaborators [Chen et al. 1995; Chen and Manning 1996], especially since my good friend, Yang Chen, is the one who introduced me to this subject, and one of the co-authors of the cited work was a co-organizer of the MSRI program on Random Matrices.

This approach can be made rigorous by using potential theory [Saff and Totik 1997] and Riemann–Hilbert problems starting from the pioneering work of Bleher and Its [1999] and cumulating in the recent series of monumental papers of Deift, Kriecherbauer, McLaughlin, Zhou, and others. First one starts with polynomials $\{p_n(x) : 0 \leq n\}$ orthogonal with respect to a weight function $w(x)$ on an interval K and associate with it an external field $v(x) := -\ln w(x)$. The weight function $w(x)$ is assumed to be positive on the interior of K. Let $x_{1,N} > x_{2,N} > \cdots > x_{N,N}$ be the zeros of $p_N(x)$. Define a probability measure ν_N to have masses $1/N$ at $x_{k,N}$, for $1 \leq k \leq N$. When this sequence of measures converge in the weak-$*$ topology to a probability measure ν then ν will be called the equilibrium measure associated with the external field $v(x)$. In general the equilibrium position of N movable charges in the external field $v(x)$ is not at the zeros of $p_N(x)$ but at what is called the Fekete points, say $y_{1,N} > y_{2,N} > \cdots > y_{N,N}$. One can think of the equilibrium measure as $N^{-1}\sigma(x)\,dx$, where σ is Dyson's fluid density. The absolute continuity of ν was not proved till very recently by Deift, Kriecherbauer, and McLaughlin [Deift et al. 1998], for real analytic external fields, thus confirming Dyson's intuition. Potential theory [Saff and Totik 1997] also confirms that the largest $(y_{1,N})$ and smallest $(y_{N,N})$ Fekete points are asymptotically equivalent to the largest $(x_{1,N})$ and smallest $(x_{N,N})$ zero of $p_N(x)$, respectively, that is,

$$\lim_{N\to\infty} \frac{x_{N,N}}{y_{N,N}} = \lim_{N\to\infty} \frac{x_{1,N}}{y_{1,N}} = 1.$$

4. Differential Equations and Discriminants

In this section, we shall assume that $\{p_n(x)\}$ is a sequence of orthonormal polynomials whose weight function is $w(x)$, that is

$$\int_a^b p_m(x)p_n(x)w(x)dx = \delta_{m,n}.$$

Throughout this section we shall assume that $w(x) > 0$ on (a, b),

$$w(x) = \exp(-v(x)), \quad x \in (a,b), \tag{4-1}$$

and that v has a continuous derivative on (a, b). We shall normalize w by setting

$$\int_a^b w(x)\,dx = 1.$$

The initial values and three-term recurrence relation of $\{p_n(x)\}$ take the form

$$\begin{aligned}&p_0(x) = 1, \quad p_1(x) = (x-b_0)/a_1,\\ &xp_n(x) = a_{n+1}p_{n+1}(x) + b_n p_n(x) + a_n p_{n-1}(x), \quad n > 0.\end{aligned} \tag{4-2}$$

Assume v has a continuous first derivative on (a, b). We define $A_n(x)$ and $B_n(x)$ via

$$A_n(x) = \frac{a_n\, w(b^-)\, p_n^2(b)}{b - x} + \frac{a_n\, w(a^+)\, p_n^2(a)}{x - a}$$
$$+ a_n \int_a^b \frac{v'(x) - v'(y)}{x - y}\, p_n^2(y)\, w(y)\, dy \quad (4\text{--}3)$$

and

$$B_n(x) = \frac{a_n\, w(a^+) p_n(a) p_{n-1}(a)}{x - a} + \frac{a_n\, w(b^-)\, p_n(b)\, p_{n-1}(b)}{b - x}$$
$$+ a_n \int_a^b \frac{v'(x) - v'(y)}{x - y}\, p_n(y) p_{n-1}(y)\, w(y)\, dy. \quad (4\text{--}4)$$

In (4–3) and (4–4) it is assumed that

$$y^n \frac{v'(x) - v'(y)}{x - y}\, w(y), \quad n = 0, 1, \ldots,$$

are integrable over (a, b) and the boundary terms in (4–3) and (4–4) exist. Under the latter assumptions, the orthonormal polynomials p_n's satisfy the differential recurrence relation [Bauldry 1990; Bonan and Clark 1990; Chen and Ismail 1997],

$$p_n'(x) = A_n(x) p_{n-1}(x) - B_n(x) p_n(x), \quad (4\text{--}5)$$

and the second-order differential equation

$$p_n''(x) + R_n(x) p_n'(x) + S_n(x) p_n(x) = 0, \quad (4\text{--}6)$$

where

$$R_n(x) := -\left(v'(x) + \frac{A_n'(x)}{A_n(x)} \right), \quad (4\text{--}7)$$

and

$$S_n(x) := B_n'(x) - B_n(x) \frac{A_n'(x)}{A_n(x)} - B_n(x)\bigl(v'(x) + B_n(x)\bigr)$$
$$+ \frac{a_n}{a_{n-1}} A_n(x) A_{n-1}(x). \quad (4\text{--}8)$$

Bonan and Clark [1990] and Bauldry [1990] were the first to establish (4–5) and (4–6), with the boundary terms assumed to vanish. Chen and Ismail [1997] rediscovered these results, and proved several others. The form of $R_n(x)$ of (4–7) in [Bonan and Clark 1990] and [Bauldry 1990] was more complicated, but Chen and Ismail [1997] observed that

$$B_{n+1}(x) + B_n(x) = \frac{x - b_n}{a_n} A_n(x) - v'(x), \quad (4\text{--}9)$$

which simplified $R_n(x)$ to the form in (4.10). Ismail and Wimp [1998] further proved that

$$B_{n+1}(x) - B_n(x) = \frac{a_{n+1}}{x - b_n} A_{n+1}(x) - \frac{a_n^2 A_{n-1}(x)}{a_{n-1}(x - b_n)} - \frac{1}{x - b_n}. \quad (4\text{--}10)$$

The recurrence relations (4–9) and (4–10) are curious because by solving them for $B_n(x)$ or $B_{n+1}(x)$ we find expressions for $B_n(x)$ and $B_{n+1}(x)$ whose consistency lead to five term inhomogeneous recursion relations for $A_n(x)$. Furthermore (4–9) gives $A_n(x)$ in terms of $B_n(x)$ and $B_{n+1}(x)$ then (4–10) yields a five-term inhomogeneous recursion relations for $B_n(x)$. The details are in [Ismail and Wimp 1998]. They have another implication. It is clear from (4–9) and (4–10) that if for some n, $A_n(x)$, $B_n(x)$, and $B_{n+1}(x)$ are polynomial (or rational) functions then $v'(x)$ is a polynomial (or rational) function. Furthermore the relationships (4–3) and (4–4) show that if $v'(x)$ is a polynomial (or rational) function then both $A_n(x)$ and $B_n(x)$ are polynomial (or rational, respectively) functions for all n.

For completeness, we indicate how differential equation (4–6) follows from (4–5). Eliminate $p_{n-1}(x)$ between (4–5) and the second line in (4–2) to get

$$-p'_n(x) + \left(\frac{(x-b_n)}{a_n}A_n(x) - B_n(x)\right)p_n(x) = \frac{a_{n+1}}{a_n}A_n(x)p_{n+1}(x).$$

In view of (4–9), we see that the polynomials $\{p_n(x)\}$ have the raising and lowering operators $L_{n,1}$ and $L_{n,2}$

$$L_{1,n} = \frac{d}{dx} + B_n(x), \quad L_{2,n} = -\frac{d}{dx} + B_n(x) + v'(x).$$

Indeed

$$L_{1,n}p_n(x) = A_n(x)p_n(x), \quad L_{2,n}p_{n-1}(x) = \frac{a_n}{a_{n-1}}A_{n-1}(x)p_n(x).$$

The differential equation (4–6) is the expanded form of

$$L_{2,n}\left(\frac{1}{A_n(x)}(L_{1,n}p_n(x))\right) = \frac{a_n}{a_{n-1}}A_{n-1}(x)p_n(x).$$

The differential equation (4–6) and the creation and annihilation operators $L_{2,n}$ and $L_{1,n}$ have many applications. First Chen and Ismail [1997] pointed out that $L_{2,n}$ and $L_{1,n}$ are adjoints in the Hilbert space $L_2(\mathbb{R}, w(x))$. Chen and Ismail also proved that the Lie algebras generated by $L_{1,n}$ and $L_{2,n}$ are finite dimensional when $v(x)$ is a polynomial. The rest of this section will cover two applications of (4–5) and (4–6), where we generalize Stieltjes' electrostatic problem and the evaluation of discriminants of orthogonal polynomials.

We now indicate how (4–6) leads to a generalization of the Stieltjes problem. This material is from [Ismail 2000b]. We propose that a weight function $w(x)$ creates two external fields. One is an external field whose potential at a point x is $v(x)$, $v(x)$ is as in (4–1). In addition in the presence of N unit charges w produces a second field whose potential is $\ln(A_N(x)/a_N)$. Thus the total external potential $V(x)$ is

$$V(x) = v(x) + \ln(A_N(x)/a_N). \tag{4–11}$$

Consider the system of N movable unit charges in $[a,b]$ in the presence of the external potential $V(x)$ of (4–11). Let $\boldsymbol{x}$ be as in (2–1) where $x_1,\ldots,x_N$ are the positions of the particles arranged in decreasing order. The total energy of the system is

$$E_N(\boldsymbol{x}) = \sum_{k=1}^{N} V(x_k) - 2 \sum_{1 \le j < k \le N} \ln|x_j - x_k|.$$

Let

$$T_N(\boldsymbol{x}) := \exp(-E_N(\boldsymbol{x})).$$

THEOREM 4.1 [Ismail 2000b]. *Assume $w(x) > 0$, $x \in (a,b)$ and let $v(x)$ of (4–1) and $v(x) + \ln A_N(x)$ be twice continuously differentiable functions whose second derivative is nonnegative on (a,b). Then the equilibrium position of N movable unit charges in $[a,b]$ in the presence of the external potential $V(x)$ of (4–11) is unique and attained at the zeros of $p_N(x)$, provided that the particle interaction obeys a logarithmic potential and that $T_N(\boldsymbol{x}) \to 0$ as $\boldsymbol{x}$ tends to any boundary point of $[a,b]^N$, where*

$$T_N(\boldsymbol{x}) = \left(\prod_{j=1}^{N} \frac{\exp(-v(x_j))}{A_N(x_j)/a_N}\right) \prod_{1 \le l < k \le N} (x_l - x_k)^2.$$

PROOF. Since T_N is symmetric in $x_1, \ldots, x_N$ and $T_N(\boldsymbol{x})$ vanishes when two of the x's coincide, we may assume that

$$x_1 > x_2 > \cdots > x_N. \qquad (4\text{–}12)$$

The assumption $v''(x) > 0$ ensures the positivity of $A_N(x)$. To find an equilibrium position we solve

$$\frac{\partial}{\partial x_j} \ln T_N(\boldsymbol{x}) = 0, \quad j = 1, 2, \ldots, N.$$

This system is

$$-v'(x_j) - \frac{A'_N(x_j)}{A_N(x_j)} + 2 \sum_{\substack{1 \le k \le N \\ k \ne j}} \frac{1}{x_j - x_k} = 0, \quad j = 1, 2, \ldots, N. \qquad (4\text{–}13)$$

The system of equations (4–13) is nonlinear in the unknowns $x_1, \ldots, x_N$. To change this to a linear system we set

$$f(x) := \prod_{j=1}^{N} (x - x_j),$$

and turn the system (4–13) to a differential equality in $f(x)$ satisfied at the points $x_j, 1 \leq j \leq N$. To see this, first observe that

$$\sum_{\substack{1 \leq k \leq N \\ k \neq j}} \frac{1}{x_j - x_k} = \lim_{x \to x_j} \left(\frac{f'(x)}{f(x)} - \frac{1}{x - x_j} \right)$$

$$= \lim_{x \to x_j} \left(\frac{(x - x_j) f'(x) - f(x)}{(x - x_j) f(x)} \right),$$

which implies, via L'Hôpital's rule,

$$2 \sum_{\substack{1 \leq k \leq N \\ k \neq j}} \frac{1}{x_j - x_k} = \frac{f(x_j)}{f'(x_j)}.$$

Now this and (4–13)

$$-v'(x_j) - \frac{A'_N(x_j)}{A_N(x_j)} + \frac{f''(x_j)}{f'(x_j)} = 0,$$

or equivalently

$$f''(x) + R_N(x) f'(x) = 0, \quad x = x_1, \ldots, x_N,$$

with R_N as in (4–7). In other words

$$f''(x) + R_N(x) f'(x) + S_N(x) f(x) = 0, \quad x = x_1, \ldots, x_N. \quad (4\text{–}14)$$

To check for local maxima and minima consider the Hessian matrix

$$H = (h_{ij}), \quad h_{ij} = \frac{\partial^2 \ln T_N(\boldsymbol{x})}{\partial x_i \partial x_j}. \quad (4\text{–}15)$$

It readily follows that

$$h_{ij} = 2(x_i - x_j)^{-2}, \quad i \neq j,$$

$$h_{ii} = -v''(x_i) - \frac{\partial}{\partial x_i} \left(\frac{A'_n(x_i)}{A_n(x_i)} \right) - 2 \sum_{\substack{1 \leq j \leq N \\ j \neq i}} \frac{1}{(x_i - x_j)^2}.$$

This shows that the matrix $-H$ is positive definite because it is real, symmetric, strictly diagonally dominant and its diagonal terms are positive [Horn and Johnson 1992, Cor. 7.2.2]. Therefore, $\ln T_N(\boldsymbol{x})$ has no relative minima nor saddle points. Thus any solution of (4–14) will provide a local maximum of $\ln T_N(\boldsymbol{x})$ or $T_N(\boldsymbol{x})$. There cannot be more than one local maximum since $T_N(\boldsymbol{x}) \to 0$ as $\boldsymbol{x} \to$ any boundary point along a path in the region defined in (4–12). Thus the system (4–14) has at most one solution. On the other hand, (4–6)–(4–8) show that the zeros of

$$f(x) = a_1 \, a_2 \, \ldots \, a_N \, p_N(x),$$

satisfy (4–14), hence the zeros of $p_N(x)$ solve (4–14). This completes the proof of Theorem 4.1. $\square$

Observe that the convexity of v in Theorem 4.1 can be replaced by requiring that $A_n(x) > 0$ for $a < x < b$.

The next result is a generalization of the Stieltjes–Hilbert evaluation of the discriminant of Jacobi polynomials (2–3). Observe that (4–5) is exactly the assumption (2–8) of Theorem 2.3. Thus we have established the following result.

THEOREM 4.2 [Ismail 1998]. *Let $\{p_N(x)\}$ be orthonormal polynomials and let $\{a_N\}$ and $\{b_N\}$ be the recursion coefficients in (4–2). Let*

$$x_{1,N} > x_{2,N} > \cdots > x_{N,N},$$

be the zeros of $p_N(x)$. Then the discriminant of $\{p_N(x)\}$ is given by

$$D(p_N) = \left(\prod_{j=1}^{N} \frac{A_N(x_{j,N})}{a_N}\right)\left(\prod_{k=1}^{N} a_k^{2k-2N+2}\right).$$

We next give a representation for the maximum value of $T_N(\boldsymbol{x})$ or the minimum value of $E_N(\boldsymbol{x})$ in terms of the recursion coefficients $\{a_n\}$.

THEOREM 4.3. *Let T_N and E_N be the maximum value of $T_N(\boldsymbol{x})$ and the equilibrium energy of the N particle system in Theorem 4.1. Then*

$$T_N = \left(\prod_{j=1}^{N} w(x_{j,N})\right)\left(\prod_{k=1}^{N} a_k^{2k}\right), \qquad E_N = \sum_{j=1}^{n} v(x_{j,N}) - 2\sum_{j=1}^{N} j \ln a_j.$$

This follows from Theorems 4.1 and 4.2.

Theorem 4.3 extends the Stieltjes results from Jacobi polynomials to general orthogonal polynomials, so it would be of interest to explore the analogue of the Selberg integral. This means replace the L_∞ norm in Theorem 4.1 by the L_p norm and evaluate the integral

$$\int_{[a,b]^N} \left(\prod_{j=1}^{N} \frac{\exp(-v(t_j))}{A_N(t_j)/a_N}\right)^p \prod_{1 \le i < k \le N}(t_i - t_k)^{2p}\, dt_1 \cdots dt_N.$$

Chihara [1985] studied orthogonal polynomials which result from modifying the orthogonality measure of a given set orthogonal by adding a one-point mass at the end of the spectral interval. He also considered specific cases of adding two point masses in certain special cases. Chihara's construction uses the kernel polynomials. Recently, Kiesel and Wimp [1995] found a different approach which avoids the use of kernel polynomials and in their later work [Kiesel and Wimp 1996], applied their results to the Koornwinder polynomials [Koornwinder 1984]. They derived closed form expressions for the recursion coefficients and the coefficients in the differential equation satisfied by the Koornwinder polynomials.

Grünbaum [1998] described an electrostatic interpretation for the zeros of the Koornwinder polynomials [Koornwinder 1984], which are orthogonal on $[-1, 1]$ with respect to measure with the absolutely continuous component $(1 - x)^\alpha \times (1 + x)^\beta$ on $[-1, 1]$ and two discrete masses at ± 1. This motivated us to write

[Ismail 2000c] where we derived second order differential equations for general polynomials orthogonal with respect a measure with a non-trivial absolutely continuous part supported on an interval and a finite discrete part outside the interval. We also extended the electrostatic models of Stieltjes and Hilbert to polynomials orthogonal with respect to a weight function supported on an interval $[a,b]$ with at most two discrete mass points at the finite end points of the interval. So far the only interesting example of this type is the Koornwinder polynomials and their special cases. Kiesel and Wimp studied the Kornwinder polynomials extensively in [Kiesel and Wimp 1996; Wimp and Kiesel 1995]. Grünbaum [≥ 2001], however, continued his research employing the Darboux transformation and described electrostatic models where the movable charges are restricted to an interval but the external field is generated by fixed charges in the plane.

5. Generalized and Quantized Discriminants

Motivated by (1–5) we were led in [Ismail 2000a] to define discriminants associated with linear operators, which reduce to (1–5) when the linear operator is the derivative operator.

DEFINITION 5.1. Let T be a linear degree reducing operator, that is $(Tf)(x)$ is a polynomial of exact degree $n-1$ whenever f has precise degree n and the leading terms in f and Tf have the same sign. We define the (generalized) discriminant relative to T by

$$D(g_n, T) := (-1)^{n(n-1)/2} \gamma_n^{-1} R\{g_n, Tg_n\} = (-1)^{n(n-1)/2} \gamma_n^{n-2} \prod_{j=1}^{n} (Tg_n)(x_j),$$

for g_n as in (1–3), and $R\{g_n, f_m\}$ is the resultant

$$R\{g_n, f_m\} := \gamma_n^m \prod_{j=1}^{n} f_m(x_j), \quad \text{for} \quad f_m(x) = \alpha_m x^m + \text{lower order terms}.$$

In this set up Theorem 2.3 becomes

THEOREM 5.2. Let T be a linear degree reducing operator and assume that $\{p_n(x)\}$ is a system of polynomials generated by (2–4) and (2–5) and assume that the zeros of $p_n(x)$ be arranged as in (2–6). If $\{p_n(x)\}$ satisfies the functional recurrence relation

$$Tp_n(x) = A_n(x)p_{n-1}(x) - B_n(x)p_n(x), \tag{5-1}$$

then the discriminant of $p_n(x)$ relative to T is given by

$$D(p_n) = \left(\prod_{j=1}^{n} A_n(x_{j,n}) \right) \prod_{k=1}^{n} \xi_k^{2n-2k-1} \zeta_k^{k-1}.$$

For particular T the above definition provides a quantization of the concept of a discriminant and leads to what we call q-discriminants which correspond to the case $T = D_q$,

$$(D_q f)(x) = \frac{f(x) - f(qx)}{x - qx}.$$

The operator D_q is called the q-difference operator [Andrews et al. 1999; Gasper and Rahman 1990]. Furthermore, $D_q f \to f'$ as $q \to 1$, for differentiable functions f. An easy calculation gives

$$D(g_n; q) := D(g_n, D_q)$$
$$= \gamma^{2n-2} q^{n(n-1)/2} \prod_{1 \leq i < j \leq n} (q^{-1/2} x_i - q^{1/2} x_j)(q^{1/2} x_i - q^{-1/2} x_j).$$

In other words,

$$D(g_n; q) := \gamma^{2n-2} q^{n(n-1)/2} \prod_{1 \leq i < j \leq n} \left(x_i^2 + x_j^2 - (q^{-1} + q) x_i x_j \right). \tag{5-2}$$

The representations (1–4) and (5–2) reaffirm the fact that $D(g_n; q) \to D(g_n)$ as $q \to 1$. We shall refer to $D(g_n; q)$ as the quantized discriminant. For $n = 2$, that is $g(x) = Ax^2 + Bx + C$, the quantized discriminant is $qB^2 - (1+q)^2 AC$, to be contrasted with the usual discriminant $B^2 - 4AC$.

In [Ismail 2000a] we extended many of the results of in §4 to polynomials orthogonal with respect to a discrete measure whose masses are at the union of at most two geometric progressions. To state these results we need the concept of a q-integral [Andrews et al. 1999; Gasper and Rahman 1990]:

$$\int_a^b f(x) d_q x := b(1-q) \sum_{n=0}^{\infty} q^n f(bq^n) - a(1-q) \sum_{n=0}^{\infty} q^n f(aq^n), \tag{5-3}$$

with

$$\int_0^{\infty} f(x) d_q x := (1-q) \sum_{-\infty}^{\infty} q^n f(q^n).$$

The orthogonality relation of discrete q-orthonormal polynomials is of the form

$$\int_a^b p_m(x) p_n(x) w(x) d_q x = \delta_{m,n}. \tag{5-4}$$

In [Ismail 2000a] we proved the following extension of (4–3)–(4–5) to discrete q-orthonormal polynomials.

THEOREM 5.3 [Ismail 2000a]. *Let $\{p_n(x)\}$ be a sequence of discrete q-orthonormal polynomials satisfying equalities (4–2). Then they have a lowering (annihilation) operator of the form*

$$D_q p_n(x) = A_n(x) p_{n-1}(x) - B_n(x) p_n(x), \tag{5-5}$$

where $A_n(x)$ and $B_n(x)$ are given by

$$A_n(x) = a_n \frac{w(y/q)p_n(y)p_n(y/q)}{x - y/q}\bigg]_a^b + a_n \int_a^b \frac{u(qx)-u(y)}{qx-y} p_n(y)\, p_n(y/q)\, w(y)\, d_q y,$$

and

$$B_n(x) = a_n \frac{w(y/q)p_n(y)p_{n-1}(y/q)}{x - y/q}\bigg]_a^b$$
$$+ a_n \int_a^b \frac{u(qx) - u(y)}{qx - y} p_n(y) p_{n-1}(y/q)\, w(y)\, d_q y,$$

where u is defined by

$$D_q w(x) = -u(qx) w(qx).$$

Furthermore the corresponding raising operator is

$$\frac{x - b_n}{a_n} A_n(x)\, p_n(x) - B_n(x)\, p_n(x) - D_q\, p_n(x) = \frac{a_{n+1}}{a_n} A_n(x)\, p_{n+1}(x).$$

We used (5–4) and (5–5) to evaluate the functions the functions $A_n(x)$ and $B_n(x)$ explicitly in [Ismail 2000a]. In [Ismail 2000a] we also computed the quantized discriminants of the big q-Jacobi polynomials. It is interesting to note that by experimentation we found that the usual discriminants for these polynomials for degrees 5 and 6 do not factor nicely.

We now proceed to study the generalized discriminants when T is the Askey–Wilson operator. Given a polynomial f we set $\check{f}(e^{i\theta}) := f(x)$ for $x = \cos\theta$; that is,

$$\check{f}(z) = f((z + 1/z)/2), \quad z = e^{i\theta}.$$

In other words, we think of $f(\cos\theta)$ as a function of $e^{i\theta}$. In this notation, the Askey–Wilson divided difference operator $\mathcal{D}_q$ is defined by

$$(\mathcal{D}_q f)(x) := \frac{\check{f}(q^{1/2}e^{i\theta}) - \check{f}(q^{-1/2}e^{i\theta})}{\check{e}(q^{1/2}e^{i\theta}) - \check{e}(q^{-1/2}e^{i\theta})}, \quad x = \cos\theta, \qquad (5\text{–}6)$$

with

$$e(x) = x.$$

A calculation reduces (5–6) to

$$(\mathcal{D}_q f)(x) = \frac{\check{f}(q^{1/2}e^{i\theta}) - \check{f}(q^{-1/2}e^{i\theta})}{(q^{1/2} - q^{-1/2})\, i\, \sin\theta}, \quad x = \cos\theta.$$

It is not difficult to see that

$$\lim_{q \to 1} (\mathcal{D}_q f)(x) = f'(x).$$

It is not difficult to compute the action of Askey–Wilson operator on the Chebyshev polynomials of the first kind, $T_n(x)$,

$$T_n(\cos\theta) = \cos(n\theta). \qquad (5\text{–}7)$$

Indeed,
$$\mathcal{D}_q T_n(x) = \frac{\left(q^{n/2} - q^{-n/2}\right)}{\left(q^{1/2} - q^{-1/2}\right)} U_{n-1}(x). \tag{5-8}$$

In [Ismail 2000a] we showed how that the appropriate generalized discriminant for the continuous q-Hermite and continuous q-Jacobi polynomials is $D(f, \mathcal{D}_q)$ and we applied Theorem 5.2 because (5–1) is known when $T = \mathcal{D}_q$ and $p_n(x)$ is any of the above mentioned polynomials.

It may be of interest to compute $D(f, \mathcal{D}_q)$ for $f(x) = Ax^2 + Bx + C$ and compare it with the familiar $B^2 - 4AC$. Since, with $x = \cos\theta$,
$$Ax^2 + Bx + C = \frac{A}{2}\cos(2\theta) + B\cos\theta + C + \frac{A}{2},$$
then (5–7) and (5–8) give
$$(\mathcal{D}_q f)(x) = \frac{A(q - q^{-1})}{q^{1/2} - q^{-1/2}} x + B,$$
so that $D(f, \mathcal{D}_q)$ is $\left(q^{1/2} + q^{-1/2} - 1\right) B^2 - \left(q^{1/2} + q^{-1/2}\right)^2 AC$.

One advantage of visiting an institute like MSRI is getting the opportunity to meet very bright young mathematicians with different backgrounds. I had the good fortune of meeting Naihuan Jing and talking to him about the different discriminants I encountered. He found interpretation of $D(f) = D(f, \frac{d}{dx})$, $D(f; \mathcal{D}_q)$ and $D(f; \Delta_h)$ as expectation values of vertex operators, where
$$(\Delta_h f)(x) := f(x + h) - f(x).$$
Our conversations led to [Ismail and Jing $\geq$ 2001].

Acknowledgments

The author is very grateful to Drs. B. H. Yoo and G. J. Yoon for their careful proofreading of this paper. We gratefully acknowledge partial research support of MSRI and the Department of Mathematics at the University of California, Berkeley.

I thank Denise Marks for the excellent job she did in putting this paper together.

References

[Andrews et al. 1999] G. E. Andrews, R. A. Askey, and R. Roy, *Special Functions*, Cambridge University Press, Cambridge, 1999.

[Bauldry 1990] W. Bauldry, "Estimates of asymmetric Freud polynomials on the real line", *J. Approximation Theory* **63** (1990), 225–237.

[Bleher and Its 1999] P. Bleher and A. Its, "Semiclassical asymptotics of orthogonal polynomials, Riemann-Hilbert problem, and universality in the matrix model", *Ann. of Math.* **150** (1999), 185–266.

[Bonan and Clark 1990] S. S. Bonan and D. S. Clark, "Estimates of the Hermite and the Freud polynomials", *J. Approximation Theory* **63** (1990), 210–224.

[Chen and Ismail 1997] Y. Chen and M. E. H. Ismail, "Ladder operators and differential equations for orthogonal polynomials", *J. Phys. A* **30** (1997), 7817–7829.

[Chen and Manning 1996] Y. Chen and S. M. Manning, "Some distribution functions of the Laguerre ensmebles", *J. Phys. A* **29** (1996), 1–19.

[Chen et al. 1995] Y. Chen, K. J. Eriksen, and C. A. Tracy, "Largest eigenvalue distribution in the double scaling limit of matrix models: a Coulomb fluid approach", *J. Phys. A* **28** (1995), L207–211.

[Chihara 1985] T. S. Chihara, "Orthogonal polynomials and measures with end point masses", *Rocky Mountain J. Math.* **15** (1985), 705–719.

[Deift et al. 1998] P. Deift, T. Kriecherbauer, and K. McLaughlin, "New results on the equilibrium measure for logarithmic potentials in the presence of an external field", *J. Approx. Theory* **95** (1998), 388–475.

[Dickson 1939] L. E. Dickson, *New Course on the Theory of Equations*, Wiley, New York, 1939.

[Dyson 1962a] F. J. Dyson, "Statistical theory of energy levels of complex systems I", *J. Math. Phys.* **3** (1962), 140–156.

[Dyson 1962b] F. J. Dyson, "Statistical theory of energy levels of complex systems II", *J. Math. Phys.* **3** (1962), 157–165.

[Dyson 1962c] F. J. Dyson, "Statistical theory of energy levels of complex systems III", *J. Math. Phys.* **3** (1962), 166–175.

[Gakhov 1990] F. D. Gakhov, *Boundary Value Problems*, Dover, New York, 1990.

[Gasper and Rahman 1990] G. Gasper and M. Rahman, *Basic Hypergeometric Series*, Cambridge University Press, Cambridge, 1990.

[Grünbaum 1998] F. A. Grünbaum, "Variations on a theme of Stieltjes and Heine: an electrostatic interpretation of zeros of certain polynomials", *J. Comp. Appl. Math.* **99** (1998), 189–194.

[Grünbaum $\geq$ 2001] F. A. Grünbaum, "Papers on electrostatics". (in preparation).

[Hilbert 1888] D. Hilbert, "Über die Discriminante der in Endlichen abbrechenden hypergeometrischen Reihe", *J. für die reine und angewandte Matematik* **103** (1888), 337–345.

[Horn and Johnson 1992] R. A. Horn and C. Johnson, *Matrix Analysis*, Cambridge University Press, Cambridge, 1992.

[Ismail 1998] M. E. H. Ismail, "Discriminants and functions of the second kind of orthogonal polynomials", *Results in Math.* **34** (1998), 132–149.

[Ismail 2000a] M. E. H. Ismail, "Difference equations and quantized discriminants for q-orthogonal polynomials", *Adv. Appl. Math.* (2000), (to appear).

[Ismail 2000b] M. E. H. Ismail, "An electrostatics model for zeros of general orthogonal polynomials", *Pacific J. Math.* **193**:2 (2000), 355–369.

[Ismail 2000c] M. E. H. Ismail, "More on electrostatic models for zeros of orthogonal polynomials", *Numer. Funct. Anal. Optim.* **21**:1-2 (2000), 191–204.

[Ismail and Jing ≥ 2001] M. E. H. Ismail and N. Jing, "q-discriminants and Vertex Operators". (to appear).

[Ismail and Wimp 1998] M. E. H. Ismail and J. Wimp, "On differential equations for orthogonal polynomials", *Methods and Applications of Analysis* **5** (1998), 439–452.

[Kiesel and Wimp 1996] H. Kiesel and J. Wimp, "A note on Koornwinder's polynomials with weight function $(1-x)^\alpha(1+x)^\beta+M\delta(x+1)+N\delta(x-1)$", *Numerical Algorithms* **11** (1996), 229–241.

[Koornwinder 1984] T. H. Koornwinder, "Orthogonal polynomials with weight function $(1-x)^\alpha(1+x)^\beta+M\delta(x+1)+N\delta(x-1)$", *Canadian Math. Bull.* **27** (1984), 205–214.

[Muskhelishvili 1953] N. I. Muskhelishvili, *Some Basic Problems of The Mathematical Theory of Elasticity*, Third ed., Noordhoff, Groningen, 1953.

[Rainville 1960] E. D. Rainville, *Special Functions*, Macmillan, New York, 1960.

[Saff and Totik 1997] E. B. Saff and V. Totik, *Logarithmic Potentials With External Fields*, Springer-Verlag, New York, 1997.

[Schur 1931] I. Schur, "Affektlose Gleichungen in der Theorie der Laguerreschen und Hermiteschen Polynomes", *J. für die Reine und Angewandte Matematik* **165** (1931), 52–58.

[Selberg 1944] A. Selberg, "Bemerkninger om et multiplet integral", *Norsk Mat. Tidsskr.* **26** (1944), 71–78.

[Stieltjes 1885a] T. J. Stieltjes, "Sur quelques théorèmes d'algèbre", *Comptes Rendus de l'Academie des Sciences, Paris* **100** (1885), 439–440. Reprinted in his *Œuvres Complètes*, v. 1, pp. 440–441, Noordhoff, Groningen, 1914 (reprinted by Springer, Berlin, 1993).

[Stieltjes 1885b] T. J. Stieltjes, "Sur les polynômes de Jacobi", *Comptes Rendus de l'Academie des Sciences, Paris* **100** (1885), 620–622. Reprinted in his *Œuvres Complètes*, v. 1, pp. 442–444, Noordhoff, Groningen, 1914 (reprinted by Springer, Berlin, 1993).

[Szegő 1975] G. Szegő, *Orthogonal polynomials*, 4th ed., Colloquium Publications **23**, Amer. Math. Soc., Providence, 1975.

[Wimp and Kiesel 1995] J. Wimp and H. Kiesel, "Non-linear recurrence relations and some derived orthogonal polynomials", *Annals of Numerical Mathematics* **2** (1995), 169–180.

MOURAD E. H. ISMAIL
DEPARTMENT OF MATHEMATICS
UNIVERSITY OF SOUTH FLORIDA
TAMPA, FL 33620-5700
UNITED STATES
ismail@math.usf.edu

Random Words, Toeplitz Determinants, and Integrable Systems I

ALEXANDER R. ITS, CRAIG A. TRACY, AND HAROLD WIDOM

ABSTRACT. It is proved that the limiting distribution of the length of the longest weakly increasing subsequence in an inhomogeneous random word is related to the distribution function for the eigenvalues of a certain *direct sum* of Gaussian unitary ensembles subject to an overall constraint that the eigenvalues lie in a hyperplane.

1. Introduction

A class of problems — important for their applications to computer science and computational biology as well as for their inherent mathematical interest — is the statistical analysis of a string of random symbols. The symbols, called *letters*, are assumed to belong to an alphabet $\mathcal{A}$ of fixed size k. The set of all such strings (or *words*) of length N, $\mathcal{W}(\mathcal{A}, N)$, forms the sample space in the statistical analysis of these strings. A natural measure on $\mathcal{W}$ is to assign each letter equal probability, namely $1/k$, and to define the probability measure on words by the product measure. Thus each letter in a word occurs independently and with equal probability. We call such random word models *homogeneous*.

Of course, for some applications, each letter in the alphabet does not occur with the same frequency and it is therefore natural to assign to each letter i a probability p_i. If we again use the product measure for the words (letters in a word occur independently), then the resulting random word models are called *inhomogeneous*.

Fixing an ordering of the alphabet $\mathcal{A}$, a *weakly increasing subsequence* of a word
$$w = \alpha_1 \alpha_2 \ldots \alpha_N \in \mathcal{W}$$
is a subsequence $\alpha_{i_1} \alpha_{i_2} \ldots \alpha_{i_m}$ such that $i_1 < i_2 < \cdots < i_m$ and $\alpha_{i_1} \leq \alpha_{i_2} \leq \cdots \leq \alpha_{i_m}$. The positive integer m is called the *length* of this weakly increasing

subsequence. For each word $w \in \mathcal{W}$ we define $l_N(w)$ to equal the *length of the longest weakly increasing subsequence* in w. We now define the fundamental object of this paper:
$$F_N(n) := \text{Prob}\,(l_N(w) \leq n)$$
where Prob is the inhomogeneous measure on random words. Of course, Prob depends upon N and the probabilities p_i.

Our results are of two types. To state our first results, we order the p_i so that
$$p_1 \geq p_2 \geq \cdots \geq p_k$$
and decompose out alphabet $\mathcal{A}$ into subsets $\mathcal{A}_1$, $\mathcal{A}_2$, ... such that $p_i = p_j$ if and only if i and j belong to the same $\mathcal{A}_\alpha$. Setting $k_\alpha = |\mathcal{A}_\alpha|$, we show that the limiting distribution function as $N \to \infty$ for the appropriately centered and normalized random variable l_N is related to the distribution function for the eigenvalues ξ_i in the *direct sum* of mutually independent $k_\alpha \times k_\alpha$ Gaussian unitary ensembles (GUE), conditional on the eigenvalues ξ_i satisfying $\sum \sqrt{p_i}\,\xi_i = 0$. (See [13], for example, for the notion of a GUE and other concepts in random matrix theory.) In the case when one letter occurs with greater probability than the others, this result implies that the limiting distribution of $(l_N - Np_1)/\sqrt{N}$ is Gaussian with variance equal to $p_1(1-p_1)$. In the case when all the probabilities p_i are distinct, we compute the next correction in the asymptotic expansion of the mean of l_N and find that
$$\mathrm{E}(l_N) = Np_1 + \sum_{j>1} \frac{p_j}{p_1 - p_j} + O(N^{-1/2}), \quad N \to \infty.$$

This last formula agrees quite well with finite N simulations. We expect this asymptotic formula remains valid when one letter occurs with greater probability than the others.

These results generalize work on the homogeneous model by Johansson [11] and by Tracy and Widom [19]. Since all the probabilities p_i are equal in the homogeneous model, the underlying random matrix model is $k \times k$ traceless GUE. That is, the direct sum reduces to just one term. In [19] the integrable system underlying the finite N homogeneous model was shown to be related to Painlevé V. In the isomonodromy formulation of Painlevé V [9], the associated 2×2 matrix linear ODE has two simple poles in the finite complex plane and one Poincaré index 1 irregular singular point at infinity. In Part II we will show that the finite N inhomogeneous model is represented by the isomonodromy deformations of the 2×2 matrix linear ODE which has $m+1$ simple poles in the finite complex plane and, again, one Poincaré index 1 irregular singular point at infinity. The number m is the total number of the subsets $\mathcal{A}_\alpha$, and the poles are located at zero point and at the points $-p_{i_\alpha}$ ($i_\alpha = \max \mathcal{A}_\alpha$). The integers k_α appear as the formal monodromy exponents at the respective points $-p_{i_\alpha}$. We will also analyse the monodromy meaning of the asymptotic results obtained in this part.

The results presented here are part of the recent flurry of activity centering around connections between combinatorial probability of the Robinson–Schensted–Knuth type on the one hand and random matrices and integrable systems on the other. From the point of view of probability theory, the quite surprising feature of these developments is that the methods came from Toeplitz determinants, integrable differential equations of the Painlevé type and the closely related Riemann–Hilbert techniques. The first to discover this connection at the level of distribution functions was Baik, Deift and Johansson [1] who showed that the limiting distribution of the length of the longest increasing subsequence in a random permutation is equal to the limiting distribution function of the appropriately centered and normalized largest eigenvalue in the GUE [17]. This result has been followed by a number of developments relating random permutations, random words and more generally random Young tableaux to the distribution functions of random matrix theory [2; 3; 4; 6; 8; 10; 12; 14; 18].

After the completion of this paper, Stanley [16] showed that the measure (2–1) also underlies the analysis of certain (generalized) riffle shuffles of Bayer and Diaconis [5]. Stanley relates this measure to quasisymmetric functions and does not require that p have finite support. (Many of our results generalize to the case when p does not have finite support, but we do not consider this here.)

2. Random Words

2.1. Probability Measure on Words and Partitions.

The Robinson–Schensted–Knuth (RSK) algorithm is a bijection between two-line arrays w_A (or generalized permutation matrices) and ordered pairs (P, Q) of semistandard Young tableaux (SSYT). For a detailed account, see [15, Chapter 7], for example; we will use without further reference various results from symmetric function theory, which can be found in the same reference.

When the two-line arrays have the special form

$$w_A = \begin{pmatrix} 1 & 2 & \cdots & N \\ \alpha_1 & \alpha_2 & \cdots & \alpha_N \end{pmatrix},$$

with $\alpha_i \in \mathcal{A} = \{1, 2, \ldots, k\}$, we identify each w_A with a word $w = \alpha_1 \alpha_2 \cdots \alpha_N$ of length N composed of letters from the alphabet $\mathcal{A}$; furthermore, in this case the insertion tableaux P have shape $\lambda \vdash N$, $l(\lambda) \leq k$, with entries coming from $\mathcal{A}$ and the recording tableaux Q are standard Young tableau (SYT) of the same shape λ. As usual, f^λ denotes the number of SYT of shape λ and $d_\lambda(k)$ the number of SSYT of shape λ whose entries come from $\mathcal{A}$.

We define a probability measure, Prob, on $\mathcal{W}(\mathcal{A}, N)$, the set of all words w of length N formed from the alphabet $\mathcal{A}$, by the two requirements:

1. For each word w consisting of a single letter $i \in \mathcal{A}$, $\text{Prob}(w = i) = p_i$, $0 < p_i < 1$, with $\sum p_i = 1$.

2. For each $w = \alpha_1\alpha_2\ldots\alpha_N \in \mathcal{W}$ and any $i_j \in \mathcal{A}$, $j = 1, 2, \ldots, N$,

$$\text{Prob}\,(\alpha_1\alpha_2\ldots\alpha_N = i_1i_2\ldots i_N) = \prod_{j=1}^{N} \text{Prob}\,(\alpha_j = i_j) \qquad \text{(independence)}.$$

Of course, Prob depends both on N and the probabilities $\{p_i\}$.

Under the RSK correspondence, the probability measure Prob induces a probability measure on partitions $\lambda \vdash N$, which we will again denote by Prob. This induced measure is expressed in terms of f^λ and the Schur function. To see this we first recall that a tableau T has *type* $\alpha = (\alpha_1, \alpha_2, \ldots)$, denoted $\alpha = \text{type}(T)$, if T has $\alpha_i = \alpha_i(T)$ parts equal to i. We write

$$x^T = x_1^{\alpha_1(T)} x_2^{\alpha_2(T)} \ldots$$

The combinatorial definition of the Schur function of shape λ in the variables $x = (x_1, x_2, \ldots)$ is the formal power series

$$s_\lambda(x) = \sum_T x^T$$

summed over all SSYT of shape λ. The $p = \{p_1, \ldots, p_k\}$ specialization of $s_\lambda(x)$ is $s_\lambda(p) = s_\lambda(p_1, p_2, \ldots, p_k, 0, 0, \ldots)$.

For each word $w \leftrightarrow (P, Q)$, the N entries of P consist of the N letters of w since P is formed by successive row bumping the letters from w. Because of the independence assumption,

$$p^P = p_1^{\alpha_1(P)} p_2^{\alpha_2(P)} \ldots p_k^{\alpha_k(P)}$$

gives the weight assigned to word w. From the combinatorial definition of the Schur function, we observe that its p specialization is summing the weights of words w that under RSK have shape $\lambda \vdash N$. The recording tableau Q keeps track of the *order* of the letters in the word. The weights of any words with the same number of letters of each type are equal (independence), so we need merely count the number of such Q, i.e. f^λ, and multiply this by the weight of any given such word to arrive at the induced measure on partitions,

$$\text{Prob}\,(\{\lambda\}) = s_\lambda(p)\, f^\lambda, \qquad (2\text{--}1)$$

which satisfies the normalization $\sum_{\lambda \vdash N} \text{Prob}(\lambda) = 1$. For the homogeneous case $p_i = 1/k$, the measure reduces to

$$\text{Prob}(\lambda) = s_\lambda(1/k, 1/k, \ldots, 1/k)\, f^\lambda = \frac{d_\lambda(k)\, f^\lambda}{k^N}, \quad \lambda \vdash N.$$

The Poissonization of this homogeneous measure is called the Charlier ensemble in [11].

If $l_N(w)$ equals the length of the longest *weakly* increasing subsequence in the word $w \in \mathcal{W}(\mathcal{A}, N)$, then by the RSK correspondence $w \leftrightarrow (P,Q)$, the number of boxes in the first row of P, λ_1, equals $l_N(w)$. Hence,

$$\text{Prob}\,(l_N(w) \leq n) = \sum_{\substack{\lambda \vdash N \\ \lambda_1 \leq n}} s_\lambda(p)\, f^\lambda. \tag{2-2}$$

2.2. Toeplitz Determinant Representation. Gessel's theorem [7] — more precisely, its dual version (see [19, §II], whose notation we follow) — is the formal power series identity

$$\sum_{\substack{\lambda \vdash N \\ \lambda_1 \leq n}} s_\lambda(x) s_\lambda(y) = \det(T_n(\varphi))$$

where $T_n(\varphi)$ is the $n \times n$ Toeplitz matrix whose i,j entry is φ_{i-j}, where φ_i is the ith Fourier coefficient of

$$\varphi(z) = \prod_{n=1}^\infty (1 + y_n z^{-1}) \prod_{n=1}^\infty (1 + x_n z), \quad z = e^{i\theta}.$$

If we define the (exponential) generating function

$$G_I(n; \{p_i\}, t) = \sum_{N=0}^\infty \text{Prob}\,(l_N(w) \leq n) \frac{t^N}{N!},$$

then an immediate consequence of Gessel's identity with p specialization of the x variables and exponential specialization of the y variables and the RSK correspondence is

$$G_I(n; \{p_i\}, t) = \det(T_n(f_I)) \tag{2-3}$$

where

$$f_I(z) = e^{t/z} \prod_{j=1}^k (1 + p_j z). \tag{2-4}$$

3. Limiting Distribution

We start with the probability distribution (2–1) on the set of partitions $\lambda = \{\lambda_1, \lambda_2, \ldots, \lambda_k\} \vdash N$. For f^λ we use the formula

$$f^\lambda = \frac{N!\,\Delta(h)}{h_1!\,h_2! \ldots h_k!}$$

where

$$h_i = \lambda_j + k - i$$

and

$$\Delta(h) = \Delta(h_1, h_2, \ldots, h_k) = \prod_{1 \leq i < j \leq k} (h_i - h_j). \tag{3-1}$$

Equivalently,
$$f^\lambda = \frac{\Delta(h)}{\prod_{i=1}^{k-1} \prod_{j=i}^{k-1} (\lambda_i + k - j)} \begin{pmatrix} & & N & \\ \lambda_1 & \lambda_2 & \cdots & \lambda_k \end{pmatrix}.$$

The (classical) definition of the Schur function is

$$s_\lambda(p) = \frac{\det(p_i^{h_j})}{\Delta(p)} = \frac{1}{\Delta(p)} \sum_{\sigma \in S_k} (-1)^\sigma p_1^{h_{\sigma(1)}} p_2^{h_{\sigma(2)}} \cdots p_k^{h_{\sigma(k)}}. \qquad (3\text{-}2)$$

This holds when all the p_i are distinct but in general the two determinants require modification, which we now describe. We order the p_i so that

$$p_1 \geq p_2 \geq \cdots \geq p_k \qquad (3\text{-}3)$$

and decompose our alphabet $\mathcal{A} = \{1, 2, \ldots, k\}$ into subsets $\mathcal{A}_1, \mathcal{A}_2, \ldots$ such that $p_i = p_j$ if and only if i and j belong to the same $\mathcal{A}_\alpha$. Set $i_\alpha = \max \mathcal{A}_\alpha$. Think of the p_i as indeterminates and for all indices i differentiate the determinant $i_\alpha - i$ times with respect to p_i if $i \in \mathcal{A}_\alpha$. Then replace the p_i by their given values. (That this is correct follows from l'Hôpital's rule.) If we set $k_\alpha = |\mathcal{A}_\alpha|$ and write p_α for p_{i_α} then we see that $\Delta(p)$ becomes

$$\Delta'(p) = \prod_\alpha (1!\, 2! \cdots (k_\alpha - 1)!) \prod_{\alpha < \beta} (p_\alpha - p_\beta)^{k_\alpha k_\beta} \qquad (3\text{-}4)$$

and (after performing row operations) that the ith row of $\det(p_i^{h_j})$ becomes $\left(h_j^{i_\alpha - i} p_i^{h_j - i_\alpha + i}\right)$. Equivalently, the partial product $\prod_{i \in \mathcal{A}_\alpha} p_i^{h_{\sigma(i)}}$ from the summand in (3-2) gets multiplied by

$$\prod_{i \in \mathcal{A}_\alpha} \left(h_{\sigma(i)}^{i_\alpha - i} p_i^{-i_\alpha + i}\right) = \left(\prod_{i \in \mathcal{A}_\alpha} h_{\sigma(i)}^{i_\alpha - i}\right) p_\alpha^{-k_\alpha(k_\alpha - 1)/2}. \qquad (3\text{-}5)$$

In the case of distinct p_i we write our formula as

$$\text{Prob}(\lambda) = s_\lambda(p_1, \ldots, p_k)\, f^\lambda$$
$$= \frac{\Delta(h)}{\Delta(p)} \frac{1}{\prod_{i=1}^{k-1} \prod_{j=i}^{k-1} (\lambda_i + k - j)}$$
$$\times \sum_{\sigma \in S_k} (-1)^\sigma p_1^{k-\sigma(1)} \cdots p_k^{k-\sigma(k)} p_1^{\lambda_{\sigma(1)}} \cdots p_k^{\lambda_{\sigma(k)}} \begin{pmatrix} & & N & \\ \lambda_1 & \lambda_2 & \cdots & \lambda_k \end{pmatrix}.$$

Let $M_q(\lambda)$ denote the multinomial distribution associated with a sequence $q = \{q_1, \ldots, q_k\}$,

$$M_q(\lambda) = q_1^{\lambda_1} \cdots q_k^{\lambda_k} \begin{pmatrix} & & N & \\ \lambda_1 & \lambda_2 & \cdots & \lambda_k \end{pmatrix}.$$

If p_σ denotes the sequence $\{p_{\sigma^{-1}(1)}, \ldots, p_{\sigma^{-1}(k)}\}$, we may write

$$\text{Prob}(\lambda) = \frac{\Delta(h)}{\Delta(p)} \frac{1}{\prod_{i=1}^{k-1}\prod_{j=i}^{k-1}(\lambda_i + k - j)} \sum_{\sigma \in S_k} (-1)^\sigma p_1^{k-\sigma(1)} \cdots p_k^{k-\sigma(k)} M_{p_\sigma}(\lambda). \tag{3-6}$$

This is the formula for distinct p_i. In the general case we must replace $\Delta(p)$ by $\Delta'(p)$ and each partial product $\prod_{i \in A_\alpha} p_i^{k-\sigma(i)}$ appearing in the sum on the right must be multiplied by the factor (3–5).

The multinomial distribution $M_q(\lambda)$ has the property that the total measure of any region where $|\lambda_i - Nq_i| > \varepsilon N$ for some i and some $\varepsilon > 0$ tends exponentially to zero as $N \to \infty$. All the other terms appearing in (3–6) or its modification are uniformly bounded by a power of N. Since $\lambda_{i+1} \leq \lambda_i$ for all i it follows that the contribution of the terms involving $M_q(\lambda)$ in (3–6) will tend exponentially to zero unless $q_{i+1} \leq q_i$ for all i. Since $q_i = p_{\sigma^{-1}(i)}$ this shows that the contribution to (3–6) of the summand corresponding to σ is exponentially small unless σ leaves each of the sets A_α invariant. It follows that if we denote the set of such permutations by S'_k then we may restrict the sum in (3–6) to the $\sigma \in S'_k$ without affecting the limit. Observe that when $\sigma \in S'_k$ all the $M_{p_\sigma}(\lambda)$ appearing in (3–6) equal $M_p(\lambda)$.

Write
$$\lambda_i = Np_i + \sqrt{Np_i}\,\xi_i.$$

In terms of the ξ_i the multinomial distribution $M_p(\lambda)$ converges to

$$(2\pi)^{-(k-1)/2} e^{-\sum \xi_i^2/2} \delta(\sum \sqrt{q_i}\,\xi_i). \tag{3-7}$$

(See Section 3.1.) Here $\delta(\sum \sqrt{q_i}\,\xi_i)$ denotes Lebesgue measure on the hyperplane $\sum \sqrt{q_i}\,\xi_i = 0$.

We now consider the contribution of the other terms in (3–6) as modified. Again, they are uniformly bounded by a power of N and the total measure of any region where $|\lambda_i - Np_i| > \varepsilon N$ for some i and some $\varepsilon > 0$ tends exponentially to zero as $N \to \infty$. Thus in determining the asymptotics of the other terms we may assume that $\lambda_i \sim Np_i$ for all i.

The constant $\Delta'(p)$ is given by (3–4). As for $\Delta(h)$, observe that the factor

$$h_i - h_j = \lambda_i - \lambda_j - i + j$$

in the product in (3–1) is asymptotically equal to $N(p_i - p_j)$ when i and j do not belong to the same A_α and to $\sqrt{Np_\alpha}\,(\xi_i - \xi_j)$ if $i, j \in A_\alpha$. It follows that

$$\Delta(h) \sim N^{k(k-1)/2 - \sum_\alpha k_\alpha(k_\alpha-1)/4} \prod p_\alpha^{k_\alpha(k_\alpha-1)/4} \prod_{\alpha<\beta}(p_\alpha - p_\beta)^{k_\alpha k_\beta} \prod_\alpha \Delta_\alpha(\xi),$$

where $\Delta_\alpha(\xi)$ is the Vandermonde determinant of those ξ_i with $i \in A_\alpha$.

The next factor in (3–6), the reciprocal of the double product, is asymptotically

$$N^{-k(k-1)/2} \prod_{i=1}^{k-1} p_i^{i-k}.$$

As for the sum in (3–6) as modified, observe that since each σ now belongs to S'_k each product appearing there is equal to $\prod p_i^{k-i}$. Each such product is to be multiplied by

$$\prod_\alpha \left(\left(\prod_{i \in \mathcal{A}_\alpha} h_{\sigma(i)}^{i_\alpha - i} \right) p_\alpha^{-k_\alpha(k_\alpha-1)/2} \right).$$

(See (3–5).) Hence the sum itself is equal to

$$\prod_i p_i^{k-i} \prod_\alpha p_\alpha^{-k_\alpha(k_\alpha-1)/2} \sum_{\sigma \in S'_k} (-1)^\sigma \prod_\alpha \prod_{i \in \mathcal{A}_\alpha} h_{\sigma(i)}^{i_\alpha - i}.$$

Since each $\sigma \in S'_k$ is uniquely expressible as a product of $\sigma_\alpha \in S(\mathcal{A}_\alpha)$ (where $S(\mathcal{A}_\alpha)$ is the group of permutations of $\mathcal{A}_\alpha$) we have

$$\sum_{\sigma \in S'_k} (-1)^\sigma \prod_\alpha \prod_{i \in \mathcal{A}_\alpha} h_{\sigma(i)}^{i_\alpha - i} = \prod_\alpha \sum_{\sigma_\alpha \in S(\mathcal{A}_\alpha)} (-1)^{\sigma_\alpha} \prod_{i \in \mathcal{A}_\alpha} h_{\sigma_\alpha(i)}^{i_\alpha - i} = \prod_\alpha \Delta_\alpha(h)$$

$$\sim N^{\sum k_\alpha(k_\alpha-1)/4} \prod_\alpha \left(p_\alpha^{k_\alpha(k_\alpha-1)/4} \Delta_\alpha(\xi) \right).$$

Putting all this together shows that the limiting distribution is

$$(2\pi)^{-(k-1)/2} \prod_\alpha (1!\,2!\cdots(k_\alpha - 1)!)^{-1} \prod_\alpha \Delta_\alpha(\xi)^2 \, e^{-\sum \xi_i^2/2} \, \delta\!\left(\sum \sqrt{p_i}\,\xi_i \right). \quad (3\text{–}8)$$

This has a random matrix interpretation. It is the distribution function for the eigenvalues in the direct sum of mutually independent $k_\alpha \times k_\alpha$ Gaussian unitary ensembles, conditional on the eigenvalues ξ_i satisfying $\sum \sqrt{p_i}\,\xi_i = 0$.

It remains to determine the support of the limiting distribution. In terms of the ξ_i the inequalities $\lambda_{i+1} \leq \lambda_i$ are equivalent to

$$\xi_{i+1} \leq \frac{N(p_i - p_{i+1})}{\sqrt{N p_i}} + \sqrt{\frac{p_i}{p_{i+1}}}\,\xi_i.$$

In the limit $N \to \infty$ this becomes no restriction if $p_{i+1} < p_i$ but becomes $\xi_{i+1} \leq \xi_i$ if $p_{i+1} = p_i$. Otherwise said, the support of the limiting distribution is restricted to those $\{\xi_i\}$ for which $\xi_{i+1} \leq \xi_i$ whenever i and $i+1$ belong to the same $\mathcal{A}_\alpha$. (In the random matrix interpretation it means that the eigenvalues within each GUE are ordered.) We denote this set of ξ_i by Ξ.

It now follows from (2–2) and (3–8) (also recall the ordering (3–3)) that

$$\lim_{N \to \infty} \mathrm{Prob}\!\left(\frac{l_N - N p_1}{\sqrt{N p_1}} \leq s \right) = (2\pi)^{-(k-1)/2} \prod_\alpha (1!\,2!\cdots(k_\alpha - 1)!)^{-1}$$

$$\times \int \cdots \int_{\xi_1 \leq s}^{\xi_i \in \Xi} \prod_\alpha \Delta_\alpha(\xi)^2 \, e^{-\sum \xi_i^2/2} \, \delta\!\left(\sum \sqrt{p_i}\,\xi_i \right) d\xi_1 \cdots d\xi_k. \quad (3\text{–}9)$$

When the probabilities are not all equal this may be reduced to a k_1-dimensional integral as follows. Let i denote the indices in $\mathcal{A}_1$ and j the other indices. We have to integrate

$$\prod_\alpha \Delta_\alpha(\xi)^2 \, e^{-\frac{1}{2}\sum \xi_i^2 - \frac{1}{2}\sum \xi_j^2} \, \delta\left(\sum \sqrt{p_i}\xi_i + \sum \sqrt{p_j}\xi_j\right)$$

over the subset of Ξ where $\xi_1 \leq s$. Since $\xi_1 = \max \xi_i$ and since the integrand is symmetric in the ξ_i and the ξ_j within their groups we may (by changing the normalization constant) integrate over all $\xi_i \leq s$ and all ξ_j. We first fix the ξ_i and integrate over the ξ_j. These have to satisfy

$$\sum \sqrt{p_j}\xi_j = -\sum \sqrt{p_i}\xi_i = -\sqrt{p_1}\sum \xi_i.$$

If we write

$$\xi_j = \eta_j + x\sqrt{p_j}, \qquad (3\text{--}10)$$

where $\{\eta_j\}$ is orthogonal to $\{\sqrt{p_j}\}$, then

$$x = \frac{\sum \sqrt{p_j}\xi_j}{\sum p_j} = -\frac{\sqrt{p_1}}{1 - k_1 p_1}\sum \xi_i.$$

(Recall that $\mathcal{A}_1$ has k_1 indices.) For each $\alpha > 1$ we have $\Delta_\alpha(\xi) = \Delta_\alpha(\eta)$ since the p_j within groups are equal and

$$\sum \xi_j^2 = \sum \eta_j^2 + x^2 \sum p_j = \sum \eta_j^2 + \frac{p_1}{1 - k_1 p_1}\left(\sum \xi_i\right)^2.$$

So the distribution function is equal to a constant times

$$\int_{-\infty}^{s} \cdots \int_{-\infty}^{s} \Delta(\xi)^2 \, e^{-\frac{1}{2}\left(\sum \xi_i^2 + \frac{p_1}{1-k_1 p_1}(\sum \xi_i)^2\right)} d\xi_1 \cdots d\xi_{k_1} \int \prod_{\alpha>1} \Delta_\alpha(\eta)^2 \, e^{-\frac{1}{2}\sum \eta_j^2} d\eta,$$

where the η integration is over the orthogonal complement of $\{\sqrt{p_j}\}$. The η integral is just another constant. Therefore the distribution function equals

$$\frac{1}{c_{k_1, p_1}} \int_{-\infty}^{s} \cdots \int_{-\infty}^{s} \Delta(\xi)^2 \, e^{-\frac{1}{2}\left(\sum \xi_i^2 + \frac{p_1}{1-k_1 p_1}(\sum \xi_i)^2\right)} d\xi_1 \cdots d\xi_{k_1},$$

where c_{k_1, p_1} is the integral over all of $\mathbb{R}^{k_1}$.

To evaluate this we make the substitution (3–10), but with j replaced by i and each p_j replaced by $1/\sqrt{k}$. The integral becomes

$$\int \prod_j \Delta(\eta)^2 \, e^{-\frac{1}{2}\sum \eta_j^2} d\eta \int e^{-\frac{x^2}{2}\left(\frac{1}{k_1} + \frac{p_1}{1-k_1 p_1}\right)} dx,$$

taken over $x \in \mathbb{R}$ and η in hyperplane $\sum \eta_i = 0$ with Lebesgue measure. The x integral equals $\sqrt{2\pi k_1(1 - k_1 p_1)}$ while the first integral equals

$$(2\pi)^{(k_1-1)/2} \, 1! \, 2! \cdots k_1!.$$

(For the last, observe that the right side of (3–9) must equal 1 when $s = \infty$.) Hence

$$c_{k_1,p_1} = (2\pi)^{k_1/2}\, 1!\, 2! \cdots k_1!\, \sqrt{k_1(1 - k_1 p_1)}.$$

3.1. Distinct Probabilities: the Next Approximation.
If all the p_i are different then $P(\lambda) := \text{Prob}(\lambda)$ equals

$$\frac{\Delta(h)}{\Delta(p)} \frac{1}{\prod_{i=1}^{k-1} \prod_{j=i}^{k-1}(\lambda_i + k - j)} \prod_{i=1}^{k} p_i^{k-i}\, M_p(\lambda) \qquad (3\text{–}11)$$

plus an exponentially small correction. We recall that

$$\lambda_j = N p_j + \sqrt{N p_j}\, \xi_j$$

and compute the Fourier transform of the measure P with respect to the ξ variables. Beginning with M_p, we have

$$\widehat{M}_p(x) = \int e^{i \sum x_j \xi_j}\, dM_p(\lambda) = e^{-i \sum \sqrt{N p_j}\, x_j} \int e^{i \sum x_j \lambda_j / \sqrt{N p_j}}\, dM_p(\lambda)$$

$$= e^{-i \sum \sqrt{N p_j}\, x_j} \left(\sum p_j\, e^{i x_j / \sqrt{N p_j}} \right)^N,$$

since M_p is the multinomial distribution. An easy computation gives

$$\widehat{M}_p(x) = \left(1 + \frac{i}{\sqrt{N}} Q(x) + O\!\left(\frac{1}{N}\right) \right) e^{-\frac{1}{2} \sum x_j^2 + \frac{1}{2}(\sum \sqrt{p_j}\, x_j)^2},$$

where $Q(x)$ is a homogeneous polynomial of degree three. (In particular the limit of M_p is the inverse Fourier transform of the exponential in the above formula, which equals (3–7).)

As for the other nonconstant factors in (3–11), we have

$$\prod_{i=1}^{k-1} \prod_{j=i}^{k-1} (\lambda_i + k - j) = \prod_{i=1}^{k-1} \left(N p_i + \sqrt{N p_i}\, \xi_i + O(1) \right)^{k-i}$$

$$= N^{k(k-1)/2} \prod_{i=1}^{k-1} p_i^{k-i} \left(1 + \frac{1}{\sqrt{N}} \sum_{i=1}^{k-1} (k - i) \frac{\xi_i}{\sqrt{p_i}} + O\!\left(\frac{1}{N}\right) \right)$$

and

$$\Delta(h) = \prod_{i<j} \left(N(p_i - p_j) + \sqrt{N}(\sqrt{p_i}\, \xi_i - \sqrt{p_j}\, \xi_j) + O(1) \right)$$

$$= N^{k(k-1)/2} \Delta(p) \left(1 + \frac{1}{\sqrt{N}} \sum_{i<j} \frac{\sqrt{p_i}\, \xi_i - \sqrt{p_j}\, \xi_j}{p_i - p_j} + O\!\left(\frac{1}{N}\right) \right).$$

Thus the factors in (3–11) aside from M_p contribute

$$1 + \frac{1}{\sqrt{N}} \left(\sum_{i<j} \frac{\sqrt{p_i}\xi_i - \sqrt{p_j}\xi_j}{p_i - p_j} - \sum_{i<j} \frac{\xi_i}{\sqrt{p_i}} \right) + O\left(\frac{1}{N}\right)$$

$$= 1 + \frac{1}{\sqrt{N}} \left(\sum_{i<j} \sqrt{\frac{p_j}{p_i}} \frac{\sqrt{p_j}\xi_i - \sqrt{p_i}\xi_j}{p_i - p_j} \right) + O\left(\frac{1}{N}\right).$$

Using the fact that multiplication by ξ_j corresponds, after taking Fourier transforms, to $-i\partial_{x_j}$ and combining this with the preceding we deduce that $\widehat{P}(x)$, the Fourier transform of $P(\lambda)$ with respect to the ξ variables, equals

$$\left(1 + \frac{i}{\sqrt{N}} \sum_{i<j} \sqrt{\frac{p_j}{p_i}} \frac{\sqrt{p_j}x_i - \sqrt{p_i}x_j}{p_i - p_j} + \frac{i}{\sqrt{N}} Q(x) + O\left(\frac{1}{N}\right) \right) e^{-\frac{1}{2}\sum x_j^2 + \frac{1}{2}(\sum \sqrt{p_j}x_j)^2}$$

plus a correction which is exponentially small in N.

The Mean. We have

$$E(\xi_1) = \int \xi_1 \, dP(\lambda) = -i \, \partial_{x_1} \widehat{P}(x) \big|_{x=0}.$$

From the preceding discussion we see that this equals

$$\frac{1}{\sqrt{Np_1}} \sum_{j>1} \frac{p_j}{p_1 - p_j} + O\left(\frac{1}{N}\right).$$

Hence

$$E(l_N) = E(\lambda_1) = Np_1 + \sum_{j>1} \frac{p_j}{p_1 - p_j} + O\left(\frac{1}{\sqrt{N}}\right), \quad N \to \infty. \tag{3–12}$$

This last formula is, in fact, an accurate approximation for $E(l_N)$ (for distinct p_i) for moderate values of N. Table 1 summarizes various simulations of l_N and compares the means of these simulated values with the asymptotic formula. We remark that even though the proof assumed distinct p_i, we expect the asymptotic formula to remain valid for $p_1 > p_2 \geq \cdots \geq p_k$. (See the last set of simulations in Table 1.)

The Variance. We write our approximation as $P = P_0 + N^{-1/2}P_1 + O(N^{-1})$ with corresponding expected values $E = E_0 + N^{-1/2}E_1 + O(N^{-1})$. (In fact P_1 is a distribution, not a measure, but the meaning is clear.) Then the variance of λ_1 is equal to

$$Np_1 \left(E(\xi_1^2) - E(\xi_1)^2 \right)$$

$$= Np_1 \left(E_0(\xi_1^2) - E_0(\xi_1)^2 + \frac{1}{\sqrt{N}} E_1(\xi_1^2) - \frac{2}{\sqrt{N}} E_0(\xi_1) E_1(\xi_1) + O\left(\frac{1}{N}\right) \right).$$

Of course $E_0(\xi_1) = 0$, but also

$$E_1(\xi_1^2) = -\partial^2_{x_1, x_1} \widehat{P}_1(x) \big|_{x=0} = 0.$$

k	Probabilities of $\{1,\ldots,k\}$	N	N_S	Mean	$E(l_N)$
2	$\{\frac{5}{7}, \frac{2}{7}\}$	50	20 000	36.37	36.38
		100	20 000	72.12	72.10
		500	20 000	357.73	357.81
2	$\{\frac{6}{11}, \frac{5}{11}\}$	50	20 000	30.54	32.27
		100	20 000	58.52	59.55
		200	20 000	113.71	114.09
		400	20 000	223.16	223.18
3	$\{\frac{1}{2}, \frac{5}{14}, \frac{1}{7}\}$	50	10 000	27.53	27.90
		100	10 000	52.79	52.90
		500	10 000	252.80	252.90
		1000	10 000	502.78	502.90
3	$\{\frac{3}{8}, \frac{1}{3}, \frac{7}{24}\}$	50	10 000	23.96	30.25
		100	10 000	44.33	49.00
		500	10 000	197.65	199.00
		1000	2 000	386.08	386.50
3	$\{\frac{3}{8}, \frac{5}{16}, \frac{5}{16}\}$	50	10 000	23.92	28.75
		100	10 000	44.16	47.50
		200	10 000	83.15	85.00
		400	10 000	159.30	160.00
		800	10 000	310.08	310.00

Table 1. Simulations of the length of the longest weakly increasing subsequence in inhomogeneous random words of length N for two- and three-letter alphabets. N_S is the sample size. The last column gives the asymptotic expected value (3–12).

Since $E_0(\xi_1^2) - E_0(\xi_1)^2 = 1 - p_1$, we find that the variance of λ_1 equals

$$Np_1(1-p_1) + O(1)$$

and so its standard deviation equals $\sqrt{Np_1(1-p_1)} + O(N^{-1/2})$.

Acknowledgments

This work was begun during the MSRI Semester Random Matrix Models and Their Applications. We wish to thank D. Eisenbud and H. Rossi for their support during this semester. This work was supported in part by the National Science Foundation through grants DMS–9801608, DMS–9802122 and DMS–9732687. The last two authors thank Y. Chen for his kind hospitality at Imperial College where part of this work was done as well as the EPSRC for the award of a Visiting Fellowship, GR/M16580, that made this visit possible.

References

[1] J. Baik, P. Deift, and K. Johansson, "On the distribution of the length of the longest increasing subsequence of random permutations", *J. Amer. Math. Soc.* **12** (1999), 1119–1178.

[2] J. Baik, P. Deift, and K. Johansson, "On the distribution of the length of the second row of a Young diagram under Plancherel measure", *Geom. Funct. Anal.* **10** (2000), 702–731.

[3] J. Baik and E. M. Rains, "The asymptotics of monotone subsequences of involutions", preprint (arXiv: math.CO/9905084).

[4] J. Baik and E. M. Rains, "Algebraic aspects of increasing subsequences", preprint (arXiv: math.CO/9905083).

[5] D. Bayer and P. Diaconis, "Trailing the dovetail shuffle to its lair", *Ann. Applied Prob.* **2** (1992), 294–313.

[6] A. Borodin, A. Okounkov and G. Olshanski, "Asymptotics of Plancherel measures for symmetric groups", *J. Amer. Math. Soc.* **13** (2000), 481–515.

[7] I. M. Gessel, "Symmetric functions and P-recursiveness", *J. Comb. Theory, Ser. A,* **53** (1990), 257–285.

[8] C. Grinstead, unpublished notes on random words, $k = 2$.

[9] M. Jimbo and T. Miwa, "Monodromy preserving deformation of linear ordinary differential equations with rational coefficients, II", *Physica D* **2** (1981), 407–448.

[10] K. Johansson, "Shape fluctuations and random matrices", *Commun. Math. Phys.* **209** (2000), 437–476.

[11] K. Johansson, "Discrete orthogonal polynomial ensembles and the Plancherel measure", preprint (arXiv: math.CO/9906120).

[12] G. Kuperberg, "Random words, quantum statistics, central limits, random matrices", preprint (arXiv: math.PR/9909104).

[13] M. L. Mehta, *Random matrices*, 2nd ed., Academic Press, San Diego, 1991.

[14] A. Okounkov, "Random matrices and random permutations", preprint (arXiv: math.CO/9903176).

[15] R. P. Stanley, *Enumerative combinatorics*, Vol. 2, Cambridge University Press, Cambridge, 1999.

[16] R. P. Stanley, "Generalized riffle shuffles and quasisymmetric functions", preprint (arXiv: math.CO/9912025).

[17] C. A. Tracy and H. Widom, "Level-spacing distributions and the Airy kernel", *Commun. Math. Phys.* **159** (1994), 151–174.

[18] C. A. Tracy and H. Widom, "Random unitary matrices, permutations and Painlevé", *Commun. Math. Phys.* **207** (1999), 665–685.

[19] C. A. Tracy and H. Widom, "On the distributions of the lengths of the longest monotone subsequences in random words", preprint (arXiv: math.CO/9904042).

ALEXANDER R. ITS
DEPARTMENT OF MATHEMATICS
INDIANA UNIVERSITY–PURDUE UNIVERSITY INDIANAPOLIS
INDIANAPOLIS, IN 46202
UNITED STATES
  itsa@math.iupui.edu

CRAIG A. TRACY
DEPARTMENT OF MATHEMATICS
INSTITUTE OF THEORETICAL DYNAMICS
UNIVERSITY OF CALIFORNIA
DAVIS, CA 95616
UNITED STATES
  tracy@itd.ucdavis.edu

HAROLD WIDOM
DEPARTMENT OF MATHEMATICS
UNIVERSITY OF CALIFORNIA
SANTA CRUZ, CA 95064
UNITED STATES
  widom@math.ucsc.edu

Random Permutations and the Discrete Bessel Kernel

KURT JOHANSSON

ABSTRACT. Let l_N denote the length of a longest increasing subsequence of a random permutation from S_N. If we write $l_N = 2\sqrt{N} + N^{1/6}\chi_N$, then χ_N converges in distribution to a random variable χ with the Tracy–Widom distribution of random matrix theory. We give an outline of the basic steps in a proof of this result which does not use the asymptotics of Toeplitz determinants, and which, in a sense, explain why the largest eigenvalue distribution occurs.

1. Introduction

Consider the length $l_N(\sigma)$ of a longest increasing subsequence in a permutation $\sigma \in S_N$; if $\sigma = i_1 i_2 \ldots i_N$ and $i_{k_1} < \cdots < i_{k_r}$, then $i_{k_1}, \ldots, i_{k_r}$ is an increasing subsequence of length r. If we give S_N the uniform probability distribution, $l_N(\sigma)$ becomes a random variable and we want to investigate its distribution. This problem was first addressed by Ulam [1961], who made Monte Carlo simulations and concluded that the expectation $E[l_N]$ seems to be of order $\sqrt{N}$. The first rigorous result was obtained by Hammersley [1972], who considered the following variant of the problem. Consider a Poisson process in the square $[0,1] \times [0,1]$ with intensity α, so that the number M of points in the square is Poisson distributed with mean α. Let $x_1 < x_2 < \cdots < x_M$ and $y_1 < y_2 < \cdots < y_N$ be the x- and y-coordinates of the points $(x_j, y_{\sigma(j)})$, $1 \le j \le M$, in the square. This associates a permutation $\sigma \in S_M$ with each point configuration, and if we condition M to be fixed, equal to N say, we get the uniform distribution on S_N. We see that $l_M(\sigma)$ equals the number of points, $L(\alpha)$, in an up/right path from $(0,0)$ to $(1,1)$ through the points, and containing as many points as possible. The random variable $L(\alpha)$ is the Poissonization of l_N,

$$P[L(\alpha) \le n] = e^{-\alpha} \sum_{N=0}^{\infty} \frac{\alpha^N}{N!} P[l_N(\sigma) \le n]. \tag{1.1}$$

Using subadditivity Hammersley showed that $E[L(\alpha)]/\sqrt{\alpha} \to c$ as $\alpha \to \infty$ with a positive constant c. Numerical simulations [Baer and Brock 1968] indicated that $c = 2$, and this was proved by Vershik and Kerov [1977]. That $c = 2$ has also been proved using Hammersley's picture in [Aldous and Diaconis 1995; Seppäläinen 1996]. Large deviation results have been obtained in [Deuschel and Zeitouni 1999; Seppäläinen 1998; Johansson 1998]. For more background on the problem see [Aldous and Diaconis 1999]. Hammersley's Poisson model also has an interesting interpretation as a certain 1+1-dimensional random growth model called polynuclear growth; see [Prähofer and Spohn 2000]. The fluctuations around the mean has been an open problem for a long time. Numerical simulations by Odlyzko and Rains [1998], indicate that the standard deviation for l_N is like a constant times $N^{1/6}$, and this can also be seen heuristically from the large deviation formulas. A precise result for the fluctuations was proved by Baik, Deift and Johansson in [Baik et al. 1999]. To state the result we need some definitions. Let u be the unique solution to the Painlevé II equation

$$u'' = 2u^3 + xu,$$

which satisfies $u(x) \sim \text{Ai}(x)$, as $x \to \infty$, where $\text{Ai}(x)$ is the Airy function. Such a solution exits [Hastings and McLeod 1980; Deift and Zhou 1995]. Put

$$F(t) = \exp\Bigl(-\int_t^\infty (x-t)u(x)^2\,dx\Bigr). \tag{1.2}$$

The result in [Baik et al. 1999] is:

THEOREM. *The random variable $(l_N - 2\sqrt{N})/N^{1/6}$ converges in distribution to a random variable with distribution function* (1.2),

$$\lim_{N \to \infty} P[l_N \leq 2\sqrt{N} + tN^{1/6}] = F(t). \tag{1.3}$$

Also, we have convergence of all moments.

The proof of (1.3) in [Baik et al. 1999] is based on the following formula due to Gessel [1990], which expresses the probability in (1.1) as a Toeplitz determinant,

$$P[L(\alpha) \leq n] = e^{-\alpha}D_n(e^{2\sqrt{\alpha}\cos\theta})$$
$$= \det\Bigl(\frac{1}{2\pi}\int_{-\pi}^\pi e^{2\sqrt{\alpha}\cos\theta - i(j-k)\theta}\,d\theta\Bigr)_{j,k=1}^n. \tag{1.4}$$

The Toeplitz determinant can be expressed in terms of the leading coefficients of the normalized orthogonal polynomials on the unit circle, $\mathbb{T}$, with respect to the weight $e^{2\sqrt{\alpha}\cos\theta}$. These orthogonal polynomials can be obtained as the solution of a certain matrix Riemann–Hilbert problem on $\mathbb{T}$, [Fokas et al. 1991], and using the powerful steepest descent argument for Riemann–Hilbert problems developed by Deift and Zhou [1993], it is possible to show that

$$\lim_{\alpha \to \infty} P[L(\alpha) \leq 2\sqrt{\alpha} + t\alpha^{1/6}] = F(t) \tag{1.5}$$

for each ∈ ℝ. This is closely related to the so called double scaling limit in unitary random matrix models [Periwal and Shevitz 1990]. From (1.5) the result (1.3) can be deduced by a de-Poissonization argument [Johansson 1998].

The distribution function $F(t)$ also has a different expression. Let

$$A(x,y) = \int_0^\infty \mathrm{Ai}(x+t)\,\mathrm{Ai}(y+t)\,dt,$$

be the *Airy kernel*. Then,

$$F(t) = \det(I - A)_{L^2[t,\infty)}, \qquad (1.6)$$

where the right hand side is the Fredholm determinant

$$\det(I - A)_{L^2[t,\infty)} = \sum_{m=0}^\infty \frac{(-1)^m}{m!} \int_{[t,\infty)^m} \det(A(x_i,x_j))_{i,j=1}^m d^m x. \qquad (1.7)$$

The fact that (1.2) and (1.7) are equal was proved by Tracy and Widom [1994], and $F(t)$ is often referred to as the *Tracy–Widom distribution*. The interesting thing about (1.6) is that the right hand side of (1.6) is the asymptotic distribution for the appropriately scaled largest eigenvalue of a random matrix from the Gaussian Unitary Ensemble (GUE) [Mehta 1991], as the size of the matrix goes to infinity [Tracy and Widom 1994]. Thus $l_N(\sigma)$ behaves, for large N, as the largest eigenvalue of a large random hermitian matrix!

In this paper we will outline a proof of (1.5) that does not use Gessel's formula (1.4) and which makes it clearer why the largest eigenvalue distribution appears. The presentation is based on [Johansson 2000; 2001], to which we refer for more details. Below we will not give all the technical details, in particular those concerning the precise justification of different limits. A closely related proof appears in [Borodin et al. 2000], see the remark at the end of section 3.

In the next section we will outline the main ingredients that go into the proof. The actual argument for (1.5) will be given in the last section.

2. Main Ingredients

2a. The Robinson–Schensted–Knuth Correspondence. General references for this subsection and the next are [Fulton 1997; Sagan 1991; Stanley 1999]. A *partition* of K is a sequence $\lambda = (\lambda_1, \ldots, \lambda_N)$, $\lambda_1 \geq \cdots \geq \lambda_N \geq 0$, of integers, such that $\lambda_1 + \cdots + \lambda_N = K$, and can be illustrated by a *Young diagram* with λ_j left justified boxes in row j. If we write numbers from $\{1, \ldots, N\}$ in the boxes in such away that the numbers in each row are weakly increasing and the numbers in each cloumn are strictly increasing, we get a *semistandard Young tableaux* T of shape λ, $\mathrm{sh}(T) = \lambda$, with entries in $\{1, \ldots, N\}$. Let $m_i(T)$ denote the number of i's in T.

An integer $N \times N$ matrix $A = (w_{ij})$, $w_{ij} \in \mathbb{N}$, can be described by a *generalized permutation*

$$\sigma = \begin{pmatrix} i_1 & \cdots & i_K \\ j_1 & \cdots & j_K \end{pmatrix}, \quad i_r, j_r \in \{1, \ldots, N\},$$

where $i_r \leq i_{r+1}$ and if $i_r = i_{r+1}$, then $j_r \leq j_{r+1}$. The matrix A is mapped bijectively to the σ in which a pair $\binom{i}{j}$ occurs w_{ij} times.

The *Robinson–Schensted–Knuth correspondence* [Knuth 1970], or RSK correspondence, maps σ bijectively to a pair of semistandard tableaux (P, Q) of the same shape λ with entries in $\{1, \ldots, N\}$. The numbers $j_1, \ldots, j_K$ go into P and $i_1, \ldots, i_K$ go into Q, so λ is a partition of $K = \sum_{i,j} w_{ij}$. Note that $m_i(Q) = \sum_j w_{ij}$ and $m_j(P) = \sum_i w_{ij}$. This correspondence has the property that λ_1, the length of the first row, equals the length $l(\sigma)$ of the longest weakly increasing subsequence in $j_1, \ldots, j_K$. In terms of the matrix A this has the following interpretation. Let $\pi = \{(m_r, n_r)\}_{r=1}^{2N-1}$ be an *up/right path* from $(1,1)$ to (N,N), i.e. $(m_{r+1}, n_{r+1}) - (m_r, n_r) = (1,0)$ or $(0,1)$, $(m_1, n_1) = (1,1)$ and $(m_{2N-1}, n_{2N-1}) = (N, N)$. Set

$$L_N(A) = \max_\pi \sum_{(i,j) \in \pi} w_{ij}. \tag{2.1}$$

It is not difficult to see that $L_N(A)$ is equal to $l(\sigma)$ and hence $L_N(A) = \lambda_1$.

2b. The Schur Polynomial. Let $\lambda = (\lambda_1, \ldots, \lambda_N)$ be a partition of K as above. Then the *Schur polynomial* $s_\lambda(x_1, \ldots, x_N)$ is a certain symmetric, homogeneous polynomial of degree K. Let $\Delta_N(x) = \prod_{1 \leq i < j \leq N}(x_i - x_j)$ denote the Vandermonde determinant. We have two different expressions for $s_\lambda(x)$,

$$s_\lambda(x) = \sum_{T: \text{sh}(T) = \lambda} \prod_{i=1}^N x_i^{m_i(T)} = \frac{1}{\Delta_N(x)} \det(x_j^{\lambda_k + N - k})_{j,k=1}^N, \tag{2.2}$$

where the sum is over all semistandard tableaux with shape λ and entries in $\{1, \ldots, N\}$. The second formula is known as the Jacobi–Trudi identity.

2c. Orthogonal Polynomial Ensembles. Consider the probability measure

$$\frac{1}{Z_N} \Delta_N(x)^2 \prod_{j=1}^N w(x_j) d\mu(x_j) \tag{2.3}$$

on $\mathbb{R}^N$, where $w(x)$ is a nonnegative weight function and the measure μ typically is the Lebesgue measure on $\mathbb{R}$, some interval in $\mathbb{R}$ or the counting measure on $\mathbb{N}$. These type of measures occur as eigenvalue measures in invariant ensembles of hermitian matrices. GUE for example has the eigenvalue measure (2.3) with $w(x) = e^{-x^2}$ and μ the Lebesgue measure on $\mathbb{R}$. An important property of measures of the form (2.3) is that all marginal distributions (correlation functions) are given by certain determinants. The k-th marginal probability is

$$\frac{(N-k)!}{N!}\det(K_N(x_i,x_j))_{i,j=1}^k d\mu(x_1)\ldots d\mu(x_k), \tag{2.4}$$

where

$$K_N(x,y) = \sum_{n=0}^{N-1} p_k(x)p_k(y)(w(x)w(y))^{1/2},$$

and $p_k(x)$, $k=0,1,\ldots$, are the normalized orthogonal polynomials with respect to the measure $w(x)d\mu(x)$,

$$\int_{\mathbb{R}} p_j(x)p_k(y)w(x)d\mu(x) = \delta_{jk},$$

see [Mehta 1991] or [Tracy and Widom 1998]. For GUE the relevant orthogonal polynomials are the Hermite polynomials. Using the fact that the marginal probabilities are given by (2.4) we see that

$P[\max_{1\le j\le N} x_j \le t]$

$$= \frac{1}{Z_N}\int_{\mathbb{R}^N} \prod_{j=1}^N (1-\chi_{(t,\infty)}(x_j))\Delta_N(x)^2 \prod_{j=1}^N w(x_j)d\mu(x_j)$$

$$= \sum_{m=0}^N \frac{(-1)^m}{m!}\int_{\mathbb{R}^m} \det(K_N(x_i,x_j)\chi_{(t,\infty)}(x_j))_{i,j=1}^m d\mu(x_1)\ldots d\mu(x_m)$$

$$= \det(I - K_N\chi_{(t,\infty)})_{L^2(\mathbb{R},d\mu)}. \tag{2.5}$$

The last expression is the Fredholm determinant with kernel $K_N(x,y)\chi_{(t,\infty)}(y)$ on $L^2(\mathbb{R},d\mu)$. By using asymptotic formulas for Hermite polynomials, (2.5) can be used to show that the appropriately scaled largest eigenvalue of a GUE-matrix has a limiting distribution given by (1.6).

In our proof of (1.5) we will get (2.3) with $d\mu(x)$ counting measure on $\mathbb{N}$ and $w(x) = q^x$, for $0 < q < 1$ a fixed parameter. The orthogonal polynomials are then the Meixner polynomials, and we turn to them next.

2d. Asymptotics for Meixner Polynomials and the Discrete Bessel Kernel. The polynomials $M_n^q(x)$ (which are multiples of the standard Meixner polynomials $m_n^{1,q}(x)$; see [Chihara 1978; Nikiforov et al. 1991]), satisfy

$$\sum_{x=0}^\infty M_n^q(x)M_m^q(x)q^x = \delta_{nm},$$

and have the integral representation [Chihara 1978]

$$M_n^q(x) = \frac{q^{n/2}\sqrt{1-q}}{2\pi i}\int_{\Gamma_{\sqrt{q},q}} \frac{(1+w/q)^x}{(1+w)^{x+1}w^{n+1}}dw, \tag{2.6}$$

where $\Gamma_{r,q}$ is the circle $re^{i\theta}$, $-\pi < \theta < \pi$, together with the line segments from $-r + i0$ to $-rq + i0$ and from $-rq - i0$ to $-r - i0$. The integral formula is a consequence of the generating function for the Meixner polynomials. The kernel we need in (2.4) is the *Meixner kernel*

$$K_N^q(x,y) = \sum_{n=0}^{N-1} M_n^q(x) M_n^q(y) q^{(x+y)/2}. \qquad (2.7)$$

If we make the change of variables $w = z\sqrt{q}$ in (2.6) we see that

$$\lim_{N\to\infty} M_{N-n}^{\alpha/N^2}(x+N) \left(\frac{\alpha}{N^2}\right)^{(x+N)/2}$$

$$= \lim_{N\to\infty} \frac{\sqrt{1-\alpha/N^2}}{2\pi i} \int_{\Gamma_{1,\alpha/N^2}} \frac{(1+\sqrt{\alpha}/(zN))^N}{(1+z\sqrt{\alpha}/N)^N} z^{x+n} \frac{(1+\sqrt{\alpha}/zN)^x}{(1+z\sqrt{\alpha}/N)^{x+1}} \frac{dz}{z}$$

$$= \frac{1}{2\pi i} \int_{\Gamma_{1,0}} e^{\sqrt{\alpha}(1/z-z)} z^{x+n} \frac{dz}{z} = J_{x+n}(2\sqrt{\alpha}), \qquad (2.8)$$

where $J_n(t)$ is the standard Bessel function. Using (2.7) and (2.8) we see that

$$\lim_{N\to\infty} K_N^{\alpha/N^2}(x+N, y+N)$$

$$= \lim_{N\to\infty} \sum_{n=1}^{N} M_{N-n}^{\alpha/N^2}(x+N) \left(\frac{\alpha}{N^2}\right)^{(x+N)/2} M_{N-n}^{\alpha/N^2}(y+N) \left(\frac{\alpha}{N^2}\right)^{(y+N)/2}$$

$$= \sum_{k=1}^{\infty} J_{x+k}(2\sqrt{\alpha}) J_{y+k}(2\sqrt{\alpha}) =: B^{\alpha}(x,y). \qquad (2.9)$$

We call $B^{\alpha}(x,y)$ the *discrete Bessel kernel*. The Bessel function $J_n(t)$ has the following asymptotics for n and t of the same order,

$$\lim_{\alpha\to\infty} \alpha^{1/6} J_{2\sqrt{\alpha}+\xi\alpha^{1/6}}(2\sqrt{\alpha}) = \operatorname{Ai}(\xi). \qquad (2.10)$$

Note the similarity with (1.5)! Combining the definition (2.9) of $B^{\alpha}(x,y)$ and (2.10) we see that we should have

$$\lim_{\alpha\to\infty} \alpha^{1/6} B^{\alpha}(2\sqrt{\alpha}+\xi\alpha^{1/6}, 2\sqrt{\alpha}+\eta\alpha^{1/6})$$

$$= \lim_{\alpha\to\infty} \sum_{k=1}^{\infty} \alpha^{1/6} J_{2\sqrt{\alpha}+\xi\alpha^{1/6}+k}(2\sqrt{\alpha}) \alpha^{1/6} J_{2\sqrt{\alpha}+\eta\alpha^{1/6}+k}(2\sqrt{\alpha}) \alpha^{-1/6}$$

$$= \lim_{\alpha\to\infty} \sum_{k=1}^{\infty} \operatorname{Ai}(\xi + k\alpha^{-1/6}) \operatorname{Ai}(\eta + k\alpha^{-1/6}) \alpha^{-1/6}$$

$$= \int_0^{\infty} \operatorname{Ai}(\xi+t) \operatorname{Ai}(\eta+t) dt = A(\xi, \eta). \qquad (2.11)$$

3. The Proof

Let w_{ij}, $1 \leq i,j \leq N$, be independent, geometrically distributed random variables, $P[w_{ij} = x] = (1-q)q^x$, $x \in \mathbb{N}$, and let $A = (w_{ij})_{i,j=1}^N$. We let $L_N(A)$ denote the maximal sum along an up/right path as in (2.1). Let P_N^q denote the measure we get on the set of integer matrices. If q is small then most of the w_{ij}'s will be zero and a few will be equal to one. In particular, if we take $q = \alpha/N^2$, then with probability going to 1 as $N \to \infty$, there will be at most one w_{ij} equal to 1 in each row and each column, and no w_{ij} will be ≥ 2. Put a particle in the square $[(i-1)/N, i/N] \times [(j-1)/N, j/N]$ if $w_{ij} = 1$ and no particle if $w_{ij} = 0$. As $N \to \infty$ this will converge to a Poisson process in $[0,1] \times [0,1]$ with intensity α, so the number of particles in the unit square will be Poisson distributed with mean α. Some thought shows that $L_N(A)$ will converge to the maximal number of points in an up/right path from $(0,0)$ to $(1,1)$ through the Poisson points, that is to the random variable $L(\alpha)$ that we defined in the introduction. Thus we have,

$$P[L(\alpha) \leq n] = \lim_{N \to \infty} P_N^{\alpha/N^2}[L_N(A) \leq n]. \tag{3.1}$$

Next, we will see that we can use the RSK correspondence and the Schur polynomial to derive an expression for the probability in the right hand side of (3.1). If A is mapped to (P,Q) by the RSK correspondence, then

$$P_N^q[A = (a_{ij})] = (1-q)^{N^2} q^{\sum_{i,j} a_{ij}}$$

$$= (1-q)^{N^2} \prod_{i=1}^N (\sqrt{q})^{\sum_j a_{ij}} \prod_{j=1}^N (\sqrt{q})^{\sum_i a_{ij}}$$

$$= (1-q)^{N^2} \prod_{i=1}^N (\sqrt{q})^{m_i(Q)} \prod_{j=1}^N (\sqrt{q})^{m_j(P)}.$$

Consequently, by the RSK correspondence and the first equality in (2.2),

$$P_N^q[L_N(A) \leq n]$$

$$= \sum_{A:L_N(A) \leq n} P_N^q[A]$$

$$= \sum_{\lambda:\lambda_1 \leq n} \sum_{P:\text{sh}(P)=\lambda} \sum_{Q:\text{sh}(Q)=\lambda} (1-q)^{N^2} \prod_{i=1}^N (\sqrt{q})^{m_i(Q)} \prod_{j=1}^N (\sqrt{q})^{m_j(P)} \tag{3.2}$$

$$= (1-q)^{N^2} \sum_{\lambda:\lambda_1 \leq n} s_\lambda(\sqrt{q},\ldots,\sqrt{q}) s_\lambda(\sqrt{q},\ldots,\sqrt{q})$$

$$= (1-q)^{N^2} \sum_{\lambda:\lambda_1 \leq n} s_\lambda(1,\ldots,1)^2 q^{\sum_j \lambda_j}.$$

Since semistandard Young tableaux are strictly increasing in columns, λ can have at most N nonzero parts, $\lambda = (\lambda_1, \ldots, \lambda_N)$. We can now use the second equality in (2.2) to get

$$s_\lambda(1,\ldots,1) = \lim_{r \to 1} s_\lambda(1,r,\ldots,r^{N-1}) = \prod_{1 \le i < j \le N} \frac{\lambda_i - \lambda_j + j - i}{j - i}. \quad (3.3)$$

If we introduce the new variables $h_j = \lambda_j + N - j$, $h_1 > h_2 > \ldots > h_N \ge 0$, we see that (3.2) and (3.3) give

$$P_N^q[L_N(A) \le n] = \frac{1}{Z_N} \sum_{h \in \{0,\ldots,n+N-1\}^N} \Delta_N(h)^2 \prod_{j=1}^N q^{h_j}, \quad (3.4)$$

where $Z_N = q^{N(N-1)/2}(1-q)^{-N^2} \prod_{j=1}^{N-1} j!^2$. Note that (3.4) has exactly the form of an orthogonal polynomial ensemble. The relevant orthogonal polynomials are the Meixner polynomials (2.6). The computation (2.5) gives

$$P_N^q[L_N(A) \le n] = \det(I - \mathcal{K}_N^q)_{\ell^2(\{n,n+1,\ldots\})}, \quad (3.5)$$

where we have the kernel

$$\mathcal{K}_N^q(x,y) = K_N^q(x+N, x+N),$$

and K_N^q is the Meixner kernel (2.7).

We can now combine (3.1) and (3.5) and use (2.9) to see that

$$P[L(\alpha) \le n] = \lim_{N \to \infty} \det(I - \mathcal{K}_N^{\alpha/N^2})_{\ell^2(\{n,n+1,\ldots\})}$$
$$= \det(I - B^\alpha)_{\ell^2(\{n,n+1,\ldots\})}. \quad (3.6)$$

Note that combining (1.4) and (3.6) gives an interesting equality between a Toeplitz and a Fredholm determinant, which has been generalized by Borodin and Okounkov [2000]; see also [Basor and Widom 2000].

When we have the formula (3.6) we are not far from a proof of (1.5). Write the Fredholm expansion of the right hand side in (3.6)

$$P[L(\alpha) \le 2\sqrt{\alpha} + t\alpha^{1/6}] = \sum_{m=0}^\infty \frac{(-1)^m}{m!}$$
$$\times \sum_{h \in \mathbb{N}^m} \det(\alpha^{1/6} B^\alpha(2\sqrt{\alpha} + t\alpha^{1/6} + h_i, 2\sqrt{\alpha} + t\alpha^{1/6} + h_j))_{i,j=1}^m (\alpha^{-1/6})^m.$$

By (2.11) this is approximately

$$\sum_{m=0}^\infty \frac{(-1)^m}{m!} \sum_{h \in \mathbb{N}^m} \det(A(t + h_i \alpha^{-1/6}, t + h_j \alpha^{-1/6}))_{i,j=1}^m (\alpha^{-1/6})^m \quad (3.7)$$

for large α, and in the limit $\alpha \to \infty$, the Riemann sums in (3.7) converge to integrals and we get

$$P[L(\alpha) \leq 2\sqrt{\alpha} + t\alpha^{1/6}] = \sum_{m=0}^{\infty} \frac{(-1)^m}{m!} \int_{[t,\infty)^m} \det(A(x_i, x_j))_{i,j=1}^m d^m x$$

$$= \det(I - A)_{L^2[t,\infty)} = F(t),$$

by (1.6). This establishes (1.5).

The RSK correspondence actually transforms the Poissonization of the uniform measure on S_N (that is, we regard N as a Poisson random variable) to a measure on partitions $\lambda = (\lambda_1, \lambda_2, \ldots)$, called the Poissonized Plancherel measure. Above we have studied the asymptotic distribution of λ_1, but the analysis can be extended to the lengths of the other rows as well [Johansson 2001]. The Poissonized Plancherel measure has also been investigated by Borodin, Okounkov and Olshanski [Borodin et al. 2000], coming from measures on partitions motivated by representation theory. They also obtain the discrete Bessel kernel and are able to give a closely related proof of (1.5).

Acknowledgement

I thank A. R. Its and P. Bleher for inviting me to the program on Random Matrices and Applications held at MSRI during the spring of 1999.

References

[Aldous and Diaconis 1995] D. Aldous and P. Diaconis, "Hammersley's interacting particle process and longest increasing subsequences", *Probab. Theory Related Fields* **103**:2 (1995), 199–213.

[Aldous and Diaconis 1999] D. Aldous and P. Diaconis, "Longest increasing subsequences: from patience sorting to the Baik-Deift-Johansson theorem", *Bull. Amer. Math. Soc. (N.S.)* **36**:4 (1999), 413–432.

[Baer and Brock 1968] R. M. Baer and P. Brock, "Natural sorting over permutation spaces", *Math. Comp.* **22** (1968), 385–410.

[Baik et al. 1999] J. Baik, P. Deift, and K. Johansson, "On the distribution of the length of the longest increasing subsequence of random permutations", *J. Amer. Math. Soc.* **12**:4 (1999), 1119–1178.

[Basor and Widom 2000] E. L. Basor and H. Widom, "On a Toeplitz determinant identity of Borodin and Okounkov", *Integral Equations Operator Theory* **37**:4 (2000), 397–401.

[Borodin and Okounkov 2000] A. Borodin and A. Okounkov, "A Fredholm determinant formula for Toeplitz determinants", *Integral Equations Operator Theory* **37**:4 (2000), 386–396.

[Borodin et al. 2000] A. Borodin, A. Okounkov, and G. Olshanski, "Asymptotics of Plancherel measures for symmetric groups", *J. Amer. Math. Soc.* **13** (2000), 491–515. Preprint at http://arXiv.org/abs/math/9905032.

[Chihara 1978] T. S. Chihara, *An introduction to orthogonal polynomials*, vol. 13, Mathematics and its Applications, Gordon and Breach Science Publishers, New York, 1978.

[Deift and Zhou 1993] P. Deift and X. Zhou, "A steepest descent method for oscillatory Riemann-Hilbert problems. Asymptotics for the MKdV equation", *Ann. of Math.* (2) **137**:2 (1993), 295–368.

[Deift and Zhou 1995] P. A. Deift and X. Zhou, "Asymptotics for the Painlevé II equation", *Comm. Pure Appl. Math.* **48**:3 (1995), 277–337.

[Deuschel and Zeitouni 1999] J.-D. Deuschel and O. Zeitouni, "On increasing subsequences of I.I.D. samples", *Combin. Probab. Comput.* **8**:3 (1999), 247–263.

[Fokas et al. 1991] A. S. Fokas, A. R. Its, and A. V. Kitaev, "Discrete Painlevé equations and their appearance in quantum gravity", *Comm. Math. Phys.* **142**:2 (1991), 313–344.

[Fulton 1997] W. Fulton, *Young tableaux*, London Math. Soc. Student texts **35**, Cambridge University Press, Cambridge, 1997.

[Gessel 1990] I. M. Gessel, "Symmetric functions and P-recursiveness", *J. Combin. Theory Ser. A* **53**:2 (1990), 257–285.

[Hammersley 1972] J. M. Hammersley, "A few seedlings of research", pp. 345–394 in *Proceedings of the Sixth Berkeley Symposium on Mathematical Statistics and Probability*, Vol. I: *Theory of statistics* (Berkeley, 1970/1971), edited by L. M. LeCam et al., Univ. California Press, Berkeley, Calif., 1972.

[Hastings and McLeod 1980] S. P. Hastings and J. B. McLeod, "A boundary value problem associated with the second Painlevé transcendent and the Korteweg–de Vries equation", *Arch. Rational Mech. Anal.* **73**:1 (1980), 31–51.

[Johansson 1998] K. Johansson, "The longest increasing subsequence in a random permutation and a unitary random matrix model", *Math. Res. Lett.* **5**:1-2 (1998), 63–82.

[Johansson 2000] K. Johansson, "Shape fluctuations and random matrices", *Comm. Math. Phys.* **209**:2 (2000), 437–476.

[Johansson 2001] K. Johansson, "Discrete orthogonal polynomial ensembles and the Plancherel measure", *Ann. of Math.* (2001). To appear; preprint version available at www.arxiv.org/abs/math.CO/9906120.html.

[Knuth 1970] D. E. Knuth, "Permutations, matrices, and generalized Young tableaux", *Pacific J. Math.* **34** (1970), 709–727.

[Mehta 1991] M. L. Mehta, *Random matrices*, 2nd ed., Academic Press, Boston, 1991.

[Nikiforov et al. 1991] A. F. Nikiforov, S. K. Suslov, and V. B. Uvarov, *Classical orthogonal polynomials of a discrete variable*, Series in Computational Physics, Springer, Berlin, 1991.

[Odlyzko and Rains 1998] A. M. Odlyzko and E. M. Rains, "On longest increasing subsequences in random permutations", preprint, 1998.

[Periwal and Shevitz 1990] V. Periwal and D. Shevitz, "Unitary-matrix models as exactly solvable string theories", *Phys. Rev. Lett.* **64** (1990), 1326–1329.

[Prähofer and Spohn 2000] M. Prähofer and H. Spohn, "Universal distributions for growth processes in 1+1 dimensions and random matrices", *Phys. Rev. Lett.* **84** (2000), 4882–.

[Sagan 1991] B. E. Sagan, *The symmetric group: representations, combinatorial algorithms, and symmetric functions*, Wadsworth & Brooks/Cole, Pacific Grove, CA, 1991.

[Seppäläinen 1996] T. Seppäläinen, "A microscopic model for the Burgers equation and longest increasing subsequences", *Electron. J. Probab.* **1**:5 (1996), 1–51.

[Seppäläinen 1998] T. Seppäläinen, "Large deviations for increasing sequences on the plane", *Probab. Theory Related Fields* **112**:2 (1998), 221–244.

[Stanley 1999] R. P. Stanley, *Enumerative combinatorics*, vol. 2, Cambridge Univ. Press, Cambridge, 1999. With a foreword by Gian-Carlo Rota and appendix 1 by Sergey Fomin.

[Tracy and Widom 1994] C. A. Tracy and H. Widom, "Level-spacing distributions and the Airy kernel", *Comm. Math. Phys.* **159**:1 (1994), 151–174.

[Tracy and Widom 1998] C. A. Tracy and H. Widom, "Correlation functions, cluster functions, and spacing distributions for random matrices", *J. Statist. Phys.* **92**:5-6 (1998), 809–835.

[Ulam 1961] S. M. Ulam, "Monte Carlo calculations in problems of mathematical physics", pp. 261–281 in *Modern mathematics for the engineer: Second series*, edited by E. F. Beckenbach, McGraw-Hill, New York, 1961.

[Versik and Kerov 1977] A. M. Versik and S. V. Kerov, "Asymptotics of the Plancherel measure of the symmetric group and the limiting form of Young tables", *Dokl. Akad. Nauk SSSR* **233** (1977), 1024–1027. In Russian; translated in *Sov. Math. Dokl.* **18** (1977), 527–531.

KURT JOHANSSON
DEPARTMENT OF MATHEMATICS
ROYAL INSTITUTE OF TECHNOLOGY
S-100 44 STOCKHOLM
SWEDEN
kurtj@math.kth.se

Solvable Matrix Models

VLADIMIR KAZAKOV

ABSTRACT. We review some old and new methods of reduction of the number of degrees of freedom from $\sim N^2$ to $\sim N$ in the multimatrix integrals.

1. Introduction

Multimatrix integrals of various types appear in many mathematical and physical applications, such as combinatorics of graphs, topology, integrable systems, string theory, theory of mesoscopic systems or statistical mechanics on random surfaces.

A general Q-matrix integral of the form

$$Z = \int \prod_{q=1}^{Q} d^{N^2} M_q \exp S(M_1, \ldots, M_Q)$$

usually goes over the $N \times N$ hermitian, real symmetric or symplectic matrices M_q with the action S and the measure symmetric under the simultaneous group rotation: $M_q \to \Omega^+ M_q \Omega$. Some other multimatrix integrals, such as these with complex matrices or with general real matrices, can be reduces to those three basic cases.

We will consider here only the case of hermitian matrices for which Ω belongs to the $U(N)$-group.

In many applications "to solve" the corresponding matrix model usually means to reduce the number of variables by explicit integrations over most of the variables in such a way that instead of QN^2 original integrations (matrix elements) one would be left in the large N limit only with $\sim N$ integration variables. In this case the integration over the rest of the variables can be performed, at least in the widely used large N limit, by means of the saddle point approximation. A more sophisticated double scaling limit [1] is also possible (if possible at all) only after such a reduction. The key of success is in the fact that after reduction the effective action at the saddle point is still of the order $\sim N^2$ whereas the corrections given by the logarithm of determinant of the second variation of the action cannot be bigger than $\sim N$ (the "entropy" of the remaining variables). The problem is thus reduced to the solution of the "classical" saddle point equations, instead of the "quantum" problem of functional (in the large N limit) integration over the original matrix variables.

Such an explicit reduction of the number of "degrees of freedom" is in general possible only for a few rather restricted, though physically and mathematically interesting, classes of multimatrix integrals. The purpose of our present notes is to review the basic old and new methods of such a reduction. Before going to the particular cases let us stress the importance of the search for new methods of such a reduction: any nontrivial finding on this way leads immediately to numerous fruitful applications.

2. Some Old Examples

The best known example of such a reduction of the number of degrees of freedom is the one-matrix integral

$$Z = \int d^{N^2} M \exp N \operatorname{Tr} S(M),$$

where $S(M)$ is an arbitrary function of one variable. We use the decomposition:

$$M = \Omega^+ x \Omega,$$

where $x = diag(x_1, \ldots, x_N)$ is a diagonal matrix of the eigenvalues and Ω is the $U(N)$ group variable. The corresponding (Dyson) measure can be written as

$$d^{N^2} M = d[\Omega]_{U(N)} \Delta^2(x) \prod_{k=1}^{N} dx_k \tag{2-1}$$

where $\Delta(x) = \prod_{i>j}(x_i - x_j)$ is the Vandermonde determinant. The integrand as an invariant function does not depend at all on Ω (the integration over it produces just a group volume factor which we will always omit). The remaining integral over the eigenvalues reads

$$Z = \int \prod_{k=1}^{N} dx_k \exp(NS(x_k)) \Delta^2(x).$$

In the large N limit the corresponding saddle point equation takes the form

$$\frac{1}{N} \frac{\partial S}{\partial x_k} = S'(x_k) + \frac{1}{N} \sum_{j \neq k} \frac{1}{x_k - x_j} = 0.$$

These arguments were successfully used for an interesting combinatorial problem: enumeration of graphs of fixed two dimensional topologies [14; 11]. There exist powerful methods to analyze this equation but it is not our present goal to review them here.

The next fruitful example is the so called two matrix model:

$$Z = \int d^{N^2}A\, d^{N^2}B \exp N \operatorname{Tr}\left(-A^2 - B^2 + cAB + U(A) + V(B)\right), \qquad (2\text{--}2)$$

where U and V are some arbitrary functions of one variable. After the decomposition $A = \Omega_1^+ x \Omega_2$, $B = \Omega_2^+ y \Omega_2$ we are left, due to the term $\operatorname{Tr}(AB)$ in the action, with one nontrivial unitary integral over the variable $\Omega = \Omega_1 \Omega_2^+$. Fortunately, this integral was explicitly calculated by Harish-Chandra [10] and by Itzykson and Zuber [11]:

$$\int d[\Omega]_{U(N)} \exp \operatorname{Tr}(\Omega^+ x \Omega y) = \prod_{k=1}^{N-1} k!\, \frac{\det_{ij} e^{x_i y_j}}{\Delta(x)\Delta(y)}. \qquad (2\text{--}3)$$

Substituting (2–3) and the Dyson measure (2–1) into (2–2) we are left again with only $2N$ variables x_k and y_k and we can write again the saddle point equations in the large N limit. They are more complicated than in the one matrix integral but can be nevertheless solved quite explicitly. The first solution of that kind was found in [16] in an indirect way, using the method of orthogonal polynomials, but the direct solution is also possible; see [18].

This model was used in [17] to solve exactly the first example of new statistical mechanical models of interacting spins on random planar graphs: in this case it was a model of Ising spins on random planar graphs.

An obvious generalization of the two matrix model is the matrix chain model:

$$Z = \int \prod_{q=1}^Q d^{N^2} M_q \, \exp \operatorname{Tr}\left(\sum_{p=1}^Q V_q(M_q) + \sum_{p=1}^{Q-1} M_{p-1} M_p\right). \qquad (2\text{--}4)$$

One easily notices that the same unitary decomposition $M_q = \Omega_q^+ x_q \Omega_q$ leads to $Q-1$ independent integrals over the variables $U_q = \Omega_{q-1}^+ \Omega_q$ of the type (2–3). We are left again with only QN eigenvalues instead of QN^2 matrix elements and are ready to apply the saddle point approximation to this integral. This model was first analyzed by the method of orthogonal polynomials by [15]. It was shown in [19] that by special choices of the potential V the model can be described by the KP integrable flow with respect to the coupling constant of the potential.

Note that if we imposed the periodicity condition $M_1 = M_q$ on this matrix chain and add the term $M_1 M_q$ to the action the problem would become much more complicated (and actually not solved so far), since this would give an extra condition $\prod_q U_q = I$ making the variables U_q not independent.

Another solvable matrix chain describing the statistical RSOS RSOS models on random planar graphs was proposed and solved in [20]. Similar models were considered in [32].

Some multimatrix models can be reduced to the solvable ones by means of simple matrix integral transformations. The first example of such transformation was described in [21] for the matrix integral describing the Q-state Potts model on random dynamical planar graphs. Its partition function is

$$Z = \int \prod_{q=1}^{Q} d^{N^2} M_q \exp \mathrm{Tr} \left(\sum_{q=1}^{Q} V_q(M_q) + \sum_{p,q=1}^{Q} M_p M_q \right).$$

One can represent the last factor under the integral as

$$\int d^{N^2} X \exp \mathrm{Tr} \left(-\tfrac{1}{2} X^2 + X \sum_{q=1}^{Q} M_q \right).$$

We consider the case $V_1 = \cdots = V_Q = V$. Then the whole integral can be expressed as

$$Z = \int d^{N^2} X \exp(-\tfrac{1}{2} \mathrm{Tr}\, X^2) \left(\int d^{N^2} M \exp \mathrm{Tr}\, (XM + V(M)) \right)^Q.$$

The integrals in this expression can be reduced to the eigenvalues: in the integral under the power the only nontrivial "angular" integration over the relative $U(N)$-"angle" can be done by means of the formula (2–3) and the external one will also depend only on the eigenvalues of X. The solution of the corresponding saddle point equations was found in [22] and analyzed in [23] and [24].

Combining these methods in the obvious ways one can generalize the large N solvability on a certain larger class of multimatrix models.

3. Matrix Quantum Mechanics

In the limit when $Q \to \infty$ and with the special scaling of coupling constants the matrix chain (2–4) becomes matrix quantum mechanics. It is defined by the Hamiltonian

$$\hat{H}_M = -\Delta_M + \mathrm{Tr}\, V(M), \qquad (3\text{–}1)$$

where Δ_M is the usual $U(N)$ invariant Laplacian on the homogeneous space of hermitian matrices and the potential $V(M)$ can actually explicitly depend on time t.

The Schrödinger equation can be written in the form of a minimization principle:

$$min_\Psi \int d^{N^2} M \,\mathrm{Tr}\, \left(\tfrac{1}{2} |\partial_M \Psi(M)|^2 + V(M)|\Psi(M)|^2 \right) \qquad (3\text{–}2)$$

To reduce this problem to the eigenvalues we use the $U(N)$ symmetry of our model and look for a wave function $\Psi(M)$ transforming according to a certain irreducible representation R of $U(N)$:

$$\Psi_R^I(\Omega^+ M \Omega) = \sum_J \Omega_R^{IJ} \Psi_R^J(M),$$

where Ω_R is a group element Ω in representation R and I, J are the indices of the representation. Such a function may be decomposed as

$$\Psi_R^I(M) = \sum_J \Omega_R^{IJ} \psi_R^J(x). \tag{3-3}$$

Here $\psi_R^I(x_1,\ldots,x_N)$ is a vector in the representation R.

Near the unity element on the group space $\Omega \simeq I + \omega$ we have $\Omega_R \simeq P_R + \sum_{ij} \omega_{ij} T_{ij}^R$ where P_R is a projector (unity element) in the R space, ω is a small deviation from it and T_{ij}^R are the $u(N)$ algebra generators. This gives

$$\frac{\partial}{\partial M_{ij}} = \delta_{kj}\frac{\partial}{\partial x_k} + \sum_{m=1}^N \frac{1}{x_k - x_m}\frac{\partial}{\partial \omega_{mj}},$$

and we finally obtain from (3-2) the variational principle

$$\min_{\psi_R} \int \prod_k dx_k \Delta^2(x)$$
$$\times \mathrm{Tr}_R\left(\frac{1}{2}\sum_j \left|\frac{\partial}{\partial x_j}\psi_R(x)\right|^2 + \frac{1}{2}\sum_{i\neq j}|T_{ij}^R \psi_R|^2 + \sum_m V(x_m)|\psi_R|^2\right),$$

where all the quantities and operators with the subscript R are subjected to the corresponding matrix operations in the matrix space of representation.

The Schrödinger equation now reads:

$$-\sum_k \Delta^{-2}(x)\frac{\partial}{\partial x_k}\Delta^2(x)\frac{\partial}{\partial x_k}\psi_R(x) - \sum_{i\neq j} T_{ij}^R T_{ji}^R \psi_R(x) = \left(E - \sum_k V(x_k)\right)\psi_R(x). \tag{3-4}$$

It is useful to introduce a new function $\phi_R(x) = \frac{1}{\Delta(x)}\psi_R(x)$ obeying the equation

$$-\sum_k \left(\frac{\partial}{\partial x_k}\right)^2 \phi_R(x) - \sum_{i\neq j}\frac{T_{ij}^R T_{ji}^R}{(x_i - x_j)^2}\phi_R(x) = \left(E - \sum_i V(x_i)\right)\phi_R(x) \tag{3-5}$$

Note that any translation $\omega_{ij} \to \omega_{ij} + \delta_{ij}\varepsilon$ does not change the wave function Ψ_R. That means that we are looking only for the states on which the condition

$$T_{kk}^R \psi_R = 0, \quad \text{for } k = 1,\ldots,N$$

is imposed.

At first sight, we fulfilled our main task for the matrix quantum mechanics: we reduced it to an eigenvalue problem and are now dealing with only N variables. But the Schrödinger equation (3-4) contains the Hamiltonian which is a matrix in the representation space acting on the wave function which is a vector in this space. For small representations whose Young tableaux contain $\ll N^2$ boxes the problem is still solvable in the large N limit (as we will demonstrate below). For a very interesting case of big representations ($\sim N^2$ boxes in the Young tableaux) the problem remains a serious challenge.

In the simplest case of singlet representation (solved long ago in [14]) the wave function is a scalar and the last term in the right-hand side of the Schrödinger equation (3–5) drops out. The problem appears to be equivalent to the quantum mechanical system of N non-interacting fermions (due to the antisymmetry of $\phi(x)$) in a potential $V(x)$. It was used in many applications, including the solution of the non-critical string theory in 1+1 dimensions [9].

The next smallest representation is adjoint. The adjoint wave function satisfying the relation (3–3) should be a function of the type

$$\Psi(M;x) = \sum_{a=0}^{N-1} C_a(x) M^a$$

where the coefficients C_A possibly depend on the invariants (eigenvalues). If we denote $\phi_{adj}(x_i;x) \equiv \phi_i(x)$ (depending of course on all N x_i) we can write the Schrödinger equation for the adjoint wave function in the following form [13]:

$$\sum_i \left(-\left(\frac{\partial}{\partial x_i}\right)^2 + V(x_i)\right)\phi_k(x) - \frac{1}{N^2}\sum_{i(\neq k)} \frac{\phi_i(x) - \phi_k(x)}{(x_i - x_k)^2} = E\phi_k(x).$$

One can see that the last term in the left-hand side of this equation is $\sim N^2$ smaller than the other terms and can be regarded as a small perturbation on the background of the free fermion solution of the singlet sector.

For one of physically most interesting applications, the 1+1 dimensional string theory, we need to solve the model in the inverted oscillatory potential $V(M) = -M^2$. The model is unstable and one needs to specify the boundary conditions for big M's. Usually one considers the boundary conditions when the absolute value of any of the eigenvalues of M cannot exceed some maximum value Λ (a cut-off wall). In the case of the large N limit one takes $\Lambda \sim N$ and it happens that the spectrum density of the model depends in a very universal (logarithmic) way on Λ. In the singlet state the spectrum is that of N independent fermions (eigenvalues) in the same potential and the eigenfunctions are the Slater determinants of the parabolic cylinder functions [25]. In the non-singlet sectors the eigenvalues start interacting and obey a more complicated statistics corresponding to the symmetry of the Young tableau of representation (see the review [30] for details). Although the problem is clearly integrable the spectrum of the non-singlet sectors of the inverted matrix quantum oscillator is still unknown. For the large N estimates of the mass gap of adjoint representation see [13; 12; 8].

It was conjectured in [12] and shown in [8] that the adjoint representation describe the vortex anti-vortex sector in the 1+1 dimensional string theory with one compact dimension. Higher representation describe higher numbers of vortex anti-vortex pairs (corresponding to the number of boxes in the Young tableau of the representation).

4. Character Expansion and New Solvable (Multi) Matrix Models

The group character expansion has shown its power in the lattice gauge theory long time ago, starting from the work of A. Migdal [26].

The character expansion method proposed in the papers [2]–[4] and inspired by the result of paper [27] is the most general approach for the reduction of the number of degrees of freedom from $\sim N^2$ to $\sim N$ in a new big class of (multi) matrix integrals. The matrix integral considered in these papers looks as follows:

$$Z = \int d^{N^2}M \exp\bigl(-\operatorname{Tr} M^2 + \operatorname{Tr} V(AM)\bigr), \qquad (4\text{–}1)$$

where $V(y) = \sum_{k>2} t_k y^k$ is an arbitrary potential and A is an arbitrary hermitian matrix (which can be taken diagonal without a loss of generality). We again diagonalize the matrix M as

$$M = \Omega^+ X \Omega. \qquad (4\text{–}2)$$

The integral over the $U(N)$ variable Ω looks difficult to do directly since the Itzykson-Zuber formula (2–3) seems to be of little use here. Instead, we expand $\exp\bigl(\operatorname{Tr} V(AM)\bigr)$ as an invariant function of the variable AM in terms of the characters $\chi_R(AM)$ of irreducible representations R of the $GL(N)$ group:

$$\exp\bigl(\operatorname{Tr} V(AM)\bigr) = \sum_R f_R \chi_R(AM), \qquad (4\text{–}3)$$

where the coefficients f_R are the functions of N highest weight components of a representation

$$R = \{0 \leq m_N \leq m_{N-1} \leq \cdots \leq m_1 < \infty\}.$$

The sum $\sum_R$ is nothing but the sum over N ordered integers. They can be calculated due to the orthogonality of characters as the following unitary integrals:

$$f_R = \int d[\Omega]_{U(N)} \exp\bigl(\operatorname{Tr} V(\Omega)\bigr) \chi_R(\Omega^+). \qquad (4\text{–}4)$$

This integral can be represented as an explicit integrals only over the Cartan subgroup $\Omega = \{e^{i\omega_1}, \ldots, e^{i\omega_N}\}$ and thus contains only N integration variables. We have $\operatorname{Tr} V(\Omega) = \sum_k V(e^{i\omega_k})$ and $d[\Omega]_{U(N)} \to \prod_k d\theta_k \prod_{i>j} \sin^2 \frac{1}{2}(\theta_i - \theta_j)$. Now if we plug (4–3) into (4–1) we realize that the decomposition (4–2) is actually useful and we can integrate over Ω using the following orthogonality relation between matrix elements of representation R:

$$\int d[\Omega]_{U(N)} \chi_R(A\Omega^+ X\Omega) = \frac{1}{\dim_R} \chi_R(A) \chi_R(X), \qquad (4\text{–}5)$$

where $\dim_R$ is the dimension of a representation R. We see that we achieved our main goal: due to the formulas (4–3), (4–4) and (4–5) we reduced the original matrix integral (4–1) to an integral over only N eigenvalues $x_1, \ldots, x_N$ of the matrix M and the sum over N highest weight components $m_1, \ldots, m_N$. In the large N limit, if we scale appropriately the constants in the potential $V(M)$, the sums over m's can be replaced by integrals and we can again apply the saddle point approximation in all $2N$ integration variables. To get explicitly the right large N scaling of the couplings one usually changes $e^V \to e^{NV}$. Then the effective action at the saddle point is always of the order $1/N^2$ and the new couplings of the potential V can be kept finite in this limit.

As was shown in [27] (see also [2]), the integral over $x_1, \ldots, x_N$ can be calculated exactly and the remaining sum over strictly ordered nonnegative integers $h_i = -m_i + N - i$ (shifted highest weights) reads

$$Z = \sum_{h_1 < h_2 < \cdots < h_N} \frac{\prod (h^e - 1)!! \, h^o!!}{\prod (h^e - h^o)} \chi_R(A) \chi_R(t), \qquad (4\text{-}6)$$

where $\{h^e\}$ and $\{h^o\}$ are the collections of even and odd integers h_k (their number is equal). Only the representations with equal amounts of even and odd h's contribute to (4-6). The products in the numerator go over all even and odd h's and the product in the denominator goes over all couples h^e, h^o. $\chi_R(t)$ is a character of the coupling constants t_k written in the Schur form

$$\chi_R = \det_{ij} P_{h_i - j}(t),$$

and the Schur polynomials $P_k(t)$ are defined as usually: $\sum_n P_n(t) z^n = e^{\sum_k t_k z^k}$.

So in the large N limit we have to do the saddle point calculation only with respect to N summation variables $h_1, \ldots, h_N$.

The details of these formulas can be found in [27], [2]–[4]. One can also find in these papers the geometrical interpretation of the integral (4-1) in terms of the so called dually weighted planar graphs. It gives the generating function of planar graphs where both vertices and faces are weighted by the generating parameters depending on their orders. In [2]–[4] one can find the solutions of some combinatorial problems related to the enumeration of planar graphs which were possible only due to the power of the character expansion method. The particular solutions of the saddle point equations could be very tricky but it is already a "classical" problem of solution of various integral equations rather than a "quantum" problem of functional integration over infinite matrices. In that sense this model is solvable.

It is obvious that there exist many ways to generalize the model (4-1) to other matrix integrals. An immediate generalization is to substitute the $\operatorname{Tr} M^2$ term in (4-1) by an arbitrary function $W(M)$. In that case we cannot calculate explicitly the coefficients f_R (except when W is a monomial: $W(M) = M^k$) but we still get an explicit integral over $3N$ variables x_i, ω_i and m_i. So the model is again solvable.

Another solvable matrix model of this kind involving general complex matrices was proposed and investigated in [33; 28]. Its free energy gives a generating functional counting branched coverings of two dimensional surfaces.

The most general solvable two matrix model reads as

$$Z = \int d^{N^2}A \, d^{N^2}B \exp N \operatorname{Tr}\left(U(AB) + V(A) + W(B)\right), \qquad (4\text{--}7)$$

where U, V and W are arbitrary functions. The way to reduce it to $\sim N$ degrees of freedom is again to expand in characters

$$\exp \operatorname{Tr} U(AB) = \sum_R u_R \chi_R(AB),$$

diagonalize the matrices A and B and integrate over the $U(N)$ variable between them by means of (4–5). In the particular case

$$Z = \int d^{N^2}A \, d^{N^2}B \exp N \operatorname{Tr}\left(\tfrac{1}{2}(A^2 + B^2) - \tfrac{1}{4}\alpha(A^4 + B^4) - \tfrac{1}{4}\beta(AB)^2\right),$$

the model describes a special trajectory of the 8-vertex model on random graphs. It was completely solved in [29]. Again it was possible, using character orthogonality relations, to integrate over the relative angle between A and B; this leads to separation into one-matrix integrals:

$$Z(\alpha, \beta) \sim \sum_{\{h\}} (N\beta/2)^{\#h/2} c_{\{h\}} (P_{\{h\}}(\alpha))^2, \qquad (4\text{--}8)$$

where

$$c_{\{h\}} = \frac{1}{\prod_i \lfloor h_i/2 \rfloor! \prod_{i,j}(h_i^e - h_j^o)}$$

is a coefficient and the one-matrix integral

$$S_{\{h\}}(\alpha) = \int d^{N^2}M \, \chi_{\{h\}}(M) \exp N\left(-\tfrac{1}{2}\operatorname{Tr} M^2 + \tfrac{1}{4}\alpha \operatorname{Tr} M^4\right)$$

appears squared in (4–8) because the contributions from the two matrices A and B are identical.

Now we can reduce the calculation of the one-matrix integral $S_{\{h\}}$ to eigenvalue integrations:

$$S_{\{h\}}(\alpha) = \int \prod_k d\lambda_k \, \Delta(\lambda) \det\left(\lambda_k^{h_j}\right) \exp N\left(-\frac{1}{2}\sum_k \lambda_k^2 + \frac{\alpha}{4}\sum_k \lambda_k^4\right),$$

where $\Delta(\lambda) = \det\left(\lambda_k^{N-j}\right) = \prod_{j<k}(\lambda_j - \lambda_k)$.

We are left only with N degrees of freedom and the action of the order N^2, so the integration is reduced to the saddle point calculation with respect to the eigenvalues λ_k (see [29] for the details).

We can immediately propose some solvable generalization of the general two matrix model (4–7) to a multimatrix chain

$$Z = \int \prod_{q=1}^{Q} d^{N^2}M_q \exp \operatorname{Tr}\left(\sum_{q=1}^{Q} V_q(M_q) + \sum_{p=1}^{Q-1} W(M_{p-1}M_p)\right),$$

where V and W are arbitrary functions.

Another interesting model solvable by the character expansion method can be written in the general form

$$Z = \int \prod_{q=1}^{Q} d^{N^2} M_q \exp\left(\operatorname{Tr} V_q(M_q) + V\left(\prod_{p=1}^{Q} M_p \right) \right).$$

It can be solved by character expansion with respect to the last factor by means of the formulas (4-3), (4-4) and the multiple application of the formula (4-5) by induction.

The results in terms of the sum over N highest weight components of the representations R reads

$$Z = \sum_R \frac{f_R}{\dim_R^{Q-1}} \prod_{q=1}^{Q} [S_{\{h\}}^{(q)}],$$

where

$$S_{\{h\}}^{(q)} = \int d^{N^2} M \, \chi_{\{h\}}(M) \exp \operatorname{Tr} V_q(M).$$

The last integral can be immediately reduced to the integrations of a type (4) over eigenvalues of the matrix M.

5. Comments and Unsolved Problems

1. The non-singlet sectors in the matrix quantum mechanics (3–1) can be effectively studied for the oscillatory potential $V(M) = M^2$. In this case the Hamilton-Ian is a collection of N^2 independent oscillators represented by the matrix elements of M. The spectrum of Hamilton-Ian of this model in a given irreducible representation of $U(N)$, is encoded into the partition functions $Z_R(q)$ for finite inverse temperature β (where $q = e^\beta$) in a given representation R. The effective way to study $Z_R(q)$ can be found in [8] or [5].
2. The character expansion is nothing but the Fourier expansion on a group manifold. As trivial as it looks for us now the Fourier transform was always a powerful method of solving problems using their symmetries. Many of the matrix models presented in the previous section and solved by this method seemed hopeless just a few years ago.
3. One of the interesting and not well studied questions is how to classify all the matrix integrals which can be reduced from $\sim N^2$ to $\sim N$ integrals or sums by use of the character expansion.
4. In many physically interesting cases we don't need a general form of potentials $V(M)$ or $W(M)$ mentioned through this paper. For example, as it was mentioned, to study the universal behavior of the large N matrix quantum mechanics near the instability point we need to know only the solution in the vicinity of a quadratic top of the potential $V(M) \simeq - \operatorname{Tr}(M - M_0)^2$. The rest

of the potential $V(M)$ has small influence on the behavior of the eigenvalues and serves only as a $U(N)$ invariant cutoff wall. It simplifies greatly the problem. For instance, all the applications in string theory, two-dimensional quantum gravity and most of statistical-mechanical applications need only the analyses of the vicinity of such critical points. The lesson to draw from it is that for some physically most interesting regimes the seemingly hopeless matrix integrals become not so hopeless and look "almost Gaussian". May be a general method of the investigation of these instability points can be worked out.

5. Another question is related to the integrability properties of sums and integrals after such reduction. The partition functions of some of them (such as the old one matrix and two matrix models) are known to be τ-functions of some integrable hierarchies of classical differential or difference equations, like Toda hierarchy [31; 5] or KP hierarchy [6] (see also [34; 35]). But many others, like the model (4–1), cannot be represented by free fermions. On the other hand, the Itzykson–Di Francesco formula (4–6) suggests that it might exist some interacting fermion representation of the partition function of the model of dually weighted graphs (4–1).

The method of character expansion as well as all other methods of calculation of the large N matrix integrals presented here represent just another refining and generalization of the usual method of reduction to the matrix eigenvalues invented long ago by Dyson. Its range of applicability is quite limited although it includes quite a few important matrix integrals known from physics and mathematics. Many more interesting matrix integrals look not hopeless for the investigation in the large N limit. The search for new tricks of integration over matrices is a fascinating and potentially extremely rewarding research direction.

Acknowledgments

I am grateful to the organizers of the MSRI Workshop "Matrix models and Painlevé equations" P. Bleher and A. Its for the kind hospitality and fruitful discussion during the Workshop.

I also thank I. K. Kostov for useful comments and M. Fukuma for the illuminating discussions concerning the matrix quantum mechanics in adjoint representation.

References

[1] E. Brézin, V. A. Kazakov, Exactly solvable field theories of closed strings, Phys. Lett. B236 (1990) 144; M. R. Douglas, S. Shenker, Strings in less than one dimension, Nucl. Phys. B335 (1990) 635; D. Gross, A. Migdal, Non-perturbative two-dimensional quantum gravity, Phys. Rev. Lett. 64 (1990) 127.

[2] V. A. Kazakov, M. Staudacher and T. Wynter, Exact Solution of Discrete Two-Dimensional R^2 Gravity, hep-th/9601069, Nucl. Phys. B471 (1996) 309-333.

[3] V. A. Kazakov, M. Staudacher and T. Wynter, Character expansion methods for matrix models of dually weighted graphs, hep-th/9502132, Comm. Math. Phys. 177 (1996) 451-468.

[4] V. A. Kazakov, M. Staudacher, T. Wynter, Almost flat planar graphs, hep-th/9506174, Comm. Math. Phys. 179 (1996) 235-256.

[5] J. Hoppe, V. Kazakov and I. Kostov, Dimensionally Reduced SYM_4 as Solvable Matrix Quantum Mechanics, hep-th/9907058, to be published in Nucl. Phys. B.

[6] V. Kazakov, I. Kostov, N. Nekrasov, D-particles, Matrix Integrals and KP hierarchy, hep-th/9810035, Nucl. Phys. B557 (1999) 413-442.

[7] M. R. Douglas and V. A. Kazakov, Large N Phase Transition in Continuum QCD_2, Phys. Lett. B319 (1993) 219 hep-th/9305047.

[8] D. Boulatov and V. Kazakov, One dimensional string theory with vortices as an upside down matrix oscillator, J. Mod. Phys A8 (1993) 809.

[9] V. Kazakov and A. Migdal, Recent Progress in the Theory of Non-critical Strings, Nucl. Phys. B 311 (1988) 171.

[10] Harish-Chandra, Amer. J. Math., 79 (1957) 87.

[11] C. Itzykson and J. B. Zuber, Planar approximation II, J. Math. Phys. 21 (1980) 411.

[12] D. Gross and I. Klebanov, Vortices and the non-singlet sector of the c=1 matrix model, Nucl. Phys. B354 (1990) 459.

[13] G. Marchesini and E. Onofri, Planar limit for $SU(N)$ symmetric quantum dynamical systems, J. Math. Phys. 21 (1980) 1103.

[14] E. Brezin, C. Itzykson, G. Parisi and J.-B. Zuber, Planar approximation, Comm. Math. Phys. 59 (1978) 35.

[15] S. Chadha, G. Mahoux and M. L. Mehta, A method of integration over matrix variables 2, J. Phys. A: Math. Gen. 14 (1981) 579.

[16] M. L. Mehta, A method of integration over matrix variables, Comm. Math. Phys. 79 (1981) 327.

[17] V. A. Kazakov, Ising model on dynamical planar random lattice: exact solution, Phys. Lett. 119A (1986) 140; D. V. Boulatov and V. A. Kazakov, The Ising model on a random planar lattice : the structure of phase transition and the exact critical exponents, Phys. Lett. B186 (1987) 379.

[18] V. A. Kazakov, D-dimensional induced gauge theory as a solvable matrix model, Proc. Intern. Symp. Lattice '92 on Lattice gauge theory, Amsterdam, Sept. 1992, Nucl. Phys. B (Proc. Suppl.) 30 (1993) 149.

[19] M. R. Douglas, Strings in less than one dimension and generalized KdV hierarchies, Phys. Lett. 238B (1990) 176.

[20] I. K. Kostov, Gauge Invariant Matrix Model for the ADE Closed Strings, hep-th/9208053, Phys. Lett. B297 (1992) 74-81.

[21] V. A. Kazakov, Exactly solvable Potts models, bond and tree-like percolation on dynamical (random) planar lattice, Nucl. Phys. B 4 (Proc. Supp.) (1988) 93.

[22] V. A. Kazakov and I. K. Kostov, published in the review of I. K. Kostov, Random surfaces, solvable matrix models and discrete quantum gravity in two dimensions, Lecture given at GIFT Int. Seminar on Nonperturbative Aspects of the Standard Model, Jaca, Spain, Jun 6-11, 1988. Published in GIFT Seminar 0295 (1988) 322.

[23] J-M. Daul, Q-states Potts model on a random planar lattice, hep-th/9502014.

[24] B. Eynard, G. Bonnet, The Potts-q random matrix model : loop equations, critical exponents, and rational case, Phys. Lett. B463 (1999) 273-279.

[25] V. A. Kazakov, Bosonic strings and string field theories in one-dimensional target space, in Random surfaces and quantum gravity, Cargèse 1990, O. Alvarez, E. Marinari, P. Windey eds., (1991).

[26] A. A. Migdal, Recursion equations in lattice gauge theories, Sov. Phys. JETP 42 (1975) 413, Zh. Eksp. Teor. Fiz. 69 (1975) 810-822.

[27] Ph. DiFrancesco and C. Itzykson, Fat graphs, Ann. Inst. Henri Poincarè, 59(2) (1993) 117.

[28] I. K. Kostov, M. Staudacher, T. Wynter, Complex Matrix Models and Statistics of Branched Coverings of 2D Surfaces, hep-th/9703189, Comm. Math. Phys. 191 (1998) 283-298.

[29] V. A. Kazakov, P. Zinn-Justin, Two-Matrix model with ABAB interaction, hep-th/9808043, Nucl. Phys. B546 (1999) 647-668.

[30] A. P. Polychronakos, Generalized statistics in one dimension, hep-th/9902157, Les Houches 1998 Lectures; 54 pages.

[31] S. Kharchev, A. Marshakov, A. Mironov, A. Morozov, Generalized Kontsevich Model Versus Toda Hierarchy and Discrete Matrix Models, hep-th/9203043, Nucl. Phys. B397 (1993) 339.

[32] S. Kharchev, A. Marshakov, A. Mironov, A. Morozov, S. Pakuliak, Conformal Matrix Models as an Alternative to Conventional Multi-Matrix Models, hep-th/9208044, Nucl. Phys. B404 (1993) 717.

[33] I. K. Kostov and M. Staudacher, Two-Dimensional Chiral Matrix Models and String Theories, hep-th/9611011, Phys. Lett. B394 (1997) 75-81.

[34] I. K. Kostov, Bilinear Functional Equations in 2D Quantum Gravity, hep-th/9602117, Talk delivered at the Workshop on New Trends in Quantum Field Theory, 28 August - 1 September 1995, Razlog, Bulgaria.

[35] V. A. Kazakov, A matrix model solution of Hirota equation, hep-th/9711019.

Vladimir Kazakov
Laboratoire de Physique Théorique
École Normale Supérieure
75231 Paris
France
kazakov@physique.ens.fr

The τ-Function for Analytic Curves

I. K. KOSTOV, I. KRICHEVER, M. MINEEV-WEINSTEIN,
P. B. WIEGMANN, AND A. ZABRODIN

ABSTRACT. We review the concept of the τ-function for simple analytic curves. The τ-function gives a formal solution to the two-dimensional inverse potential problem and appears as the τ-function of the integrable hierarchy which describes conformal maps of simply-connected domains bounded by analytic curves to the unit disk. The τ-function also emerges in the context of topological gravity and enjoys an interpretation as a large N limit of the normal matrix model.

1. Introduction

Recently, it has been realized [1; 2] that conformal maps exhibit an integrable structure: conformal maps of compact simply connected domains bounded by analytic curves provide a solution to the dispersionless limit of the two-dimensional Toda hierarchy. As is well known from the theory of solitons, solutions of an integrable hierarchy are represented by τ-functions. The dispersionless limit of the τ-function emerges as a natural object associated with the curves. In this paper we discuss the τ-function for simple analytic curves and its connection to the inverse potential problem, area preserving diffeomorphisms, the Dirichlet boundary problem, and matrix models.

2. The Inverse Potential Problem

Define a *closed analytic curve* as a curve that can be parametrized by a function $z \equiv x + iy = z(w)$, analytic in a domain that includes the unit circle $|w| = 1$. Consider a closed analytic curve γ in the complex plane and denote by D_+ and D_- the interior and exterior domains with respect to the curve. The point $z = 0$ is assumed to be in D_+. Assume that the domain D_+ is filled homogeneously with electric charge, with a density that we set to 1. The potential Φ created by the charge obeys the equation

$$-\partial_z \partial_{\bar{z}} \Phi(z, \bar{z}) = \begin{cases} 1 & \text{if } z = x + iy \in D_+, \\ 0 & \text{if } z = x + iy \in D_-. \end{cases} \quad (2\text{--}1)$$

The potential Φ can be written as an integral over the domain D^+:

$$\Phi(z, \bar{z}) = -\frac{2}{\pi} \int_{D_+} d^2 z' \log |z - z'| \quad (2\text{--}2)$$

In the exterior domain D_-, the potential is the harmonic function whose asymptotic expansion as $z \to \infty$ is given by

$$\Phi^-(z, \bar{z}) = -2t_0 \log |z| + 2 \operatorname{Re} \sum_{k>0} \frac{v_k}{k} z^{-k}, \quad (2\text{--}3)$$

where

$$v_k = \frac{1}{\pi} \int_{D_+} z^k d^2 z \quad (k > 0) \quad (2\text{--}4)$$

are the harmonic moments of the interior domain D_+ and

$$\pi t_0 = \int_{D_+} d^2 z$$

is its area. In the interior domain D_+, the potential (2-2) is equal to a function Φ^+, which is harmonic up to the term $-|z|^2$. The expansion of this function around $z = 0$ is

$$\Phi^+(z, \bar{z}) = -|z|^2 - v_0 + 2 \operatorname{Re} \sum_{k>0} t_k z^k. \quad (2\text{--}5)$$

Here

$$t_k = -\frac{1}{\pi k} \int_{D_-} z^{-k} d^2 z \quad (k > 0) \quad (2\text{--}6)$$

are the harmonic moments of the exterior domain D_- and

$$v_0 = \frac{2}{\pi} \int_{D_+} \log |z| \, d^2 z.$$

The two sets of moments (2-4) and (2-6) are related by the conditions $\Phi^+ = \Phi^-$ and $\partial_z \Phi^+ = \partial_z \Phi^-$ on the curve γ.

The inverse potential problem is to determine the form of the curve γ given one of the functions Φ^+ or Φ^-, i.e., given one of the infinite sets of moments. We will choose as independent variables the area πt_0 and the moments of the exterior t_k, for $k \geq 1$. Under certain conditions, they completely determine the form of the curve as well as the moments v_k, for $k \geq 0$ [3]. More precisely, $\{t_k\}_{k=0}^{\infty}$ is a good set of local coordinates in the space of analytic curves. For simplicity we assume in this paper that only a finite number of t_k are nonzero. In this case the series (2-5) is a polynomial in z, $\bar{z}$ and, therefore, it gives the function Φ^+ for $z \in D_+$. Note that t_0, v_0 are real quantities while all other moments are in general complex variables.

3. Variational Principle

Consider the energy functional describing a charge with a density $\rho(z,\bar{z})$ in the background potential created by the homogeneously distributed charge with the density $+1$ inside the domain D_+ (2–1):

$$\mathcal{E}\{\rho\} = -\frac{1}{\pi^2}\iint d^2z\, d^2z'\, \rho(z,\bar{z})\log|z-z'|\rho(z',\bar{z}') - \frac{1}{\pi}\int d^2z\, \rho(z,\bar{z})\,\Phi(z,\bar{z}).$$

The first term is the two-dimensional "Coulomb" energy of the charge while the second one is the energy due to the background charge. Clearly, the distribution of the charge neutralizing the background charge gives the minimum to the functional: $\rho_0 = -1$ inside the domain and $\rho_0 = 0$ outside. At the minimum the functional is equal to minus electrostatic energy E of the background charge :

$$-E = \min_\rho \mathcal{E}\{\rho\} = \frac{1}{\pi^2}\int_{D_+} d^2z \int_{D_+} d^2z'\, \log|z-z'| = -\frac{1}{2\pi}\int_{D_+} d^2z\, \Phi(z,\bar{z}).$$

Varying over ρ and then setting $\rho = -1$ inside the domain, we obtain (2–5).

The first corollary of the variational principle is that the E is a potential function for the moments. Equation (2–5) suggests treating v_0 and t_k as independent variables, so moments of the interior, v_k, $k \geq 1$, and t_0 are functions of v_0 and t_k. Differentiate E or $-\mathcal{E}\{\rho\}$ at the extremum with respect to the parameters v_0, t_k. Since ρ_0 minimizes the functional, the derivative is equivalent to the partial derivative of $\mathcal{E}$ at the fixed extremum ρ. This gives

$$\frac{\partial E}{\partial t_k} = v_k, \quad \frac{\partial E}{\partial \bar{t}_k} = \bar{v}_k, \quad \frac{\partial E}{\partial v_0} = -t_0, \tag{3-1}$$

where the partial derivative with respect to t_k is taken at fixed v_0 and t_j, for $j \neq 0, k$. Therefore the differential dE reads

$$dE = \sum_{k>0}(v_k dt_k + \bar{v}_k d\bar{t}_k) - t_0 dv_0.$$

The variational principle may be formulated in a number of different ways. One particular variational principle is suggested by the matrix model discussed in Section 9. In this case one considers a charged liquid in the potential

$$V(z,\bar{z}) = z\bar{z} + v_0 - \sum_{k>0}\left(t_k z^k + \bar{t}_k \bar{z}^k\right) \tag{3-2}$$

defined everywhere on the plane and v_0 and t_k are parameters. The energy of the charged liquid,

$$\mathcal{E}\{\rho, V\} = -\frac{1}{\pi^2}\int d^2z \int d^2z'\, \rho(z,\bar{z})\rho(z',\bar{z}')\log|z-z'| + \frac{1}{\pi}\int d^2z\, \rho(z,\bar{z})\, V(z,\bar{z}), \tag{3-3}$$

reaches its minimum if the liquid forms a drop with the density $\rho_0 = -1$ bounded by the curve determined by parameters of the potential v_0 and t_k. For another version of the variational principle see [4].

4. The τ-Function

It is more natural to treat the total charge t_0 rather than v_0 as an independent variable, i.e., to consider the variational principle at a fixed total charge $t_0 = \int \rho d^2 z$. This is achieved via the Legendre transformation. Introduce the function $F = E + t_0 v_0$, whose differential is

$$dF = \sum_{k>0}(v_k dt_k + \bar{v}_k d\bar{t}_k) + v_0 dt_0.$$

We define the τ-function as $\tau = e^F$, so that

$$\log \tau = \frac{1}{2\pi}\int_{D_+} d^2 z\, \Phi(z,\bar{z}) + t_0 v_0 = -\frac{1}{\pi^2}\int\int_{D_+} \log\left|\frac{1}{z} - \frac{1}{z'}\right| d^2 z\, d^2 z'. \quad (4\text{–}1)$$

The τ-function is a real function of the moments $\{t_0, t_1, t_2, \ldots\}$. Under the assumption that only a finite number of them are nonzero, we can substitute (2–5) into (4–1) and perform the term-wise integration. Taking into account that $\frac{1}{\pi}\int_{D_+} |z|^2 d^2 z = \frac{1}{2}t_0^2 + \frac{1}{2}\sum_{k>0} k(t_k v_k + \bar{t}_k \bar{v}_k)$ (a simple consequence of the Stokes formula), we get the expression for the τ-function in terms of t_k and v_k:

$$2\log \tau = -\frac{1}{2}t_0^2 + t_0 v_0 - \frac{1}{2}\sum_{k>0}(k-2)(t_k v_k + \bar{t}_k \bar{v}_k).$$

Rephraising (3–1) we get the main property of the τ-function, which was used in [2] as its definition:

$$\frac{\partial \log \tau}{\partial t_k} = v_k, \quad \frac{\partial \log \tau}{\partial \bar{t}_k} = \bar{v}_k, \quad \frac{\partial \log \tau}{\partial t_0} = v_0, \quad (4\text{–}2)$$

where the derivative with respect to t_k is taken at fixed t_j ($j \neq k$).

Two immediate consequences of the very existence of the potential function are symmetry relations for the moments

$$\frac{\partial v_k}{\partial t_n} = \frac{\partial v_n}{\partial t_k}, \quad \frac{\partial v_k}{\partial \bar{t}_n} = \frac{\partial \bar{v}_n}{\partial t_k},$$

and the quasi-homogeneity condition for the τ-function:

$$4\log \tau = -t_0^2 + 2t_0 \frac{\partial \log \tau}{\partial t_0} - \sum_{n>0}(n-2)\left(t_n \frac{\partial \log \tau}{\partial t_n} + \bar{t}_n \frac{\partial \log \tau}{\partial \bar{t}_n}\right).$$

Apart from the term $-t_0^2$, this formula reflects the scaling of moments as $z \to \lambda z$: $t_k \to \lambda^{2-k} t_k$ ($k \geq 0$), $v_k \to \lambda^{2+k} v_k$ ($k \geq 1$).

As an illustration we present the τ-function of ellipse [2]. In this case only the first two moments t_1 and t_2 are nonzero:

$$\log\tau = -\frac{3}{4}t_0^2 + \frac{1}{2}t_0^2 \log\left(\frac{t_0}{1-4|t_2|^2}\right) + \frac{t_0}{1-4|t_2|^2}\left(|t_1|^2 + t_1^2\bar{t}_2 + \bar{t}_1^2 t_2\right).$$

(The τ-function for the ellipse (at $t_1 = 0$) appeared in [5] as the limit of the Laughlin wave function or a planar limit of the free energy of normal matrix models; see Section 9.)

5. The Schwarz Function and the Generating Function of the Conformal Map

Consider a univalent conformal map of the exterior domain D_- to the exterior of the unit disk and expand it in a Laurent series:

$$w(z) = \frac{1}{r}z + \sum_{j=0}^{\infty} p_j z^{-j},$$

where the coefficient r is chosen to be real and positive. The series for the inverse map (from the exterior of the unit disk to D_-) has a similar form:

$$z(w) = rw + \sum_{j=0}^{\infty} u_j w^{-j}. \tag{5-1}$$

Chosen w on the unit circle, (5-1) gives a parametrization of the curve. By the definition of an analytic curve, the map can be analytically continued to a strip-like neighborhood of the curve belonging to D_+. The continuation is given by the Riemann–Schwarz reflection principle (see [6], for example):

$$w = (\bar{w}(S(z)))^{-1},$$

where $S(z)$ is the point reflected relative to the curve, and where the bar notation has the following meaning: Given an analytic function $f(z) = \sum_j f_j z^j$, we set $\bar{f}(z) = \sum_j \bar{f}_j z^j$. Following [7], we call $S(z)$ the *Schwarz function* of the curve. We recall its construction. Write the equation for the curve $F(x,y) = 0$ in complex coordinates, $F(\frac{1}{2}(z+\bar{z}), \frac{1}{2i}(z-\bar{z})) = 0$, and solve it with respect to $\bar{z}$. One gets the Schwarz function: $\bar{z} = S(z)$. The Schwarz function is analytic in a strip-like domain that includes the curve. On the curve the Schwarz function is equal to the complex conjugate argument. The main property of the Schwarz function is the obvious but important *unitarity condition*

$$\bar{S}(S(z)) = z$$

(the inverse function coincides with the complex conjugate function). In terms of a conformal map the Schwarz function is

$$S(z) = rw^{-1}(z) + \sum_{j=0}^{\infty} \bar{u}_j w^j(z). \tag{5-2}$$

Using the Schwarz function one can write the moments of the exterior and the interior domains (2–4), (2–6) as contour integrals

$$t_n = \frac{1}{2\pi i n} \oint_\gamma z^{-n} S(z)\, dz, \quad v_n = \frac{1}{2\pi i} \oint_\gamma z^n S(z)\, dz. \tag{5-3}$$

(This follows from the more general statement

$$\int_{D_\pm} f(z)\, d^2z = \pm \frac{1}{2i} \oint_\gamma f(z) S(z)\, dz,$$

where $f(z)$ is an analytic function in the domain $D_\pm$.) Equation (5–3) yields the Laurent expansion of the Schwarz function

$$S(z) = \sum_{k=1}^{\infty} k t_k z^{k-1} + \frac{t_0}{z} + \sum_{k=1}^{\infty} v_k z^{-k-1}. \tag{5-4}$$

We now define the *generating function* $\Omega(z)$, related to the Schwarz function by

$$S(z) = \partial_z \Omega(z).$$

The latter is given, according to (5–4), by the Laurent series

$$\Omega(z) = \sum_{k=1}^{\infty} t_k z^k - \frac{1}{2} v_0 + t_0 \log z - \sum_{k=1}^{\infty} \frac{v_k}{k} z^{-k}.$$

It can be represented as $\Omega(z) = \Omega^{(+)}(z) + \Omega^{(-)}(z) - \frac{1}{2} v_0$, where $\Omega^{(\pm)}(z)$ are analytic in $D_\pm$:

$$\Omega^{(+)}(z) = \frac{1}{\pi} \int_{D_-} \log\left(1 - \frac{z}{z'}\right) d^2 z' = \sum_{k=1}^{\infty} t_k z^k,$$

$$\Omega^{(-)}(z) = \frac{1}{\pi} \int_{D_+} \log(z - z')\, d^2 z' = t_0 \log z - \sum_{k=1}^{\infty} \frac{v_k}{k} z^{-k}.$$

From (2–3) and (2–5) we see that $\Phi^-(z,\bar{z}) = -2\,\mathrm{Re}\,\Omega^{(-)}(z)$ and $\Phi^+(z,\bar{z}) = 2\,\mathrm{Re}\,\Omega^{(+)}(z) - v_0 - |z|^2$. Contrary to the potentials $\Phi^\pm$, the analytical functions Ω^+ and $-\Omega^-$ do not match each other on the curve. The discontinuity gives the value of the generating function restricted to the curve

$$\Omega(z) = \frac{1}{2}|z|^2 + 2iA(z), \quad z \in \gamma,$$

where $A(z)$ is the area of the interior domain bound by the ray $\varphi = \arg z$ and the real axis. As a corollary, it is easy to show that variations of the $\Omega(z)$ on the curve with respect to the *real* parameters t_0, $\operatorname{Re} t_k$ and $\operatorname{Im} t_k$ are purely imaginary. This allows one to apply the Riemann–Schwarz reflection principle to analytical continuation of

$$H_k(z) = \partial_{t_k} \Omega(z), \quad \bar{H}_k(z) = -\partial_{\bar{t}_k} \Omega(z),$$

and to prove the fundamental relations

$$\partial_{t_0} \Omega(z) = \log w(z), \tag{5-5}$$

$$\partial_{t_k} \Omega(z) = \left(z^k(w)\right)_+ + \tfrac{1}{2}\left(z^k(w)\right)_0, \tag{5-6}$$

$$\partial_{\bar{t}_k} \Omega(z) = \left(S^k(z(w))\right)_- + \tfrac{1}{2}\left(S^k(z(w))\right)_0. \tag{5-7}$$

The symbols $(f(w))_\pm$ stand for the truncated Laurent series, preserving only terms with positive or negative powers of w, as the case may be; $(f(w))_0$ is the constant term of the series. The derivatives in (5-5) and (5-7) are taken at fixed z.

To prove (5-5), we first notice that

$$\partial_{t_0}\Omega(z(w)) = \log z - \tfrac{1}{2}\partial_{t_0} v_0 + \text{negative powers in } z$$
$$= \log wr - \tfrac{1}{2}\partial_{t_0} v_0 + \text{negative powers in } w.$$

Independently, one can show that $\partial_{t_0} v_0 = 2\log r$.

Then, using the Riemann–Schwarz reflection principle, we may also write $\partial_{t_0}\Omega(z(w))$ in the form $\partial_{t_0}\bar{\Omega}(S(z(w)))$. Expanding this in $S(z)$ and then using the expansion of (5-2) in w, we have

$$\partial_{t_0}\bar{\Omega}(S(z(w))) = \log S(z) - \tfrac{1}{2}\partial_{t_0} v_0 + \text{negative powers in } S(z)$$
$$= \log w + \text{positive powers in } w.$$

Comparing both expansions, we conclude that $\partial_{t_0}\Omega(z) = \log w(z)$. Similar arguments are used in the proof of (5-6) and (5-7).

6. Dispersionless Hirota Equation and the Dirichlet Boundary Problem

Using the representation (4-2) of the moments v_k as derivatives of the τ-function, one can express the conformal map $w(z)$ (5-5) through the τ-function:

$$\log w = \log z - \partial_{t_0}\left(\tfrac{1}{2}\partial_{t_0} + \sum_{k\geq 1} \frac{z^{-k}}{k}\partial_{t_k}\right)\log \tau. \tag{6-1}$$

With the help of the τ-function, equations (5-6) and (5-7) can be similarly encoded as follows:

$$\partial_z \partial_\zeta \log\left(w(z)-w(\zeta)\right) = \frac{1}{(z-\zeta)^2} + \left(\sum_{k\geq 1} z^{-k-1}\partial_{t_k}\right)\left(\sum_{n\geq 1} \zeta^{-n-1}\partial_{t_n}\right)\log \tau, \tag{6-2}$$

$$-\partial_z \partial_{\bar{\zeta}} \log \left(w(z) \bar{w}(\bar{\zeta}) - 1 \right) = \left(\sum_{k \geq 1} z^{-k-1} \partial_{t_k} \right) \left(\sum_{n \geq 1} \bar{\zeta}^{-n-1} \partial_{\bar{t}_n} \right) \log \tau. \quad (6\text{--}3)$$

The derivation is similar to the one given in [8; 9] for the case of the KP hierarchy. Moreover, these equations in the integrated form are most conveniently written in terms of the differential operators

$$D(z) = \sum_{k \geq 1} \frac{z^{-k}}{k} \partial_{t_k},, \quad \bar{D}(\bar{z}) = \sum_{k \geq 1} \frac{\bar{z}^{-k}}{k} \partial_{\bar{t}_k}. \quad (6\text{--}4)$$

From (6–2) and (6–3) one obtains:

$$\log \frac{w(z) - w(\zeta)}{z - \zeta} = -\tfrac{1}{2} \partial_{t_0}^2 \log \tau + D(z) D(\zeta) \log \tau, \quad (6\text{--}5)$$

$$-\log \left(1 - \frac{1}{w(z)\bar{w}(\bar{\zeta})} \right) = D(z) \bar{D}(\bar{\zeta}) \log \tau. \quad (6\text{--}6)$$

Combining (6–1) and (6–5), one obtains the dispersionless Hirota equation (or the dispersionless Fay identity) for two-dimensional Toda lattice hierarchy [2]:

$$(z - \zeta) e^{D(z) D(\zeta) \log \tau} = z e^{-\partial_{t_0} D(z) \log \tau} - \zeta e^{-\partial_{t_0} D(\zeta) \log \tau} \quad (6\text{--}7)$$

This equation, after being expanded in powers of z and ζ, generates an infinite set of relations between the second derivatives $\partial_{t_n} \partial_{t_m} \log \tau$ of the τ-function. Using (6–6) instead of (6–5), a similar equation for the mixed derivatives $\partial_{t_n} \partial_{\bar{t}_m} \log \tau$ can be written:

$$1 - e^{-D(z)\bar{D}(\bar{\zeta}) \log \tau} = \frac{1}{z\bar{\zeta}} e^{\partial_{t_0}(\partial_{t_0} + D(z) + \bar{D}(\bar{\zeta})) \log \tau}$$

We conclude this section with two other forms of the dispersionless Hirota equation for the conformal map. They emphasize a relation between the Hirota equation and two fundamental objects of the classical analysis: the Green function of the Dirichlet problem (which was pointed out to us by L. Takhtajan) and the Schwarz derivative.

The Green function of the Dirichlet boundary problem for the Laplace operator in D_- expressed through the conformal map $w(z)$ is:

$$G(z, \zeta) = \log \left| \frac{w(z) - w(\zeta)}{w(z) \bar{w}(\bar{\zeta}) - 1} \right|.$$

Combining (6–5) and (6–6), and using the notation (6–4), we represent the Green function as follows:

$$2G(z, \zeta) = 2\log |z^{-1} - \zeta^{-1}| + (\partial_{t_0} + D(z) + \bar{D}(\bar{z}))(\partial_{t_0} + D(\zeta) + \bar{D}(\bar{\zeta})) \log \tau. \quad (6\text{--}8)$$

This formula generalizes (6–1), since (6–8) becomes the real part of (6–1) as $\zeta \to \infty$. (As ζ approaches infinity, $G(z, \zeta)$ tends to $-\log |w(z)|$.) The real part of (6–1) can be written in the form

$$\Phi(z,\bar{z}) = -2t_0 \log|z| + \left(D(z) + \bar{D}(\bar{z})\right) \log \tau,$$

where Φ is the potential (2–2) and $z \in D_-$.

The left-hand side of (6–5) generalizes the Schwarz derivative of the conformal map

$$T(z) \equiv \frac{w'''(z)}{w'(z)} - \frac{3}{2}\left(\frac{w''(z)}{w'(z)}\right)^2 = 6 \lim_{z \to \zeta} \partial_z \partial_\zeta \log \frac{w(z) - w(\zeta)}{z - \zeta}.$$

Taking the limit $\zeta \to z$ of both sides of (6–5), we get a relation between the Schwarz derivative and the τ-function:

$$T(z) = 6\, z^{-2} \sum_{k,n \geq 1} z^{-k-n} \frac{\partial^2 \log \tau}{\partial t_k \partial t_n}$$

This can be used as an alternative definition of the τ-function.

7. Integrable Structure of Conformal Maps and Area-Preserving Diffeomorphisms

Equations (5–5)–(5–7) allow one to say that the differential

$$d\Omega = S\, dz + \log w\, dt_0 + \sum_{k=1}^{\infty} (H_k\, dt_k - \bar{H}_k\, d\bar{t}_k)$$

generates the set of Hamiltonian equations for deformations of the curve due to variation of t_k:

$$\partial_{t_k} S(z) = \partial_z H_k(z), \quad \partial_{\bar{t}_k} S(z) = -\partial_z \bar{H}_k(z), \qquad (7\text{--}1)$$

where we set $H_0(z) = \log w(z)$. The equations are consistent due to commutativity of the flows:

$$\left(\partial_{t_j} H_k\right)_z = \left(\partial_{t_k} H_j\right)_z = \partial_{t_j} \partial_{t_k} \Omega(z).$$

Equations (7–1) are more transparent when written in terms of canonical variables. The differential $d\Omega$ suggests that the pairs $\log w$, t_0 and $z(w)$, $S(z(w))$ are canonical and establishes the *symplectic structure for conformal maps*. Indeed, treating w as an independent variable, one rewrites (5–5) as

$$\{z(w), S(z(w))\} = 1, \qquad (7\text{--}2)$$

where the Poisson bracket $\{\,\cdot\,,\,\cdot\,\}$ is with respect to $\log w$ and the area t_0 is defined as

$$\{f, g\} = w\frac{\partial f}{\partial w}\frac{\partial g}{\partial t_0} - w\frac{\partial g}{\partial w}\frac{\partial f}{\partial t_0},$$

where the derivatives with respect to t_0 are taken at fixed t_k and w.

The other flows read

$$\frac{\partial z(w)}{\partial t_k} = \{H_k, z(w)\}, \tag{7-3}$$

$$\frac{\partial S(z(w))}{\partial t_k} = \{H_k, S(z(w))\}, \tag{7-4}$$

and similarly for the flows with respect to $\bar{t}_k$. Now the Hamiltonian functions H_k and $\bar{H}_k$ are degree k polynomials of w and w^{-1} respectively.

The consistency conditions (7–1) now take the form of the zero-curvature conditions

$$\partial_{t_j} H_i - \partial_{t_i} H_j + \{H_i, H_j\} = 0, \tag{7-5}$$

$$\partial_{t_j} \bar{H}_i + \partial_{\bar{t}_i} H_j + \{\bar{H}_i, H_j\} = 0. \tag{7-6}$$

The infinite set of Poisson-commuting flows forms a *Whitham integrable hierarchy* [10]. Equations (7–3) and (7–4) are the Lax–Sato equations for the hierarchy. They generate an infinite set of differential equations for the coefficients (potentials) u_j of the inverse conformal map (5–1). The first equation of the hierarchy is

$$\partial^2_{t_1 \bar{t}_1} \phi = \partial_{t_0} \exp(\partial_{t_0} \phi), \quad \partial_{t_0} \phi = \log r^2.$$

The integrable hierarchy describing conformal maps is also known in the soliton literature as the *dispersionless Toda lattice hierarchy*, or SDiff(2) Toda hierarchy [11]; see the next section. The algebra Sdiff(2) of area-preserving diffeomorphisms is the symmetry algebra of this hierarchy [11]. Equations (7–3)–(7–6) describe infinitesimal deformations of the curve such that the area t_0 is kept fixed.

(A relation between conformal maps of slit domains and special solutions to equations of hydrodynamic type, namely the Benney equations, was first observed by Gibbons and Tsarev [12].)

The integrable hierarchy possesses many solutions. The particular solution relevant to conformal maps is selected by the subsidiary condition (7–2). This condition, known as *dispersionless string equation*, has already appeared in the study of the $c = 1$ topological gravity [11; 13; 14] and in the large N limit of a model of normal random matrices [15]. The latter is discussed in Section 9.

8. Toda Lattice Hierarchy and its Dispersionless Limit

We review the two-dimensional Toda lattice hierarchy and show that its dispersionless limit gives the equations describing the conformal maps (5–6), (5–7), (7–3), (7–4).

The two-dimensional Toda hierarchy is defined by two Lax operators

$$L = r(t_0)\, e^{\hbar \partial/\partial t_0} + \sum_{k=0}^{\infty} u_k(t_0)\, e^{-k\hbar \partial/\partial t_0}, \tag{8-1}$$

$$\bar{L} = e^{-\hbar\partial/\partial t_0} r(t_0) + \sum_{k=0}^{\infty} e^{k\hbar\partial/\partial t_0} \bar{u}_k(t_0), \qquad (8\text{-}2)$$

acting in the space of functions of t_0 where the coefficients u_j and $\bar{u}_j$ are functions of t_0 and also of two independent sets of parameters ("times") t_k and $\bar{t}_k$. Note that u_k and $\bar{u}_k$ as well as t_k and $\bar{t}_k$ in (8–1), (8–2) are not necessarily complex conjugate to each other, although we choose them to be so.

The dependence of the coefficient u_k and $\bar{u}_k$ on t_k and $\bar{t}_k$ are given by the Lax–Sato equations:

$$\hbar \frac{\partial L}{\partial t_k} = [H_k, L], \qquad (8\text{-}3)$$

$$\hbar \frac{\partial L}{\partial \bar{t}_k} = [L, \bar{H}_k], \qquad (8\text{-}4)$$

and similar equations for $\bar{L}$. The flows are generated by

$$H_k = \left(L^k\right)_+ + \tfrac{1}{2}\left(L^k\right)_0 \qquad (8\text{-}5)$$

and

$$\bar{H}_k = \left(\bar{L}^k\right)_- + \tfrac{1}{2}\left(\bar{L}^k\right)_0,$$

where the symbol $\left(L^k\right)_\pm$ means positive (negative) parts of the series in the shift operator $e^{\hbar \partial/\partial t_0}$. The first equation of the hierarchy is the Toda lattice equation

$$\partial^2_{t_1 \bar{t}_1} \phi(t_0) = e^{\phi(t_0+\hbar)-\phi(t_0)} - e^{\phi(t_0)-\phi(t_0-\hbar)},$$

where $r^2 = e^{\phi(t_0+\hbar)-\phi(t_0)}$.

The spectrum of the Lax operator is determined by the linear problem $L\Psi = z\Psi$. The wave function Ψ is expressed through the τ-function $\tau_\hbar$ of the dispersionful hierarchy (8–3), (8–4) by the formula

$$\Psi(z; t_0, t_1, t_2, \cdots)$$
$$= \tau_\hbar^{-1}(t_0, t_1, t_2, \ldots) z^{t_0/\hbar} e^{(1/\hbar) \sum_{k>0} t_k z^k} e^{\hbar \sum_{k>0} (z^{-k}/k) \partial/\partial t_k} \tau_\hbar(t_0, t_1, t_2, \ldots).$$

Among many solutions of the hierarchy, one is of particular interest. It is selected by the *string equation* [16]

$$[L, \bar{L}] = \hbar. \qquad (8\text{-}6)$$

This solution is known to describe the normal matrix model at finite size of matrices [15].

The dispersionless limit of the Toda hierarchy is a formal semi-classical limit $\hbar \to 0$. To proceed we notice that the shift operator $W = e^{\hbar\partial/\partial t_0}$ obeys the commutation relation $[W, t_0] = \hbar W$. In the semiclassical limit it is supposed to be replaced by the canonical variable w with the Poisson bracket $\{\log w, t_0\} = 1$. The Lax operator then becomes a c-valued function which is identified with the inverse conformal map $z(w)$ (5–1). Similarly, $\bar{L}$ is identified with $S(z(w))$. In their turn, the Lax–Sato equations (8–3) and (8–4) are identified with equations

(7–3) and (7–4) for the conformal map. In the same fashion the dispersionless limit of the string equation (8–6) is identified with (7–2). The semiclassical limits of the wave function and the τ-function give the generating function Ω and the dispersionless τ-function: $\Psi \to e^{\Omega/\hbar}$, $\tau_\hbar \to e^{(\log \tau)/\hbar^2}$. Similarly, equation (6–7) is a semiclassical limit of the Hirota equation for the τ-function of the two-dimensional Toda hierarchy.

9. The τ-Function of the Conformal Map as a Large N Matrix Integral

The integrable structure of conformal maps is identical to the one observed in a class of random matrix models related to noncritical string theories. Moreover, there exists a random matrix model whose large N limit reproduces *exactly* the τ-function for analytic curves.

Consider the partition function of the ensemble of normal random $N \times N$ matrices [15], with the potential (3–2):

$$\tau_\hbar[t,\bar{t}] = \int dM\, dM^\dagger e^{-(1/\hbar)\operatorname{Tr} V(M,M^\dagger)}.$$

(V. Kazakov pointed out to us that the Lax equations (8–3) and (8–4) are generated by the Hermitian 2-matrix model [17] with complex conjugated potentials. The latter and the normal matrix model have an identical $1/N$-expansion.)

A matrix is called normal if it commutes with its Hermitian conjugated $[M, M^\dagger] = 0$. Passing to the eigenvalues $\operatorname{diag}(z_1, \ldots, z_N)$ of the matrix M, one obtains the measure of the integral in a factorized form

$$dM\, dM^\dagger \sim \prod_{i=1}^N dz_i\, d\bar{z}_i \prod_{k<j} (z_k - z_j)(\bar{z}_k - \bar{z}_j).$$

Then the partition function is represents a two-dimensional Coulomb gas in the potential (3–2)

$$\tau_\hbar[t,\bar{t}] = \int \prod_{k=1}^N dz_k\, d\bar{z}_k\, e^{-(1/\hbar)V(z_k,\bar{z}_k)} \prod_{i<j} e^{2\log|z_i - z_j|}.$$

To proceed to the large N limit one introduces a parameter $t_0 = \hbar N$ and expresses the integrand in terms of density of eigenvalues as $e^{-\hbar^{-2}\mathcal{E}\{\rho,V\}}$, where $\mathcal{E}\{\rho,V\}$ is given by (3–3). Then the limit for large N (or $\hbar \to 0$) yields to the variational principle of Section 3. In the large N limit the eigenvalues of the matrix homogeneously fill the domain D_+ bound by the curve, characterized by the harmonic moments t_k and the area t_0 and leads to the τ-function defined by (4–1). Other objects introduced in Sections 3 to 7 can also be identified with expectation values of the matrix model. In particular the moments v_k of (2–4) are

$$v_k = \hbar \langle \operatorname{Tr} M^k \rangle$$

and $\Omega^- - \frac{1}{2}v_0 = \hbar \langle \mathrm{Tr}\, \log(z - M) \rangle$.

In order to identify the Lax operator, we follow [18; 15; 17]. Introduce the basis of orthogonal polynomials $P_n(z) = h_n z^n + \cdots$ ($n \geq 0$), by the orthonormality relations

$$\langle m | n \rangle \equiv \int d^2 z \, \overline{P_n(z)} \, e^{-\frac{1}{\hbar} V(z, \bar{z})} \, P_m(z) = \delta_{m,n}.$$

The polynomials are uniquely defined by the potential V up to phase factors. It is easy to see that the τ-function is given by the product of the coefficients $N! \, |h_n h_{n-1} \cdots h_0|^2$ of the highest powers of the polynomials $P_n(z) = h_n z^n + \cdots$. Then Lax operators L and $\bar{L}$ appear as the operators $\langle m|z|n \rangle$ and $\langle m|\bar{z}|n \rangle$. Since $zP_n(z)$ can be expressed through polynomials of the degree not grater than n, one may represent $\langle m|z|n \rangle$ and $\langle m|\bar{z}|n \rangle$ in terms of shifts operators $W = e^{\hbar \frac{\partial}{\partial t_0}}$ in the form of (8–1), (8–2), where $r(t_0 = \hbar n) = h_n / h_{n+1}$.

Similar arguments allow one to identify the flows. Consider a variation of some operator $\langle m|O|n \rangle$ under a variation of t_k. We have $\hbar \partial_{t_k} \langle m|O|n \rangle = \langle m|[H_k, O]|n \rangle$, where $H_k = A_k - A^\dagger{}_k$ and $\langle m|A_k|n \rangle = \langle m|\partial_{t_k}|n \rangle$. Obviously $H_k = -L^k(W) +$ negative powers of W. Choosing O to be $\bar{L}$ (see (8–2)) which consists on W^{-1} and positive powers of W, one concludes that H_k does not consists of negative powers of W. This brings us to (8–5).

Finally, the operator $D = \langle m| \hbar \partial_z | n \rangle$ is equal to

$$D = \bar{L} - \sum_{k \geq 1} k t_k L^{k-1}.$$

The Heisenberg relation $[D, L] = \hbar$ prompts the string equation (8–6).

The matrix model also offers an effective method to derive equations (6–1)–(6–7); see [17], for example.

Acknowledgements

We thank M. Brodsky, V. Kazakov, S. P. Novikov, and L. Takhtajan for valuable comments and interest to this work.

Kostov's work is supported in part by European TMR contract ERBFM-RXCT960012 and EC Contract FMRX-CT96-0012.

Krichever's work is supported in part by NSF grant DMS-98-02577.

Wiegmann would like to thanks P. Bleher and A. Its for the hospitality in MSRI during the workshop on Random Matrices in spring 1999.

Krichever and Zabrodin have been partially supported by CRDF grant 6531.

Wiegmann and Zabrodin have been partially supported by grants NSF DMR 9971332 and MRSEC NSF DMR 9808595.

Zabrodin's work was supported in part by grant INTAS-99-0590 and RFBR grant 00-02-16477. He also thanks for hospitality the Erwin Schrödinger Institute in Vienna, where this work was completed.

References

[1] M. Mineev-Weinstein, P. B. Wiegmann and A. Zabrodin, Phys. Rev. Lett. **84** (2000) 5106–5109.

[2] P. B. Wiegmann and A. Zabrodin, hep-th/9909147, Commun. Math. Phys. to appear.

[3] P. S. Novikov, C. R. (Dokl.) Acad. Sci. URSS (N. S.) **18** (1938) 165–168; M. Sakai, Proc. Amer. Math. Soc. **70** (1978) 35–38; V. Strakhov and M. Brodsky, SIAM J. Appl. Math. **46** (1986) 324–344.

[4] V. K. Ivanov, Soviet Doklady, Ser. Math. **105** (1955) 409–414.

[5] P. Di Francesco, M. Gaudin, C. Itzykson and F. Lesage, Int. J. Mod. Phys. **A9** (1994) 4257–4351.

[6] A. Hurwitz and R. Courant, The theory of functions, Springer-Verlag, 1964.

[7] P. J. Davis, The Schwarz function and its applications, The Carus Mathematical Monographs, No. 17, The Math. Assotiation of America, Buffalo, N. Y., 1974.

[8] J. Gibbons and Y. Kodama, Phys. Lett. **135 A** (1989) 167–170; in Proceedings of NATO ASI, Singular Limits of Dispersive Waves, ed. N. Ercolani, Plenum 1994.

[9] R. Carroll and Y. Kodama, J. Phys. A: Math. Gen. **A28** (1995) 6373–6388.

[10] I. M. Krichever, Function. Anal. Appl. **22** (1989) 200–213; Commun. Math. Phys. **143** (1992) 415–429; Commun. Pure. Appl. Math. **47** (1992) 437–476.

[11] K. Takasaki and T. Takebe, Lett. Math. Phys. **23** (1991) 205–214; Rev. Math. Phys. **7** (1995) 743–808.

[12] J. Gibbons and S. P. Tsarev, Phys. Lett. **211A** (1996) 19–24; ibid **258A** (1999) 263–271.

[13] R. Dijkgraaf, G. Moore and R. Plesser, Nucl. Phys. **B394** (1993) 356–382; A Hanany, Y. Oz and R. Plesser, Nucl. Phys. **B425** (1994) 150–172; K. Takasaki, Commun. Math. Phys. **170** (1995) 101–116; T. Eguchi and H. Kanno, Phys. Lett. **331B** (1994) 330.

[14] R. Dijkgraaf and E. Witten, Nucl. Phys. **B342** (1990) 486–522;A. Losev and I. Polyubin, Int. J. Mod. Phys. **A10** (1995) 4161–4178;S. Aoyama and Y. Kodama, Commun. Math. Phys. **182** (1996) 185–220.

[15] Ling-Lie Chau and Y. Yu Phys. Lett. **167A** (1992) 452,Ling-Lie Chau and O. Zaboronsky, Commun. Math. Phys. **196** (1998) 203.

[16] M. Douglas, Phys. Lett. **238B** (1990) 176; in Proceedings of the 1990 Cargèse Workshop on Random Surfaces and Quantum Gravity, NATO ASI Series, Plenum Press, New York.

[17] J. M. Daul, V. A. Kazakov and I. K. Kostov, Nucl. Phys. **B409** (1993) 311–338;L. Bonora and C. S. Xiong, Phys. Lett. **B347** (1995) 41–48.

[18] M. L. Mehta, Commun. Math. Phys. **79** (1981) 327; S. Chadha, G. Mahoux and M. L. Mehta, J. Phys. A: Math. Gen. **14** (1981) 579.

I. K. Kostov
SERVICE DE PHYSIQUE THÉORIQUE
CEA-SACLAY
91191 GIF SUR YVETTE
FRANCE
    kostov@spht.saclay.cea.fr

I. KRICHEVER
LANDAU INSTITUTE FOR THEORETICAL PHYSICS
and
DEPARTMENT OF MATHEMATICS
COLUMBIA UNIVERSITY
NEW YORK, NY 10027
UNITED STATES
    krichev@math.columbia.edu

M. MINEEV-WEINSTEIN
THEORETICAL DIVISION, MS-B213
LOS ALAMOS NATIONAL LABORATORIES
LOS ALAMOS, NM 87545
UNITED STATES
    mineev@t13.lanl.gov

P. B. WIEGMANN
LANDAU INSTITUTE FOR THEORETICAL PHYSICS
and
JAMES FRANCK INSTITUTE AND ENRICO FERMI INSTITUTE
UNIVERSITY OF CHICAGO
5640 S. ELLIS AVENUE
CHICAGO, IL 60637
UNITED STATES
    wiegmann@uchicago.edu

A. ZABRODIN
JOINT INSTITUTE OF CHEMICAL PHYSICS
KOSYGINA STR. 4
117334, MOSCOW
and
ITEP
117259, MOSCOW
RUSSIA
    zabrodin@heron.itep.ru

Integration over Angular Variables for Two Coupled Matrices

G. MAHOUX, M. L. MEHTA, AND J.-M. NORMAND

ABSTRACT. An integral over the angular variables for two coupled $n \times n$ real symmetric, complex hermitian or quaternion self-dual matrices is expressed in terms of the eigenvalues and eigenfunctions of a hamiltonian closely related to the Calogero hamiltonian. This generalizes the known result for the complex hermitian matrices. The integral can thus be evaluated for $n = 2$ and reduced to a single sum for $n = 3$.

1. Introduction

The remarkable and useful formula

$$\int dU \, \exp\left(-\frac{1}{2t} \, \mathrm{tr}(A - UA'U^{-1})^2\right) = t^{n(n-1)/2} \left(\prod_{j=0}^{n-1} j!\right)$$
$$\times \left(\Delta(\boldsymbol{x})\Delta(\boldsymbol{x}')\right)^{-1} \det\left[\exp\left(-\frac{1}{2t}(x_j - x'_k)^2\right)\right]_{j,k=1,\ldots,n} \quad (1\text{-}1)$$

has been known for the last two decades; see [Itzykson and Zuber 1980; Mehta 1981; Mehta 1991, Appendix A.5]. Here A and A' are $n \times n$ complex hermitian matrices having eigenvalues $\boldsymbol{x} := \{x_1, \ldots, x_n\}$ and $\boldsymbol{x}' := \{x'_1, \ldots, x'_n\}$ respectively, integration is over the $n \times n$ complex unitary matrices U with the invariant Haar measure dU normalized such that $\int dU = 1$. The function $\Delta(\boldsymbol{x})$ is the product of differences of the x_j:

$$\Delta(\boldsymbol{x}) := \begin{cases} 1 & \text{if } n = 1, \\ \prod_{1 \le j < k \le n}(x_k - x_j) & \text{if } n \ge 2. \end{cases}$$

We would like to have a similar formula when A and A' are $n \times n$ real symmetric or quaternion self-dual matrices and the integration is over $n \times n$ real orthogonal or quaternion symplectic matrices U, a formula not presently known. These three cases are usually denoted by a parameter β taking values 1, 2 and 4 corresponding

---

Mehta is a member of C.N.R.S. France.

respectively to the integration over $n \times n$ real orthogonal, complex unitary and quaternion symplectic matrices U. We will show that the integral in (1–1) with a measure dU invariant under the appropriate group can be expressed in terms of the eigenvalues and eigenfunctions of a particular hamiltonian. This hamiltonian is closely related to the Calogero model [1969a; 1969b; 1971] where one considers the quantum n-body problem with the hamiltonian

$$H := -\sum_{j=1}^{n} \frac{\partial^2}{\partial x_j^2} + \frac{1}{2}\beta(\beta - 2) \sum_{1 \leq j < k \leq n} (x_j - x_k)^{-2}. \qquad (1\text{–}2)$$

For $n = 2$ and $n = 3$ the complete set of the relevant eigenfunctions and eigenvalues are known. The three integrals (namely, for $\beta = 1, 2, 4$) can thus be explicitly computed for $n = 2$ and reduced to a single infinite sum for $n = 3$ and $\beta = 1$ or 4. For $n > 3$ and $\beta = 1$ or $\beta = 4$, the question remains open.

Integral (1–1) over the orthogonal group has also been of interest. Muirhead [1982, Chapters 7 and 9; 1975] gives this integral in terms of a hypergeometric function of two matrix variables, expressed itself as a series of matrix zonal polynomials. But no general formula for these zonal polynomials is known.

2. The Diffusion Equation

Recall that, D_j being constants, the partial differential (or diffusion) equation

$$\frac{\partial \xi}{\partial t} = \frac{1}{2} \sum_{j=1}^{n} D_j \frac{\partial^2 \xi}{\partial x_j^2} \qquad (2\text{–}1)$$

with the initial condition

$$\xi(\boldsymbol{x}; 0) := \eta(\boldsymbol{x})$$

has, for $t \geq 0$, the unique solution

$$\xi(\boldsymbol{x}; t) = \int d\boldsymbol{x}' \, K(\boldsymbol{x}, \boldsymbol{x}'; t) \eta(\boldsymbol{x}'), \quad \text{with } d\boldsymbol{x}' := dx'_1 \ldots dx'_n,$$

$$K(\boldsymbol{x}, \boldsymbol{x}'; t) := \prod_{j=1}^{n} \left((2\pi D_j t)^{-1/2} \exp\left(-\frac{1}{2 D_j t}(x_j - x'_j)^2\right) \right).$$

Indeed, for $t > 0$, $K(\boldsymbol{x}, \boldsymbol{x}'; t)$ satisfies (2–1) and it reduces to $\prod_{j=1}^{n} \delta(x_j - x'_j)$ as $t \to 0$.

Now, in these formulae, we use as variables x_j the $n + \beta n(n-1)/2$ real variables that determine the matrix A, namely, the n real diagonal elements A_{jj} and the β components $A_{jk(r)}$ (with $r = 1, \ldots, \beta$) of each of the $n(n-1)/2$ nondiagonal elements A_{jk}. For real symmetric matrices, only one component A_{jk} is present; for complex hermitian matrices, the two components are the real and imaginary parts of A_{jk}; for the quaternion self-dual matrices, A_{jk} has four components, one scalar and three vectorial parts. Similarly, the x'_j are the $n + \beta n(n-1)/2$ real variables that determine the matrix A'. The kernel $K(x, x'; t)$ then becomes

$$K(A, A'; t) := \prod_{j=1}^{n}\left((2\pi t)^{-1/2} \exp\left(-\frac{1}{2t}(A_{jj} - A'_{jj})^2\right)\right)$$

$$\times \prod_{1 \le j < k \le n} \prod_{r=1}^{\beta}\left((\pi t)^{-1/2} \exp\left(-\frac{1}{t}(A_{jk(r)} - A'_{jk(r)})^2\right)\right)$$

$$= (2\pi t)^{-n/2}(\pi t)^{-\beta n(n-1)/4} \exp\left(-\frac{1}{2t} \operatorname{tr}(A - A')^2\right). \qquad (2\text{-}2)$$

The integral

$$\xi(A; t) = \int dA'\, K(A, A'; t)\eta(A'), \qquad (2\text{-}3)$$

with the measure

$$dA' := \left(\prod_{j=1}^{n} dA'_{jj}\right) \prod_{1 \le j < k \le n} \prod_{r=1}^{\beta} dA'_{jk(r)} \qquad (2\text{-}4)$$

satisfies the diffusion equation

$$\frac{\partial \xi}{\partial t} = \frac{1}{2}\nabla_A^2 \xi, \quad \nabla_A^2 := \sum_{j=1}^{n} \frac{\partial^2}{\partial A_{jj}^2} + \frac{1}{2} \sum_{1 \le j < k \le n} \sum_{r=1}^{\beta} \frac{\partial^2}{\partial A_{jk(r)}^2}, \qquad (2\text{-}5)$$

and the initial condition

$$\xi(A; 0) = \eta(A).$$

The dA' in (2-4) is a measure invariant under the automorphism $A' \to UA'U^{-1}$ for any U in the group $\mathcal{G}_\beta$ of $n \times n$ real orthogonal, complex unitary or quaternion symplectic matrices, respectively for $\beta = 1$, 2 or 4. Let us assume from now on that $\eta(A')$ is invariant under the same transformation; that is, $\eta(A') = \eta(UA'U^{-1})$ for any U in $\mathcal{G}_\beta$. From the invariance of the measure dA' in (2-3), (2-4) and the cyclic invariance of the trace in (2-2), it follows that $\xi(A; t)$ is also invariant under the same transformation; that is, $\xi(A; t) = \xi(UAU^{-1}; t)$. We can choose a matrix U_A in $\mathcal{G}_\beta$ to diagonalize A, [1]

$$A = U_A X U_A^{-1}, \quad X := [x_i \delta_{ij}], \qquad (2\text{-}6)$$

and similarly for A'. The invariance of η and ξ implies that $\eta(A')$ and $\xi(A; t)$ are symmetric functions of the eigenvalues of A' and A respectively; we denote them as

$$\eta(x') := \eta(A') \qquad \xi(x; t) := \xi(A; t).$$

The hyperplanes $x_j = x_k$ for $1 \le j < k \le n$ divide the n-dimensional space into $n!$ sectors (sometimes called Weyl chambers). Taking advantage of the symmetry

---

[1] Diagonalization of quaternion matrices is not as well known as that of real or complex matrices. See [Mehta 1989, Chapters 4 and 8], for example.

property of $\xi(\boldsymbol{x};t)$, from now on we will restrict our attention to one of the sectors where $\Delta(\boldsymbol{x}) \geq 0$, namely the sector S defined by the conditions

$$S := \{\boldsymbol{x}; x_1 \leq x_2 \leq \cdots \leq x_n\}. \tag{2-7}$$

Changing the variables from matrix element components to the n (real) eigenvalues and the $\beta n(n-1)/2$ real "angle" variables on which U_A and U'_A depend, we have as usual [Mehta 1991, Chapter 3][2]

$$dA' = \pi^{\beta n(n-1)/4} \left(\prod_{j=1}^{n} \frac{\Gamma(1+\beta/2)}{\Gamma(1+\beta j/2)} \right) |\Delta(\boldsymbol{x}')|^\beta \, d\boldsymbol{x}' \, dU'_A, \tag{2-8}$$

where dU'_A is the invariant measure over the group $\mathcal{G}_\beta$ normalized such that $\int dU'_A = 1$. Hence equation (2–3) reads, for any $\boldsymbol{x}$ in S,

$$\xi(\boldsymbol{x};t) = \int_{R^n} d\boldsymbol{x}' \, |\Delta(\boldsymbol{x}')|^\beta \, \mathcal{K}(\boldsymbol{x},\boldsymbol{x}';t) \, \eta(\boldsymbol{x}'), \tag{2-9}$$

where

$$\mathcal{K}(\boldsymbol{x},\boldsymbol{x}';t) = (2\pi t)^{-n/2} t^{-\beta n(n-1)/4} \left(\prod_{j=1}^{n} \frac{\Gamma(1+\beta/2)}{\Gamma(1+\beta j/2)} \right)$$
$$\times \int dU \, \exp\left(-\frac{1}{2t} \mathrm{tr}(\boldsymbol{X} - U\boldsymbol{X}'U^{-1})^2 \right). \tag{2-10}$$

It follows from the integration over the group $\mathcal{G}_\beta$ with the invariant measure dU, and from the cyclic invariance of the trace, that $\mathcal{K}(\boldsymbol{x},\boldsymbol{x}';t)$ is a function symmetric in the x_j and symmetric in the x'_j. Since the measure $d\boldsymbol{x}' \, |\Delta(\boldsymbol{x}')|^\beta$ and $\eta(\boldsymbol{x}')$ are symmetric in the x'_j, the integral in (2–9) can be restricted to the sector S and multiplying it by a factor $n!$; thus

$$\xi(\boldsymbol{x};t) = n! \int_S d\boldsymbol{x}' \, (\Delta(\boldsymbol{x}'))^\beta \, \mathcal{K}(\boldsymbol{x},\boldsymbol{x}';t) \eta(\boldsymbol{x}'). \tag{2-11}$$

We recognize in the right-hand side of (2–10) precisely the quantity we are interested in. Namely, apart from explicitly known constant factors, this is the left-hand side of (1–1), including now the cases where A, A' are $n \times n$ real symmetric or quaternion self-dual matrices and the integration is over $n \times n$ real orthogonal or quaternion symplectic matrices. Our problem thus amounts to constructing the kernel $\mathcal{K}(\boldsymbol{x},\boldsymbol{x}';t)$ of the evolution operator of the diffusion equation (2–5).

Separating the laplacian ∇_A^2 into parts depending on $\boldsymbol{x}$ and on U_A, we get (see Appendix A for proof)

$$\nabla_A^2 = (\Delta(\boldsymbol{x}))^{-\beta} \sum_{j=1}^{n} \frac{\partial}{\partial x_j} (\Delta(\boldsymbol{x}))^\beta \frac{\partial}{\partial x_j} + \mathcal{D}_{U_A}^2, \tag{2-12}$$

---

[2]There the constant factor in (2–8) is not evaluated. The evaluation can be done by computing $\int dA e^{-\mathrm{tr}\, A^2}$ both directly and using (3.3.10) of the reference.

where $\mathcal{D}^2_{U_A}$ involves derivatives with respect to the angle variables entering U_A, one sees that $\xi(\boldsymbol{x};t)$ satisfies the diffusion equation, for all $\boldsymbol{x}$ in $\mathcal{S}$:

$$\frac{\partial \xi}{\partial t} = -\frac{1}{2}\mathcal{H}\xi, \tag{2-13}$$

$$\mathcal{H} := -\left(\Delta(\boldsymbol{x})\right)^{-\beta} \sum_{j=1}^{n} \frac{\partial}{\partial x_j}\left(\Delta(\boldsymbol{x})\right)^{\beta}\frac{\partial}{\partial x_j}$$

$$= -\sum_{j=1}^{n}\frac{\partial^2}{\partial x_j^2} - \beta \sum_{1 \leq j < k \leq n}\frac{1}{x_j - x_k}\left(\frac{\partial}{\partial x_j} - \frac{\partial}{\partial x_k}\right), \tag{2-14}$$

with the initial value $\xi(\boldsymbol{x};0) = \eta(\boldsymbol{x})$. Equation (2–13) is similar to a Schrödinger equation with the hamiltonian $\mathcal{H}$ and a purely imaginary time $-i\hbar t/2$.

We now have to specify the space of functions ξ and η, in which we solve the preceding equations.

3. The Evolution Operator

To construct the evolution operator of equation (2–13), we build a Hilbert space $\mathcal{L}^2(\mathcal{S},\mu)$ of functions $f(\boldsymbol{x})$. These functions are supposed to be the restrictions to the sector $\mathcal{S}$, defined in (2–7), of symmetric functions of the variables x_j in R^n. The scalar product using Dirac's notation is defined as

$$\langle f|g \rangle := \int_{\mathcal{S}} d\boldsymbol{x}\, \mu(\boldsymbol{x})\, f^*(\boldsymbol{x}) g(\boldsymbol{x})$$

and we choose the weight $\mu(\boldsymbol{x})$ in such a way that $\mathcal{H}$ be hermitian. From (2–14) one deduces for twice differentiable functions f and g the identity

$$\mu\left(f\mathcal{H}g - g\mathcal{H}f\right) = -\sum_{j=1}^{n}\frac{\partial}{\partial x_j}\left(\mu W_j(f,g)\right) + \sum_{j=1}^{n}\left(\frac{\partial \mu}{\partial x_j} - \beta\mu\sum_{k \neq j}\frac{1}{x_j - x_k}\right)W_j(f,g), \tag{3-1}$$

where W_j is the wronskian $W_j(f,g) := f\,\partial g/\partial x_j - g\,\partial f/\partial x_j$. By integrating both sides over the sector $\mathcal{S}$, the first term in the right-hand side of (3–1), which is a divergence, gives an integral over the boundary $\partial\mathcal{S}$ of $\mathcal{S}$. The points at finite distance of $\partial\mathcal{S}$ do not contribute: because of the symmetry of f and g extended to R^n, the vector with components $W_j(f,g)$ has no component normal to $\partial\mathcal{S}$. Also the points at infinity of $\partial\mathcal{S}$ do not contribute if f and g vanish fast enough at infinity; that is, if they belong to the domain of $\mathcal{H}$. As for the second term, it vanishes identically if and only if μ satisfies n linear first order partial differential equations, with the solution unique up to a constant c,

$$\mu(\boldsymbol{x}) = c\left(\Delta(\boldsymbol{x})\right)^{\beta}. \tag{3-2}$$

With such a weight μ, $\mathcal{H}$ defines a hermitian operator in $\mathcal{L}^2(\mathcal{S},\mu)$.

Looking at factorized solutions of the form $\varphi_\alpha(x)\exp(-\mathcal{E}_\alpha t/2)$ of (2–13), $\varphi_\alpha(x)$ is the eigenfunction with the eigenvalue $\mathcal{E}_\alpha$ of the t-independent hamiltonian $\mathcal{H}$. In Dirac's bracket notation $\varphi_\alpha(x) := \langle x|\varphi_\alpha\rangle$ satisfies the "Schrödinger equation"

$$\mathcal{H}|\varphi_\alpha\rangle = \mathcal{E}_\alpha|\varphi_\alpha\rangle. \tag{3-3}$$

Here, α denotes a convenient set of indices. The orthogonality and closure relations for the basis $\{|x\rangle\}$ are

$$\langle x|x'\rangle = (\mu(x))^{-1}\delta(x-x'), \quad \int_S dx\,\mu(x)|x\rangle\langle x| = 1 \tag{3-4}$$

and those for the basis $\{|\varphi_\alpha\rangle\}$ are

$$\langle\varphi_\alpha|\varphi_{\alpha'}\rangle = (\rho(\alpha))^{-1}\delta(\alpha-\alpha'), \quad \int d\alpha\,\rho(\alpha)|\varphi_\alpha\rangle\langle\varphi_\alpha| = 1, \tag{3-5}$$

where the positive weight $\rho(\alpha)$ can be chosen at will. With these notations, it follows from (2–11) and (3–2)–(3–5) that the kernel $\mathcal{K}$ equals

$$\mathcal{K}(x,x';t) = \frac{c}{n!}\langle x|e^{-\mathcal{H}t/2}|x'\rangle = \frac{c}{n!}\int d\alpha\,\rho(\alpha)\varphi_\alpha(x)e^{-\mathcal{E}_\alpha t/2}\varphi_\alpha^*(x'). \tag{3-6}$$

Using (2–10), this last equation allows us to extend (1–1) to the cases $\beta = 1$ and $\beta = 4$:

$$\int dU\,\exp\left(-\frac{1}{2t}\mathrm{tr}(X-UX'U^{-1})^2\right) = \frac{c}{n!}(2\pi t)^{n/2}t^{\beta n(n-1)/4}\left(\prod_{j=1}^n \frac{\Gamma(1+\beta j/2)}{\Gamma(1+\beta/2)}\right)$$

$$\times \int d\alpha\,\rho(\alpha)\varphi_\alpha(x)e^{-\mathcal{E}_\alpha t/2}\varphi_\alpha^*(x'). \tag{3-7}$$

4. Connection with the Calogero Model

If we wanted to eliminate the linear derivative terms in (2–13) and (2–14), we would change the unknown function for all x in the sector S defined in (2–7), as follows

$$\psi(x;t) := (\Delta(x))^{\beta/2}\xi(x;t). \tag{4-1}$$

A straightforward calculation shows that this $\psi(x;t)$ satisfies the partial differential equation

$$\frac{\partial\psi}{\partial t} = -\tfrac{1}{2}H\psi \tag{4-2}$$

where H is precisely the Calogero hamiltonian (1–2). Looking at factorized solutions of the form $\phi_\alpha(x)\exp(-E_\alpha t/2)$ of (4–2), $\phi_\alpha(x)$ is the eigenfunction with the eigenvalue E_α of the t-independent hamiltonian H. In Dirac's bracket notation $\phi_\alpha(x) := \langle x|\phi_\alpha\rangle$ satisfies the "Schrödinger equation"

$$H|\phi_\alpha\rangle = E_\alpha|\phi_\alpha\rangle. \tag{4-3}$$

The discussion of the hermitian character of $\mathcal{H}$, (3–1) and what follows, can be applied to H. For the function ψ of the form (4–1), with $\xi(x;t)$ analytic in x, one can check that H is hermitian in S with $\mu(x) = 1$. Hence finding a solution ϕ of (4–3) in $\mathcal{L}^2(S,1)$ is equivalent to calculating the integral (1–1). Indeed the considerations leading from (2–13) to (3–6) when applied to (4–2) yields for all x in S

$$\psi(x;t) = \int_S dx' \langle x|e^{-\frac{1}{2}Ht}|x'\rangle (\Delta(x'))^{\beta/2} \eta(x') \tag{4-4}$$

with

$$\langle x|e^{-\frac{1}{2}Ht}|x'\rangle = \int d\alpha \phi_\alpha(x) e^{-\frac{1}{2}E_\alpha t} \phi_\alpha^*(x'). \tag{4-5}$$

Here the orthogonality and closure relations for both bases $\{|x\rangle\}$ and $\{|\phi_\alpha\rangle\}$ are

$$\langle x|x'\rangle = \delta(x - x'), \qquad \int_S dx \, |x\rangle\langle x| = 1,$$

$$\langle \phi_\alpha|\phi_{\alpha'}\rangle = \delta(\alpha - \alpha'), \qquad \int d\alpha \, |\phi_\alpha\rangle\langle \phi_\alpha| = 1.$$

The $\xi(x;t)$ defined in S from (4–1) and (4–4) is then extended to R^n by the requirement that it is a symmetric function of the x_j.

In spite of a slight difference of the point of view, our problem is similar to that of Calogero. In sections 2 and 3 we deal with either one particle in n dimensions and completely symmetric wave functions in $(x_1, \ldots, x_n)$ or with n bosons in one dimension. Calogero considers n particles in one dimension which can be bosons, fermions or boltzmannions. Calogero's particles, all on the real line, cannot cross each other due to the singular potential (see Appendix B). The phase space is thus naturally divided in $n!$ sectors, each sector corresponding to a certain order of the n particles. For Calogero it is sufficient to find the eigenfunctions and eigenvalues in any one sector, say when $x_1 \leq x_2 \leq \cdots \leq x_n$, the solutions in other sectors being obtained by the proper symmetry according as the particles satisfy Boltzmann, Bose or Fermi statistics [Calogero 1969a]. In our case, the "Schrödinger equation" (4–3) for a single particle in n dimensions is well defined only in S and the singular potential requires a special treatment as detailed in Appendix B for the case $n = 2$.

When $\beta = 2$, the hamiltonian (1–2) reduces to the "kinetic energy"

$$-\sum_{j=1}^n \partial^2/\partial x_j^2,$$

each variable x_j in the "Schrödinger equation" (4–3) separates, and the normalized solutions with the corresponding eigenvalues are

$$\phi_\alpha(x) = \prod_{j=1}^n (2\pi)^{-1/2} \exp(ik_j x_j),$$

$$E_\alpha = \sum_{j=1}^n k_j^2, \qquad \alpha = \{k_1, \ldots, k_n\},$$

where k_j are real varying from $-\infty$ to ∞.

Gaussian integrals in (4–5) over the k_j can be performed and on extending it by symmetry to the whole of R^n [Itzykson and Zuber 1980] one gets back (1–1), as one should.

When $\beta = 1$ or $\beta = 4$, the solutions of (3–3) or of (4–3) are not completely known for general n.

5. The Case $n = 2$

The Hilbert space considered in section 3 is $\mathcal{L}^2\big(x_1 \leq x_2, 2^{-\beta/2}(x_2 - x_1)^\beta\big)$ and the hermitian hamiltonian reads

$$\mathcal{H} := -\frac{\partial^2}{\partial x_1^2} - \frac{\partial^2}{\partial x_2^2} - \frac{\beta}{(x_1 - x_2)}\left(\frac{\partial}{\partial x_1} - \frac{\partial}{\partial x_2}\right).$$

Following Calogero [1969a] with a slight modification, we change the variables to

$$X := \frac{x_1 + x_2}{\sqrt{2}}, \qquad x := \frac{x_2 - x_1}{\sqrt{2}} = \frac{\Delta(x)}{\sqrt{2}},$$

where X varies from $-\infty$ to ∞ and x from 0 to ∞. This is an orthogonal change of variables, and (3–3) becomes

$$\mathcal{H}\varphi = -\left(\frac{\partial^2}{\partial X^2} + \frac{\partial^2}{\partial x^2} + \frac{\beta}{x}\frac{\partial}{\partial x}\right)\varphi = \mathcal{E}\varphi.$$

We look for solutions of the form

$$\varphi(X, x) = f(X)g(x).$$

The variables can be separated. Letting primes denote differentiation, we get

$$f''(X) = -K^2 f(X), \qquad (5\text{–}1)$$

$$g''(x) + \frac{\beta}{x}g'(x) + k^2 g(x) = 0, \qquad (5\text{–}2)$$

and

$$\mathcal{E} \equiv \mathcal{E}_{K,k} := K^2 + k^2,$$

where $\mathcal{E}_{K,k}$ is real. Since $-d^2/dX^2$ is the square of the hermitian operator $-id/dX$, K^2 is real nonnegative. As a consequence k^2 is also real. Then (5–1) has only a continuous spectrum labelled by K real and a complete set of normalized solutions

$$f_K(X) = \frac{1}{\sqrt{2\pi}} \exp(iKX) \qquad K \text{ real}. \qquad (5\text{–}3)$$

Setting $g(x) := x^{-\nu}J(x)$ and $\nu := (\beta - 1)/2$, equation (5–2) changes to Bessel's differential equation

$$x^2 J''(x) + x J'(x) + \left(k^2 x^2 - \nu^2\right) J(x) = 0,$$

having two linearly independent solutions. They are: for $k^2 \geq 0$ the Bessel functions $J_\nu(kx)$ and $Y_\nu(kx)$ with $k \geq 0$, and for $k^2 = -\kappa^2 < 0$ the modified

Bessel functions $I_\nu(\kappa x)$ and $K_\nu(\kappa x)$ with $\kappa > 0$. Only the solution J_ν provides a function
$$g_k(x) = (kx)^{-\nu} J_\nu(kx), \quad \nu = (\beta - 1)/2 \tag{5-4}$$
which is square-integrable for $x \geq 0$ with the measure x^β. Then (5–2) has only a continuous spectrum labelled by k real nonnegative. The orthogonality and closure relations (3–5) for the $g_k(x)$ with $k \geq 0$ read
$$\int_0^\infty dx\, x^\beta g_k(x) g_{k'}(x) = k^{-\beta} \delta(k - k')$$
$$\int_0^\infty dk\, k^\beta g_k(x) g_k(x') = x^{-\beta} \delta(x - x').$$
They can be verified by taking the limit $\rho \to 0$ in [Gradshteyn and Ryzhik 1991, page 718, formula 6.633.2]:
$$\int_0^\infty dk\, k\, e^{-\rho^2 k^2} J_\nu(kx) J_\nu(kx') = \frac{1}{2\rho^2} e^{-(x^2 + x'^2)/(4\rho^2)} I_\nu\left(\frac{xx'}{2\rho^2}\right). \tag{5-5}$$
Thus
$$\langle x | e^{-\mathcal{H}t/2} | x' \rangle = \int_{-\infty}^{\infty} dK \int_0^\infty dk\, k^\beta e^{-(K^2 + k^2)t/2} f_K(X) f_K^*(X') g_k(x) g_k(x').$$
The integration over K is a gaussian integral while for the integration over k one can use (5–5). One finally gets from (3–7)
$$\int dU\, \exp\left(-\frac{1}{2t} \operatorname{tr}(A - UA'U^{-1})^2\right) = \frac{\sqrt{\pi}}{2} t^{(\beta-1)/2} \frac{\Gamma(1+\beta)}{\Gamma(1+\beta/2)}$$
$$\times \left((x_2 - x_1)(x_2' - x_1')\right)^{(1-\beta)/2} I_{(\beta-1)/2}\left(\frac{(x_2 - x_1)(x_2' - x_1')}{2t}\right)$$
$$\times \exp\left(-\frac{1}{2t}\left(x_1^2 + x_2^2 + x_1'^2 + x_2'^2 - (x_1 + x_2)(x_1' + x_2')\right)\right). \tag{5-6}$$
This result can be directly verified for $\beta = 1$. For $\beta = 2$, we have
$$I_{1/2}(z) = \sqrt{\frac{2}{\pi z}} \sinh z,$$
and (5–6) gives back the known result (1–1), as it should. More effort is needed to verify directly the result for $\beta = 4$.

6. The Case $n = 3$

The Hilbert space considered in section 3 is $\mathcal{L}^2(x_1 \leq x_2 \leq x_3, 2^{\beta/2}(\Delta(\boldsymbol{x}))^\beta)$, and the hermitian hamiltonian reads
$$\mathcal{H} := -\left(\frac{\partial^2}{\partial x_1^2} + \frac{\partial^2}{\partial x_2^2} + \frac{\partial^2}{\partial x_3^2}\right) - \beta\left(\frac{1}{(x_2 - x_3)}\left(\frac{\partial}{\partial x_2} - \frac{\partial}{\partial x_3}\right)\right.$$
$$\left. + \frac{1}{(x_3 - x_1)}\left(\frac{\partial}{\partial x_3} - \frac{\partial}{\partial x_1}\right) + \frac{1}{(x_1 - x_2)}\left(\frac{\partial}{\partial x_1} - \frac{\partial}{\partial x_2}\right)\right).$$

Following Calogero [1969a] with some slight modifications, we change the variables to

$$X := \frac{1}{\sqrt{3}}(x_1 + x_2 + x_3), \tag{6-1}$$

$$x := \frac{1}{\sqrt{2}}(-x_1 + x_2) =: r\cos\theta,$$

$$y := \frac{1}{\sqrt{6}}(-x_1 - x_2 + 2x_3) := r\sin\theta.$$

It is convenient to make this change of variables in two steps: from x_1, x_2, x_3 to X, x, y and then to X, r, θ. Thereby one gets

$$\Delta(x) = -\frac{1}{\sqrt{2}}r^3\cos(3\theta).$$

Therefore, in the sector $x_1 \leq x_2 \leq x_3$, X varies from $-\infty$ to ∞; x, y and r vary from 0 to ∞; and θ varies from $\pi/6$ to $\pi/2$. Using the identity

$$\tan\theta + \tan\left(\theta + \frac{\pi}{3}\right) + \tan\left(\theta + \frac{2\pi}{3}\right) = 3\tan 3\theta,$$

one has

$$\mathcal{H} = -\left(\frac{\partial^2}{\partial X^2} + \frac{\partial^2}{\partial r^2} + \frac{1}{r}\frac{\partial}{\partial r} + \frac{1}{r^2}\frac{\partial^2}{\partial \theta^2}\right) - 3\beta\left(\frac{1}{r}\frac{\partial}{\partial r} - \frac{1}{r^2}\tan 3\theta \frac{\partial}{\partial \theta}\right).$$

Now we look for factorized solutions of the form

$$\xi(X, r, \theta) = f(X)g(r)h(\theta).$$

The variables can again be separated. The equation $(\mathcal{H} - \mathcal{E})\xi = 0$ splits into three equations

$$f''(X) = -K^2 f(X), \tag{6-2}$$

$$r^2 g''(r) + (1 + 3\beta)rg'(r) + (k^2 r^2 - L^2)g(r) = 0, \tag{6-3}$$

$$h''(\theta) - 3\beta\tan(3\theta)h'(\theta) + L^2 h(\theta) = 0, \tag{6-4}$$

where we have introduced three constants K^2, k^2 and L^2, with

$$\mathcal{E} \equiv \mathcal{E}_{K,k} := K^2 + k^2.$$

Like (5–1), equation (6–2) has a continuous spectrum labelled by K real, a complete set of normalized solutions given by (5–3) and k^2 is real.

Setting $g(r) := r^{-3\beta/2} J(r)$ in (6–3) one gets Bessel's differential equation for J:

$$z^2 J''(z) + z J'(z) + (z^2 - \nu^2)J(z) = 0, \quad z = kr, \quad \nu^2 = L^2 + (3\beta/2)^2.$$

As in (5–4), only the Bessel function $J_\nu(z)$ with k real nonnegative and $\nu = l + 3\beta/2$ with l an integer, that is,

$$L^2 = \nu^2 - (3\beta/2)^2 = l(l + 3\beta) \tag{6-5}$$

gives for (6–3) a square integrable solution $g(r)$ over $x_1 \leq x_2 \leq x_3$ with the measure $r^{3\beta+1}$.

The singularities of (6–4) are at the points θ with $\cos 3\theta = 0$, that is, the end points of the interval $[\pi/6, \pi/2]$. It is convenient to shift from the variable θ to the new variable $z = \frac{1}{2}(1 - \sin 3\theta)$, which maps this interval $[\pi/6, \pi/2]$ on the interval $[0, 1]$. The function $F(z) := h(\theta)$ satisfies the hypergeometric differential equation

$$z(1-z)F'' + (1+\beta)\left(\frac{1}{2}-z\right)F' + \frac{L^2}{9}F = 0.$$

A careful examination shows that any solution of this equation has necessarily a singularity in the variable z, either at $z = 0$ or at $z = 1$, unless $L^2/9$ has the form $m(m+\beta)$, where m is an integer. This leads, using (6–5), to the only acceptable integer values of l, namely $l = 3m$, for m a nonnegative integer. The unique regular solution $h(\theta)$ is then a Gegenbauer polynomial in $\sin 3\theta$.

Finally,

$$f_K(X) = \frac{1}{\sqrt{2\pi}} e^{iKX},$$

$$g_{k,m}(r) = (kr)^{-3\beta/2} J_{3m+3\beta/2}(kr),$$

$$h_m(\theta) = \text{const}\, F\big(-m,\, m+\beta;\, \tfrac{1}{2}(1+\beta);\, \tfrac{1}{2}(1-\sin 3\theta)\big)$$

$$= \Gamma\left(\frac{\beta}{2}\right)\left(\frac{2^\beta(3m+3\beta/2)m!}{2\pi\Gamma(m+\beta)}\right)^{1/2} C_m^{(\beta/2)}(\sin 3\theta).$$

The orthogonality and closure relations (3–5) for $g_{k,m}(r)h_m(\theta)$ read

$$\int_0^\infty dr\, r^{3\beta+1} \int_{\pi/6}^{\pi/2} d\theta (-\cos 3\theta)^\beta g_{k,m}(r)h_m(\theta)g_{k',m'}(r)h_{m'}(\theta)$$
$$= k^{-3\beta-1}\delta(k-k')\delta_{mm'},$$

$$\int_0^\infty dk\, k^{3\beta+1} \sum_{m=0}^\infty g_{k,m}(r)h_m(\theta)g_{k,m}(r')h_m(\theta')$$
$$= r^{-3\beta-1}(-\cos 3\theta)^{-\beta}\delta(r-r')\delta(\theta-\theta').$$

Therefore

$$\langle \boldsymbol{x}|e^{-\mathcal{H}t/2}|\boldsymbol{x}'\rangle = \int_{-\infty}^\infty dK \int_0^\infty dk\, k^\beta \sum_{m=0}^\infty e^{-(K^2+k^2)t/2}$$
$$\times f_K(X)f_K^*(X')g_{k,m}(r)h_m(\theta)g_{k,m}(r')h_m(\theta'),$$

where we recall that the variables are: X given by (6–1) and r and θ such that

$$r^2 = \tfrac{1}{3}\big((x_1-x_2)^2 + (x_2-x_3)^2 + (x_3-x_1)^2\big), \quad \tan\theta = \frac{x_1+x_2-2x_3}{\sqrt{3}(x_1-x_2)},$$

and X', r', θ' have the same expressions in terms of x'_1, x'_2, x'_3. The integrations over K and k can be done as before and yield

$$\int_{-\infty}^{\infty} dK \, e^{iK(X-X') - \frac{t}{2}K^2} = \sqrt{\frac{2\pi}{t}} \exp\left(-\frac{1}{2t}(X-X')^2\right),$$

$$\int_0^{\infty} dk \, k \, e^{-\frac{t}{2}k^2} J_\nu(kr) J_\nu(kr') = \frac{1}{t} \exp\left(-\frac{1}{2t}(r^2 + r'^2)\right) I_\nu\left(\frac{rr'}{t}\right).$$

Collecting the constants one gets from (3–7)

$$\int dU \exp\left(-\frac{1}{2t} \operatorname{tr}(A - UA'U^{-1})^2\right) = 3(2t)^{3\beta/2} \, \Gamma(\beta) \, \Gamma\left(\tfrac{3}{2}\beta\right)$$

$$\times (rr')^{3\beta/2} \exp\left(-\frac{1}{2t}\left((X-X')^2 + r^2 + r'^2\right)\right) S,$$

with

$$S = \sum_{m=0}^{\infty} \frac{(m+\beta/2)m!}{\Gamma(m+\beta)} \, C_m^{(\beta/2)}(\sin 3\theta) \, C_m^{(\beta/2)}(\sin 3\theta') \, I_{3m+3\beta/2}\left(\frac{rr'}{t}\right). \quad (6\text{–}6)$$

Therefore the integration over the 3×3 orthogonal, unitary or symplectic matrices is effectively replaced by the infinite sum S of (6–6).

For $\beta = 2$, the Gegenbauer polynomials reduce to Chebyshev polynomials of the second kind:

$$C_m^{(1)}(\cos\theta) = U_m(\cos\theta) := \frac{\sin(m+1)\theta}{\sin\theta}.$$

Using the integral representation

$$I_n(z) = \frac{1}{2\pi} \int_0^{2\pi} d\phi \, \exp(z\cos\phi \pm in\phi) \quad (6\text{–}7)$$

one can write the sum (6–6) as

$$S = \sum_{m=0}^{\infty} \frac{\sin\left((m+1)\left(\frac{\pi}{2} - 3\theta\right)\right)}{\sin\left(\frac{\pi}{2} - 3\theta\right)} \frac{\sin\left((m+1)\left(\frac{\pi}{2} - 3\theta'\right)\right)}{\sin\left(\frac{\pi}{2} - 3\theta'\right)}$$

$$\times \frac{1}{2\pi} \int_0^{2\pi} d\phi \, \exp\left(\frac{rr'}{t}\cos\phi \pm i(3m+3)\phi\right)$$

$$= \frac{1}{8\pi \cos 3\theta \cos 3\theta'} \sum_{m=-\infty}^{\infty} \int_0^{2\pi} d\phi \, e^{(rr'/t)\cos\phi} \left(e^{im(3\theta-3\theta'+3\phi)} - e^{im(\pi-3\theta-3\theta'+3\phi)}\right).$$

Interchanging the order of summation and integration and using the identity

$$\sum_{m=-\infty}^{\infty} e^{imz} = 2\pi \sum_{m=-\infty}^{\infty} \delta(z + 2m\pi),$$

one gets

$$\cos(3\theta)\cos(3\theta')\,S = \tfrac{1}{12}\Big(e^{(rr'/t)\cos(\theta-\theta')} + e^{(rr'/t)\cos(\theta-\theta'+2\pi/3)}$$
$$+ e^{(rr'/t)\cos(\theta-\theta'-2\pi/3)} - e^{-(rr'/t)\cos(\theta+\theta')}$$
$$- e^{-(rr'/t)\cos(\theta+\theta'+2\pi/3)} - e^{-(rr'/t)\cos(\theta+\theta'-2\pi/3)}\Big).$$

The six terms in the right-hand side correspond to the six terms in the expansion of the 3×3 determinant $\det\big[\exp(-(1/2t)(x_j - y_k)^2)\big]_{j,k=1,2,3}$ of (1–1), and the correctness of the multiplying factors can be easily checked.

For $\beta = 1$ the Gegenbauer polynomials reduce to Legendre polynomials, $C_m^{(1/2)}(z) = P_m(z)$; while for $\beta = 4$ we have from [Bateman 1953, p. 176, § 10.9, eq. (23)]

$$\frac{d}{dz}C_{m+1}^{(1)}(z) = 2C_m^{(2)}(z),$$

or more explicitely

$$C_m^{(2)}(\cos\theta) = \frac{\sin\big((m+1)\theta\big)}{2\sin^3\theta} - (m+1)\frac{\cos\big((m+2)\theta\big)}{2\sin^2\theta}.$$

But we do not know how to evaluate the sum S in (6–6).

A sum similar to (6–6) is known [Watson 1952, p. 370, eq. (9)], namely

$$\sum_{m=0}^{\infty} \frac{(\pm 1)^m (m+\nu)m!}{\Gamma(m+2\nu)} C_m^{(\nu)}(\cos\theta) C_m^{(\nu)}(\cos\theta') I_{m+\nu}(z)$$
$$= \frac{\sqrt{2\pi}}{2^{2\nu}\Gamma(\nu)^2} \frac{I_{\nu-1/2}(z\sin\theta\sin\theta')}{(z\sin\theta\sin\theta')^{\nu-1/2}} \exp(\pm z\cos\theta\cos\theta').$$

The only significant difference with our sum is the index $m + \nu$ of the Bessel function I instead of $3(m + \nu)$.

7. The Case $n > 3$

When $n > 3$, following Calogero [1971], with slight modifications we can again change variables to

$$X := \frac{1}{\sqrt{n}}(x_1 + \ldots + x_n)$$
$$z_j := \frac{1}{\sqrt{j(j+1)}}(-x_1 - \cdots - x_j + jx_{j+1}), \quad j = 1, 2, \ldots, n-1.$$

This is an orthogonal change of variables, so that

$$\sum_{j=1}^{n} \frac{\partial^2}{\partial x_j^2} = \frac{\partial^2}{\partial X^2} + \sum_{j=1}^{n-1} \frac{\partial^2}{\partial z_j^2}.$$

The "linear derivative terms"

$$\sum_{1 \leq j < k \leq n} (x_j - x_k)^{-1}\left(\frac{\partial}{\partial x_j} - \frac{\partial}{\partial x_k}\right)$$

being independent of the "center of mass" X, one can separate the variable X. One can then change to "polar coordinates"

$$z_1 := r\cos\theta_1,$$
$$z_j := r\sin\theta_1 \ldots \sin\theta_{j-1}\cos\theta_j, \quad 2 \le j \le n-2,$$
$$z_{n-1} := r\sin\theta_1 \ldots \sin\theta_{n-2},$$
$$\sum_{j=1}^{n-1} \frac{\partial^2}{\partial z_j^2} = \frac{\partial^2}{\partial r^2} + \frac{n-2}{r}\frac{\partial}{\partial r} + \frac{1}{r^2}\nabla^2_{(\theta)},$$
$$r^2 = \frac{1}{n} \sum_{1 \le j < k \le n} (x_j - x_k)^2,$$

and separate the variable r. The linear derivative terms having a complicated expression in terms of the remaining variables θ_j, it seems difficult to say something more.

A. Expression of the Laplacian. Proof of Equation (2–12)

For completeness we give here a derivation of the splitting of the laplacian of a matrix in terms of its eigenvalues and the "angle variables".

We first recall some well known results of the general tensor analysis on a riemannian manifold. The line element ds in terms of the nondegenerate positive definite metric tensor $g = [g_{jk}]$ is

$$ds^2 := \sum_{j,k} g_{jk}\, dx_j\, dx_k.$$

Then the volume element or measure $d\mu(\boldsymbol{x})$ and the laplacian ∇^2 read as follows (see [Gouyons 1963, p. 91, § 103], for example):

$$d\mu(\boldsymbol{x}) = \sqrt{\det g} \prod_j dx_j, \tag{A–1}$$

$$\nabla^2 = \sum_{j,k} \frac{1}{\sqrt{\det g}} \frac{\partial}{\partial x_j} \sqrt{\det g}\, (g^{-1})_{jk} \frac{\partial}{\partial x_k}. \tag{A–2}$$

As in (2–6), an $n \times n$ real symmetric, complex hermitian or quaternion self-dual matrix A can be diagonalized:

$$A = U\boldsymbol{X}U^{-1}, \tag{A–3}$$

with $\boldsymbol{X}$ real diagonal and U in $\mathcal{G}_\beta$, for $\beta = 1$, 2 or 4 respectively. These matrices A depend on the n real diagonal elements A_{jj} and the β real components of each nondiagonal element A_{jk} for $j < k$. These later parameters, a total of $\beta n(n-1)/2$, will be denoted by A_μ in what follows. The latin indices $j, k, \ldots$

will vary from 1 to n, while the greek indices μ, ν, ... will vary from 1 to $\beta n(n-1)/2$. The line element is

$$ds^2 := \text{tr } dA^2 \tag{A-4}$$

where dA denotes the variation of a matrix A in the corresponding set of matrices according as $\beta = 1$, 2 or 4. In terms of the variables A_{jj}, A_μ, the line element reads

$$ds^2 = \sum_j dA_{jj}^2 + 2\sum_\mu dA_\mu^2. \tag{A-5}$$

Thus the metric tensor is diagonal and it follows from (A–1) and (A–2) that the measure and the laplacian are respectively

$$d\mu(A) = 2^{\beta n(n-1)/4}\left(\prod_j dA_{jj}\right)\left(\prod_\mu dA_\mu\right) \tag{A-6}$$

and

$$\nabla^2 = \sum_j \frac{\partial^2}{\partial A_{jj}^2} + \frac{1}{2}\sum_\mu \frac{\partial^2}{\partial A_\mu^2}.$$

Notice that the measure $d\mu(A)$ considered in (2–4) was denoted by dA for briefness (it must not be confused with dA in (A–4)), and furthermore the normalization is different.

Now to specify any matrix A of the corresponding set according as $\beta = 1$, 2 or 4, we take the n real eigenvalues x_j and $\beta n(n-1)/2$ additional real parameters p_μ, i.e., the "angle variables" entering the definition of U in (A–3). Notice that these matrices U, according as $\beta = 1$, 2 or 4, actually depend on $n(n-1)/2$, n^2 or $n(2n-1)$ real parameters respectively. Thus, for $\beta = 2$ or 4—that is, for the unitary or symplectic group—U depends on n more real parameters than the collection of the variables p_μ. These n parameters correspond to the possibility of multiplying each column k of U by a phase factor $\exp(i\phi_k)$ without changing A in (A–3). From (A–3) the variation of the matrix A reads

$$dA = U\bigl(d\boldsymbol{X} + (U^{-1}dU)\boldsymbol{X} - \boldsymbol{X}(U^{-1}dU)\bigr)U^{-1}.$$

Then, using the cyclic invariance of the trace and the diagonal character of $\boldsymbol{X}$ and $d\boldsymbol{X}$, the line element (A–4) is

$$ds^2 = \sum_j dx_j^2 + \text{tr}\bigl((U^{-1}dU)\boldsymbol{X} - \boldsymbol{X}(U^{-1}dU)\bigr)^2.$$

Notice that

$$dU = \begin{cases} \displaystyle\sum_\mu \frac{\partial U}{\partial p_\mu} dp_\mu & \text{for } \beta = 1, \\ \displaystyle\sum_\mu \frac{\partial U}{\partial p_\mu} dp_\mu + \sum_k \frac{\partial U}{\partial \phi_k} d\phi_k & \text{for } \beta = 2 \text{ or } 4, \end{cases}$$

where for the unitary and symplectic groups, the dependence of the matrix element U_{jk} on the parameter ϕ_k is through the phase factor $\exp(i\phi_k)$ already mentioned. It follows that

$$\left(\frac{\partial U}{\partial \phi_k}\right)_{jl} = iU_{jl}\delta_{lk}$$

and then, a straightforward calculation shows that the variations $d\phi_k$ do not contribute, as expected, to $(U^{-1}dU)\boldsymbol{X} - \boldsymbol{X}(U^{-1}dU)$. Finally, in terms of the variables x_j and p_μ the metric tensor has the block diagonal structure

$$g_{jk} = \delta_{jk},$$
$$g_{j\mu} = g_{\mu j} = 0,$$
$$g_{\mu\nu} = \operatorname{tr}\left(\left(U^{-1}\frac{\partial U}{\partial p_\mu}\boldsymbol{X} - \boldsymbol{X}U^{-1}\frac{\partial U}{\partial p_\mu}\right)\left(U^{-1}\frac{\partial U}{\partial p_\nu}\boldsymbol{X} - \boldsymbol{X}U^{-1}\frac{\partial U}{\partial p_\nu}\right)\right)$$
$$= -\sum_{j,k}(x_j - x_k)^2 \left(U^{-1}\frac{\partial U}{\partial p_\mu}\right)_{jk}\left(U^{-1}\frac{\partial U}{\partial p_\nu}\right)_{kj}.$$

Consequently, the inverse matrix g^{-1} of g, occuring in (A–2), also has a block diagonal structure

$$\left.\begin{array}{l} (g^{-1})_{jk} = \delta_{jk}, \\ (g^{-1})_{j\mu} = (g^{-1})_{\mu j} = 0 \\ (g^{-1})_{\mu\nu} = \text{a complicated expression not needed here.} \end{array}\right\} \quad \text{(A–7)}$$

We now show that the determinant of the metric tensor, which is positive from (A–5), satisfies

$$\sqrt{\det g} = |\Delta(\boldsymbol{x})|^\beta f(\boldsymbol{p}), \quad \text{(A–8)}$$

where f depends only on the p_μ, denoted collectively $\boldsymbol{p}$. Indeed, one shows that the measure $d\mu(A)$, (A–6), is given by

$$d\mu(A) = \left|\frac{\partial(A_{jj}, A_\mu)}{\partial(x_j, p_\mu)}\right| \left(\prod_j dx_j\right)\left(\prod_\mu dp_\mu\right)$$

(see [Mehta 1991, Chapter 3]), where the absolute value of the jacobian is

$$\left|\frac{\partial(A_{jj}, A_\mu)}{\partial(x_j, p_\mu)}\right| = |\Delta(\boldsymbol{x})|^\beta f(\boldsymbol{p}).$$

Using (A–1) this ends the proof of (A–8). Finally, from (A–2), (A–7), and (A–8) one gets the expression (2–12) for the laplacian assuming we restrict our attention to one of the sectors where $\Delta(\boldsymbol{x}) \geq 0$ (the absolute value of $\Delta(\boldsymbol{x})$ no longer occurs).

B. The Schrödinger Equation with the Singular Calogero Potential

We consider the time independent Schrödinger equation (4–3), with Calogero's hamiltonian H defined in (1–2), interpreted as describing a system of n one-dimensional particles. Our aim is to investigate the validity of the assertion that two such particles cannot cross each other. For that, it is sufficient to study the simple case $n = 2$.

Separating the center of mass motion and using the notations of section 5, we write the wave function $\phi(\boldsymbol{x})$ solution of (4–3) as

$$\phi(\boldsymbol{x}) = f(X)\, u(x),$$

where $u(x)$ satisfies the equation

$$u''(x) + \left(\frac{\gamma}{4x^2} + E\right) u(x) = 0 \tag{B–1}$$

and $\gamma = \beta(2 - \beta)$. When $\beta = 2$, the singular potential disappears ($\gamma = 0$); when $\beta = 1$, it is attractive ($\gamma = 1$), and when $\beta = 4$, it is repulsive ($\gamma = -8$). In the following, γ is allowed to take any real value.

We are interested in the neighbourhood of the singularity at $x = 0$, so in (B–1) we drop the energy term $E\, u(x)$, negligible in comparison to the potential term. Furthermore, following Landau [1958, § 35, p. 118], we regularize (B–1) by replacing the potential $\gamma/4x^2$ by the constant $\gamma/4x_0^2$ in a small interval $[-x_0, x_0]$ around the origin:

$$\begin{aligned} u''(x) + \frac{\gamma}{4x^2}\, u(x) &= 0, \quad \text{for } |x| > x_0, \\ u''(x) + \frac{\gamma}{4x_0^2}\, u(x) &= 0, \quad \text{for } |x| < x_0. \end{aligned} \tag{B–2}$$

Ultimately, we will let x_0 go to zero.

In the outer region $|x| > x_0$, when $\gamma \neq 1$, we define two linearly independent solutions $u_\pm(x)$ by setting

$$u_+(x) := |x|^{s_+} \quad \text{and} \quad u_-(x) := \pm |x|^{s_-} \quad \text{if } \pm x > x_0, \tag{B–3}$$

where the indices $s_\pm$ are equal to $(1 \pm \sqrt{1-\gamma})/2$ or $(1 \pm i\sqrt{\gamma-1})/2$, according to whether $\gamma < 1$ or $\gamma > 1$. When $\gamma = 1$, the two indices coincide, and we choose the following two linearly independent solutions

$$u_+(x) := |x|^{1/2} \quad \text{and} \quad u_-(x) := \pm |x|^{1/2} \log |x| \quad \text{if } \pm x > x_0, \tag{B–4}$$

(the signs $\pm$ in (B–3) and (B–4) ensure that the wronskian $W(u_+, u_-)$ takes on the same value for $x > x_0$ and $x < -x_0$).

In the inner region $|x| < x_0$, when $\gamma \neq 0$, we define two linearly independent solutions $v_\pm(x)$

$$v_\pm(x) := e^{\pm \kappa x/x_0}, \quad \text{where } \kappa = \begin{cases} \sqrt{-\gamma}/2 & \text{if } \gamma < 0, \\ i\sqrt{\gamma}/2 & \text{if } \gamma > 0. \end{cases}$$

The general solution of (B–2) reads

$$u(x) = \begin{cases} A_+ u_+(x) + A_- u_-(x) & \text{for } x > x_0, \\ B_+ v_+(x) + B_- v_-(x) & \text{for } -x_0 < x < x_0, \\ C_+ u_+(x) + C_- u_-(x) & \text{for } x < -x_0. \end{cases}$$

The constants $A_\pm$, $B_\pm$ and $C_\pm$ are related by the continuity conditions of $u(x)$ and $u'(x)$ at both points x_0 and $-x_0$. The only relations of interest are the two connecting $A_\pm$ to $C_\pm$, which we write

$$\begin{pmatrix} C_+ \\ C_- \end{pmatrix} = M \begin{pmatrix} A_+ \\ A_- \end{pmatrix}. \tag{B–5}$$

Here, M is a 2×2 unimodular matrix which depends on x_0. Straightforward calculations lead to the following expression, valid when $\gamma \neq 0$ and 1:

$$M = \begin{pmatrix} a & b x_0^{s_- - s_+} \\ c x_0^{s_+ - s_-} & a \end{pmatrix},$$

where a, b and c do not depend on x_0:

$$\begin{pmatrix} a & b \\ c & a \end{pmatrix} = \frac{1}{2\kappa(s_+ - s_-)} \begin{pmatrix} s_- + \kappa & s_- - \kappa \\ s_+ + \kappa & s_+ - \kappa \end{pmatrix} \begin{pmatrix} e^{-2\kappa} & 0 \\ 0 & e^{2\kappa} \end{pmatrix} \begin{pmatrix} -s_+ - \kappa & -s_- - \kappa \\ s_+ - \kappa & s_- - \kappa \end{pmatrix}. \tag{B–6}$$

Equation (B–5) can be rewritten conveniently as

$$\begin{pmatrix} A_- \\ C_- \end{pmatrix} = \frac{1}{b} x_0^{s_+ - s_-} \begin{pmatrix} -a & 1 \\ -1 & a \end{pmatrix} \begin{pmatrix} A_+ \\ C_+ \end{pmatrix}. \tag{B–7}$$

The x_0 dependence is now contained in a global factor $x_0^{s_+ - s_-}$. Indeed, a simple dimensional argument leads directly to this result.

When $\gamma = 1$, similar calculations lead to the following formula (we have dropped terms which are negligible when $x_0 \to 0$)

$$\begin{pmatrix} A_- \\ C_- \end{pmatrix} \simeq \begin{pmatrix} -1/\log x_0 & 1/(\log x_0)^2 \cos 1 \\ 1/\log x_0 & -1/(\log x_0)^2 \cos 1 \end{pmatrix} \begin{pmatrix} A_+ \\ C_+ \end{pmatrix}. \tag{B–8}$$

We now let x_0 go to zero.

When $\gamma < 1$, the difference $s_+ - s_- = \sqrt{1-\gamma}$ is positive, and the right-hand side of (B–7) goes to zero with x_0. Similarly, the right-hand side of (B–8) vanishes in the same limit. Thus, when $\gamma \leq 1$, the constants A_- and C_- vanish with x_0, and A_+ and C_+ are the two arbitrary integration constants of the problem. Consequently, two particular linearly independent solutions are, in this limit, $\theta(x)u_+(x)$ and $\theta(-x)u_+(x)$, which are localised respectively in the sectors $x \geq 0$

and $x \leq 0$. This proves that *when the singular interaction $\gamma/(x_1 - x_2)^2$ between two one-dimensional particles is either repulsive or weakly attractive ($\gamma \leq 1$), the particles cannot cross each other.*

When $\gamma > 1$, the difference $s_+ - s_- = i\sqrt{\gamma - 1}$ is pure imaginary, and the right-hand side of (B-7) has no limit when $x_0 \to 0$: it oscillates indefinitely. Indeed, in that case, namely *when the interaction $\gamma/(x_1 - x_2)^2$ is strongly attractive, the two particles collapse.* The argument is exactly the one developed in reference [Landau and Lifshitz 1958], for a three-dimensional particle in a central potential const$/r^2$, and it will not be reproduced here.

As a final remark, we raise the question of the independence of these results with respect to the regularization process. We just note that if, choosing a different regularization, we replace in the small interval $[-x_0, x_0]$ the potential $\gamma/4x^2$ by a constant τ^2 independent of x_0 (which induces discontinuities in the potential), (B-6) and (B-7) are still valid once κ has been replaced by τx_0. The quantities a, b and c now depend on x_0, but they have a finite limit when $x_0 \to 0$, and nothing is changed in the above conclusions.

Acknowledgements

We are thankful to R. Balian for indicating us the use of (6–7) to evaluate the sum in (6–6) for the case $\beta = 2$.

References

[Bateman 1953] H. Bateman, *Higher transcendental functions*, vol. 2, edited by A. Erdélyi, Mc Graw-Hill, New York, 1953.

[Calogero 1969a] F. Calogero, "Solution of a three-body problem in one dimension", *J. Math. Phys.* **10** (1969), 2191–2196.

[Calogero 1969b] F. Calogero, "Ground state of a one-dimensional N-body system", *J. Math. Phys.* **10** (1969), 2197–2200.

[Calogero 1971] F. Calogero, "Solution of the one-dimensional N-body problems with quadratic and/or inversely quadratic pair potentials", *J. Math. Phys.* **12** (1971), 419–436.

[Gouyons 1963] R. Gouyons, *Calcul tensoriel*, Vuibert, Paris, 1963.

[Gradshteyn and Ryzhik 1991] I. S. Gradshteyn and I. M. Ryzhik, *Table of integrals, series, and products*, Academic Press, Boston, 1991.

[Itzykson and Zuber 1980] C. Itzykson and J. B. Zuber, "The planar approximation, II", *J. Math. Phys.* **21**:3 (1980), 411–421.

[Landau and Lifshitz 1958] L. D. Landau and E. M. Lifshitz, *Quantum mechanics: non-relativistic theory*, Pergamon Press, London and Paris, 1958.

[Mehta 1981] M. L. Mehta, "A method of integration over matrix variables", *Comm. Math. Phys.* **79**:3 (1981), 327–340.

[Mehta 1989] M. L. Mehta, *Matrix theory*, Les Éditions de Physique, Z. I. Courtabeuf, 91944 Les Ulis, France, 1989.

[Mehta 1991] M. L. Mehta, *Random matrices*, 2nd ed., Academic Press, Boston, 1991. There is an error in the value of the constant c given there: the product over j should go from 1 to $n-1$, not from 1 to n, when dU is a normalized Haar measure.

[Muirhead 1975] R. J. Muirhead, "Expressions for some hypergeometric functions of matrix argument with applications", *J. Multivariate Anal.* **5**:3 (1975), 283–293.

[Muirhead 1982] R. J. Muirhead, *Aspects of multivariate statistical theory*, Wiley, New York, 1982.

[Watson 1952] G. N. Watson, *A treatise on the theory of Bessel functions*, 2nd ed., Cambridge Univ. Press, Cambridge, 1952.

G. MAHOUX
CEA/SACLAY
SERVICE DE PHYSIQUE THÉORIQUE
F-91191 GIF-SUR-YVETTE CEDEX
FRANCE
   mahoux@spht.saclay.cea.fr

M. L. MEHTA
CEA/SACLAY
SERVICE DE PHYSIQUE THÉORIQUE
F-91191 GIF-SUR-YVETTE CEDEX
FRANCE
   mehta@spht.saclay.cea.fr

J.-M. NORMAND
CEA/SACLAY
SERVICE DE PHYSIQUE THÉORIQUE
F-91191 GIF-SUR-YVETTE CEDEX
FRANCE
   norjm@spht.saclay.cea.fr

Integrable Lattices: Random Matrices and Random Permutations

PIERRE VAN MOERBEKE

CONTENTS

Introduction	322
1. Matrix Integrals, Random Matrices and Permutations	323
1.1. Tangent Space to Symmetric Spaces and Associated Random Matrix Ensembles	323
1.2. Infinite Hermitian Matrix Ensembles	330
1.3. Integrals over Classical Groups	333
1.4. Permutations and Integrals over Groups	335
2. Integrals, Vertex Operators and Virasoro Relations	341
2.1. β-Integrals	342
2.2. Double Matrix Integrals	350
2.3. Integrals over the Unit Circle	351
3. Integrable Systems and Associated Matrix Integrals	353
3.1. Toda lattice and Hermitian matrix integrals	353
3.2. Pfaff Lattice and symmetric/symplectic matrix integrals	361
3.3. 2d-Toda Lattice and Coupled Hermitian Matrix Integrals	369
3.4. The Toeplitz Lattice and Unitary Matrix Integrals	372
4. Ensembles of Finite Random Matrices	377
4.1. PDEs Defined by the Probabilities in Hermitian, Symmetric and Symplectic Random Ensembles	377
4.2. ODEs When E Has One Boundary Point	380
4.3. Proof of Theorems 4.1, 4.2 and 4.3	382
5. Ensembles of Infinite Random Matrices: Fredholm Determinants As τ-Functions of the KdV Equation	386
6. Coupled Random Hermitian Ensembles	395
7. Random Permutations	396
8. Random Involutions	399
Appendix: Chazy Classes	401
Acknowledgment	402
References	403

The support of National Science Foundation grant #DMS-98-4-50790, a NATO grant, a FNRS grant and a Francqui Foundation grant is gratefully acknowledged.

Introduction

The purpose of this article is to survey recent interactions between statistical questions and integrable theory. Two types of questions will be tackled here:

(i) Consider a random ensemble of matrices, with certain symmetry conditions to guarantee the reality of the spectrum and subjected to a given statistics. What is the probability that all its eigenvalues belong to a given subset E? What happens, when the size of the matrices gets very large? The probabilities here are functions of the boundary points c_i of E.

(ii) What is the statistics of the length of the largest increasing sequence in a random permutation, assuming each permutation is equally probable? Here, one considers generating functions (over the size of the permutations) for the probability distributions, depending on the variable x.

The main emphasis of this article is to show that integrable theory serves as a useful tool for finding equations satisfied by these functions of x, and conversely the probabilities point the way to new integrable systems.

These questions are all related to integrals over spaces of matrices. Such spaces can be classical Lie groups or algebras, symmetric spaces or their tangent spaces. In infinite-dimensional situations, the "∞-fold" integrals get replaced by Fredholm determinants.

During the last decade, astonishing discoveries have been made in a variety of directions. A first striking feature is that these probabilities are all related to Painlevé equations or interesting generalizations. In this way, new and unusual distributions have entered the statistical world.

Another feature is that each of these problems is related to some integrable hierarchy. Indeed, by inserting an infinite set of time variables $t_1, t_2, t_3, \ldots$ in the integrals or Fredholm determinants — e.g., by introducing appropriate exponentials $e^{\sum_1^\infty t_i y^i}$ in the integral — this probability, as a function of $t_1, t_2, t_3, \ldots$, satisfies an integrable hierarchy. Korteweg-de Vries, KP, Toda lattice equations are only a few examples of such integrable equations.

Typically integrable systems can be viewed as isospectral deformations of differential or difference operators $\mathcal{L}$. Perhaps, one of the most startling discoveries of integrable theory is that $\mathcal{L}$ can be expressed in terms of a single "τ-function" $\tau(t_1, t_2, \ldots)$ (or vector of τ-functions), which satisfy an infinite set of nonlinear equations, encapsulated in a single "*bilinear identity*". The t_i account for the commuting flows of this integrable hierarchy. In this way, many interesting classical functions live under the same hat: characters of representations, Θ-functions of algebraic geometry, hypergeometric functions, certain integrals over classical Lie algebras or groups, Fredholm determinants, arising in statistical mechanics, in scattering and random matrix theory! They are all special instances of "τ-*functions*".

The point is that the probabilities or generating functions above, as functions of $t_1, t_2, \ldots$ (after some minor renormalization) are precisely such τ-functions for the corresponding integrable hierarchy and thus automatically satisfy a large set of equations.

These probabilities are very special τ-functions: they happen to be a solution of yet another hierarchy of (linear) equations in the variables t_i and the boundary points c_i, namely $\mathbb{J}_k^{(2)}\tau(t;c) = 0$, where the $\mathbb{J}_k^{(2)}$ form — roughly speaking — a Virasoro-like algebra:

$$[\mathbb{J}_k^{(2)}, \mathbb{J}_l^{(2)}] = (k-l)\mathbb{J}_{k+l}^{(2)} + \cdots.$$

Each integrable hierarchy has a natural *"vertex operator"*, which automatically leads to a natural Virasoro algebra. Then, eliminating the partial derivatives in t from the two hierarchy of equations, the integrable and the Virasoro hierarchies, and finally setting $t = 0$, lead to PDEs or ODEs satisfied by the probabilities.

Table 1 gives an overview of the different problems discussed in this article, the relevant integrals in the second column and the different hierarchies satisfied by the integrals. To fix notation, $\mathcal{H}_l$, $\mathcal{S}_l$, $\mathcal{T}_l$ refer to the Hermitian, symmetric and symplectic ensembles, populated respectively by $l \times l$ Hermitian matrices, symmetric matrices and self-dual Hermitian matrices, with quaternionic entries. $\mathcal{H}_l(E)$, $\mathcal{S}_l(E)$, $\mathcal{T}_l(E)$ are the corresponding set of matrices, with all spectral points belonging to E. $U(l)$ and $O(l)$ are the unitary and orthogonal groups respectively. In Table 1 we have $V_t(z) := V_0(z) + \sum t_i z^i$, where $V_0(z)$ stands for the unperturbed problem; in the last integral $\tilde{V}_t(z)$ is a more complicated function of $t_1, t_2, \ldots$ and z, to be specified later.

1. Matrix Integrals, Random Matrices and Permutations

1.1. Tangent Space to Symmetric Spaces and Associated Random Matrix Ensembles.
Random matrices provided a model for excitation spectra of heavy nuclei at high excitations (Wigner [74], Dyson [27] and Mehta [49]), based on the nuclear experimental data by Porter and Rosenzweig [56]; they observed that the occurrence of two levels, close to each other, is a rare event (level repulsion), showing that the spacing is not Poissonian, as one might expect from a naive point of view.

Random matrix ideas play an increasingly prominent role in mathematics: not only have they come up in the spacings of the zeroes of the Riemann zeta function, but their relevance has been observed in the chaotic Sinai billiard and, more generally, in chaotic geodesic flows. Chaos seems to lead to the "spectral rigidity", typical of the spectral distributions of random matrices, whereas the spectrum of an integrable system is random (Poisson)! (e.g., see Odlyzko [53] and Sarnak [59]).

All these problems have led to three very natural random matrix ensembles: Hermitian, symmetric and symplectic ensembles. The purpose of this section is to show that these three examples appear very naturally as tangent spaces to symmetric spaces.

Probability problem	underlying t-perturbed integral, τ-function of $\longrightarrow$	corresponding integrable hierarchies
$P(M \in \mathcal{H}_n(E))$	$\int_{\mathcal{H}_n(E)} e^{\text{Tr}(-V(M)+\sum_1^\infty t_i M^i)} dM$	Toda lattice KP hierarchy
$P(M \in \mathcal{S}_n(E))$	$\int_{\mathcal{S}_n(E)} e^{\text{Tr}(-V(M)+\sum_1^\infty t_i M^i)} dM$	Pfaff lattice Pfaff-KP hierarchy
$P(M \in \mathcal{T}_n(E))$	$\int_{\mathcal{T}_n(E)} e^{\text{Tr}(-V(M)+\sum_1^\infty t_i M^i)} dM$	Pfaff lattice Pfaff-KP hierarchy
$P((M_1, M_2) \in \mathcal{H}_n(E_1) \times \mathcal{H}_n(E_2))$	$\int_{\mathcal{H}_n^2(E)} dM_1\, dM_2$ $e^{-\text{Tr}(V_t(M_1) - V_s(M_2) - c M_1 M_2)}$	2d-Toda lattice KP-hierarchy
$P(M \in \mathcal{H}_\infty(E))$	$\det(I - K_t(y,z) I_{E^c}(z))$ (Fredholm determinant)	KdV equation
longest increasing sequence in random permutations	$\int_{U(l)} e^{\text{Tr} \sum_1^\infty (t_i M^i - s_i \bar{M}^i)} dM$	Toeplitz lattice 2d-Toda lattice
longest increasing sequence in random involutions	$\int_{O(l)} e^{\text{Tr}(xM + \tilde{V}_t(M))} dM$	Toda lattice KP-hierarchy

Table 1. Overview of the article: problems discussed, relevant integrals (second column), and the hierarchies satisfied by the integrals (last column).

A symmetric space G/K is given by a semisimple Lie group G and a Lie group involution $\sigma : G \to G$ such that

$$K = \{x \in G, \, \sigma(x) = x\}.$$

Then the following identification holds:

$$G/K \cong \{g\sigma(g)^{-1} \quad \text{with } g \in G\},$$

and the involution σ induces a map of the Lie algebra,

$$\sigma_* : \mathfrak{g} \longrightarrow \mathfrak{g}, \quad \text{such that } \sigma_*^{\,2} = 1,$$

where

$$\mathfrak{g} = \mathfrak{k} \oplus \mathfrak{p} \quad \text{with } \mathfrak{k} = \{a \in \mathfrak{g} \mid \sigma_*(a) = a\} \text{ and } \mathfrak{p} = \{a \in \mathfrak{g} \mid \sigma_*(a) = -a\},$$

and

$$[\mathfrak{k}, \mathfrak{k}] \subset \mathfrak{k}, \quad [\mathfrak{k}, \mathfrak{p}] \subset \mathfrak{p}, \quad [\mathfrak{p}, \mathfrak{p}] \subset \mathfrak{k}.$$

Then K acts on $\mathfrak{p}$ by conjugation: $k\mathfrak{p}k^{-1} \subset \mathfrak{p}$ for all $k \in K$ and $\mathfrak{p}$ is the tangent space to G/K at the identity. The action of K on $\mathfrak{p}$ induces a root space decomposition, with $\mathfrak{a}$ being a maximal abelian subalgebra in $\mathfrak{p}$:

$$\mathfrak{p} = \mathfrak{a} + \sum_{\alpha \in \Delta} \mathfrak{p}_\alpha, \quad \text{with } m_\alpha = \dim \mathfrak{p}_\alpha.$$

Then, according to Helgason [35], the volume element on $\mathfrak{p}$ is

$$dV = \left(\prod_{\alpha \in \Delta_+} \alpha(z)^{m_\alpha} \right) dz_1 \ldots dz_n,$$

where Δ_+ is the set of positive roots; see [35; 36; 60; 61]. This will subsequently be worked out for the three so-called A_n-symmetric spaces. I like to thank Chuu-Lian Terng for very helpful conversations on these matters.

Examples

(i) Hermitian ensemble. Consider the *non-compact symmetric space*[1]

$$\mathrm{SL}(n,\mathbb{C})/\mathrm{SU}(n)$$

with $\sigma(g) = \bar{g}^{\top -1}$. Then

$$\mathrm{SL}(n,\mathbb{C})/\mathrm{SU}(n) = \{g\bar{g}^\top \mid g \in \mathrm{SL}(n,\mathbb{C})\}$$
$$= \{\text{positive definite matrices with det} = 1\}$$

with

$$K = \{g \in \mathrm{SL}(n,\mathbb{C}) \mid \sigma(g) = g\} = \{g \in \mathrm{SL}(n,\mathbb{C}) \mid g^{-1} = \bar{g}^\top\} = \mathrm{SU}(n).$$

Then $\sigma_*(a) = -\bar{a}^\top$ and the tangent space to G/K is then given by the space $\mathfrak{p} = \mathcal{H}_n$ of Hermitian matrices

$$\mathrm{sl}(n,\mathbb{C}) = \mathfrak{k} \oplus \mathfrak{p} = \mathrm{su}(n) \oplus \mathcal{H}_n \quad (\text{i.e., } a = a_1 + a_2 \text{ with } a_1 \in \mathrm{su}(n), a_2 \in \mathcal{H}_n).$$

If $M \in \mathcal{H}_n$, then the M_{ii}, $\mathrm{Re}\, M_{ij}$ and $\mathrm{Im}\, M_{ij}$ ($1 \leq i < j \leq n$) are free variables, so that Haar measure on $M \in \mathcal{H}_n$ takes on the form

$$dM := \prod_1^n dM_{ii} \prod_{1 \leq i < j \leq n} (d\,\mathrm{Re}\, M_{ij}\, d\,\mathrm{Im}\, M_{ij}). \tag{1.1.1}$$

A maximal abelian subalgebra $a \subset \mathfrak{p} = \mathcal{H}_n$ is given by real diagonal matrices $z = \mathrm{diag}(z_1, \ldots, z_n)$. Each $M \in \mathfrak{p} = \mathcal{H}_n$ can be written as

$$M = e^A z e^{-A}, \quad e^A \in K = \mathrm{SU}(n),$$

with[2]

$$A = \sum_{1 \leq k \leq l \leq n} (a_{kl}(e_{kl} - e_{lk}) + ib_{kl}(e_{kl} + e_{lk})) \in \mathfrak{k} = \mathrm{su}(n), \quad a_{ll} = 0. \tag{1.1.2}$$

Notice that $e_{kl} - e_{lk}$ and $i(e_{kl} + e_{lk})$ belong to $\mathfrak{k} = \mathrm{su}(n)$ and that

$$[e_{kl} - e_{lk}, z] = (z_l - z_k)(e_{kl} + e_{lk}) \in \mathfrak{p} = \mathcal{H}_n,$$
$$[i(e_{kl} + e_{lk}), z] = (z_l - z_k)i(e_{kl} - e_{lk}) \in \mathfrak{p} = \mathcal{H}_n. \tag{1.1.3}$$

---
[1]The corresponding compact symmetric space is $(\mathrm{SU}(n) \times \mathrm{SU}(n))/\mathrm{SU}(n)$.
[2]e_{kl} is the $n \times n$ matrix with all zeroes, except for 1 at the (k,l)-th entry.

Incidentally, this implies that $e_{kl} + e_{lk}$ and $i(e_{kl} + e_{lk})$ are two-dimensional eigenspaces of $(\text{ad } z)^2$ (where ad is defined by ad $x(y) := [x, y]$) with eigenvalue $(z_l - z_k)^2$. From (1.1.2) and (1.1.3) it follows that

$$[A, z] = (z_l - z_k) \sum_{1 \leq k < l \leq n} (a_{kl}(e_{kl} + e_{lk}) + ib_{kl}(e_{kl} - e_{lk})) \in \mathfrak{p} = \mathcal{H}_n \quad (1.1.4)$$

and thus, for small A, we have[3]

$$dM = d(e^A z e^{-A}) = d(z + [A, z] + \cdots)$$
$$= \prod_1^n dz_i \prod_{1 \leq k < l \leq n} d((z_l - z_k)a_{kl}) \, d((z_l - z_k)b_{kl}) \quad \text{using (1.1.4) and (1.1.1)}$$
$$= \prod_1^n dz_i \Delta_n^2(z) \prod_{1 \leq k < l \leq n} da_{kl} \, db_{kl}. \quad (1.1.5)$$

Therefore $\Delta^2(z)$ is also the Jacobian determinant of the map $M \to (z, U)$, such that $M = UzU^{-1} \in \mathcal{H}_n$, and thus dM admits the decomposition in polar coordinates:

$$dM = \Delta_n^2(z) \, dz_1 \ldots dz_n \, dU, \quad U \in \text{SU}(n). \quad (1.1.6)$$

In random matrix theory, $\mathcal{H}_n$ is endowed with the following probability,

$$P(M \in dM) = c_n e^{-\text{tr} V(M)} dM, \quad \rho(dz) = e^{-V(z)} dz, \quad (1.1.7)$$

where dM is Haar measure (1.1.6) on $\mathcal{H}_n$ and c_n is the normalizing factor. Since dM as in (1.1.6) contains dU and since the probability measure (1.1.7) only depends on the trace of $V(M)$, dU completely integrates out. Given $E \subset \mathbb{R}$, define

$$\mathcal{H}_n(E) := \{M \in \mathcal{H}_n \text{ with all spectral points} \in E \subset \mathbb{R}\} \subset \mathcal{H}_n. \quad (1.1.8)$$

Then

$$P(M \in \mathcal{H}_n(E)) = \int_{\mathcal{H}_n(E)} c_n e^{-\text{Tr } V(M)} dM = \frac{\int_{E^n} \Delta^2(z) \prod_1^n \rho(dz_k)}{\int_{\mathbb{R}^n} \Delta^2(z) \prod_1^n \rho(dz_k)}. \quad (1.1.9)$$

As explained in the excellent book by Mehta [49], $V(z)$ is quadratic (Gaussian ensemble) if the probability $P(M \in dM)$ satisfies

(i) invariance under conjugation by unitary transformations $M \mapsto UMU^{-1}$, and
(ii) the condition that the random variables M_{ii}, $\text{Re } M_{ij}$, and $\text{Im } M_{ij}$, for $1 \leq i < j \leq n$, be independent.

---

[3] $\Delta_n(z) = \prod_{1 \leq i < j \leq n} (z_i - z_j)$ is the Vandermonde determinant.

(ii) Symmetric ensemble. Here we consider the *non-compact symmetric space*[4] $\mathrm{SL}(n,\mathbb{R})/\mathrm{SO}(n)$ with $\sigma(g) = g^{\top -1}$. Then

$$\mathrm{SL}(n,\mathbb{R})/\mathrm{SO}(n) = \{gg^\top \mid g \in \mathrm{SL}(n,\mathbb{R})\}$$
$$= \{\text{positive definite matrices with det} = 1\}$$

with

$$K = \{g \in \mathrm{SL}(n,\mathbb{R}) \mid \sigma(g) = g\} = \{g \in \mathrm{SL}(n,\mathbb{R}) \mid g^\top = g^{-1}\} = \mathrm{SO}(n).$$

Then $\sigma_*(a) = -a^\top$ and the tangent space to G/K is then given by the space $\mathfrak{p} = \mathcal{S}_n$ of symmetric matrices appearing in the decomposition of $\mathrm{sl}(n,\mathbb{R})$,

$$\mathrm{sl}(n,\mathbb{R}) = \mathfrak{k} \oplus \mathfrak{p} = \mathrm{so}(n) \oplus \mathcal{S}_n \quad (\text{i.e., } a = a_1 + a_2 \text{ with } a_1 \in \mathrm{so}(n), a_2 \in \mathcal{S}_n),$$

with Haar measure $dM = \prod_{1 \leq i \leq j \leq n} dM_{ij}$ on $\mathcal{S}_n$.

A maximal abelian subalgebra $a \subset \mathfrak{p} = \mathcal{S}_n$ is given by real traceless diagonal matrices $z = \mathrm{diag}(z_1, \ldots, z_n)$. Each $M \in \mathfrak{p} = \mathcal{S}_n$ conjugates to a diagonal matrix z:

$$M = e^A z e^{-A}, \quad e^A \in K = \mathrm{SO}(n), \quad A \in \mathrm{so}(n).$$

A calculation analogous to example (1.1.5)(i) leads to

$$dM = |\Delta_n(z)|\, dz_1 \ldots dz_n\, dU, \quad U \in \mathrm{SO}(n).$$

Random matrix theory deals with the following probability on $\mathcal{S}_n$:

$$P(M \in dM) = c_n e^{-\mathrm{tr}V(M)}\, dM, \quad \rho(dz) = e^{-V(z)}\, dz, \tag{1.1.10}$$

with normalizing factor c_n. Setting as in (1.1.8): $\mathcal{S}_n(E) \subset \mathcal{S}_n$ is the subset of matrices with spectrum $\in E$. Then

$$P(M \in \mathcal{S}_n(E)) = \int_{\mathcal{S}_n(E)} c_n e^{-\mathrm{Tr}\,V(M)}\, dM = \frac{\int_{E^n} |\Delta(z)| \prod_1^n \rho(dz_k)}{\int_{\mathbb{R}^n} |\Delta(z)| \prod_1^n \rho(dz_k)}. \tag{1.1.11}$$

As in the Hermitian case, $P(M \in dM)$ is Gaussian, if $P(M \in dM)$ satisfies

(i) invariance under conjugation by orthogonal conjugation $M \to OMO^{-1}$, and

(ii) the condition that M_{ii}, M_{ij} ($i < j$) be independent random variables.

---

[4] The compact version is given by $\mathrm{SU}(n)/\mathrm{SO}(n)$.

(iii) Symplectic ensemble. Consider the *non-compact symmetric space*[5]

$$\mathrm{SU}^*(2n)/\mathrm{USp}(n)$$

with $\sigma(g) = Jg^{\top-1}J^{-1}$, where J is the $2n \times 2n$ matrix

$$J := \begin{pmatrix} \boxed{\begin{matrix} 0 & 1 \\ -1 & 0 \end{matrix}} & & & 0 \\ & \boxed{\begin{matrix} 0 & 1 \\ -1 & 0 \end{matrix}} & & \\ & & \boxed{\begin{matrix} 0 & 1 \\ -1 & 0 \end{matrix}} & \\ 0 & & & \ddots \end{pmatrix} \quad \text{with } J^2 = -I, \qquad (1.1.12)$$

and

$$G = \mathrm{SU}^*(2n) = \{g \in \mathrm{SL}(2n,\mathbb{C}) \mid g = J\bar{g}J^{-1}\},$$

$$K = \{g \in \mathrm{SU}^*(2n) \mid \sigma(g) = g\} := \mathrm{Sp}(n,\mathbb{C}) \cap U(2n)$$
$$= \{g \in \mathrm{SL}(2n,\mathbb{C}) \mid g^\top J g = J\} \cap \{g \in \mathrm{SL}(2n,\mathbb{C}) \mid g^{-1} = \bar{g}^\top\}$$
$$= \{g \in \mathrm{SL}(2n,\mathbb{C}) \mid g^{-1} = \bar{g}^\top \quad \text{and} \quad g = J\bar{g}J^{-1}\}$$
$$=: \mathrm{USp}(n).$$

Then $\sigma_*(a) = -Ja^\top J^{-1}$ and

$$\mathfrak{k} = \{a \in \mathrm{su}^*(2n) \mid \sigma_*(a) = a\} = \mathrm{sp}(n,\mathbb{C}) \cap u(2n)$$
$$= \{a \in \mathbb{C}^{2n \times 2n} \mid a^\top = -\bar{a},\ a = J\bar{a}J^{-1}\},$$

$$\mathfrak{p} = \{a \in \mathrm{su}^*(2n) \mid \sigma_*(a) = -a\} = \mathrm{su}^*(2n) \cap iu(2n)$$
$$= \{a \in \mathbb{C}^{2n \times 2n} \mid a^\top = \bar{a},\ a = J\bar{a}J^{-1}\}$$
$$= \left\{ M = (M_{kl})_{1 \le k,l \le n},\ M_{kl} = \begin{pmatrix} M_{kl}^{(0)} & M_{kl}^{(1)} \\ -\bar{M}_{kl}^{(1)} & \bar{M}_{kl}^{(0)} \end{pmatrix} \text{ with } M_{lk} = \bar{M}_{kl}^\top \in \mathbb{C}^{2 \times 2} \right\}$$
$$\cong \{\text{self-dual } n \times n \text{ Hermitian matrices, with quaternionic entries}\}$$
$$=: \mathcal{T}_{2n}.$$

The condition on the 2×2 matrices M_{kl} implies that $M_{kk} = M_k I$, with $M_k \in \mathbb{R}$ and the 2×2 identity I. Notice that $\mathrm{USp}(n)$ acts naturally by conjugation on the tangent space $\mathfrak{p}$ to G/K. Haar measure on $\mathcal{T}_{2n}$ is given by

$$dM = \prod_{1}^{n} dM_k \prod_{1 \le k < l \le n} dM_{kl}^{(0)}\, d\bar{M}_{kl}^{(0)}\, dM_{kl}^{(1)}\, d\bar{M}_{kl}^{(1)}, \qquad (1.1.13)$$

---

[5] The corresponding compact symmetric space is $\mathrm{SU}(2n)/\mathrm{Sp}(n)$.

since these M_{ij} are the only free variables in the matrix $M \in \mathcal{T}_{2n}$. A maximal abelian subalgebra in $\mathfrak{p}$ is given by real diagonal matrices of the form $z = \operatorname{diag}(z_1, z_1, z_2, z_2, \ldots, z_n, z_n)$. Each $M \in \mathfrak{p} = \mathcal{T}_{2n}$ can be written as

$$M = e^A z e^{-A}, \quad e^A \in K = \operatorname{USp}(n), \tag{1.1.14}$$

with $(a_{kl}, b_{kl}, c_{kl}, d_{kl} \in \mathbb{R})$

$$A = \sum_{1 \leq k \leq l \leq n} a_{kl}(e^{(0)}_{kl} - e^{(0)}_{lk}) + b_{kl}(e^{(1)}_{kl} + e^{(1)}_{lk}) + c_{kl}(e^{(2)}_{kl} - e^{(2)}_{lk}) + d_{kl}(e^{(3)}_{kl} + e^{(3)}_{lk}) \in \mathfrak{k} \tag{1.1.15}$$

in terms of the four 2×2 matrices[6]

$$e^{(0)} = \begin{pmatrix} 1 & 0 \\ 0 & 1 \end{pmatrix}, \quad e^{(1)} = \begin{pmatrix} i & 0 \\ 0 & -i \end{pmatrix}, \quad e^{(2)} = \begin{pmatrix} 0 & 1 \\ -1 & 0 \end{pmatrix}, \quad e^{(3)} = \begin{pmatrix} 0 & i \\ i & 0 \end{pmatrix}.$$

Since

$$\begin{aligned}
\left[e^{(0)}_{kl} - e^{(0)}_{lk}, z\right] &= (z_l - z_k)(e^{(0)}_{kl} + e^{(0)}_{lk}) \in \mathfrak{p}, \\
\left[e^{(1)}_{kl} + e^{(1)}_{lk}, z\right] &= (z_l - z_k)(e^{(1)}_{kl} - e^{(1)}_{lk}) \in \mathfrak{p}, \\
\left[e^{(2)}_{kl} - e^{(2)}_{lk}, z\right] &= (z_l - z_k)(e^{(2)}_{kl} + e^{(2)}_{lk}) \in \mathfrak{p}, \\
\left[e^{(3)}_{kl} + e^{(3)}_{lk}, z\right] &= (z_l - z_k)(e^{(3)}_{kl} - e^{(3)}_{lk}) \in \mathfrak{p},
\end{aligned} \tag{1.1.16}$$

$[A, z] \in \mathfrak{p}$ has 2×2 zero blocks along the diagonal, and from (1.1.16) and (1.1.15),

$$((k,l)\text{-th block in } [A,z]) = (z_l - z_k) \begin{pmatrix} a_{kl} + ib_{kl} & c_{kl} + id_{kl} \\ -c_{kl} + id_{kl} & a_{kl} - ib_{kl} \end{pmatrix}, \quad k < l. \tag{1.1.17}$$

Therefore, using (1.1.17), Haar measure dM on $\mathcal{T}_{2n}$ equals

$$\begin{aligned}
dM &= d(e^A z e^{-A}) = d(I + A + \cdots)z(I - A + \cdots) = d(z + [A,z] + \cdots) \\
&= \prod_{1 \leq k \leq n} dz_k \prod_{1 \leq k < l \leq n} d((z_l - z_k)(a_{kl} + ib_{kl})) d((z_l - z_k)(a_{kl} - ib_{kl})) \\
&\quad \times d((z_l - z_k)(c_{kl} + id_{kl})) d((z_l - z_k)(-c_{kl} + id_{kl})) \\
&= \Delta^4(z) \, dz_1 \cdots dz_n \prod_{1 \leq k < l \leq n} 4 \, da_{kl} \, db_{kl} \, dc_{kl} \, dd_{kl}.
\end{aligned}$$

As before, define $\mathcal{T}_{2n}(E) \subset \mathcal{T}_{2n}$ as the subset of matrices with spectrum $\in E$ and define the probability

$$P(M \in \mathcal{T}_{2n}(E)) = \int_{\mathcal{T}_{2n}(E)} c_n e^{-\operatorname{Tr} V(M)} \, dM = \frac{\int_{E^n} \Delta^4(z) \prod_1^n \rho(dz_k)}{\int_{\mathbb{R}^n} \Delta^4(z) \prod_1^n \rho(dz_k)}. \tag{1.1.18}$$

REMARK. $\mathcal{T}_{2n}$ is called the symplectic ensemble, although the matrices in $\mathfrak{p} = \mathcal{T}_{2n}$ are not at all symplectic; rather, it's the matrices in $\mathfrak{k}$ that are.

---

[6]The notation $e^{(i)}_{kl}$ in (1.1.15) refers to putting the 2×2 matrix $e^{(i)}$ at place (k,l).

1.2. Infinite Hermitian Matrix Ensembles.
Now consider the limit of the probability

$$P(M \in \mathcal{H}_n(E)) = \frac{\int_{E^n} \Delta^2(z) \prod_1^n \rho(dz_k)}{\int_{\mathbb{R}^n} \Delta^2(z) \prod_1^n \rho(dz_k)} \quad \text{when } n \nearrow \infty. \tag{1.2.1}$$

Dyson [27] (see also Mehta [49]) used the following trick to circumvent the problem of dealing with ∞-fold integrals. Using the orthogonality of the *monic orthogonal polynomials* $p_k = p_k(z)$ for the weight $\rho(dz)$ on $\mathbb{R}$, and the L^2-norms $h_k = \int_\mathbb{R} p_k^2(z)\rho(dz)$ of the p_k's, one finds, using $(\det A)^2 = \det(AA^\top)$,

$$\int_{\mathbb{R}^n} \Delta^2(z) \prod_1^n \rho(dz_i) = \int_{\mathbb{R}^n} \det(p_{i-1}(z_j))_{1\leq i,j\leq n} \det(p_{k-1}(z_l))_{1\leq k,l\leq n} \prod_{k=1}^n \rho(dz_k)$$

$$= \sum_{\pi,\pi' \in \sigma_n} (-1)^{\pi+\pi'} \prod_{k=1}^n \int_\mathbb{R} p_{\pi(k)-1}(z_k) p_{\pi'(k)-1}(z_k) \rho(dz_k)$$

$$= n! \prod_0^{n-1} \int_\mathbb{R} p_k^2(z)\rho(dz) = n! \prod_0^{n-1} h_k. \tag{1.2.2}$$

For the integral over an arbitrary subset $E \subset \mathbb{R}$, one stops at the second equality, since the p_n's are not necessarily orthogonal over E. This leads to the probability (1.2.1),

$$P(M \in \mathcal{H}_n(E)) = \frac{1}{n! \prod_1^n h_{i-1}} \int_{E^n} \det\left(\sum_{1\leq j\leq n} p_{j-1}(z_k)p_{j-1}(z_l)\right)_{1\leq k,l\leq n} \prod_1^n \rho(dz_i)$$

$$= \frac{1}{n!} \int_{E^n} \det(K_n(z_k, z_l))_{1\leq k,l\leq n} \prod_1^n \rho(dz_i), \tag{1.2.3}$$

in terms of the kernel

$$K_n(y,z) := \sum_{j=1}^n \frac{p_{j-1}(y)}{\sqrt{h_{j-1}}} \frac{p_{j-1}(z)}{\sqrt{h_{j-1}}}. \tag{1.2.4}$$

The orthonormality relations of the $p_k(y)/\sqrt{h_k}$ lead to the reproducing property for the kernel $K_n(y,z)$:

$$\int_\mathbb{R} K_n(y,z) K_n(z,u) \rho(dz) = K_n(y,u), \quad \int_\mathbb{R} K_n(z,z)\rho(dz) = n. \tag{1.2.5}$$

Upon replacing E^n by $\prod_1^k dz_i \times \mathbb{R}^{n-k}$ in (1.2.3), integrating out all the remaining variables $z_{k+1}, \ldots, z_n$ and using the reproducing property (1.2.5), one finds the n-point correlation function

$P(\text{one eigenvalue in each } [z_i, z_i + dz_i], \text{ for } i = 1, \ldots, k)$

$$= c_n \det(K_n(z_i, z_j))_{1\leq i,j\leq k} \prod_1^k \rho(dz_i). \tag{1.2.6}$$

Finally, by Poincaré's formula for the probability $P(\cup E_i)$, the probability that no spectral point of M belongs to E is given by a Fredholm determinant

$$P(M \in \mathcal{H}_n(E^c)) = \det(I - \lambda K_n^E)$$

$$= 1 + \sum_{k=1}^{\infty} (-\lambda)^k \int_{z_1 \leq \ldots \leq z_k} \det\left(K_n^E(z_i, z_j)\right)_{1 \leq i,j \leq k} \prod_{1}^{k} \rho(dz_i),$$

for the kernel $K_n^E(y, z) = K_n(y, z) I_E(z)$.

- *Wigner's semicircle law:* For this ensemble (defined by a large class of ρ's, in particular for the Gaussian ensemble) and for very large n, the density of eigenvalues tends to Wigner's semicircle distribution on the interval $[-\sqrt{2n}, \sqrt{2n}]$:

$$\text{density of eigenvalues} = \begin{cases} \dfrac{1}{\pi}\sqrt{2n - z^2}\, dz & \text{for } |z| \leq \sqrt{2n}, \\ 0 & \text{for } |z| > \sqrt{2n} \end{cases}$$

- *Bulk scaling limit:* From the formula above, it follows that the average number of eigenvalues per unit length near $z = 0$ ("the bulk") is given by $\sqrt{2n}/\pi$ and thus the average distance between two consecutive eigenvalues is given by $\pi/\sqrt{2n}$. Upon using this rescaling, one shows (see [43; 48; 52; 55; 39]) that

$$\lim_{n \nearrow \infty} \frac{\pi}{\sqrt{2n}} K_n\left(\frac{\pi x}{\sqrt{2n}}, \frac{\pi y}{\sqrt{2n}}\right) = \frac{\sin \pi(x - y)}{\pi(x - y)} \quad \text{(sine kernel)}$$

and

$$P(\text{exactly } k \text{ eigenvalues in } [0, a]) = \frac{(-1)^k}{k!} \left(\frac{\partial}{\partial \lambda}\right)^k \det(I - \lambda K I_{[0,a]})\Big|_{\lambda=1}$$

with

$$\det(I - \lambda K I_{[0,a]}) = \exp \int_0^{\pi a} \frac{f(x;\lambda)}{x}\, dx, \quad (1.2.7)$$

where $f(x, \lambda)$ is a solution to the following differential equation (where the prime stands for differentiation with respect to x), due to the pioneering work of Jimbo, Miwa, Mori, and Sato [39]:

$$(xf'')^2 = 4(xf' - f)(-f'^2 - xf' + f), \quad \text{with } f(x; \lambda) \cong -\frac{\lambda}{\pi} x \text{ for } x \simeq 0.$$

$$\textbf{(Painlevé V)} \quad (1.2.8)$$

- *Soft edge scaling limit:* Near the edge $\sqrt{2n}$ of the Wigner semicircle, the scaling is $\sqrt{2}n^{1/6}$ and thus the scaling is more subtle (see [21; 30; 51; 49; 63]):

$$y = \sqrt{2n} + \frac{u}{\sqrt{2}n^{1/6}}, \quad (1.2.9)$$

and so for the kernel K_n as in (1.2.4), with the p_n's being Hermite polynomials,

$$\lim_{n \nearrow \infty} \frac{1}{\sqrt{2}n^{1/6}} K_n\left(\sqrt{2n} + \frac{u}{\sqrt{2}n^{1/6}}, \sqrt{2n} + \frac{v}{\sqrt{2}n^{1/6}}\right) = K(u, v),$$

where
$$K(u,v) = \int_0^\infty A(x+u)A(x+v)\,dx, \quad A(u) = \int_{-\infty}^\infty e^{iux - x^3/3}\,dx.$$

Relating y and u by (1.2.9), the statistics of the largest eigenvalue for very large n is governed by the function

$$P(\lambda_{\max} \le y) = P\left(2n^{2/3}\left(\frac{\lambda_{\max}}{\sqrt{2n}} - 1\right) \le u\right), \text{ for } n \nearrow \infty,$$
$$= \det(I - K(y,z)I_{(-\infty, u]}(z)) = \exp\left(-\int_u^\infty (\alpha - u)g^2(\alpha)d\alpha\right),$$

with $g(x)$ a solution of

$$\begin{cases} g'' = xg + 2g^3 \\ g(x) \cong -\dfrac{e^{-(2/3)x^{3/2}}}{2\sqrt{\pi}x^{1/4}} \text{ for } x \nearrow \infty. \end{cases} \quad \textbf{(Painlevé II)} \quad (1.2.10)$$

The latter is essentially the asymptotics of the Airy function. In Section 5, I shall derive, via Virasoro constraints, not only this result, due to Tracy and Widom [63], but also a PDE for the probability that the eigenvalues belong to several intervals, due to Adler, Shiota, and van Moerbeke [11; 12].

• *Hard edge scaling limit*: Consider the ensemble of $n \times n$ random matrices for the Laguerre probability distribution, thus corresponding to (1.1.9) with $\rho(dz) = z^{\nu/2}e^{-z/2}\,dz$. One shows the density of eigenvalues near $z = 0$ is given by $4n$ for very large n. At this edge, one computes for the kernel (1.2.4) with Laguerre polynomials p_n [52; 30]:

$$\lim_{n \nearrow \infty} \frac{1}{4n} K_n^{(\nu)}\left(\frac{u}{4n}, \frac{v}{4n}\right) = K^{(\nu)}(u,v), \quad (1.2.11)$$

where $K^{(\nu)}(u,v)$ is the *Bessel kernel*, with Bessel functions J_ν:

$$K^{(\nu)}(u,v) = \frac{1}{2}\int_0^1 x J_\nu(xu) J_\nu(xv)\,dx$$
$$= \frac{J_\nu(u)\sqrt{u}J_\nu'(v) - J_\nu(\sqrt{v})\sqrt{v}J_\nu'(\sqrt{u})}{2(u-v)}. \quad (1.2.12)$$

Then
$$P(\text{no eigenvalues in } [0,x]) = \exp\left(\int_0^x \frac{f(u)}{u}\,du\right),$$

with f satisfying

$$(xf'')^2 - 4(xf' - f)f'^2 + ((x - \nu^2)f' - f)f' = 0. \quad \textbf{(Painlevé V)} \quad (1.2.13)$$

This result of Tracy and Widom [64] and a more general statement from [11; 12] will be shown using Virasoro constraints in Section 5.

1.3. Integrals over Classical Groups.
Integration on a compact, semisimple, simply connected Lie group G is given by the formula

$$\int_G f(M)\,dM = \frac{1}{|W|}\int_T \left|\prod_{\alpha \in \Delta} 2\sin\frac{\alpha(iH)}{2}\right| dt \int_U f(utu^{-1})\,du, \quad t = e^H \quad (1.3.1)$$

(see Helgason [36]), where $A \subset G$ is a maximal subgroup, du and dt are Haar measures on G and A respectively, satisfying $\int_A dt = \int_U du = 1$, the symbol Δ denotes the set of roots of $\mathfrak{g}$ with respect to $\mathfrak{a}$ ($\mathfrak{g}$ and $\mathfrak{a}$ being the Lie algebras of G and A), and $|W|$ is the order of the Weyl group of G.

Integration formula (1.3.1) will be applied to integrals of $f = e^{\sum_1^\infty t_i \operatorname{Tr} M^i}$ over the groups $SO(2n)$, $SO(2n+1)$ and $Sp(n)$. Their Lie algebras (over $\mathbb{C}$) are given respectively by $\mathfrak{d}_n$, $\mathfrak{b}_n$, $\mathfrak{c}_n$, with the following sets of roots (see [20], for example):

$$\Delta_n = \{\pm\varepsilon e_i \mid 1 \leq i \leq k\} \cup \{\pm(e_i+e_j),\ \pm(e_i-e_j) \mid 1 \leq i < j \leq n\},$$

where

$$\varepsilon = \begin{cases} 0 & \text{for } \mathfrak{d}_n = so(2n), \\ 1 & \text{for } \mathfrak{b}_n = so(2n+1), \\ 2 & \text{for } \mathfrak{c}_n = sp(n). \end{cases}$$

Setting $H = i\theta$, we have, in view of formula (1.3.1),

$$\left|\prod_{\alpha \in \Delta} 2\sin\frac{\alpha(iH)}{2}\right| dt$$

$$= \begin{cases} c_n \left(\displaystyle\prod_{1\leq j<k\leq n} \sin\frac{\theta_j-\theta_k}{2}\sin\frac{\theta_j+\theta_k}{2}\right)^2 \displaystyle\prod_1^n d\theta_j & \text{for } \mathfrak{d}_n \\[2mm] c_n \left(\displaystyle\prod_{1\leq j<k\leq n} \sin\frac{\theta_j-\theta_k}{2}\sin\frac{\theta_j+\theta_k}{2}\right)^2 \displaystyle\prod_1^n \sin^2\frac{\varepsilon\theta_j}{2}\,d\theta_j & \text{for } \mathfrak{b}_n,\mathfrak{c}_n \end{cases}$$

$$= c'_n \prod_{1\leq j<k'\leq n}(\cos\theta_j - \cos\theta_k)^2 \begin{cases} \displaystyle\prod_{1\leq j\leq n} d\theta_j & \text{for } \mathfrak{d}_n \\[2mm] \displaystyle\prod_{1\leq j\leq n}\left(\frac{1-\cos\theta_j}{2}\right)d\theta_j & \text{for } \mathfrak{b}_n \\[2mm] \displaystyle\prod_{1\leq j\leq n}(1-\cos^2\theta_j)\,d\theta_j & \text{for } \mathfrak{c}_n \end{cases}$$

$$= \begin{cases} c'_n \Delta^2(z) \displaystyle\prod_{1\leq j\leq n} \frac{dz_j}{\sqrt{1-z_j^2}|} & \text{for } \mathfrak{d}_n \\[2mm] c'_n \Delta^2(z) \displaystyle\prod_{1\leq j\leq n}(1-z_j)\frac{dz_j}{\sqrt{1-z_j^2}|} & \text{for } \mathfrak{b}_n \\[2mm] c'_n \Delta^2(z) \displaystyle\prod_{1\leq j\leq n}(1-z_j^2)\frac{dz_j}{\sqrt{1-z_j^2}|} & \text{for } \mathfrak{c}_n \end{cases}$$

$$= c_n'' \Delta^2(z) \prod_{1 \le j \le n} (1-z_j)^\alpha (1+z_j)^\beta \, dz_j \text{ with } \begin{cases} \alpha = \beta = -\frac{1}{2} & \text{for } \mathfrak{d}_n \\ \alpha = \frac{1}{2}, \ \beta = -\frac{1}{2} & \text{for } \mathfrak{b}_n \\ \alpha = \beta = \frac{1}{2} & \text{for } \mathfrak{c}_n \end{cases}$$

For $M \in SO(2n)$, $Sp(n)$, the eigenvalues are given by $e^{i\theta_j}$ and $e^{-i\theta_j}$, for $1 \le j \le n$; therefore, setting $f = \exp(\sum t_k \mathrm{tr} M^k)$ in formula (1.3.1) leads to

$$e^{\sum_1^\infty t_k \operatorname{Tr} M^k} = e^{\sum_1^\infty t_k \sum_{j=1}^n (e^{ik\theta_j} + e^{-ik\theta_j})} = \prod_{j=1}^n e^{2\sum_{k=1}^\infty t_k \cos k\theta_j}$$
$$= \prod_{j=1}^n e^{2\sum t_k T_k(z_j)}, \qquad (1.3.2)$$

where $T_n(z)$ are the Chebyshev polynomials, defined by $T_n(\cos\theta) := \cos n\theta$; in particular $T_1(z) = z$.

For $M \in SO(2n+1)$, the eigenvalues are given by 1, $e^{i\theta_j}$ and $e^{-i\theta_j}$, for $1 \le j \le n$, which is responsible for the extra-exponential $e^{\sum t_i}$ appearing in (1.3.2).

Before listing various integrals, define the Jacobi weight

$$\rho_{\alpha\beta}(z)\, dz := (1-z)^\alpha (1+z)^\beta dz, \qquad (1.3.3)$$

and the formal sum

$$g(z) := 2 \sum_1^\infty t_i T_i(z).$$

The arguments above lead to the following integrals, originally due to H. Weyl [73], and in its present form, due to Johansson [40]; besides the integrals over $SO(k) = O_+(k)$, the integrals over $O_-(k)$ and $U(n)$ will also be of interest in the theory of random permutations:

$$\int_{O(2n)_+} e^{\sum_1^\infty t_i \mathrm{tr} M^i}\, dM = \int_{[-1,1]^n} \Delta_n(z)^2 \prod_{k=1}^n e^{g(z_k)} \rho_{(-\frac{1}{2},-\frac{1}{2})}(z_k)\, dz_k,$$

$$\int_{O(2n+1)_+} e^{\sum_1^\infty t_i \mathrm{tr} M^i}\, dM = e^{\sum_1^\infty t_i} \int_{[-1,1]^n} \Delta_n(z)^2 \prod_{k=1}^n e^{g(z_k)} \rho_{(\frac{1}{2},-\frac{1}{2})}(z_k)\, dz_k,$$

$$\int_{Sp(n)} e^{\sum_1^\infty t_i \mathrm{tr} M^i}\, dM = \int_{[-1,1]^n} \Delta_n(z)^2 \prod_{k=1}^n e^{g(z_k)} \rho_{(\frac{1}{2},\frac{1}{2})}(z_k)\, dz_k,$$

$$\int_{O(2n)_-} e^{\sum_1^\infty t_i \mathrm{tr} M^i}\, dM = e^{\sum_1^\infty 2 t_{2i}} \int_{[-1,1]^{n-1}} \Delta_{n-1}(z)^2 \prod_{k=1}^{n-1} e^{g(z_k)} \rho_{(\frac{1}{2},\frac{1}{2})}(z_k)\, dz_k,$$

$$\int_{O(2n+1)_-} e^{\sum_1^\infty t_i \mathrm{tr} M^i}\, dM = e^{\sum_1^\infty (-1)^i t_i} \int_{[-1,1]^n} \Delta_n(z)^2 \prod_{k=1}^n e^{g(z_k)} \rho_{(-\frac{1}{2},\frac{1}{2})}(z_k)\, dz_k,$$

$$\int_{U(n)} e^{\sum_1^\infty \mathrm{tr}(t_i M^i - s_i \bar{M}^i)}\, dM = \frac{1}{n!} \int_{(S^1)^n} |\Delta_n(z)|^2 \prod_{k=1}^n e^{\sum_1^\infty (t_i z_k^i - s_i z_k^{-i})} \frac{dz_k}{2\pi i z_k}, \qquad (1.3.4)$$

1.4. Permutations and Integrals over Groups.

Let S_n be the group of permutations π and S_{2n}^0 the subset of fixed-point free involutions π^0 (this means that $(\pi^0)^2 = I$ and $\pi^0(k) \neq k$ for $1 \leq k \leq 2n$). Put the uniform distribution on S_n and S_{2n}^0 — that is, give all permutations $\pi_n \in S_n$ and all involutions $\pi_{2n}^0 \in S_{2n}^0$ equal probability:

$$P(\pi_n) = 1/n! \quad \text{and} \quad P(\pi_{2n}^0) = \frac{2^n n!}{(2n)!}. \qquad (1.4.1)$$

An *increasing subsequence* of $\pi \in S_n$ or S_n^0 is a sequence $1 \leq j_1 < \cdots < j_k \leq n$ such that $\pi(j_1) < \cdots < \pi(j_k)$. Define

$$L(\pi_n) = \text{length of the longest increasing subsequence of } \pi_n. \qquad (1.4.2)$$

Example: for $\pi = (\underline{3}, 1, \underline{4}, 2, \underline{6}, \underline{7}, 5)$, we have $L(\pi_7) = 4$.

Around 1960 and based on Monte Carlo methods, Ulam [68] conjectured that

$$\lim_{n \to \infty} \frac{E(L_n)}{\sqrt{n}} = c \text{ exists.}$$

An argument of Erdös and Szekeres [28], dating back from 1935 showed that $E(L_n) \geq \frac{1}{2}\sqrt{n-1}$, and thus $c \geq \frac{1}{2}$. In 1972, Hammersley [33] showed rigorously that the limit exists. Logan and Shepp [46] showed the limit $c \geq 2$, and finally Vershik and Kerov [72] that $c = 2$. In 1990, I. Gessel [31] showed that the following generating function is the determinant of a Toeplitz matrix:

$$\sum_{n=0}^{\infty} \frac{t^n}{n!} P(L_n \leq l) = \det \left(\int_0^{2\pi} e^{2\sqrt{t}\cos\theta} e^{i(k-m)\theta} d\theta \right)_{0 \leq k, m \leq l-1}. \qquad (1.4.3)$$

The next major contribution was due to Johansson [41] and Baik, Deift, and Johansson [17], who prove that for arbitrary $x \in \mathbb{R}$, we have a *"law of large numbers"* and a *"central limit theorem"*, where $F(x)$ is the statistics (1.2.10),

$$\lim_{n \to \infty} \frac{L_n}{2\sqrt{n}} = 1 \quad \text{and} \quad P\left(\frac{L_n - 2\sqrt{n}}{n^{1/6}} \leq x \right) \longrightarrow F(x) \text{ for } n \longrightarrow \infty.$$

The next set of ideas is due to Diaconis and Shashahani [26], Rains [57; 58], and Baik and Rains [18]. For a nice state-of-the-art account, see Aldous and Diaconis [14]. An illustration is contained in the following proposition; the first part is essentially Gessel's and the second can be found in [26; 58; 18].

PROPOSITION 1.1.

(i) $$\sum_{n=0}^{\infty} \frac{t^n}{n!} P(L(\pi_n) \leq l) = \int_{U(l)} e^{\sqrt{t}\operatorname{Tr}(M+\bar{M})} dM \qquad (1.4.4)$$

$$= \int_{[0,2\pi]^l} \prod_{1 \leq j < k \leq l} |e^{i\theta_j} - e^{i\theta_k}|^2 \prod_{1 \leq k \leq l} e^{2\sqrt{t}\cos\theta_k} \frac{d\theta_k}{2\pi}.$$

(ii) $$\sum_{n=0}^{\infty} \frac{(t^2/2)^n}{n!} P(L(\pi_{2n}^0) \le l) = \int_{O(l)} e^{t \operatorname{Tr} M} dM. \qquad (1.4.5)$$

The proof of this statement will be sketched later. The connection with integrable systems goes via this chain of ideas:

Combinatorics
↓
Robinson–Schensted–Knuth correspondence
↓
Theory of symmetric polynomials
↓
Integrals over classical groups
↓
Integrable systems

All arrows but the last will be explained in this section; the last arrow will be discussed in Sections 7 and 8. We briefly sketch a few of the basic well known facts going into these arguments. They can be found in MacDonald [47], Knuth [45], and Aldous and Diaconis [14]. Useful facts on symmetric functions, applicable to integrable theory, can be found in the appendix to [1]. To mention a few:

- A *Young diagram* λ is a finite sequence of non-increasing, non-negative integers $\lambda_1 \ge \lambda_2 \ge \cdots \ge \lambda_l \ge 0$; also called a *partition* of $n = |\lambda| := \lambda_1 + \cdots + \lambda_l$, with $|\lambda|$ being the weight. It can be represented by a diagram, having λ_1 boxes in the first row, λ_2 boxes in the second row, etc., all aligned to the left. A *dual Young diagram* $\hat{\lambda} = (\hat{\lambda}_1 \ge \hat{\lambda}_2 \ge \cdots)$ is the diagram obtained by flipping the diagram λ about its diagonal.
- A *Young tableau* of shape λ is an array of positive integers a_{ij} (at place (i,j) in the Young diagram) placed in the Young diagram λ, which are non-decreasing from left to right *and* strictly increasing from top to bottom.
- A *standard Young tableau* of shape λ is an array of integers $1, \ldots, n$ placed in the Young diagram, which are strictly increasing from left to right *and* from top to bottom. There are several formulae for the number of Young tableaux of a given shape $\lambda = (\lambda_1 \ge \cdots \ge \lambda_m)$:

$$\begin{aligned} f^\lambda &= \#\{\text{standard tableaux of shape } \lambda\} \\ &= \text{coefficient of } x_1 x_2 \ldots x_n \text{ in the Schur polynomial } s_\lambda(x) \quad \text{(see next entry)} \\ &= \frac{|\lambda|!}{\prod_{\text{all } i,j} h_{ij}^\lambda} \\ &= |\lambda|! \det\left(\frac{1}{(\lambda_i - i + j)!}\right) \quad (\text{with } h_{ij}^\lambda := \lambda_i + \hat{\lambda}_j - i - j + 1 = \text{hook length}) \\ &= |\lambda|! \prod_{1 \le i < j \le m} (h_i - h_j) \prod_1^m \frac{1}{h_i!} \quad (\text{with } h_i := \lambda_i - i + m, \ m := \hat{\lambda}_1). \end{aligned} \qquad (1.4.6)$$

- The *Schur polynomial* s_λ associated with a Young diagram λ is a symmetric function in the variables $x_1, x_2, \ldots$ (finite or infinite), defined by

$$s_\lambda(x_1, x_2, \ldots) := \sum_{\{a_{ij}\} \text{ tableaux of } \lambda} \prod_{ij} x_{a_{ij}}. \tag{1.4.7}$$

- The linear *space* Λ_n *of symmetric polynomials* in $x_1, \ldots, x_n$ with rational coefficients comes equipped with the inner product

$$\langle f, g \rangle = \frac{1}{n!} \int_{(S_1)^n} f(z_1, \ldots, z_n) g(\bar{z}_1, \ldots, \bar{z}_n) \prod_{1 \le k < l \le n} |z_k - z_l|^2 \prod_1^n \frac{dz_k}{2\pi i z_k}$$

$$= \int_{U(n)} f(M) g(\bar{M}) \, dM. \tag{1.4.8}$$

- An *orthonormal basis of the space* Λ_n is given by the Schur polynomials $s_\lambda(x_1, \ldots, x_n)$, in which the numbers a_{ij} are restricted to $1, \ldots, n$. Therefore, each symmetric function admits a *"Fourier series"*

$$f(x_1, \ldots, x_n) = \sum_{\substack{\lambda \text{ with} \\ \hat{\lambda}_1 \le n}} \langle f, s_\lambda \rangle s_\lambda(x_1, \ldots, x_n), \quad \text{with } \langle s_\lambda, s_{\lambda'} \rangle = \delta_{\lambda \lambda'}. \tag{1.4.9}$$

In particular, one proves (see (1.4.6) for the definition of f^λ)

$$(x_1 + \cdots + x_n)^k = \sum_{\substack{|\lambda| = k \\ \hat{\lambda}_1 \le n}} f^\lambda s_\lambda, \tag{1.4.10}$$

If $\lambda = (\lambda_1 \ge \cdots \ge \lambda_l > 0)$, with[7] $\hat{\lambda}_1 = l > n$, then obviously $s_\lambda = 0$.

- *Robinson–Schensted–Knuth correspondence:* There is a 1-1 correspondence

$$S_n \longrightarrow \left\{ \begin{array}{l} (P, Q), \text{two standard Young} \\ \text{tableaux from } 1, \ldots, n, \text{ where} \\ P \text{ and } Q \text{ have the same shape} \end{array} \right\}$$

Given a permutation $i_1, \ldots, i_n$, the correspondence constructs two standard Young tableaux P, Q having the same shape λ. This construction is inductive. Namely, having obtained two equally shaped Young diagrams P_k, Q_k from $i_1, \ldots, i_k$, with the numbers $(i_1, \ldots, i_k)$ in the boxes of P_k and the numbers $(1, \ldots, k)$ in the boxes of Q_k, one creates a new diagram Q_{k+1}, by putting the next number i_{k+1} *in the first row of* P, according to the following rule:

(i) if $i_{k+1} \ge$ all numbers appearing in the first row of P_k, then one creates a new box with i_{k+1} in that box to the right of the first column,

---

[7] Remember from the definition of the dual Young diagram that $\hat{\lambda}_1$ is the length of the first column of λ.

(ii) if not, place i_{k+1} in the box (of the first row) with the smallest higher number. That number then gets pushed down to the second row of P_k according to the rule (i) or (ii), as if the first row had been removed.

The diagram Q is a bookkeeping device; namely, add a box (with the number $k+1$ in it) to Q_k exactly at the place, where the new box has been added to P_k. This produces a new diagram Q_{k+1} of same shape as P_{k+1}.

The inverse of this map is constructed essentially by reversing the steps above.

EXAMPLE. We take $\pi = (5, 1, 4, 3, 2) \in S_5$ and follow the construction rules for P and Q:

$$
\begin{array}{ccccc}
5 & 1 & 1\ 4 & 1\ 3 & 1\ 2 \\
  & 5 & 5   & 4   & 3   \\
  &   &   & 5   & 4   \\
  &   &   &   & 5
\end{array}
\qquad
\begin{array}{ccccc}
1 & 1 & 1\ 3 & 1\ 3 & 1\ 3 \\
  & 2 & 2   & 2   & 2   \\
  &   &   & 4   & 4   \\
  &   &   &   & 5
\end{array}
$$

Hence $\pi \longmapsto (P(\pi), Q(\pi)) = \left(\begin{pmatrix} 1\ 2 \\ 3 \\ 4 \\ 5 \end{pmatrix}, \begin{pmatrix} 1\ 3 \\ 2 \\ 4 \\ 5 \end{pmatrix} \right)$ and so $L_5(\pi) = 2 =$ number of columns of P or Q.

The Robinson–Schensted–Knuth correspondence has these properties:

- if $\pi \mapsto (P, Q)$, then $\pi^{-1} \mapsto (Q, P)$;
- length (longest increasing subsequence of π) = # (columns in P);
- length (longest decreasing subsequence of π) = # (rows in P);
- if $\pi^2 = I$, then $\pi \mapsto (P, P)$;
- if $\pi^2 = I$ with k fixed points, then P has exactly k columns of odd length.

(1.4.11)

From representation theory (see Weyl [73] and especially Rains [57]), one proves:

LEMMA 1.2. *The following perpendicularity relations hold:*

(i) $\displaystyle\int_{U(n)} s_\lambda(M) s_\mu(\bar M) dM = \langle s_\lambda, s_\mu \rangle = \delta_{\lambda\mu}.$

(ii) $\displaystyle\int_{O(n)} s_\lambda(M)\, dM = \begin{cases} 1 & \text{for } \lambda = (\lambda_1 \geq \cdots \geq \lambda_k \geq 0),\ k \leq n,\ \lambda_i \text{ even,} \\ 0 & \text{otherwise.} \end{cases}$

(iii) $\displaystyle\int_{\mathrm{Sp}(n)} s_\lambda(M)\, dM = \begin{cases} 1 & \text{for } \hat\lambda_i \text{ even},\ \hat\lambda_1 \leq 2n, \\ 0 & \text{otherwise.} \end{cases}$ (1.4.12)

PROOF OF PROPOSITION 1.1. On the one hand,

$$\langle (x_1 + \cdots + x_n)^k, (x_1 + \cdots + x_n)^k \rangle$$
$$= \sum_{\substack{|\lambda|=|\mu|=k \\ \hat{\lambda}_1, \hat{\mu}_1 \leq n}} f^\lambda f^\mu \langle s_\lambda, s_\mu \rangle = \sum_{\substack{|\lambda|=k \\ \hat{\lambda}_1 \leq n}} (f^\lambda)^2 = \sum_{\substack{|\lambda|=k \\ \lambda_1 \leq n}} (f^\lambda)^2$$
$$= \#\{(P,Q) \text{ standard Young tableaux, each of arbitrary}$$
$$\text{shape } \lambda \text{ with } |\lambda| = k, \ \lambda_1 \leq n\}$$
$$= \#\{\pi_k \in S_k \text{ such that } L(\pi_k) \leq n\}. \qquad (1.4.13)$$

On the other hand, notice that, upon setting $\theta_j = \theta'_j + \theta_1$ for $2 \leq j \leq n$, the expression $\prod_{1 \leq j < k \leq n} |e^{i\theta_j} - e^{i\theta_k}|^2$ is independent of θ_1. Then, setting $z_k = e^{i\theta_k}$, one computes:

$$\langle (x_1 + \cdots + x_n)^k, (x_1 + \cdots + x_n)^l \rangle$$
$$= \frac{1}{n!} \int_{[0,2\pi]^n} (z_1 + \cdots + z_n)^k (\bar{z}_1 + \cdots + \bar{z}_n)^l \prod_{1 \leq j < k \leq n} |e^{i\theta_j} - e^{i\theta_k}|^2 \, d\theta_1 \ldots d\theta_n$$
$$= \frac{1}{n!} \int_{[0,2\pi]^n} e^{ik\theta_1}(1 + z'_2 + \cdots + z'_n)^k e^{-il\theta_1}(1 + \bar{z}'_2 + \cdots + \bar{z}'_n)^l$$
$$\prod_{1 \leq j < k \leq n} |e^{i\theta_j} - e^{i\theta_k}|^2 \, d\theta_1 \ldots d\theta_n$$

and upon setting $\theta_j = \theta'_j + \theta_1$, for $j \geq 2$ and $z'_k = e^{i\theta'_k}$,

$$= \frac{1}{n!} \int_0^{2\pi} e^{i(k-l)\theta_1} \, d\theta_1 \times \text{an } (n-1)\text{-fold integral}$$
$$= \delta_{kl} \langle (x_1 + \cdots + x_n)^k, (x_1 + \cdots + x_n)^k \rangle = \delta_{kl} \int_{U(n)} |\text{Tr } M|^{2k} \, dM. \qquad (1.4.14)$$

It follows that

$$\int_{U(n)} (\text{tr}(M + \bar{M}))^k \, dM = \sum_{0 \leq j \leq k} \binom{k}{j} \int_{U(n)} (\text{tr}M)^j (\overline{\text{tr}M})^{k-j} \, dM$$
$$= \begin{cases} 0 & \text{if } k \text{ is odd,} \\ \binom{k}{k/2} \int_{U(n)} |\text{tr}M|^k \, dM & \text{if } k \text{ is even.} \end{cases} \qquad (1.4.15)$$

(The equality for k odd follows because then $j \neq k - j$ for all $0 \leq j \leq k$.) Combining the three identities (1.4.13), (1.4.14) and (1.4.15) leads to

$$\#\{\pi_k \in S_k \text{ such that } L(\pi_k) \leq n\} = \binom{2k}{k}^{-1} \int_{U(n)} (\text{Tr}(M + \bar{M}))^{2k} \, dM. \qquad (1.4.16)$$

Finally

$$\sum_{n=0}^{\infty} \frac{t^n}{n!} P(L(\pi_n) \leq l) = \sum_{n=0}^{\infty} \frac{t^n}{n!} \frac{\#\{\pi_n \in S_n \mid L(\pi_n) \leq l\}}{n!}$$

$$= \sum_{n=0}^{\infty} \frac{t^n}{(n!)^2} \binom{2n}{n}^{-1} \int_{U(l)} (\operatorname{tr}(M + \bar{M}))^{2n} \, dM$$

$$= \sum_{n=0}^{\infty} \frac{(\sqrt{t})^{2n}}{(2n)!} \int_{U(l)} (\operatorname{tr}(M + \bar{M}))^{2n} \, dM$$

$$= \int_{U(l)} e^{\sqrt{t}\operatorname{Tr}(M+\bar{M})} \, dM$$

$$= \frac{1}{l!} \int_{[0,2\pi]^l} e^{\sqrt{t}(z_1 + z_1^{-1} + \cdots + z_l + z_l^{-1})} \prod_{1 \leq j < k \leq l} |e^{i\theta_j} - e^{i\theta_k}|^2 \prod_{1 \leq k \leq l} \frac{d\theta_k}{2\pi},$$

$$\text{(where } z_k = e^{i\theta_k}\text{)}$$

$$= \frac{1}{l!} \int_{[0,2\pi]^l} \prod_{1 \leq j < k \leq l} |e^{i\theta_j} - e^{i\theta_k}|^2 \prod_{k=1}^{l} e^{2\sqrt{t}\cos\theta_k} \frac{d\theta_k}{2\pi},$$

showing (1.4.4) of Proposition 1.1. The latter also equals:

$$= \frac{1}{l!} \int_{(S^1)^l} \Delta_l(z) \Delta_l(\bar{z}) \prod_{k=1}^{l} \left(e^{\sqrt{t}(z_k + \bar{z}_k)} \frac{dz_k}{2\pi i z_k} \right)$$

$$= \frac{1}{l!} \int_{(S^1)^l} \sum_{\sigma \in S_l} \det\left(z_{\sigma(m)}^{k-1} \bar{z}_{\sigma(m)}^{m-1} \right)_{1 \leq k,m \leq l} \prod_{k=1}^{l} \left(e^{\sqrt{t}(z_k + \bar{z}_k)} \frac{dz_k}{2\pi i z_k} \right)$$

$$= \frac{1}{l!} \sum_{\sigma \in S_l} \det\left(\int_{S^1} z_k^{k-1} \bar{z}_k^{m-1} e^{\sqrt{t}(z_k + \bar{z}_k)} \frac{dz_k}{2\pi i z_k} \right)_{1 \leq k,m \leq l}$$

$$= \det\left(\int_0^{2\pi} e^{2\sqrt{t}\cos\theta} e^{i(k-m)\theta} d\theta \right)_{1 \leq k,m \leq l},$$

confirming Gessel's result (1.4.3).

The proof of the second relation (1.4.5) of Proposition 1.1 is based on the following computation:

$$\int_{O(n)} (\operatorname{Tr} M)^k \, dM = \sum_{\substack{|\lambda|=k, \\ \hat{\lambda}_1 \leq n}} f^\lambda \int_{O(n)} s_\lambda(M) \, dM \qquad \text{using (1.4.10)}$$

$$= \sum_{\substack{|\lambda|=k, \, \hat{\lambda}_1 \leq n \\ \lambda_i \text{ even}}} f^\lambda \qquad \text{using Lemma 1.2}$$

$$= \sum_{\substack{|\lambda|=k,\ \lambda_1 \leq n, \\ \hat{\lambda}_i \text{ even}}} f^\lambda \qquad \text{using duality}$$

$$= \#\left\{ (P,P) \mid P \text{ standard Young tableau of shape } \lambda \text{ with } |\lambda| = k,\ \lambda_1 \leq n,\ \hat{\lambda}_i \text{ even} \right\}$$

$$= \#\{\pi_k^0 \in S_k^0,\ \text{no fixed points and } L(\pi_k^0) \leq n\}. \qquad (1.4.17)$$

In the last equality, we have used property (1.4.11): an involution has no fixed points if and only if all columns of P have even length. Since all columns $\hat{\lambda}_i$ have even length, it follows that $|\lambda| = k$ is even and then only is $\int_{O(n)} (\operatorname{Tr} M)^k dM > 0$; otherwise this integral equals 0. Finally, one computes

$$\sum_{k=0}^{\infty} \frac{(t^2/2)^k}{k!} P\left(L(\pi_{2k}^0) \leq n,\ \pi_{2k}^0 \in S_{2k}^0\right)$$

$$= \sum_{k=0}^{\infty} \frac{t^{2k}}{2^k k!} \frac{2^k k!}{(2k)!} \#\{\pi_{2k}^0 \in S_{2k}^0,\ L(\pi_{2k}^0) \leq n\} \qquad \text{using (1.4.1)}$$

$$= \sum_{k=0}^{\infty} \frac{t^k}{k!} \#\{\pi_k^0 \in S_k^0,\ L(\pi_k^0) \leq n\}$$

$$= \sum_{k=0}^{\infty} \frac{t^k}{k!} \int_{O(n)} (\operatorname{Tr} M)^k dM \qquad \text{using (1.4.17)}$$

$$= \int_{O(n)} e^{t \operatorname{Tr} M} dM,$$

ending the proof of Proposition 1.1. $\square$

2. Integrals, Vertex Operators and Virasoro Relations

In Section 1, we discussed random matrix problems over different finite and infinite matrix ensembles, generating functions for the statistics of the length of longest increasing sequences in random permutations and involutions. One can also consider two Hermitian random matrix ensembles, coupled together. All those problems lead to matrix integrals or Fredholm determinants, which we list in the following formulas (where $\beta = 2, 1, 4$):

- $\int_{\mathcal{H}_n(E),\ \mathcal{S}_n(E)\ \text{or}\ \mathcal{T}_n(E)} e^{-\operatorname{Tr} V(M)} dM = c_n \int_{E^n} |\Delta_n(z)|^\beta \prod_1^n \rho(z_k)\, dz_k$

- $\int\int_{\mathcal{H}_n^2(E_1 \times E_2)} dM_1\, dM_2\, e^{-\frac{1}{2}\operatorname{Tr}(M_1^2 + M_2^2 - 2cM_1 M_2)}$

- $\int_{O(n)} e^{x \operatorname{Tr} M} dM$

- $\int_{U(n)} e^{\sqrt{x}\,\mathrm{Tr}(M+\bar{M})}\, dM$

- $\det\bigl(I - \lambda K(y,z) I_E(z)\bigr)$, with $K(y,z)$ as in (1.2.4). (2.0.1)

Each of these quantities admits a natural deformation, by inserting time variables $t_1, t_2, \ldots$ and possibly a second set $s_1, s_2, \ldots$, seemingly *ad hoc*. Each of these integrals or Fredholm determinant is then a fixed point for a natural *vertex operator*, which generates a Virasoro-like algebra. These new integrals in $t_1, t_2, \ldots$ are all annihilated by the precise subalgebra of the Virasoro generators, which annihilates τ_0. This will be the topic of this section.

2.1. β-Integrals

2.1.1. Virasoro constraints for β-integrals.

Consider weights of the form
$$\rho(z)\, dz := e^{-V(z)}\, dz$$
on an interval $F = [A, B] \subseteq \mathbb{R}$, with rational logarithmic derivative and subjected to the boundary conditions
$$-\frac{\rho'}{\rho} = V' = \frac{g}{f} = \frac{\sum_0^\infty b_i z^i}{\sum_0^\infty a_i z^i}, \quad \lim_{z \to A, B} f(z)\rho(z) z^k = 0 \text{ for } k \geq 0, \qquad (2.1.1)$$
and a disjoint union of intervals,
$$E = \bigcup_1^r [c_{2i-1}, c_{2i}] \subset F \subset \mathbb{R}. \qquad (2.1.2)$$

These data define an algebra of differential operators
$$\mathcal{B}_k = \sum_1^{2r} c_i^{k+1} f(c_i) \frac{\partial}{\partial c_i}. \qquad (2.1.3)$$

Take the first type of integrals in the list (2.0.1) for general $\beta > 0$, thus generalizing the integrals appearing in the probabilities (1.1.9), (1.1.11) and (1.1.18). Consider t-deformations of such integrals, for general (fixed) $\beta > 0$; they can be written as follows, with $t := (t_1, t_2, \ldots)$, $c = (c_1, c_2, \ldots, c_{2r})$ and $z = (z_1, \ldots, z_n)$:
$$I_n(t, c; \beta) := \int_{E^n} |\Delta_n(z)|^\beta \prod_{k=1}^n \left(e^{\sum_1^\infty t_i z_k^i} \rho(z_k)\, dz_k\right) \quad \text{for } n > 0. \qquad (2.1.4)$$

The main statement of this section is Theorem 2.1, whose proof will be outlined in the next subsection. The central charge (2.1.9) has already appeared in the work of Awata et al. [16].

THEOREM 2.1 (Adler and van Moerbeke [3; 6]). *The multiple integrals*
$$I_n(t, c; \beta) := \int_{E^n} |\Delta_n(z)|^\beta \prod_{k=1}^n \left(e^{\sum_1^\infty t_i z_k^i} \rho(z_k)\, dz_k\right), \quad \text{for } n > 0, \qquad (2.1.5)$$

and

$$I_n\left(t,c;\frac{4}{\beta}\right) := \int_{E^n} |\Delta_n(z)|^{4/\beta} \prod_{k=1}^n \left(e^{\sum_1^\infty t_i z_k^i} \rho(z_k)\,dz_k\right), \quad \text{for } n > 0, \quad (2.1.6)$$

with $I_0 = 1$, satisfy respectively the Virasoro constraints[8]

$$\left(-\mathcal{B}_k + \sum_{i\geq 0} a_i\,{}^\beta\mathbb{J}^{(2)}_{k+i,n}(t,n) - b_i\,{}^\beta\mathbb{J}^{(1)}_{k+i+1,n}(t,n)\right) I_n(t,c;\beta) = 0,$$

$$\left(-\mathcal{B}_k + \sum_{i\geq 0} a_i\,{}^\beta\mathbb{J}^{(2)}_{k+i,n}\left(-\frac{\beta t}{2}, -\frac{2n}{\beta}\right) + \frac{\beta b_i}{2}\,{}^\beta\mathbb{J}^{(1)}_{k+i+1,n}\left(-\frac{\beta t}{2}, -\frac{2n}{\beta}\right)\right) I_n\left(t,c;\frac{4}{\beta}\right) = 0,$$

(2.1.7)

for all $k \geq -1$, in terms of the coefficients a_i, b_i of the rational function $(-\log \rho)'$ and the end points c_i of the subset E, as in (2.1.1) to (2.1.2). For all $n \in \mathbb{Z}$, the ${}^\beta\mathbb{J}^{(2)}_{k,n}(t,n)$ and ${}^\beta\mathbb{J}^{(1)}_{k,n}(t,n)$ form a Virasoro and a Heisenberg algebra respectively, interacting via the formulas

$$[{}^\beta\mathbb{J}^{(2)}_{k,n},{}^\beta\mathbb{J}^{(2)}_{l,n}] = (k-l)\,{}^\beta\mathbb{J}^{(2)}_{k+l,n} + c\left(\frac{k^3-k}{12}\right)\delta_{k,-l}$$

$$[{}^\beta\mathbb{J}^{(2)}_{k,n},{}^\beta\mathbb{J}^{(1)}_{l,n}] = -l\,{}^\beta\mathbb{J}^{(1)}_{k+l,n} + c'k(k+1)\delta_{k,-l}.$$

$$[{}^\beta\mathbb{J}^{(1)}_{k,n},{}^\beta\mathbb{J}^{(1)}_{l,n}] = \frac{k}{\beta}\delta_{k,-l}, \quad (2.1.8)$$

with central charge

$$c = 1 - 6\left(\left(\frac{\beta}{2}\right)^{1/2} - \left(\frac{\beta}{2}\right)^{-1/2}\right)^2 \quad \text{and} \quad c' = \left(\frac{1}{\beta} - \frac{1}{2}\right). \quad (2.1.9)$$

REMARK 1. The ${}^\beta\mathbb{J}^{(2)}_{k,n}$'s are defined by

$$ {}^\beta\mathbb{J}^{(2)}_{k,n} = \frac{\beta}{2}\sum_{i+j=k} :{}^\beta\mathbb{J}^{(1)}_{i,n}\,{}^\beta\mathbb{J}^{(1)}_{j,n}: + \left(1-\frac{\beta}{2}\right)\left((k+1)\,{}^\beta\mathbb{J}^{(1)}_{k,n} - k\mathbb{J}^{(0)}_{k,n}\right). \quad (2.1.10)$$

Componentwise, we have

$${}^\beta\mathbb{J}^{(1)}_{k,n}(t,n) = {}^\beta J^{(1)}_k + nJ^{(0)}_k \quad \text{and} \quad {}^\beta\mathbb{J}^{(0)}_{k,n} = nJ^{(0)}_k = n\delta_{0k}$$

and

$${}^\beta\mathbb{J}^{(2)}_{k,n}(t,n) = \frac{\beta}{2}\,{}^\beta J^{(2)}_k + \left(n\beta + (k+1)\left(1-\frac{\beta}{2}\right)\right){}^\beta J^{(1)}_k + n\left((n-1)\frac{\beta}{2}+1\right)J^{(0)}_k,$$

---

[8] When E equals the whole range F, then the $\mathcal{B}_k$'s are absent in the formulae (2.1.7).

where

$$^\beta \mathbb{J}_k^{(1)} = \frac{\partial}{\partial t_k} + \frac{1}{\beta}(-k)t_{-k},$$

$$^\beta \mathbb{J}_k^{(2)} = \sum_{i+j=k} \frac{\partial^2}{\partial t_i \partial t_j} + \frac{2}{\beta} \sum_{-i+j=k} it_i \frac{\partial}{\partial t_j} + \frac{1}{\beta^2} \sum_{-i-j=k} it_i jt_j. \quad (2.1.11)$$

We put n explicitly in $^\beta \mathbb{J}_{l,n}^{(2)}(t,n)$ to indicate the n-th component contains n explicitly, besides t. For $\beta = 2$, (2.1.10) becomes particularly simple:

$$^\beta \mathbb{J}_{k,n}^{(2)}|_{\beta=2} = \sum_{i+j=k} : {}^2\mathbb{J}_{i,n}^{(1)} \, {}^2\mathbb{J}_{j,n}^{(1)} : .$$

REMARK 2. The Heisenberg and Virasoro generators satisfy the following *duality* properties:

$$\frac{4}{\beta}\mathbb{J}_{l,n}^{(2)}(t,n) = {}^\beta\mathbb{J}_{l,n}^{(2)}\left(-\frac{\beta t}{2}, -\frac{2n}{\beta}\right), \quad n \in \mathbb{Z},$$

$$\frac{4}{\beta}\mathbb{J}_{l,n}^{(1)}(t,n) = -\frac{\beta}{2} \, {}^\beta\mathbb{J}_{l,n}^{(1)}\left(-\frac{\beta t}{2}, -\frac{2n}{\beta}\right), \quad n > 0. \quad (2.1.12)$$

In (2.1.7), $^\beta\mathbb{J}_{l,n}^{(2)}(-\beta t/2, -2n/\beta)$ means that the variable n, which appears in the n-th component, gets replaced by $2n/\beta$ and t by $-\beta t/2$.

REMARK 3. Theorem 2.1 states that the integrals (2.1.5) and (2.1.6) satisfy two sets of differential equations (2.1.7) respectively. Of course, the second integral also satisfies the first set of equations, with β replaced by $4/\beta$.

2.1.2. Proof: β-integrals as fixed points of vertex operators. Theorem 2.1 can be established by using the invariance of the integral under the transformation $z_i \mapsto z_i + \varepsilon f(z_i) z_i^{k+1}$ of the integration variables. However, the most transparent way to prove Theorem 2.1 is via vector vertex operators, for which the β-integrals are fixed points. This is a technique that we have used already in [2]. Indeed, define the (vector) vertex operator, for $t = (t_1, t_2, \dots) \in \mathbb{C}^\infty$, $u \in \mathbb{C}$, and setting $\chi(z) := (1, z, z^2, \dots)$:

$$\mathbb{X}_\beta(t,u) = \Lambda^{-1} e^{\sum_1^\infty t_i u^i} e^{-\beta \sum_1^\infty \frac{u^{-i}}{i} \frac{\partial}{\partial t_i}} \chi(|u|^\beta). \quad (2.1.13)$$

It acts on vectors $f(t) = (f_0(t), f_1(t), \dots)$ of functions, as follows:[9]

$$(\mathbb{X}_\beta(t,u)f(t))_n = e^{\sum_1^\infty t_i u^i} (|u|^\beta)^{n-1} f_{n-1}(t - \beta[u^{-1}]).$$

For the sake of these arguments, it is convenient to introduce the vector Virasoro generators $^\beta\mathbb{J}_k^{(i)}(t) := (^\beta\mathbb{J}_{k,n}^{(i)}(t,n))_{n\in\mathbb{Z}}$.

---

[9]For $\alpha \in \mathbb{C}$, define $[\alpha] := (\alpha, \frac{1}{2}\alpha^2, \frac{1}{3}\alpha^3, \dots) \in \mathbb{C}^\infty$. The operator Λ is the shift matrix with zeroes everywhere except for 1's just above the diagonal: $(\Lambda v)_n = v_{n+1}$.

PROPOSITION 2.2. *The multiplication operator z^k and the differential operators $\frac{\partial}{\partial z} z^{k+1}$ with $z \in \mathbb{C}^*$, acting on the vertex operator $\mathbb{X}_\beta(t,z)$, have realizations as commutators, in terms of the Heisenberg and Virasoro generators ${}^\beta \mathbb{J}_k^{(1)}(t,n)$ and ${}^\beta \mathbb{J}_k^{(2)}(t,n)$*:

$$z^k \mathbb{X}_\beta(t,z) = [{}^\beta \mathbb{J}_k^{(1)}(t), \mathbb{X}_\beta(t,z)]$$
$$\frac{\partial}{\partial z} z^{k+1} \mathbb{X}_\beta(t,z) = [{}^\beta \mathbb{J}_k^{(2)}(t), \mathbb{X}_\beta(t,z)]. \qquad (2.1.14)$$

PROOF. By explicit computation; see [3]. □

COROLLARY 2.3. *Given a weight $\rho(z)\,dz$ on $\mathbb{R}$ satisfying (2.1.1), we have*

$$\frac{\partial}{\partial z} z^{k+1} f(z) \mathbb{X}_\beta(t,z) \rho(z) = \left[\sum_{i \geq 0} \left(a_i\,{}^\beta \mathbb{J}_{k+i}^{(2)}(t) - b_i\,{}^\beta \mathbb{J}_{k+i+1}^{(1)}(t) \right),\ \mathbb{X}_\beta(t,z) \rho(z) \right]. \qquad (2.1.15)$$

PROOF. Using (2.1.14) on the last line below, compute

$$\frac{\partial}{\partial z} z^{k+1} f(z) \mathbb{X}_\beta(t,z) \rho(z)$$
$$= \left(\frac{\rho'(z)}{\rho(z)} f(z) \right) z^{k+1} \mathbb{X}_\beta(t,z) \rho(z) + \rho(z) \frac{\partial}{\partial z} \left(z^{k+1} f(z) \mathbb{X}_\beta(t,z) \right)$$
$$= -\left(\sum_0^\infty b_i z^{k+i+1} \mathbb{X}_\beta(t,z) \right) \rho(z) + \rho(z) \frac{\partial}{\partial z} \left(\sum_0^\infty a_i z^{k+i+1} \mathbb{X}_\beta(t,z) \right)$$
$$= -\left[\sum_0^\infty b_i\,{}^\beta \mathbb{J}_{k+i+1}^{(1)}, \mathbb{X}_\beta(t,z) \rho(z) \right] + \left[\sum_0^\infty a_i\,{}^\beta \mathbb{J}_{k+i}^{(2)}, \mathbb{X}_\beta(t,z) \rho(z) \right], \qquad (2.1.16)$$

establishing (2.1.15). □

Giving the weight $\rho_E(u)\,du = \rho(u) I_E(u)\,du$, with ρ and E as before, define the integrated vector vertex operator

$$\mathbb{Y}_\beta(t, \rho_E) := \int_E du\, \rho(u) \mathbb{X}_\beta(t, u), \qquad (2.1.17)$$

and the vector operator

$$\mathcal{D}_k := \mathcal{B}_k - \mathcal{V}_k$$
$$:= \sum_1^{2r} c_i^{k+1} f(c_i) \frac{\partial}{\partial c_i} - \sum_{i \geq 0} \left(a_i\,{}^\beta \mathbb{J}_{k+i}^{(2)}(t) - b_i\,{}^\beta \mathbb{J}_{k+i+1}^{(1)}(t) \right), \qquad (2.1.18)$$

consisting of a c-dependent boundary part $\mathcal{B}_k$ and a (t,n)-dependent Virasoro part $\mathcal{V}_k$.

PROPOSITION 2.4. *The following commutation relation holds:*

$$[\mathcal{D}_k, \mathbb{Y}_\beta(t, \rho_E)] = 0. \qquad (2.1.19)$$

PROOF. Integrating both sides of (2.1.15) over E, one computes:

$$\int_E dz \frac{\partial}{\partial z}\left(z^{k+1} f(z) \mathbb{X}_\beta(t,z)\rho(z)\right) = \sum_1^{2r}(-1)^i c_i^{k+1} f(c_i) \mathbb{X}_\beta(t,c_i)\rho(c_i)$$

$$= \sum_1^{2r} c_i^{k+1} f(c_i) \frac{\partial}{\partial c_i} \int_E \mathbb{X}_\beta(t,z)\rho(z)\,dz$$

$$= \left[\mathcal{B}_k, \mathbb{Y}_\beta(t,\rho_E)\right], \qquad (2.1.20)$$

while, on the other hand,

$$\int_E dz\left[\sum_{i\geq 0}\left(a_i{}^\beta \mathbb{J}^{(2)}_{k+i} - b_i{}^\beta \mathbb{J}^{(1)}_{k+i+1}\right), \mathbb{X}_\beta(t,z)\rho(z)\right]$$

$$= \left[\sum_{i\geq 0}\left(a_i{}^\beta \mathbb{J}^{(2)}_{k+i} - b_i{}^\beta \mathbb{J}^{(1)}_{k+i+1}\right), \int_\mathbb{R} dz\rho_E(z)\mathbb{X}_\beta(t,z)\right]$$

$$= \left[\mathcal{V}_k, \mathbb{Y}_\beta(t,\rho_E)\right]. \qquad (2.1.21)$$

Subtracting both expressions (2.1.20) and (2.1.21) yields

$$0 = \left[\mathcal{B}_k - \mathcal{V}_k, \mathbb{Y}_\beta(t,\rho_E)\right] = \left[\mathcal{D}_k, \mathbb{Y}_\beta(t,\rho_E)\right]. \qquad \square$$

PROPOSITION 2.5. *The column vector*

$$I(t) := \left(\int_{E^n} |\Delta_n(z)|^\beta \prod_{k=1}^n e^{\sum_1^\infty t_i z_k^i} \rho(z_k)\,dz_k\right)_{n\geq 0}$$

is a fixed point for the vertex operator $\mathbb{Y}_\beta(t,\rho_E)$ *(see definition (2.1.17)):*

$$(\mathbb{Y}_\beta(t,\rho_E)I)_n = I_n, \quad n \geq 1. \qquad (2.1.22)$$

PROOF. We have

$$I_n(t) = \int_{\mathbb{R}^n} |\Delta_n(z)|^\beta \prod_{k=1}^n \left(e^{\sum_1^\infty t_i z_k^i} I_E(z_k)\rho(z_k)\,dz_k\right)$$

$$= \int_\mathbb{R} du\,\rho_E(u) e^{\sum_1^\infty t_i u^i} |u|^{\beta(n-1)}$$

$$\times \int_{\mathbb{R}^{n-1}} \prod_{k=1}^{n-1}\left|1 - \frac{z_k}{u}\right|^\beta |\Delta_{n-1}(z)|^\beta \prod_{k=1}^{n-1}\left(e^{\sum_1^\infty t_i z_k^i}\rho_E(z_k)\,dz_k\right)$$

$$= \int_\mathbb{R} du\,\rho_E(u) e^{\sum_1^\infty t_i u^i} |u|^{\beta(n-1)}$$

$$\times e^{-\beta \sum_1^\infty \frac{u^{-i}}{i}\frac{\partial}{\partial t_i}} \int_{\mathbb{R}^{n-1}} |\Delta_{n-1}(z)|^\beta \prod_{k=1}^{n-1}\left(e^{\sum_1^\infty t_i z_k^i}\rho_E(z_k)\,dz_k\right)$$

$$= \int_{\mathbb{R}} du\, \rho_E(u)|u|^{\beta(n-1)} e^{\sum_1^\infty t_i u^i} e^{-\beta \sum_1^\infty \frac{u^{-i}}{i} \frac{\partial}{\partial t_i}} I_{n-1}(t)$$

$$= \left(\mathbb{Y}_\beta(t,\rho_E) I(t)\right)_n. \tag{2.1.23}$$

□

PROOF OF THEOREM 2.1. Proposition 2.4 implies that, for $n \geq 1$,

$$0 = [\mathcal{D}_k, (\mathbb{Y}_\beta(t,\rho_E))^n] I$$
$$= \mathcal{D}_k \mathbb{Y}_\beta(t,\rho_E)^n I - \mathbb{Y}_\beta(t,\rho_E)^n \mathcal{D}_k I. \tag{2.1.24}$$

Taking the n-th component for $n \geq 1$ and $k \geq -1$, setting

$$X_\beta(t,u) = e^{\sum t_i u^i} e^{-\beta \sum \frac{u^{-i}}{i} \frac{\partial}{\partial t_i}},$$

and using (2.1.22), we get

$$0 = \left(\mathcal{D}_k I - \mathbb{Y}_\beta(t,\rho_E)^n \mathcal{D}_k I\right)_n$$
$$= (\mathcal{D}_k I)_n - \int du\, \rho_E(u) X_\beta(t;u)(|u|^\beta)^{n-1} \ldots \int du\, \rho_E(u) X_\beta(t;u)(\mathcal{D}_k I)_0$$
$$= (\mathcal{D}_k I)_n.$$

Indeed $(\mathcal{D}_k I)_0 = 0$ for $k \geq -1$, since $\tau_0 = 1$ and $\mathcal{D}_k$ involves ${}^\beta\mathbb{J}_k^{(2)}$, ${}^\beta\mathbb{J}_k^{(1)}$ and $\mathbb{J}_k^{(0)}$ for $k \geq -1$:

${}^\beta\mathbb{J}_k^{(2)}$ is pure differentiation for $k \geq -1$;

${}^\beta\mathbb{J}_k^{(1)}$ is pure differentiation, except for $k = -1$; but

${}^\beta\mathbb{J}_{-1}^{(1)}$ appears with coefficient $n\beta$, which vanishes for $n = 0$;

$\mathbb{J}_k^{(0)}$ appears with coefficient $n((n-1)\frac{\beta}{2} + 1)$, vanishing for $n = 0$. □

2.1.3. Examples

Example 1: Gaussian β-integrals. The weight and the a_i and b_i, as in (2.1.1), are given by

$$\rho(z) = e^{-V(z)} = e^{-z^2}, \quad V' = g/f = 2z,$$

$a_0 = 1$, $b_0 = 0$, $b_1 = 2$, and all other $a_i, b_i = 0$.

From (2.1.7), the integrals

$$I_n = \int_{E^n} |\Delta_n(z)|^\beta \prod_{k=1}^n e^{-z_k^2 + \sum_{i=1}^\infty t_i z_k^i}\, dz_k \tag{2.1.25}$$

satisfy the Virasoro constraints, for $k \geq -1$,

$$-\mathcal{B}_k I_n = -\sum_1^{2r} c_i^{k+1} \frac{\partial}{\partial c_i} I_n = \left(-{}^\beta\mathbb{J}_{k,n}^{(2)} + 2\,{}^\beta\mathbb{J}_{k+2,n}^{(1)}\right) I_n. \tag{2.1.26}$$

Introducing the notation
$$\sigma_i = \left(n - \frac{i+1}{2}\right)\beta + i + 1 - b_0 = \left(n - \frac{i+1}{2}\right)\beta + i + 1, \qquad (2.1.27)$$
and setting $F_n = \log I_n$, the first three constraints have the form
$$-\mathcal{B}_{-1}F = \left(2\frac{\partial}{\partial t_1} - \sum_{i \geq 2} it_i \frac{\partial}{\partial t_{i-1}}\right)F - nt_1,$$
$$-\mathcal{B}_0 F = \left(2\frac{\partial}{\partial t_2} - \sum_{i \geq 1} it_i \frac{\partial}{\partial t_i}\right)F - \frac{n}{2}\sigma_1,$$
$$-\mathcal{B}_1 F = \left(2\frac{\partial}{\partial t_3} - \sigma_1 \frac{\partial}{\partial t_1} - \sum_{i \geq 1} it_i \frac{\partial}{\partial t_{i+1}}\right)F. \qquad (2.1.28)$$

For later use, take the linear combinations
$$\mathcal{D}_1 = -\tfrac{1}{2}\mathcal{B}_{-1}, \quad \mathcal{D}_2 = -\tfrac{1}{2}\mathcal{B}_0, \quad \mathcal{D}_3 = -\tfrac{1}{2}(\mathcal{B}_1 + \tfrac{1}{2}\sigma_1 \mathcal{B}_{-1}), \qquad (2.1.29)$$
such that each $\mathcal{D}_i$ contains the pure term $\partial F/\partial t_i$, i.e., $\mathcal{D}_i F = (\partial F/\partial t_i) + \cdots$.

Example 2: Laguerre β-integrals. Here the weight and the a_i and b_i, as in (2.1.1), are given by
$$e^{-V} = z^a e^{-z}, \quad V' = \frac{g}{f} = \frac{z-a}{z},$$
$a_0 = 0$, $a_1 = 1$, $b_0 = -a$, $b_1 = 1$, and all other $a_i, b_i = 0$. Thus, from Theorem 2.1, the integrals
$$I_n = \int_{E^n} |\Delta_n(z)|^\beta \prod_{k=1}^n z_k^a e^{-z_k + \sum_{i=1}^\infty t_i z_k^i} \, dz_k \qquad (2.1.30)$$
satisfy the Virasoro constraints, for $k \geq -1$:
$$-\mathcal{B}_k I_n = -\sum_1^{2r} c_i^{k+2} \frac{\partial}{\partial c_i} I_n = \left(-{}^\beta \mathbb{J}^{(2)}_{k+1,n} - a\, {}^\beta \mathbb{J}^{(1)}_{k+1,n} + {}^\beta \mathbb{J}^{(1)}_{k+2,n}\right) I_n. \qquad (2.1.31)$$

Introducing the notation
$$\sigma_i = = \left(n - \frac{i+1}{2}\right)\beta + i + 1 - b_0 = \left(n - \frac{i+1}{2}\right)\beta + i + 1 + a,$$
and setting $F = F_n = \log I_n$, the first three have the form
$$-\mathcal{B}_{-1}F = \left(\frac{\partial}{\partial t_1} - \sum_{i \geq 1} it_i \frac{\partial}{\partial t_i}\right)F - \frac{n}{2}(\sigma_1 + a),$$
$$-\mathcal{B}_0 F = \left(\frac{\partial}{\partial t_2} - \sigma_1 \frac{\partial}{\partial t_1} - \sum_{i \geq 1} it_i \frac{\partial}{\partial t_{i+1}}\right)F$$
$$-\mathcal{B}_1 F = \left(\frac{\partial}{\partial t_3} - \sigma_2 \frac{\partial}{\partial t_2} - \sum_{i \geq 1} it_i \frac{\partial}{\partial t_{i+2}} - \frac{\beta}{2}\frac{\partial^2}{\partial t_1^2}\right)F - \frac{\beta}{2}\left(\frac{\partial F}{\partial t_1}\right)^2.$$

Again, replace the operators $\mathcal{B}_i$ by linear combinations $\mathcal{D}_i$, such that $\mathcal{D}_i F = (\partial F/\partial t_i) + \cdots$:

$$\mathcal{D}_1 = -\mathcal{B}_{-1}, \quad \mathcal{D}_2 = -\mathcal{B}_0 - \sigma_1 \mathcal{B}_{-1}, \quad \mathcal{D}_3 = -\mathcal{B}_1 - \sigma_2 \mathcal{B}_0 - \sigma_1 \sigma_2 \mathcal{B}_{-1}. \quad (2.1.32)$$

Example 3: Jacobi β-integral. This case is particularly important, because it covers not only the first integral, but also the third integral in the list (2.0.1), used in the problem of random permutations. The weight and the a_i and b_i, as in (2.1.1), are given by

$$\rho(z) := e^{-V} = (1-z)^a (1+z)^b, \quad V' = \frac{g}{f} = \frac{a-b+(a+b)z}{1-z^2},$$

$a_0 = 1$, $a_1 = 0$, $a_2 = -1$, $b_0 = a-b$, $b_1 = a+b$, and all other $a_i, b_i = 0$. The integrals

$$I_n = \int_{E^n} |\Delta_n(z)|^\beta \prod_{k=1}^n (1-z_k)^a (1+z_k)^b e^{\sum_{i=1}^\infty t_i z_k^i} dz_k \quad (2.1.33)$$

satisfy the Virasoro constraints ($k \geq -1$):

$$-\mathcal{B}_k I_n = -\sum_{1}^{2r} c_i^{k+1}(1-c_i^2)\frac{\partial}{\partial c_i} I_n$$
$$= \left({}^\beta\mathbb{J}_{k+2,n}^{(2)} - {}^\beta\mathbb{J}_{k,n}^{(2)} + b_0\, {}^\beta\mathbb{J}_{k+1,n}^{(1)} + b_1\, {}^\beta\mathbb{J}_{k+2,n}^{(1)}\right) I_n. \quad (2.1.34)$$

Introducing $\sigma_i = \left(n - \frac{i+1}{2}\right)\beta + i + 1 + b_1$, the first four ($k = -1, 0, 1, 2$) have the form

$$-\mathcal{B}_{-1} F = \left(\sigma_1 \frac{\partial}{\partial t_1} + \sum_{i\geq 1} it_i \frac{\partial}{\partial t_{i+1}} - \sum_{i\geq 2} it_i \frac{\partial}{\partial t_{i-1}}\right) F + n(b_0 - t_1),$$

$$-\mathcal{B}_0 F = \left(\sigma_2 \frac{\partial}{\partial t_2} + b_0 \frac{\partial}{\partial t_1} + \sum_{i\geq 1} it_i \left(\frac{\partial}{\partial t_{i+2}} - \frac{\partial}{\partial t_i}\right) + \frac{\beta}{2}\frac{\partial^2}{\partial t_1^2}\right) F + \frac{\beta}{2}\left(\frac{\partial F}{\partial t_1}\right)^2$$
$$- \frac{n}{2}(\sigma_1 - b_1),$$

$$-\mathcal{B}_1 F = \left(\sigma_3 \frac{\partial}{\partial t_3} + b_0 \frac{\partial}{\partial t_2} - (\sigma_1 - b_1)\frac{\partial}{\partial t_1} + \sum_{i\geq 1} it_i \left(\frac{\partial}{\partial t_{i+3}} - \frac{\partial}{\partial t_{i+1}}\right) + \beta \frac{\partial^2}{\partial t_1 \partial t_2}\right) F$$
$$+ \beta \frac{\partial F}{\partial t_1} \frac{\partial F}{\partial t_2},$$

$$-\mathcal{B}_2 F = \left(\sigma_4 \frac{\partial}{\partial t_4} + b_0 \frac{\partial}{\partial t_3} - (\sigma_2 - b_1)\frac{\partial}{\partial t_2} + \sum_{i\geq 1} it_i \left(\frac{\partial}{\partial t_{i+4}} - \frac{\partial}{\partial t_{i+2}}\right)\right.$$
$$\left.+ \frac{\beta}{2}\left(\frac{\partial^2}{\partial t_2^2} - \frac{\partial^2}{\partial t_1^2} + 2\frac{\partial^2}{\partial t_1 \partial t_3}\right)\right) F + \frac{\beta}{2}\left(\left(\frac{\partial F}{\partial t_2}\right)^2 - \left(\frac{\partial F}{\partial t_1}\right)^2 + 2\frac{\partial F}{\partial t_1}\frac{\partial F}{\partial t_3}\right).$$

$$(2.1.35)$$

2.2. Double Matrix Integrals.
Consider now weights of the form

$$\rho(x,y) = e^{\sum_{i,j\geq 1} r_{ij} x^i y^j} \rho(x)\tilde{\rho}(y), \qquad (2.2.1)$$

defined on a product of intervals $F_1 \times F_2 \subset \mathbb{R}^2$, with rational logarithmic derivative

$$-\frac{\rho'}{\rho} = \frac{g}{f} = \frac{\sum_{i\geq 0} b_i x^i}{\sum_{i\geq 0} a_i x^i} \quad \text{and} \quad -\frac{\tilde{\rho}'}{\tilde{\rho}} = \frac{\tilde{g}}{\tilde{f}} = \frac{\sum_{i\geq 0} \tilde{b}_i y^i}{\sum_{i\geq 0} \tilde{a}_i y^i},$$

satisfying

$$\lim_{x\to\partial F_1} f(x)\rho(x)x^k = \lim_{y\to\partial F_2} \tilde{f}(y)\tilde{\rho}(y)y^k = 0 \quad \text{for all } k \geq 0. \qquad (2.2.2)$$

Consider subsets of the form

$$E = E_1 \times E_2 := \bigcup_{i=1}^{r}[c_{2i-1}, c_{2i}] \times \bigcup_{i=1}^{s}[\tilde{c}_{2i-1}, \tilde{c}_{2i}] \subset F_1 \times F_2 \subset \mathbb{R}^2. \qquad (2.2.3)$$

A natural deformation of the second integral in the list (2.0.1) is given by the following integrals:

$$I_n(t,s,r;E) = \iint_{E^n} \Delta_n(x)\Delta_n(y) \prod_{k=1}^{n} e^{\sum_{i=1}^{\infty}(t_i x_k^i - s_i y_k^i)} \rho(x_k, y_k)\, dx_k\, dy_k \qquad (2.2.4)$$

In the theorem below, $\mathbb{J}_{k,n}^{(i)}$ and $\tilde{\mathbb{J}}_{k,n}^{(i)}$ are vectors of operators, whose components are given by the operators (2.1.10) for $\beta = 1$; i.e.,

$$\mathbb{J}_{k,n}^{(i)}(t) = {}^{\beta}\mathbb{J}_{k,n}^{(i)}(t)\big|_{\beta=1}, \quad \tilde{\mathbb{J}}_{k,n}^{(i)}(s) := {}^{\beta}\mathbb{J}_{k,n}^{(i)}(t)\big|_{\beta=1,\, t\mapsto -s};$$

thus, from (2.1.10) and (2.1.11), one finds

$$\mathbb{J}_{k,n}^{(2)}(t) = \frac{1}{2}\left(J_k^{(2)}(t) + (2n+k+1)J_k^{(1)}(t) + n(n+1)J_k^{(0)}\right), \qquad (2.2.5)$$

satisfying the Heisenberg and Virasoro relations (2.1.8), with *central charge* $c = -2$ and $c' = \frac{1}{2}$.

The $a_i, \tilde{a}_i, b_i, \tilde{b}_i, c_i, \tilde{c}_i, r_{ij}$ given by (2.2.1), (2.2.2) and (2.2.3) define differential operators

$$\mathcal{D}_{k,n} := \sum_{1}^{2r} c_i^{k+1} f(c_i) \frac{\partial}{\partial c_i} - \sum_{i\geq 0}\left(a_i\left(\mathbb{J}_{k+i,n}^{(2)} + \sum_{m,l\geq 1} m r_{ml} \frac{\partial}{\partial r_{m+k+i,l}}\right) - b_i \mathbb{J}_{k+i+1,n}^{(1)}\right),$$

$$\tilde{\mathcal{D}}_{k,n} := \sum_{1}^{2r} \tilde{c}_i^{k+1} \tilde{f}(\tilde{c}_i) \frac{\partial}{\partial \tilde{c}_i} - \sum_{i\geq 0}\left(\tilde{a}_i\left(\tilde{\mathbb{J}}_{k+i,n}^{(2)} + \sum_{m,l\geq 1} l r_{ml} \frac{\partial}{\partial r_{m,l+k+i}}\right) - \tilde{b}_i \tilde{\mathbb{J}}_{k+i+1,n}^{(1)}\right).$$

$$(2.2.6)$$

THEOREM 2.6 (Adler and van Moerbeke [3; 4]). *Given $\rho(x,y)$ as in (2.2.1), the integrals*

$$I_n(t,s,r;E) := \iint_{E^n} \Delta_n(x)\Delta_n(y) \prod_{k=1}^n e^{\sum_{i=1}^\infty (t_i x_k^i - s_i y_k^i)} \rho(x_k, y_k)\, dx_k\, dy_k \quad (2.2.7)$$

satisfy two families of Virasoro equations for $k \geq -1$:

$$\mathcal{D}_{k,n} I_n(t,s,r;E) = 0 \quad \text{and} \quad \tilde{\mathcal{D}}_{k,n} I_n(t,s,r;E) = 0. \quad (2.2.8)$$

PROOF. The proof of this statement is very similar to the one for β-integrals. Namely, define the vector vertex operator,

$$\mathbb{X}_{12}(t,s;u,v) = \Lambda^{-1} e^{\sum_1^\infty (t_i u^i - s_i v^i)} e^{-\sum_1^\infty (\frac{u^{-i}}{i} \frac{\partial}{\partial t_i} - \frac{v^{-i}}{i} \frac{\partial}{\partial s_i})} \chi(uv), \quad (2.2.9)$$

which, as a consequence of Proposition 2.2 for $\beta = 1$, interacts with the operators $\mathbb{J}_k^{(i)}(t) = \left(\mathbb{J}_{k,n}^{(i)}(t,n)\right)_{n \in \mathbb{Z}}$ as follows:

$$u^k \mathbb{X}_{12}(t,s;u,v) = [\mathbb{J}_k^{(1)}(t), \mathbb{X}_{12}(t,s;u,v)]$$

$$\frac{\partial}{\partial u} u^{k+1} \mathbb{X}_{12}(t,s;u,v) = [\mathbb{J}_k^{(2)}(t), \mathbb{X}_{12}(t,s;u,v)]. \quad (2.2.10)$$

A similar statement can be made, upon replacing the operators u^k and $\frac{\partial}{\partial u} u^{k+1}$ by v^k and $\frac{\partial}{\partial v} v^{k+1}$, and upon using the $\tilde{\mathbb{J}}_k^{(i)}(s)$'s.

Finally, one checks that the integral vertex operator

$$\mathbb{Y}(t,s;\rho_E) := \iint_E dx\, dy\, \rho(x,y) \mathbb{X}_{12}(t,s;x,y) \quad (2.2.11)$$

commutes with the two vectors of differential operators $\mathcal{D}_k = (\mathcal{D}_{k,n})_{n \in \mathbb{Z}}$, as in (2.2.6):

$$[\mathcal{D}_k, \mathbb{Y}(t,s;\rho_E)] = [\tilde{\mathcal{D}}_k, \mathbb{Y}(t,s;\rho_E)] = 0,$$

and that the vector $I = (I_0 = 1, I_1, \ldots)$ of integrals (2.2.7) is a fixed point for $\mathbb{Y}(t,s;\rho_E)$,

$$\mathbb{Y}(t,s;\rho_E) I(t,s,r;E) = I(t,s,r;E).$$

Then, as before, the proof of Theorem 2.6 hinges ultimately on the fact that $\mathcal{D}_{k,0}$ annihilates $I_0 = 1$. □

2.3. Integrals over the Unit Circle. We now deal with the fourth type of integral in the list (2.0.1), which we deform, this time, by inserting two sequences of times $t_1, t_2, \ldots$ and $s_1, s_2, \ldots$. The following theorem holds:

THEOREM 2.7 (Adler and van Moerbeke [7]). *The multiple integrals over the unit circle S^1,*

$$I_n(t,s) = \int_{(S^1)^n} |\Delta_n(z)|^2 \prod_{k=1}^n e^{\sum_1^\infty (t_i z_k^i - s_i z_k^{-i})} \frac{dz_k}{2\pi i z_k}, \quad n > 0, \quad (2.3.1)$$

with $I_0 = 1$, satisfy an $SL(2,\mathbb{Z})$-algebra of Virasoro constraints:

$$\mathcal{D}_{k,n}^\theta I_n(t,s) = 0, \quad \text{only for} \quad \begin{cases} k = -1, & \theta = 0, \\ k = 0, & \theta \text{ arbitrary}, \\ k = 1, & \theta = 1, \end{cases} \quad (2.3.2)$$

where the operators $\mathcal{D}_{k,n}^\theta := \mathcal{D}_{k,n}^\theta(t,s,n)$, $k \in \mathbb{Z}$, $n \geq 0$ are given by

$$\mathcal{D}_{k,n}^\theta := \mathbb{J}_{k,n}^{(2)}(t,n) - \mathbb{J}_{-k,n}^{(2)}(-s,n) - k\bigl(\theta \mathbb{J}_{k,n}^{(1)}(t,n) + (1-\theta)\mathbb{J}_{-k,n}^{(1)}(-s,n)\bigr), \quad (2.3.3)$$

with $\mathbb{J}_{k,n}^{(i)}(t,n) := {}^\beta\mathbb{J}_{k,n}^{(i)}(t,n)\bigr|_{\beta=1}$, as in (2.1.11).

The explicit expressions are

$$\mathcal{D}_{-1} I_n = \left(\sum_{i \geq 1}(i+1)t_{i+1}\frac{\partial}{\partial t_i} - \sum_{i \geq 2}(i-1)s_{i-1}\frac{\partial}{\partial s_i} + n\left(t_1 + \frac{\partial}{\partial s_1}\right)\right) I_n = 0$$

$$\mathcal{D}_0 I_n = \sum_{i \geq 1}\left(it_i\frac{\partial}{\partial t_i} - is_i\frac{\partial}{\partial s_i}\right) I_n = 0 \quad (2.3.4)$$

$$\mathcal{D}_1 I_n = \left(-\sum_{i \geq 1}(i+1)s_{i+1}\frac{\partial}{\partial s_i} + \sum_{i \geq 2}(i-1)t_{i-1}\frac{\partial}{\partial t_i} + n\left(s_1 + \frac{\partial}{\partial t_1}\right)\right) I_n = 0.$$

Here the key vertex operator is a reduction of $\mathbb{X}_{12}(t,s;u,v)$, defined in the previous section (formula (2.2.9)). For all $k \in \mathbb{Z}$, the vector of operators

$$\mathcal{D}_k^\theta(t,s) = \bigl(\mathcal{D}_{k,n}^\theta(t,s,n)\bigr)_{n \in \mathbb{Z}}$$

forms a realization of the first order differential operators $(d/du)u^{k+1}$, using the vertex operator $\mathbb{X}_{12}(t,s;u,u^{-1})$, namely

$$\frac{d}{du}u^{k+1}\frac{\mathbb{X}_{12}(t,s;u,u^{-1})}{u} = \left[\mathcal{D}_k^\theta(t,s), \frac{\mathbb{X}_{12}(t,s;u,u^{-1})}{u}\right]. \quad (2.3.5)$$

Indeed,

$$u\frac{d}{du}u^k \mathbb{X}_{12}(t,s;u,u^{-1})$$

$$= \left(\frac{\partial}{\partial u}u^{k+1} - \frac{\partial}{\partial v}v^{1-k} - k\theta u^k - k(1-\theta)v^{-k}\right)\mathbb{X}_{12}(t,s;u,v)\Bigr|_{v=-u}$$

$$= \bigl[\mathbb{J}_k^{(2)}(t) - \mathbb{J}_{-k}^{(2)}(-s) - k\bigl(\theta \mathbb{J}_k^{(1)}(t) + (1-\theta)\mathbb{J}_k^{(1)}(-s)\bigr), \mathbb{X}_{12}(t,s;u,-u)\bigr]$$

$$= \bigl[\mathcal{D}_k^\theta(t,s), \mathbb{X}_{12}(t,s;u,u^{-1})\bigr].$$

The $\mathcal{D}_k^\theta := \mathcal{D}_k^\theta(t,s)$ satisfy Virasoro relations with central charge zero:

$$[\mathcal{D}_k^\theta, \mathcal{D}_l^\theta] = (k-l)\mathcal{D}_{k+l}^\theta; \quad (2.3.6)$$

thus, from (2.3.5) we have the commutation relation

$$[\mathcal{D}_k^\theta(t,s), \mathbb{Y}(t,s)] = 0, \quad \text{with } \mathbb{Y}(t,s) := \int_{S^1} \frac{du}{2\pi i u}\mathbb{X}_{12}(t,s;u,u^{-1}). \quad (2.3.7)$$

The point is that the column vector $I(t,s) = (I_0, I_1, \dots)$ of integrals (2.3.1) is a fixed point for $\mathbb{Y}(t,s)$:

$$(\mathbb{Y}(t,s)I)_n = I_n, \quad n \geq 1, \tag{2.3.8}$$

which is shown in a way similar to Proposition 2.5.

PROOF OF THEOREM 2.7. Here again the proof is similar to the one of Theorem 2.1. Taking the n-th component and the n-th power of $\mathbb{Y}(t,s)$, with $n \geq 1$, and noticing from the explicit formulae (2.3.4) that $\left(\mathcal{D}_k^\theta(t,s)I\right)_0 = 0$, we have, by means of a calculation similar to the proof of Theorem 2.1,

$$\begin{aligned} 0 &= \left([\mathcal{D}_k^\theta, \mathbb{Y}(t,s)^n]I\right)_n \\ &= \left(\mathcal{D}_k^\theta \mathbb{Y}(t,s)^n I - \mathbb{Y}(t,s)^n \mathcal{D}_k^\theta I\right)_n \\ &= \left(\mathcal{D}_k^\theta I - \mathbb{Y}(t,s)^n \mathcal{D}_k^\theta I\right)_n = \left(\mathcal{D}_k^\theta I\right)_n. \end{aligned}$$

$\square$

3. Integrable Systems and Associated Matrix Integrals

3.1. Toda lattice and Hermitian matrix integrals

3.1.1. Toda lattice, factorization of symmetric matrices and orthogonal polynomials. Given a weight $\rho(z) = e^{-V(z)}$ defined as in (2.1.1), the inner product over $E \subseteq \mathbb{R}$,

$$\langle f, g \rangle_t = \int_E f(z)g(z)\rho_t(z)\,dz, \quad \text{with } \rho_t(z) := e^{\sum t_i z^i}\rho(z), \tag{3.1.1}$$

leads to a moment matrix

$$m_n(t) = (\mu_{ij}(t))_{0 \leq i,j < n} = (\langle z^i, z^j \rangle_t)_{0 \leq i,j < n}, \tag{3.1.2}$$

which is a *Hänkel matrix*[10], thus symmetric. This is tantamount to $\Lambda m_\infty = m_\infty \Lambda^\top$, where Λ denotes the shift matrix; see footnote 9. As easily seen, the semi-infinite moment matrix m_∞ evolves in t according to the equations

$$\frac{\partial \mu_{ij}}{\partial t_k} = \mu_{i+k,j}, \text{ and thus } \frac{\partial m_\infty}{\partial t_k} = \Lambda^k m_\infty. \quad \begin{pmatrix} \text{commuting} \\ \text{vector fields} \end{pmatrix} \tag{3.1.3}$$

Another important ingredient is the factorization of m_∞ into a lower-triangular times an upper-triangular matrix[11]

$$m_\infty(t) = S(t)^{-1} S(t)^{\top -1},$$

where $S(t)$ is lower triangular with non-zero diagonal elements.

The main ideas of the following theorem can be found in [2; 5]. Remember that $c = (c_1, \dots, c_{2r})$ denotes the boundary points of the set E; further, dM refers to properly normalized Haar measure on $\mathcal{H}_n$.

---

[10] Hänkel means that μ_{ij} depends on $i+j$ only.

[11] This factorization is possible for those t's for which $\tau_n(t) := \det m_n(t) \neq 0$ for all $n > 0$.

THEOREM 3.1. *The determinants of the moment matrices*

$$\tau_n(t,c) := \det m_n(t,c) = \frac{1}{n!} \int_{E^n} \Delta_n^2(z) \prod_{k=1}^n \rho_t(z_k)\, dz_k$$

$$= \int_{\mathcal{H}_n(E)} e^{tr(-V(M) + \sum_1^\infty t_i M^i)}\, dM, \qquad (3.1.4)$$

satisfy the following relations:

(i) *Virasoro constraints* (2.1.7) *for* $\beta = 2$,

$$\left(-\sum_1^{2r} c_i^{k+1} f(c_i) \frac{\partial}{\partial c_i} + \sum_{i \geq 0} \left(a_i \, \mathbb{J}^{(2)}_{k+i,n} - b_i \, \mathbb{J}^{(1)}_{k+i+1,n} \right) \right) \tau_n(t,c) = 0. \qquad (3.1.5)$$

(ii) *The KP-hierarchy*[12] ($k = 0,1,2,\dots$)

$$\left(\mathbf{s}_{k+4} \left(\frac{\partial}{\partial t_1}, \frac{1}{2}\frac{\partial}{\partial t_2}, \frac{1}{3}\frac{\partial}{\partial t_3}, \dots \right) - \frac{1}{2}\frac{\partial^2}{\partial t_1 \partial t_{k+3}} \right) \tau_n \circ \tau_n = 0,$$

of which the first equation reads:

$$\left(\left(\frac{\partial}{\partial t_1}\right)^4 + 3\left(\frac{\partial}{\partial t_2}\right)^2 - 4\frac{\partial^2}{\partial t_1 \partial t_3} \right) \log \tau_n + 6 \left(\frac{\partial^2}{\partial t_1^2} \log \tau_n \right)^2 = 0. \qquad (3.1.6)$$

(iii) *The standard Toda lattice, i.e., the symmetric tridiagonal matrix*

$$L(t) := S(t)\Lambda S(t)^{-1} = \begin{pmatrix} \frac{\partial}{\partial t_1} \log \frac{\tau_1}{\tau_0} & \left(\frac{\tau_0 \tau_2}{\tau_1^2}\right)^{1/2} & 0 & \\ \left(\frac{\tau_0 \tau_2}{\tau_1^2}\right)^{1/2} & \frac{\partial}{\partial t_1} \log \frac{\tau_2}{\tau_1} & \left(\frac{\tau_1 \tau_3}{\tau_2^2}\right)^{1/2} & \\ 0 & \left(\frac{\tau_1 \tau_3}{\tau_2^2}\right)^{1/2} & \frac{\partial}{\partial t_1} \log \frac{\tau_3}{\tau_2} & \\ & & & \ddots \end{pmatrix} \qquad (3.1.7)$$

satisfies the commuting equations[13]

$$\frac{\partial L}{\partial t_k} = [\tfrac{1}{2}(L^k)_s, L]. \qquad (3.1.8)$$

(iv) *Eigenvectors of* L: *The tridiagonal matrix* L *admits two independent eigenvectors:*

---

[12] Given a polynomial $p(t_1, t_2, \dots)$, define the customary Hirota symbol
$$p(\partial_t) f \circ g := p\left(\frac{\partial}{\partial y_1}, \frac{\partial}{\partial y_2}, \dots\right) f(t+y)g(t-y)\Big|_{y=0}.$$
The $\mathbf{s}_l$'s are the elementary Schur polynomials $e^{\sum_1^\infty t_i z^i} := \sum_{i \geq 0} \mathbf{s}_i(t) z^i$. For later use, set
$$\mathbf{s}_l(\tilde{\partial}) := \mathbf{s}_l\left(\frac{\partial}{\partial t_1}, \frac{1}{2}\frac{\partial}{\partial t_2}, \dots\right).$$

[13] The notation $(\ )_s$ means the skew-symmetric part of $(\ )$; for details, see Section 3.1.2.

- $p(t;z) = (p_n(t;z))_{n\geq 0}$ satisfying $(L(t)p(t;z))_n = zp_n(t;z)$, $n \geq 0$, where $p_n(t;z)$ are n-th degree polynomials in z, depending on $t \in \mathbb{C}^\infty$, orthonormal with respect to the t-dependent inner product[14] (3.1.1)

$$\langle p_k(t;z), p_l(t;z)\rangle_t = \delta_{kl};$$

they are eigenvectors of L, i.e., $L(t)p(t;z) = zp(t;z)$, and enjoy the following representations, where $\chi(z) = (1, z, z^2, \dots)^\top$:

$$p_n(t;z) := (S(t)\chi(z))_n = \frac{1}{\sqrt{\tau_n(t)\tau_{n+1}(t)}} \det \begin{pmatrix} & & & 1 \\ & m_n(t) & & z \\ & & & \vdots \\ \mu_{n,0} & \cdots & \mu_{n,n-1} & z^n \end{pmatrix}$$

$$= z^n h_n^{-1/2} \frac{\tau_n(t-[z^{-1}])}{\tau_n(t)}, \quad \text{with } h_n := \frac{\tau_{n+1}(t)}{\tau_n(t)}. \qquad (3.1.9)$$

- $q(t,z) = (q_n(t;z))_{n\geq 0}$, with

$$q_n(t;z) := z \int_{\mathbb{R}^n} \frac{p_n(t;u)}{z-u} \rho_t(u)\, du,$$

satisfying $(L(t)q(t;z))_n = zq_n(t;z)$ for $n \geq 1$; the function $q_n(t;z)$ enjoys the representations

$$q_n(t;z) = \left(S^{\top-1}(t)\chi(z^{-1})\right)_n = \left(S(t)m_\infty(t)\chi(z^{-1})\right)_n$$

$$= z^{-n} h_n^{-1/2} \frac{\tau_{n+1}(t+[z^{-1}])}{\tau_n(t)}. \qquad (3.1.10)$$

In the case $\beta = 2$, the Virasoro generators (2.1.11) take on a particularly elegant form, namely for $n \geq 0$,

$$\mathbb{J}_{k,n}^{(2)}(t) = \sum_{i+j=k} : \mathbb{J}_{i,n}^{(1)}(t)\, \mathbb{J}_{j,n}^{(1)}(t) := J_k^{(2)}(t) + 2n J_k^{(1)}(t) + n^2 \delta_{0k}$$

$$\mathbb{J}_{k,n}^{(1)}(t) = J_k^{(1)}(t) + n\delta_{0k},$$

with[15]

$$J_k^{(1)} = \frac{\partial}{\partial t_k} + \tfrac{1}{2}(-k)t_{-k},$$

$$J_k^{(2)} = \sum_{i+j=k} \frac{\partial^2}{\partial t_i \partial t_j} + \sum_{-i+j=k} it_i \frac{\partial}{\partial t_j} + \tfrac{1}{4}\sum_{-i-j=k} it_i jt_j. \qquad (3.1.11)$$

Statement (i) is already contained in Section 2, whereas statement (ii) will be established in Section 3.1.2, using elementary methods.

---

[14] The explicit dependence on the boundary points c will be omitted in this point (iv).
[15] The expression $J_k^{(1)}$ vanishes for $k = 0$.

3.1.2. Sketch of Proof

Orthogonal polynomials and τ-function representation: Representation (3.1.4) of the determinants of moment matrices as integrals follows immediately from the fact that the square of a Vandermonde determinant can be represented as a sum of determinants

$$\Delta^2(u_1,\ldots,u_n) = \sum_{\sigma \in S_n} \det\left(u_{\sigma(k)}^{l+k-2}\right)_{1\leq k,l\leq n}.$$

Indeed,

$$n!\tau_n(t) = n!\det m_n(t) = \sum_{\sigma \in S_n} \det\left(\int_E z_{\sigma(k)}^{l+k-2} \rho_t(z_{\sigma(k)})dz_{\sigma(k)}\right)_{1\leq k,l\leq n}$$

$$= \sum_{\sigma \in S_n}\int_{E^n} \det\left(z_{\sigma(k)}^{l+k-2}\right)_{1\leq k,l\leq n} \rho_t(z_{\sigma(k)}) dz_{\sigma(k)}$$

$$= \int_{E^n} \Delta_n^2(z) \prod_{k=1}^n \rho_t(z_k)\, dz_k,$$

whereas the representation (3.1.4) in terms of integrals over Hermitian matrices follows from Section 1.1.

The Borel factorization of m_∞ is responsible for the orthonormality of the polynomials $p_n(t;z) = (S(t)\chi(z))_n$; indeed,

$$\langle p_k(t;z), p_l(t;z)\rangle_{0\leq k,l<\infty} = \int_E S\chi(z)(S\chi(z))^\top \rho_t(z)\, dz = S m_\infty S^\top = I.$$

Note that $S\chi(z)(S\chi(z))^\top$ should be viewed as a semi-infinite matrix obtained by multiplying a semi-infinite column and row. The determinantal representation (3.1.9) follows at once from noticing that $\langle p_n(t;z), z^k\rangle = 0$ for $0 \leq k \leq n-1$, because taking that inner product produces two identical columns in the matrix thus obtained. From the same representation (3.1.9), one has $p_n(t;z) = h_n^{-1/2} z^n + \cdots$, where $h_n := \tau_{n+1}/\tau_n(t)$.

The "Sato" representation (3.1.9) of $p_n(t;z)$ in terms of the determinant $\tau_n(t)$ of the moment matrix can be shown by first proving the Heine representation of the orthogonal polynomials, which goes as follows:

$$h_n^{1/2} p_n(t;z) = \frac{1}{\tau_n} \det \begin{pmatrix} & & & 1 \\ & m_n(t) & & z \\ & & & \vdots \\ \mu_{n,0} & \cdots & \mu_{n,n-1} & z^n \end{pmatrix}$$

$$= \frac{1}{\tau_n} \int_{E^n} \det \begin{pmatrix} u_1^0 & u_2^1 & \cdots & u_n^{n-1} & 1 \\ u_1^1 & u_2^2 & \cdots & u_n^n & z \\ \vdots & & & \vdots & \vdots \\ u_1^{n-1} & u_2^n & \cdots & u_n^{2n-2} & z^{n-1} \\ u_1^n & u_2^{n+1} & \cdots & u_n^{2n-1} & z^n \end{pmatrix} \prod_1^n \rho_t(u_i)\, du_i$$

$$= \frac{1}{\tau_n} \int_{E^n} \det \begin{pmatrix} u_1^0 & u_2^0 & \cdots & u_n^0 & 1 \\ u_1^1 & u_2^1 & \cdots & u_n^1 & z \\ \vdots & \vdots & & \vdots & \vdots \\ u_1^{n-1} & u_2^{n-1} & \cdots & u_n^{n-1} & z^{n-1} \\ u_1^n & u_2^n & \cdots & u_n^n & z^n \end{pmatrix} u_1^0 u_2^1 \cdots u_n^{n-1} \prod_1^n \rho_t(u_i)\, du_i$$

$$= \frac{1}{\tau_n} \int_{E^n} \det \begin{pmatrix} u_{\sigma(1)}^0 & u_{\sigma(2)}^0 & \cdots & u_{\sigma(n)}^0 & 1 \\ u_{\sigma(1)}^1 & u_{\sigma(2)}^1 & \cdots & u_{\sigma(n)}^1 & z \\ \vdots & \vdots & & \vdots & \vdots \\ u_{\sigma(1)}^{n-1} & u_{\sigma(2)}^{n-1} & \cdots & u_{\sigma(n)}^{n-1} & z^{n-1} \\ u_{\sigma(1)}^n & u_{\sigma(2)}^n & \cdots & u_{\sigma(n)}^n & z^n \end{pmatrix} u_{\sigma(1)}^0 u_{\sigma(2)}^1 \cdots u_{\sigma(n)}^{n-1} \prod_1^n \rho_t(u_{\sigma(i)}) du_{\sigma(i)}$$

for any permutation $\sigma \in S_n$

$$= \frac{1}{\tau_n} \int_{E^n} \det \begin{pmatrix} u_1^0 & u_2^0 & \cdots & u_n^0 & 1 \\ u_1^1 & u_2^1 & \cdots & u_n^1 & z \\ \vdots & \vdots & & \vdots & \vdots \\ u_1^{n-1} & u_2^{n-1} & \cdots & u_n^{n-1} & z^{n-1} \\ u_1^n & u_2^n & \cdots & u_n^n & z^n \end{pmatrix} (-1)^\sigma u_{\sigma(1)}^0 u_{\sigma(2)}^1 \cdots u_{\sigma(n)}^{n-1} \prod_1^n \rho_t(u_i) du_i$$

$$= \frac{1}{n!\, \tau_n} \int_{E^n} \Delta_n^2(u) \prod_{k=1}^n (z - u_k) \rho_t(u_k)\, du_k, \qquad \text{upon summing over all } \sigma.$$

Therefore, using again the representation of $\Delta^2(z)$ as a sum of determinants, Heine's formula leads to

$$h_n^{1/2} p_n(t,z) = \frac{z^n}{n!\, \tau_n} \int_{E^n} \sum_{\sigma \in S_n} \det(u_{\sigma(k)}^{l+k-2})_{1 \le k, l \le n} \prod_{k=1}^n \left(1 - \frac{u_{\sigma(k)}}{z}\right) \rho_t(u_{\sigma(k)})\, du_{\sigma(k)},$$

$$= \frac{z^n}{n!\, \tau_n} \int_{E^n} \sum_{\sigma \in S_n} \det \left(u_{\sigma(k)}^{l+k-2} - \frac{1}{z} u_{\sigma(k)}^{l+k-1} \right)_{1 \le k, l \le n} \rho_t(u_{\sigma(k)})\, du_{\sigma(k)}$$

$$= \frac{z^n}{\tau_n} \det \left(\mu_{ij} - \frac{1}{z} \mu_{i,j+1} \right)_{0 \le i,j \le n-1}$$

$$= \frac{z^n}{\tau_n} \det \left(\mu_{ij}(t - [z^{-1}]) \right)_{0 \le i,j \le n-1}$$

$$= z^n \frac{\tau_n(t - [z^{-1}])}{\tau_n(t)}, \qquad (3.1.12)$$

invoking the fact that

$$\mu_{ij}(t-[z^{-1}]) = \int u^{i+j} e^{\sum_1^\infty \left(t_i - \frac{z^{-i}}{i}\right) u^i} \rho(u)\, du = \int u^{i+j} \left(1 - \frac{u}{z}\right) \rho(u) e^{\sum_1^\infty t_i u^i}\, du$$

$$= \mu_{i+j}(t) - \frac{1}{z} \mu_{i+j+1}(t).$$

Formula (3.1.10) follows from computing on the one hand $S(t)m_\infty \chi(z)$ using the explicit moments μ_{ij}, together with (3.1.12), and on the other hand the equivalent expression $S^{\top-1}(t)\chi(z^{-1})$. Indeed, using $(S(t)\chi(z))_n = p_n(t;z) = \sum_0^n p_{nk}(t)z^k$,

$$\sum_{j\geq 0} (Sm_\infty)_{nj} z^{-j} = \sum_{j\geq 0} z^{-j} \sum_{l\geq 0} p_{nl}(t)\mu_{lj}$$

$$= \sum_{j\geq 0} z^{-j} \sum_{l\geq 0} p_{nl}(t) \int_E u^{l+j} \rho_t(u)\, du$$

$$= \int_E \sum_{l\geq 0} p_{nl}(t) u^l \sum_{j\geq 0} \left(\frac{u}{z}\right)^j \rho_t(u)\, du$$

$$= z \int_E \frac{p_n(t,u)\rho_t(u)}{z-u}\, du.$$

Mimicking computation (3.1.12), one shows that

$$h_n^{1/2} \sum_{j\geq 0} \left(S^{\top-1}(t)\right)_{nj} z^{-j} = \frac{\tau_{n+1}(t+[z^{-1}])}{\tau_n(t)} z^{-n},$$

from which (3.1.10) follows, upon using $Sm_\infty = S^{\top-1}$. Details of this and subsequent derivation can be found in [5; 6].

The vectors p and q are eigenvectors of L. Indeed, remembering $\chi(z) = (1, z, z^2, \dots)^\top$, and the shift $(\Lambda v)_n = v_{n+1}$, we have

$$\Lambda \chi(z) = z\chi(z) \quad \text{and} \quad \Lambda^\top \chi(z^{-1}) = z\chi(z^{-1}) - ze_1, \quad \text{with } e_1 = (1,0,0,\dots)^\top.$$

Therefore, $p(z) = S\chi(z)$ and $q(z) = S^{\top-1}\chi(z^{-1})$ are eigenvectors, in the sense that

$$Lp = S\Lambda S^{-1} S\chi(z) = zS\chi(z) = zp,$$
$$L^\top q = S^{\top-1} \Lambda^\top S^\top S^{\top-1} \chi(z^{-1}) = zS^{\top-1}\chi(z^{-1}) - zS^{\top-1}e_1 = zq - zS^{\top-1}e_1.$$

Then, using $L = L^\top$, one is led to

$$((L-zI)p)_n = 0 \quad \text{for } n \geq 0,$$
$$((L-zI)q)_n = 0 \quad \text{for } n \geq 1.$$

Toda lattice and Lie algebra splitting: The Lie algebra splitting of semi-infinite matrices and the corresponding projections (used in (3.1.8)), denoted by $(\ )_\mathfrak{s}$ and $(\ )_\mathfrak{b}$, are defined as follows:

$$\mathfrak{gl}(\infty) = \mathfrak{s} \oplus \mathfrak{b} \quad \text{with} \quad \begin{cases} \mathfrak{s} = \{\text{skew-symmetric matrices}\}, \\ \mathfrak{b} = \{\text{lower-triangular matrices}\}. \end{cases}$$

Conjugating the shift matrix Λ by $S(t)$ yields a matrix

$$\begin{aligned} L(t) &= S(t)\Lambda S(t)^{-1} \\ &= S\Lambda S^{-1} S^{\top -1} S^{\top} \\ &= S\Lambda m_\infty S^{\top} && \text{using (3.1.3)} \\ &= S m_\infty \Lambda^{\top} S^{\top} && \text{using } \Lambda m_\infty = m_\infty \Lambda^{\top} \\ &= S(S^{-1} S^{\top -1})\Lambda^{\top} S^{\top} && \text{using (3.1.3) again} \\ &= (S\Lambda S^{-1})^{\top} = L(t)^{\top}, \end{aligned}$$

which is symmetric and thus tridiagonal. Moreover, from (3.1.3) one computes

$$0 = S\left(\Lambda^k m_\infty - \frac{\partial m_\infty}{\partial t_k}\right) S^{\top} = S\Lambda^k S^{-1} - S \frac{\partial}{\partial t_k}(S^{-1} S^{\top -1}) S^{\top}$$
$$= L^k + \frac{\partial S}{\partial t_k} S^{-1} + S^{\top -1} \frac{\partial S^{\top}}{\partial t_k}.$$

Taking the ( )$_-$ and ( )$_0$ parts of this equation (A_- means the lower-triangular part of the matrix A, including the diagonal and A_0 the diagonal part) leads to

$$(L^k)_- + \frac{\partial S}{\partial t_k} S^{-1} + \left(S^{\top -1} \frac{\partial S^{\top}}{\partial t_k}\right)_0 = 0 \quad \text{and} \quad \left(\frac{\partial S}{\partial t_k} S^{-1}\right)_0 = -\tfrac{1}{2}(L^k)_0.$$

Upon observing that, for any symmetric matrix

$$\begin{pmatrix} a & c \\ c & b \end{pmatrix}_\flat = \begin{pmatrix} a & 0 \\ 2c & b \end{pmatrix} = 2\begin{pmatrix} a & c \\ c & b \end{pmatrix}_- - \begin{pmatrix} a & c \\ c & b \end{pmatrix}_0,$$

it follows that the matrices $L(t)$, $S(t)$ and the vector $p(t;z) = (p_n(t;z))_{n\geq 0} = S(t)\chi(z)$ satisfy the (commuting) differential equations and the eigenvalue problem

$$\frac{\partial S}{\partial t_k} = -\tfrac{1}{2}(L^k)_\flat S, \quad L(t)p(t;z) = zp(t;z), \qquad (3.1.13)$$

and thus

$$\frac{\partial L}{\partial t_k} = -[\tfrac{1}{2}(L^k)_\flat, L], \quad \frac{\partial p}{\partial t_k} = -\tfrac{1}{2}(L^k)_\flat p \qquad \text{(Standard Toda lattice)}.$$

The bilinear identity: The functions $\tau_n(t)$ satisfy the following identity, for $n \geq m+1$, $t, t' \in \mathbb{C}$, where one integrates along a small circle about ∞,

$$\oint_{z=\infty} \tau_n(t - [z^{-1}]) \tau_{m+1}(t' + [z^{-1}]) e^{\sum(t_i - t'_i)z^i} z^{n-m-1} \, dz = 0. \qquad (3.1.14)$$

An elementary proof can be given by expressing the left hand side of (3.1.14), in terms of $p_n(t;z)$ and $p_m(t,z)$, using (3.1.9) and (3.1.10). One uses below the following identity (see [2]):

$$\int_\mathbb{R} f(z)g(z)\,dz = \left\langle f, \int_\mathbb{R} \frac{g(u)}{z-u}\,du \right\rangle_\infty, \qquad (3.1.15)$$

involving the residue pairing[16]. So, modulo terms depending on t and t' only, the left-hand side of (3.1.14) equals

$$\oint_{z=\infty} dz\, z^{-n} p_n(t;z) e^{\sum_1^\infty (t_i-t'_i)z^i} z^{n-m-1} z^{m+1} \int_\mathbb{R} \frac{p_m(t';u)}{z-u} e^{\sum_1^\infty t'_i u^i} \rho(u)\, du$$

$$= \int_\mathbb{R} p_n(t;z) e^{\sum (t_i-t'_i)z^i} p_m(t';z) e^{\sum t'_i z^i} \rho(z)\, dz \qquad \text{using (3.1.15)}$$

$$= \int_\mathbb{R} p_n(t;z) p_m(t';z) e^{\sum t_i z^i} \rho(z)\, dz = 0 \quad \text{when } m \leq n-1. \qquad (3.1.16)$$

The KP-hierarchy: Setting $n = m+1$, shifting $t \mapsto t-y, t' \mapsto t+y$, evaluating the residue and Taylor expanding in y_k and using the elementary Schur polynomials s_n, leads to (see footnote 12 for the definition of $p(\partial_t) f \circ g$ and $\tilde{\partial}$)

$$0 = \frac{1}{2\pi i} \oint dz\, e^{-\sum_1^\infty 2 y_i z^i} \tau_n(t-y-[z^{-1}])\tau_n(t+y+[z^{-1}])$$

$$= \frac{1}{2\pi i} \oint dz \left(\sum_0^\infty z^i s_i(-2y) \right) \left(\sum_0^\infty z^{-j} s_j(\tilde{\partial}) \right) e^{\sum_1^\infty y_k \frac{\partial}{\partial t_k}} \tau_n \circ \tau_n$$

$$= e^{\sum_1^\infty y_k \frac{\partial}{\partial t_k}} \sum_0^\infty s_i(-2y) s_{i+1}(\tilde{\partial}) \tau_n \circ \tau_n$$

$$= \left(1 + \sum_1^\infty y_j \frac{\partial}{\partial t_j} + O(y^2) \right) \left(\frac{\partial}{\partial t_1} + \sum_1^\infty s_{i+1}(\tilde{\partial})(-2y_i + O(y^2)) \right) \tau_n \circ \tau_n$$

$$= \left(\frac{\partial}{\partial t_1} + \sum_1^\infty y_k \left(\frac{\partial}{\partial t_k} \frac{\partial}{\partial t_1} - 2 s_{k+1}(\tilde{\partial}) \right) \right) \tau_n \circ \tau_n + O(y^2),$$

thus yielding (3.1.6), taking into account that

$$\frac{\partial}{\partial t_1} \tau \circ \tau = 0$$

and that the coefficient of y_k is trivial for $k = 1, 2$.

The Riemann–Hilbert problem: As a function of z, the integral (3.1.10) has a jump across the real axis:

$$\frac{1}{2\pi i} \lim_{\substack{z' \to z \\ \Im z' < 0}} \int_\mathbb{R} \frac{p_n(t;u)}{z'-u} \rho_t(u)\, du = p_n(t,z)\rho_t(z) + \frac{1}{2\pi i} \lim_{\substack{z' \to z \\ \Im z' > 0}} \int_\mathbb{R} \frac{p_n(t;u)}{z'-u} \rho_t(u)\, du.$$

Thus we have (see [29; 19; 5]):

---

[16]The residue pairing about $z = \infty$ between $f = \sum_{i \geq 0} a_i z^i \in \mathcal{H}^+$ and $g = \sum_{j \in \mathbb{Z}} b_j z^{-j-1} \in \mathcal{H}$ is defined as

$$\langle f, g \rangle_\infty = \oint_{z=\infty} f(z) g(z) \frac{dz}{2\pi i} = \sum_{i \geq 0} a_i b_i.$$

COROLLARY 3.2. *The matrix*

$$Y_n(z) = \begin{pmatrix} \dfrac{T_n(t-[z^{-1}])}{T_n(t)} z^n & \dfrac{T_{n+1}(t+[z^{-1}])}{T_n(t)} z^{-n-1} \\ \dfrac{T_{n-1}(t-[z^{-1}])}{T_n(t)} z^{n-1} & \dfrac{T_n(t+[z^{-1}])}{T_n(t)} z^{-n} \end{pmatrix}$$

satisfies the Riemann–Hilbert problem:

1. $Y_n(z)$ holomorphic on the upper and lower half-planes $\mathbb{C}_+$ and $\mathbb{C}_-$;
2. $Y_{n-}(z) = Y_{n+}(z) \begin{pmatrix} 1 & 2\pi i \rho_t(z) \\ 0 & 1 \end{pmatrix}$ (jump condition);
3. $Y'_n(z) \begin{pmatrix} z^{-n} & 0 \\ 0 & z^n \end{pmatrix} = 1 + O(z^{-1})$, when $z \to \infty$.

3.2. Pfaff Lattice and symmetric/symplectic matrix integrals

3.2.1. Pfaff lattice, factorization of skew-symmetric matrices and skew-orthogonal polynomials.
Consider an inner product with a skew-symmetric weight $\tilde{\rho}(y,z)$:

$$\langle f, g \rangle_t = \iint_{\mathbb{R}^2} f(y)g(z) e^{\sum t_i(y^i + z^i)} \tilde{\rho}(y,z)\, dy\, dz, \text{ with } \tilde{\rho}(z,y) = -\tilde{\rho}(y,z). \tag{3.2.1}$$

Since $\langle f, g \rangle_t = -\langle g, f \rangle_t$, the moment matrix, depending on $t = (t_1, t_2, \dots)$,

$$m_n(t) = (\mu_{ij}(t))_{0 \le i,j \le n-1} = (\langle y^i, z^j \rangle_t)_{0 \le i,j \le n-1}$$

is skew-symmetric. It is clear from formula (3.2.1) that the semi-infinite matrix m_∞ evolves in t according to the *commuting vector fields*:

$$\frac{\partial \mu_{ij}}{\partial t_k} = \mu_{i+k,j} + \mu_{i,j+k}, \quad \text{i.e.,} \quad \frac{\partial m_\infty}{\partial t_k} = \Lambda^k m_\infty + m_\infty \Lambda^{\top k}. \tag{3.2.2}$$

Since m_∞ is skew-symmetric, m_∞ does not admit a Borel factorization in the standard sense, but m_∞ admits a unique factorization, with an inserted semi-infinite, skew-symmetric matrix J, with $J^2 = -I$, of the form (1.1.12) (see [2]):

$$m_\infty(t) = Q^{-1}(t) J Q^{\top -1}(t),$$

where

$$Q(t) = \begin{pmatrix} \ddots & & & & \\ & \boxed{\begin{matrix} Q_{2n,2n} & 0 \\ 0 & Q_{2n,2n} \end{matrix}} & & 0 & \\ & & \boxed{\begin{matrix} Q_{2n+2,2n+2} & 0 \\ 0 & Q_{2n+2,2n+2} \end{matrix}} & \\ & * & & & \ddots \end{pmatrix} \in K. \tag{3.2.3}$$

K is the group of lower-triangular invertible matrices of the form above, with Lie algebra $\mathfrak{k}$. Consider the Lie algebra splitting, given by

$$\mathrm{gl}(\infty) = \mathfrak{k} \oplus \mathfrak{n} \quad \begin{cases} \mathfrak{k} = \{\text{lower-triangular matrices of the form (3.2.3)}\} \\ \mathfrak{n} = \mathrm{sp}(\infty) = \{a \text{ such that } Ja^\top J = a\}, \end{cases} \quad (3.2.4)$$

with unique decomposition[17]

$$\begin{aligned} a &= (a)_\mathfrak{k} + (a)_\mathfrak{n} \\ &= \left((a_- - J(a_+)^\top J) + \tfrac{1}{2}(a_0 - J(a_0)^\top J)\right) \\ &\quad + \left((a_+ + J(a_+)^\top J) + \tfrac{1}{2}(a_0 + J(a_0)^\top J)\right). \end{aligned} \quad (3.2.5)$$

Consider as a special skew-symmetric weight (3.2.1) (see [13]):

$$\tilde{\rho}(y,z) = 2D^\alpha \delta(y-z)\tilde{\rho}(y)\tilde{\rho}(z) \text{ with } \alpha = \mp 1, \quad \tilde{\rho}(y) = e^{-\tilde{V}(y)}, \quad (3.2.6)$$

together with the associated inner product[18] of type (3.2.1):

$$\langle f, g\rangle_t = \iint_{\mathbb{R}^2} f(y)g(z) e^{\sum t_i(y^i+z^i)} 2D^\alpha \delta(y-z)\tilde{\rho}(y)\tilde{\rho}(z)\,dy\,dz \quad (3.2.7)$$

$$= \begin{cases} \iint_{\mathbb{R}^2} f(y)g(z) e^{\sum_1^\infty t_i(y^i+z^i)} \varepsilon(y-z)\tilde{\rho}(y)\tilde{\rho}(z)\,dy\,dz & \text{for } \alpha = -1, \\ \int_{\mathbb{R}} \{f,g\}(y) e^{\sum_1^\infty 2t_i y^i} \tilde{\rho}(y)^2\,dy & \text{for } \alpha = +1, \end{cases}$$

in terms of the Wronskian $\{f,g\} := \frac{\partial f}{\partial y}g - f\frac{\partial g}{\partial y}$. The moments with regard to these inner products (with that precise definition of time t!) satisfy the differential equations $\partial \mu_{ij}/\partial t_k = \mu_{i+k,j} + \mu_{i,j+k}$, as in (3.2.2).

Now recall that the determinant of an odd skew-symmetric matrix equals 0, whereas the determinant of an even skew-symmetric matrix is the square of a polynomial in the entries, the *Pfaffian*, which is defined by this property up to sign.[19] Now introduce the *Pfaffian τ-functions*, defined with regard to the inner products (3.2.7):

$$\tau_{2n}(t) :=$$

$$\begin{cases} \mathrm{pf}\left(\iint_{\mathbb{R}^2} y^k z^l \varepsilon(y-z) e^{\sum_1^\infty t_i(y^i+z^i)} \tilde{\rho}(y)\tilde{\rho}(z)\,dy\,dz\right)_{0\le k,l\le 2n-1} & \text{if } \alpha = -1, \\ \mathrm{pf}\left(\int_{\mathbb{R}} \{y^k, y^l\} e^{\sum_1^\infty 2t_i y^i} \tilde{\rho}^2(y)\,dy\right)_{0\le k,l\le 2n-1} & \text{if } \alpha = +1. \end{cases} \quad (3.2.8)$$

Setting

$$\tilde{\rho}(z) = \begin{cases} \rho(z)I_E(z) & \text{for } \alpha = -1, \\ \rho^{1/2}(z)I_E(z), \; t \mapsto t/2 & \text{for } \alpha = +1 \end{cases}$$

---

[17]$a_\pm$ refers to projection onto strictly upper (strictly lower) triangular matrices, with all 2×2 diagonal blocks equal zero. a_0 refers to projection onto the "diagonal", consisting of 2×2 blocks.

[18]We set $\varepsilon(x) = \mathrm{sign}\, x$, so that $\varepsilon' = 2\delta(x)$.

[19]We have $(\det m_{2n}(t))^{1/2} = \mathrm{pf}(m_{2n}(t)) = \dfrac{1}{n!} \dfrac{\left(\sum_{0\le i<j\le 2n-1} \mu_{ij}(t)\,dx_i \wedge dx_j\right)^n}{dx_0 \wedge dx_1 \wedge \cdots \wedge dx_{2n-1}}$.

in the identities (3.2.8) leads to the identities (3.2.9) between integrals and Pfaffians, spelled out in Theorem 3.3 below. Recall that $c = (c_1, \ldots, c_{2r})$ stands for the boundary points of the disjoint union $E \subset \mathbb{R}$. Denote by $\mathcal{S}_n(E)$ and $\mathcal{T}_n(E)$ the set of matrices in $\mathcal{S}_n$ and $\mathcal{T}_n$ with spectrum in E.

THEOREM 3.3 (Adler, Horozov, and van Moerbeke [9]; Adler and van Moerbeke [7]). *Consider the integral*

$$I_n = I_n(t,c) := \int_{E^n} |\Delta_n(z)|^\beta \prod_{k=1}^n \left(e^{\sum_1^\infty t_i z_k^i} \rho(z_k) \, dz_k \right).$$

Then I_n is a Pfaffian in certain cases:

- $\beta = 1$, n even:

$$I_n = \int_{\mathcal{S}_n(E)} e^{\mathrm{Tr}(-V(X) + \sum_1^\infty t_i X^i)} \, dX$$

$$= n! \operatorname{pf} \left(\iint_{E^2} y^k z^l \varepsilon(y-z) e^{\sum_1^\infty t_i (y^i + z^i)} \rho(y) \rho(z) \, dy \, dz \right)_{0 \le k, l \le n-1}$$

$$= n! \, \tau_n(t,c); \qquad (3.2.9a)$$

- $\beta = 4$, n arbitrary:

$$I_n = \int_{\mathcal{T}_{2n}(E)} e^{\mathrm{Tr}(-V(X) + \sum_1^\infty t_i X^i)} \, dX$$

$$= n! \operatorname{pf} \left(\int_E \{y^k, y^l\} e^{\sum_1^\infty t_i y^i} \rho(y) \, dy \right)_{0 \le k, l \le 2n-1}$$

$$= n! \, \tau_{2n}(t/2, c). \qquad (3.2.9b)$$

The I_n and τ_n's satisfy satisfy the following relations:

(i) *The <u>Virasoro constraints</u>[20] (2.1.7) for $\beta = 1, 4$:*

$$\left(-\sum_1^{2r} c_i^{k+1} f(c_i) \frac{\partial}{\partial c_i} + \sum_{i \ge 0} \left(a_i \, {}^\beta \mathbb{J}^{(2)}_{k+i,n} - b_i \, {}^\beta \mathbb{J}^{(1)}_{k+i+1,n} \right) \right) I_n(t,c) = 0. \qquad (3.2.10)$$

(ii) *The <u>Pfaff-KP hierarchy</u> (see footnote 12 for notation)*

$$\left(\mathbf{s}_{k+4}(\tilde{\partial}) - \frac{1}{2} \frac{\partial^2}{\partial t_1 \partial t_{k+3}} \right) \tau_n \circ \tau_n = \mathbf{s}_k(\tilde{\partial}) \, \tau_{n+2} \circ \tau_{n-2} \qquad (3.2.11)$$

for n even and $k = 0, 1, 2, \ldots$. The first relation in this hierarchy reads (for n even)

$$\left(\left(\frac{\partial}{\partial t_1} \right)^4 + 3 \left(\frac{\partial}{\partial t_2} \right)^2 - 4 \frac{\partial^2}{\partial t_1 \partial t_3} \right) \log \tau_n + 6 \left(\frac{\partial^2}{\partial t_1^2} \log \tau_n \right)^2 = 12 \frac{\tau_{n-2} \tau_{n+2}}{\tau_n^2}.$$

---

[20] Here the a_i's and b_i's are defined in the usual way, in terms of $\rho(z)$; namely, $-\rho'/\rho = (\sum b_i z^i)/(\sum a_i z^i)$.

(iii) The *Pfaff Lattice*: The time-dependent matrix $L(t)$, zero above the first superdiagonal, obtained by dressing up Λ and having the general form

$$L(t) = Q(t)\Lambda Q(t)^{-1} = \begin{pmatrix} 0 & 1 & & & \\ -d_1 & (h_2/h_0)^{1/2} & & 0 & \\ & d_1 & 1 & & \\ & & -d_2 & (h_4/h_2)^{1/2} & \\ & * & & d_2 & \\ & & & & \ddots \end{pmatrix} \quad (3.2.12)$$

satisfies the Hamiltonian commuting equations

$$\frac{\partial L}{\partial t_i} = [-(L^i)_{\mathfrak{k}}, L]. \qquad \textbf{(Pfaff lattice)} \quad (3.2.13)$$

(iv) *Skew-orthogonal polynomials*: The vector of time-dependent polynomials

$$q(t;z) := (q_n(t;z))_{n \geq 0} = Q(t)\chi(z)$$

in z satisfies the eigenvalue problem

$$L(t)q(t,z) = zq(t,z) \quad (3.2.14)$$

and enjoy the following representations (with $h_{2n} = \tau_{2n+2}(t)/\tau_{2n}(t)$)

$$q_{2n}(t;z) = \frac{h_{2n}^{-1/2}}{\tau_{2n}(t)} \operatorname{pf} \begin{pmatrix} & & & & 1 \\ & m_{2n+1}(t) & & & z \\ & & & & \vdots \\ & & & & z^{2n} \\ \hline -1 & -z & \cdots & -z^{2n} & 0 \end{pmatrix}$$

$$= z^{2n} h_{2n}^{-1/2} \frac{\tau_{2n}(t - [z^{-1}])}{\tau_{2n}(t)} = z^{2n} h_{2n}^{-1/2} + \cdots,$$

$$q_{2n+1}(t;z) = \frac{h_{2n}^{-1/2}}{\tau_{2n}(t)} \operatorname{pf} \begin{pmatrix} & & & 1 & \mu_{0,2n+1} \\ & m_{2n}(t) & & z & \mu_{1,2n+1} \\ & & & \vdots & \vdots \\ & & & z^{2n-1} & \mu_{2n-1,2n+1} \\ \hline -1 & \cdots & -z^{2n-1} & 0 & -z^{2n+1} \\ \mu_{2n+1,0} & \cdots & \mu_{2n+1,2n-1} & z^{2n+1} & 0 \end{pmatrix}.$$

$$= z^{2n} h_{2n}^{-1/2} \frac{1}{\tau_{2n}(t)} \left(z + \frac{\partial}{\partial t_1}\right) \tau_{2n}(t - [z^{-1}]) = z^{2n+1} h_{2n}^{-1/2} + \cdots. \quad (3.2.15)$$

They are skew-orthogonal polynomials in z; that is, $\langle q_i(t;z), q_j(t;z)\rangle_t = J_{ij}$.

The hierarchy (3.2.11) already appears in the work of Kac and van de Leur [42] in the context of what they call the DKP-hierarchy. Interesting further work has been done by van de Leur [69].

3.2.2. Sketch of Proof.

Skew-orthogonal polynomials and the Pfaff Lattice: The equalities (3.2.9) between the Pfaffians and the matrix integrals are based on two identities [49]. The first, due to de Bruyn, reads

$$\frac{1}{n!}\int_{\mathbb{R}^n} \prod_1^n dy_i \, \det\bigl(F_i(y_1) \; G_i(y_1) \; \ldots \; F_i(y_n) \; G_i(y_n)\bigr)_{0 \leq i \leq 2n-1}$$
$$= \det{}^{1/2}\left(\int_{\mathbb{R}} (G_i(y)F_j(y) - F_i(y)G_j(y))\,dy\right)_{0 \leq i,j \leq 2n-1};$$

the second (Mehta [50]) is

$$\Delta_n^4(x) = \det\bigl(x_1^i \; (x_1^i)' \; x_2^i \; (x_2^i)' \; \ldots \; x_n^i \; (x_n^i)'\bigr)_{0 \leq i \leq 2n-1}.$$

On the one hand (see Mehta [49]), setting in the calculation below $\rho_{t,E}(z) = \rho(z)e^{\sum t_i z^i} I_E(z)$, $F_i(x) := \int_{-\infty}^x y^i \rho_{t,E}(y)\,dy$, and $G_i(x) := F_i'(x) = x^i \rho_{t,E}(x)$, one computes:

$$\frac{1}{(2n)!}\int_{\mathbb{R}^{2n}} |\Delta_{2n}(z)| \prod_{i=1}^{2n} \rho_{t,E}(z_i)\,dz_i$$

$$= \int_{\substack{-\infty < z_1 < z_2 < \\ \cdots < z_{2n} < \infty}} \det\bigl(z_{j+1}^i \rho_{t,E}(z_{j+1})\bigr)_{0 \leq i,j \leq 2n-1} \prod_{i=1}^{2n} dz_i,$$

$$= \int_{\substack{-\infty < z_2 < z_4 < \\ \cdots < z_{2n} < \infty}} \prod_{k=1}^{n} \rho_{t,E}(z_{2k})\,dz_{2k}$$
$$\det\left(\int_{-\infty}^{z_2} z_1^i \rho_{t,E}(z_1)\,dz_1, z_2^i, \ldots, \int_{z_{2n-2}}^{z_{2n}} z_{2n-1}^i \rho_{t,E}(z_{2n-1})\,dz_{2n-1}, z_{2n}^i\right)_{0 \leq i \leq 2n-1}$$

$$= \int_{\substack{-\infty < z_2 < z_4 < \\ \cdots < z_{2n} < \infty}} \prod_{k=1}^{n} \rho_{t,E}(z_{2k})\,dz_{2k}$$
$$\det\bigl(F_i(z_2), z_2^i, F_i(z_4) - F_i(z_2), z_4^i, \ldots, F_i(z_{2n}) - F_i(z_{2n-2}), z_{2n}^i\bigr)_{0 \leq i \leq 2n-1}$$

$$= \int_{\substack{-\infty < z_2 < z_4 < \\ \cdots < z_{2n} < \infty}} \prod_{1}^{n} dz_{2i}\, \det\bigl(F_i(z_2), G_i(z_2), \ldots, F_i(z_{2n}), G_i(z_{2n})\bigr)_{0 \leq i \leq 2n-1},$$

$$= \frac{1}{n!}\int_{\mathbb{R}^n} \prod_1^n dy_i\, \det\bigl(F_i(y_1), G_i(y_1), \ldots, F_i(y_n), G_i(y_n)\bigr)_{0 \leq i \leq 2n-1},$$

$$= \det^{1/2} \left(\int_{\mathbb{R}} (G_i(y)F_j(y) - F_i(y)G_j(y)) \, dy \right)_{0 \leq i,j \leq 2n-1}$$

$$= \operatorname{pf} \left(\iint_{E^2} y^k z^l \varepsilon(y-z) e^{\sum_1^\infty t_i(y^i+z^i)} \rho(y)\rho(z) \, dy \, dz \right)_{0 \leq k,l \leq 2n-1} = \tau_{2n}(t),$$

which establishes (3.2.9a).

On the other hand, upon setting

$$F_j(x) = x^j \rho(x) e^{\sum t_i x^i} \quad \text{and} \quad G_j(x) := F_j'(x) = \left(x^j \rho(x) e^{\sum t_i x^i} \right)',$$

one computes

$$\frac{1}{n!} \int_E \prod_{1 \leq i,j \leq n} (x_i - x_j)^4 \prod_{k=1}^n \left(\rho^2(x_k) e^{2 \sum_{i=1}^\infty t_i x_k^i} \, dx_k \right)$$

$$= \frac{1}{n!} \int_E \prod_{k=1}^n \left(\rho^2(x_k) e^{2 \sum t_i x_k^i} \, dx_k \right) \det\left(x_1^i \; (x_1^i)' \; x_2^i \; (x_2^i)' \; \ldots \; x_n^i \; (x_n^i)' \right)_{0 \leq i \leq 2n-1}$$

$$= \frac{1}{n!} \int_E \prod_1^n dy_i \; \det\left(F_i(y_1) \; G_i(y_1) \; \ldots \; F_i(y_n) \; G_i(y_n) \right)_{0 \leq i \leq 2n-1}$$

$$= \det^{1/2} \left(\int_E (G_i(y) F_j(y) - F_i(y) G_j(y)) \, dy \right)_{0 \leq i,j \leq 2n-1}$$

$$= \operatorname{pf} \left(\int_E \{y^k, y^l\} e^{\sum_1^\infty 2 t_i y^i} \rho^2(y) \, dy \right)_{0 \leq k,l \leq 2n-1} = \tau_{2n}(t),$$

establishing (3.2.9b).

The skew-orthogonality of the polynomials $q_k(t;z)$ follows immediately from the skew-Borel decomposition of m_∞:

$$\langle q_k(t,y), q_l(t,z) \rangle_{k,l \geq 0} = Q(\langle y^i, z^j \rangle)_{i,j \geq 0} Q^\top = Q m_\infty Q^\top = J,$$

with the q_n's admitting the representation (3.2.15) in terms of the moments.

Using $L = Q \Lambda Q^{-1}$, $m_\infty = Q^{-1} J Q^{\top-1}$ and $J^2 = -I$, one computes from the differential equations (3.2.2)

$$0 = Q \left(\Lambda^k m_\infty + m_\infty \Lambda^{\top k} - \frac{\partial m_\infty}{\partial t_k} \right) Q^\top$$

$$= (Q \Lambda^k Q^{-1}) J - (J Q^\top{}^{-1} \Lambda^{\top k} Q^\top J) J + \frac{\partial Q}{\partial t_k} Q^{-1} J - \left(J Q^{-1\top} \frac{\partial Q^\top}{\partial t_k} J \right) J$$

$$= \left(L^k + \frac{\partial Q}{\partial t_k} Q^{-1} \right) - J \left(L^k + \frac{\partial Q}{\partial t_k} Q^{-1} \right)^\top J.$$

Then computing the $+$, $-$ and the diagonal part (in the sense of footnote 17) of the expression leads to commuting Hamiltonian differential equations for Q, and thus for L and $q(t;z)$, confirming (3.2.13):

$$\frac{\partial Q}{\partial t_i} = -(L^i)_\ell Q, \quad \frac{\partial L}{\partial t_i} = [(L^i)_\mathfrak{n}, L], \quad \frac{\partial q}{\partial t_i} = -(L^i)_\ell q \quad \text{(Pfaff lattice)}. \quad (3.2.16)$$

The bilinear identities: For all $n, m \geq 0$, the τ_{2n}'s satisfy the bilinear identity

$$\oint_{z=\infty} \tau_{2n}(t - [z^{-1}])\tau_{2m+2}(t' + [z^{-1}])e^{\sum(t_i - t'_i)z^i} z^{2n-2m-2} \frac{dz}{2\pi i}$$

$$+ \oint_{z=0} \tau_{2n+2}(t + [z])\tau_{2m}(t' - [z])e^{\sum(t'_i - t_i)z^{-i}} z^{2n-2m} \frac{dz}{2\pi i} = 0. \quad (3.2.17)$$

The differential equation (3.2.2) on the moment matrix m_∞ admits the following solution, which upon using the Borel decomposition $m_\infty = Q^{-1} J Q^{\top -1}$, leads to

$$m_\infty(0) = e^{-\sum_1^\infty t_k \Lambda^k} m_\infty(t) e^{-\sum_1^\infty t_k \Lambda^{\top k}}$$
$$= \left(Q(t)e^{\sum_1^\infty t_k \Lambda^k}\right)^{-1} J \left(Q(t)e^{\sum_1^\infty t_k \Lambda^k}\right)^{\top -1}, \quad (3.2.18)$$

so the right-hand side of (3.2.18) is independent of t; equal, say, to the same expression with t replaced by t'. Upon rearrangement, one finds

$$\left(Q(t)e^{\sum t_k \Lambda^k}\right)\left(JQ(t')e^{\sum t'_k \Lambda^k}\right)^{-1} = \left(JQ(t)e^{\sum t_k \Lambda^k}\right)^{\top -1}\left(Q(t')e^{\sum t'_k \Lambda^k}\right)^\top,$$

and therefore[21]

$$\oint_{z=\infty} \left(Q(t)\chi(z) \otimes (JQ(t'))^{\top -1}\chi(z^{-1})\right) e^{\sum_1^\infty (t_k - t'_k)z^k} \frac{dz}{2\pi i z}$$
$$= \oint_{z=0} \left((JQ(t))^{\top -1}\chi(z) \otimes Q(t')\chi(z^{-1})\right) e^{\sum_1^\infty (t'_k - t_k)z^{-k}} \frac{dz}{2\pi i z}. \quad (3.2.19)$$

Setting $t - t' = [z_1^{-1}] + [z_2^{-1}]$ in the exponential leads to

$$e^{\sum_1^\infty (t_k - t'_k)z^k} = \left(1 - \frac{z}{z_1}\right)^{-1}\left(1 - \frac{z}{z_2}\right)^{-1}, \quad e^{\sum_1^\infty (t'_k - t_k)z^{-k}} = \left(1 - \frac{1}{zz_1}\right)\left(1 - \frac{1}{zz_2}\right),$$

and somewhat enlarging the integration circle about $z = \infty$ to include the points z_1 and z_2, the integrand on the left-hand side has poles at $z = z_1$ and z_2, whereas the integrand on the right-hand side is holomorphic. Combining the identity obtained and the one, with $z_2 \nearrow \infty$, one finds a functional relation involving a function $\varphi(t;z) = 1 + O(z^{-1})$:

$$\frac{\varphi(t - [z_2^{-1}]; z_1)}{\varphi(t; z_1)} = \frac{\varphi(t - [z_1^{-1}]; z_2)}{\varphi(t; z_2)}, \quad t \in \mathbb{C}^\infty, \ z \in \mathbb{C}.$$

---

[21] We use $\Lambda \chi(z) = z\chi(z)$, $\Lambda^\top \chi(z) = z^{-1}\chi(z)$ and the matrix identities (see [25])
$$U_1 V_1 = \oint_{z=\infty} U_1 \chi(z) \otimes V_1^\top \chi(z^{-1}) \frac{dz}{2\pi i z}, \quad U_2 V_2 = \oint_{z=0} U_2 \chi(z) \otimes V_2^\top \chi(z^{-1}) \frac{dz}{2\pi i z}.$$

Such an identity leads, by a standard argument (see [8, Appendix], for example) to the existence of a function $\tau(t)$ such that

$$\varphi(t; z) = \frac{\tau(t - [z^{-1}])}{\tau(t)}.$$

This, combined with the bilinear identity (3.2.19), yields the bilinear identity (3.2.17).

The Pfaff-KP hierarchy: Shifting

$$t \mapsto t - y, \quad t' \mapsto t + y$$

in (3.2.17), evaluating the residue and Taylor expanding in y_k leads to (for $\tilde{\partial}$, see footnote 12)

$$\frac{1}{2\pi i} \oint_{z=\infty} e^{-\sum_1^\infty 2y_i z^i} \tau_{2n}(t - y - [z^{-1}])\tau_{2m+2}(t + y + [z^{-1}])z^{2n-2m-2}\, dz$$

$$+ \frac{1}{2\pi i} \oint_{z=0} e^{\sum_1^\infty 2y_i z^{-i}} \tau_{2n+2}(t - y + [z])\tau_{2m}(t + y - [z])z^{2n-2m}\, dz$$

$$= \frac{1}{2\pi i} \oint_{z=\infty} \sum_{j=0}^\infty z^j s_j(-2y) e^{\sum -y_i \frac{\partial}{\partial t_i}} \sum_{k=0}^\infty z^{-k} s_k(-\tilde{\partial}) \tau_{2n} \circ \tau_{2m+2} z^{2n-2m-2}\, dz$$

$$+ \frac{1}{2\pi i} \oint_{z=0} \sum_{j=0}^\infty z^{-j} s_j(2y) e^{\sum -y_i \frac{\partial}{\partial t_i}} \sum_{k=0}^\infty z^k s_k(\tilde{\partial}) \tau_{2n+2} \circ \tau_{2m} z^{2n-2m}\, dz$$

$$= \sum_{j-k=-2n+2m+1} s_j(-2y) e^{\sum_1^\infty -y_i \frac{\partial}{\partial t_i}} s_k(-\tilde{\partial}) \tau_{2n} \circ \tau_{2m+2}$$

$$+ \sum_{k-j=-2n+2m-1} s_j(2y) e^{\sum_1^\infty -y_i \frac{\partial}{\partial t_i}} s_k(\tilde{\partial}) \tau_{2n+2} \circ \tau_{2m}$$

$$= \cdots + y_k \left(\left(\frac{1}{2}\frac{\partial}{\partial t_1}\frac{\partial}{\partial t_k} - s_{k+1}(\tilde{\partial})\right)\tau_{2n} \circ \tau_{2n} + s_{k-3}(\tilde{\partial})\tau_{2n+2} \circ \tau_{2n-2}\right) + \cdots,$$

establishing the Pfaff-KP hierarchy (3.2.11), different from the usual KP hierarchy, because of the presence of a right-hand side.

REMARK. L admits the following representation in terms of τ, much in the style of (3.1.7):

$$L = h^{-1/2} \begin{pmatrix} \hat{L}_{00} & \hat{L}_{01} & 0 & 0 & \\ \hat{L}_{10} & \hat{L}_{11} & \hat{L}_{12} & 0 & \\ * & \hat{L}_{21} & \hat{L}_{22} & \hat{L}_{23} & \\ * & * & \hat{L}_{32} & \hat{L}_{33} & \\ & & & & \ddots \end{pmatrix} h^{1/2},$$

with the 2×2 entries $\hat{L}_{ij}$ and

$$h = \mathrm{diag}(h_0, h_0, h_2, h_2, h_4, h_4, \ldots), \quad h_{2n} = \tau_{2n+2}/\tau_{2n}.$$

For example, using $\cdot$ to denote partial differentiation with respect to t_1:

$$\hat{L}_{nn} := \begin{pmatrix} -(\log \tau_{2n})^{\cdot} & 1 \\ -\dfrac{s_2(\tilde{\partial})\tau_{2n}}{\tau_{2n}} - \dfrac{s_2(-\tilde{\partial})\tau_{2n+2}}{\tau_{2n+2}} & (\log \tau_{2n+2})^{\cdot} \end{pmatrix},$$

$$\hat{L}_{n,n+1} := \begin{pmatrix} 0 & 0 \\ 1 & 0 \end{pmatrix}, \quad \hat{L}_{n+1,n} := \begin{pmatrix} * & (\log \tau_{2n+2})^{\cdot\cdot} \\ * & * \end{pmatrix}.$$

3.3. 2d-Toda Lattice and Coupled Hermitian Matrix Integrals

3.3.1. 2d-Toda lattice, factorization of moment matrices and bi-orthogonal polynomials. Consider the inner product

$$\langle f, g \rangle_{t,s} = \iint_{E \subset \mathbb{R}^2} f(y)g(z) e^{\sum_1^\infty (t_i y^i - s_i z^i) + cyz} \, dy \, dz, \tag{3.3.1}$$

on a subset $E = E_1 \times E_2 := \bigcup_{i=1}^r [c_{2i-1}, c_{2i}] \times \bigcup_{i=1}^s [\tilde{c}_{2i-1}, \tilde{c}_{2i}] \subset F_1 \times F_2 \subset \mathbb{R}^2$. Define the customary moment matrix, depending on $t = (t_1, t_2, \dots)$ and $s = (s_1, s_2, \dots)$:

$$m_n(t,s) = \bigl(\mu_{ij}(t,s)\bigr)_{0 \leq i,j \leq n-1} = \bigl(\langle y^i, z^j \rangle_{t,s}\bigr)_{0 \leq i,j \leq n-1},$$

and let its factorization in lower-triangular times upper-triangular matrices be

$$m_\infty(t,s) = S_1^{-1}(t,s) S_2(t,s). \tag{3.3.2}$$

Then m_∞ evolves in t, s according to the equations

$$\text{i.e.,} \quad \frac{\partial \mu_{ij}}{\partial t_k} = \mu_{i+k,j}, \quad \frac{\partial \mu_{ij}}{\partial s_k} = -\mu_{i,j+k},$$
$$\frac{\partial m_\infty}{\partial t_k} = \Lambda^k m_\infty, \quad \frac{\partial m_\infty}{\partial s_k} = -m_\infty \Lambda^{\top k}. \tag{3.3.3}$$

In the next integral (3.3.4), dM denotes properly normalized Haar measure on $\mathcal{H}_n$.

THEOREM 3.4 (Adler and van Moerbeke [4; 3]). *The integrals $I_n(t, s; c, \tilde{c})$, with $I_0 = 1$,*

$$\tau_n = \det m_n = \frac{1}{n!} I_n = \frac{1}{n!} \iint_{E^n} \Delta_n(x) \Delta_n(y) \prod_{k=1}^n e^{\sum_1^\infty (t_i x_k^i - s_i y_k^i) + c x_k y_k} \, dx_k \, dy_k$$

$$= \iint_{\mathcal{H}_n^2(E)} e^{c \operatorname{Tr}(M_1 M_2)} e^{\operatorname{Tr} \sum_1^\infty (t_i M_1^i - s_i M_2^i)} \, dM_1 \, dM_2, \tag{3.3.4}$$

satisfy the following relations:

(i) *Virasoro constraints*[22] (2.2.8) *for* $k \geq -1$:

$$\left(-\sum_{i=1}^{r} c_i^{k+1}\frac{\partial}{\partial c_i} + J_{k,n}^{(2)}\right)\tau_n^E + c\,\mathbf{s}_{k+n}(\tilde{\partial}_t)\mathbf{s}_n(-\tilde{\partial}_s)\tau_1^E \circ \tau_{n-1}^E = 0,$$

$$\left(-\sum_{i=1}^{s} \tilde{c}_i^{k+1}\frac{\partial}{\partial \tilde{c}_i} + \tilde{J}_{k,n}^{(2)}\right)\tau_n^E + c\,\mathbf{s}_n(\tilde{\partial}_t)\mathbf{s}_{k+n}(-\tilde{\partial}_s)\tau_1^E \circ \tau_{n-1}^E = 0, \quad (3.3.5)$$

with

$$J_{k,n}^{(2)} = \tfrac{1}{2}\left(J_k^{(2)} + (2n+k+1)J_k^{(1)} + n(n+1)J_k^{(0)}\right),$$
$$\tilde{J}_{k,n}^{(2)} = \tfrac{1}{2}\left(\tilde{J}_k^{(2)} + (2n+k+1)\tilde{J}_k^{(1)} + n(n+1)J_k^{(0)}\right).$$

(ii) *A Wronskian identity*:[23]

$$\left\{\frac{\partial^2 \log \tau_n}{\partial t_1 \partial s_2},\frac{\partial^2 \log \tau_n}{\partial t_1 \partial s_1}\right\}_{t_1} + \left\{\frac{\partial^2 \log \tau_n}{\partial s_1 \partial t_2},\frac{\partial^2 \log \tau_n}{\partial t_1 \partial s_1}\right\}_{s_1} = 0. \quad (3.3.6)$$

(iii) *The 2d-Toda lattice*: Given the factorization (3.3.2), the matrices $L_1 := S_1 \Lambda S_1^{-1}$ and $L_2 := S_2 \Lambda^\top S_2^{-1}$, with $h_n = \tau_{n+1}/\tau_n$, have the following form, where the $(k-l)$-th subdiagonal is given by the diagonal matrix in front of Λ^{k-l}:

$$L_1^k = \sum_{l=0}^{\infty} \operatorname{diag}\left(\frac{\mathbf{s}_l(\tilde{\partial}_t)\tau_{n+k-l+1} \circ \tau_n}{\tau_{n+k-l+1}\tau_n}\right)_{n\in\mathbb{Z}} \Lambda^{k-l}$$

$$hL_2^{\top k}h^{-1} = \sum_{l=0}^{\infty} \operatorname{diag}\left(\frac{\mathbf{s}_l(-\tilde{\partial}_s)\tau_{n+k-l+1} \circ \tau_n}{\tau_{n+k-l+1}\tau_n}\right)_{n\in\mathbb{Z}} \Lambda^{k-l}, \quad (3.3.7)$$

and satisfy the **2d-Toda Lattice**[24]

$$\frac{\partial L_i}{\partial t_n} = [(L_1^n)_+, L_i] \quad \text{and} \quad \frac{\partial L_i}{\partial s_n} = [(L_2^n)_-, L_i], \quad i=1,2. \quad (3.3.8)$$

(iv) *Bi-orthogonal polynomials*: The expressions

$$p_n^{(1)}(t,s;y) := (S_1(t,s)\chi(y))_n = y^n \frac{\tau_n(t-[y^{-1}],s)}{\tau_n(t,s)},$$

$$p_n^{(2)}(t,s;z) := (hS_2^{\top -1}(t,s)\chi(z))_n = z^n \frac{\tau_n(t,s+[z^{-1}])}{\tau_n(t,s)} \quad (3.3.9)$$

---

[22]For the Hirota symbol, see footnote 12. The $J_k^{(i)}$'s are as in remark 1 at the end of Theorem 2.1, for $\beta = 1$ and $\tilde{J}_k^{(i)} = J_k^{(i)}|_{t \to -s}$, with
$$J_k^{(1)} = \frac{\partial}{\partial t_k} + (-k)t_{-k}, \quad J_k^{(2)} = \sum_{i+j=k} \frac{\partial^2}{\partial t_i \partial t_j} + 2\sum_{-i+j=k} it_i \frac{\partial}{\partial t_j} + \sum_{-i-j=k} it_i jt_j.$$

[23]in terms of the Wronskian $\{f,g\}_t = \frac{\partial f}{\partial t}g - f\frac{\partial g}{\partial t}$.

[24]P_+ and P_- denote the upper (including diagonal) and strictly lower triangular parts of the matrix P, respectively.

form a system of monic bi-orthogonal polynomials in z:

$$\langle p_n^{(1)}(t,s;y), p_m^{(2)}(t,s;z)\rangle_{t,s} = \delta_{n,m} h_n \quad \text{with } h_n = \frac{\tau_{n+1}}{\tau_n}, \tag{3.3.10}$$

which are also eigenvectors of L_1 and L_2:

$$zp_n^{(1)}(t,s;z) = L_1(t,s)p_n^{(1)}(t,s;z),$$
$$zp_n^{(2)}(t,s;z) = L_2^\top(t,s)p_n^{(2)}(t,s;z). \tag{3.3.11}$$

REMARK. Each statement can be dualized via the duality $t \leftrightarrow -s$, $L_1 \leftrightarrow hL_2^\top h^{-1}$.

3.3.2. Sketch of proof. Identity (3.3.4) follows from the fact that the product of the two Vandermonde appearing in the integral (3.3.4) can be expressed as a sum of determinants:

$$\Delta_n(u)\Delta_n(v) = \sum_{\sigma \in S_n} \det\left(u_{\sigma(k)}^{l-1} v_{\sigma(k)}^{k-1}\right)_{1 \leq l,k \leq n}, \tag{3.3.12}$$

together with the Harish-Chandra, Itzykson and Zuber formula [34; 38]

$$\int_{U(n)} dU \, e^{c\,\text{Tr}\,xUy\bar{U}^\top} = \frac{(2\pi)^{n(n-1)/2}}{n!} \frac{\det(e^{cx_i y_j})_{1 \leq i,j \leq n}}{\Delta_n(x)\Delta_n(y)}. \tag{3.3.13}$$

Moreover the τ_n's satisfy the following bilinear identities, for all integer $m, n \geq 0$ and $t, s \in \mathbb{C}^\infty$:

$$\oint_{z=\infty} \tau_n(t-[z^{-1}],s)\tau_{m+1}(t'+[z^{-1}],s')e^{\sum_1^\infty (t_i-t'_i)z^i} z^{n-m-1} \, dz$$
$$= \oint_{z=0} \tau_{n+1}(t, s-[z])\tau_m(t', s'+[z])e^{\sum_1^\infty (s_i-s'_i)z^{-i}} z^{n-m-1} \, dz. \tag{3.3.14}$$

Again, the bi-orthogonal nature (3.3.10) of the polynomials (3.3.9) is tantamount to the Borel decomposition, written in the form $S_1 m_\infty (hS_2^{\top -1})^\top = h$. These polynomials satisfy the eigenvalue problem (3.3.11) and evolve in t, s according to the differential equations

$$\frac{\partial p^{(1)}}{\partial t_n} = -(L_1^n)_- p^{(1)}, \qquad \frac{\partial p^{(1)}}{\partial s_n} = -(L_2^n)_- p^{(1)},$$
$$\frac{\partial p^{(2)}}{\partial t_n} = -\left((h^{-1}L_1 h)^{\top n}\right)_- p^{(2)}, \qquad \frac{\partial p^{(2)}}{\partial s_n} = \left((h^{-1}L_2 h)^{\top n}\right)_- p^{(2)}. \tag{3.3.15}$$

From the representation (3.3.7) and the bilinear identity (3.3.14), it follows that

$$\frac{p_{k-1}(\tilde{\partial}_t)\tau_{n+2} \circ \tau_n}{\tau_{n+1}^2} = -\frac{\partial^2}{\partial s_1 \partial t_k} \log \tau_{n+1}, \tag{3.3.16}$$

and so, for $k = 1$,

$$\frac{\tau_n \tau_{n+2}}{\tau_{n+1}^2} = -\frac{\partial^2}{\partial s_1 \partial t_1} \log \tau_{n+1}. \tag{3.3.17}$$

Thus, using (3.3.7), (3.3.16) and (3.3.17), we have

$$\left(L_1^k\right)_{n,n+1} = \frac{p_{k-1}(\tilde{\partial}_t)\tau_{n+2} \circ \tau_n}{\tau_{n+2}\tau_n} = \frac{\frac{\partial^2 \log \tau_{n+1}}{\partial s_1 \partial t_k}}{\frac{\partial^2 \log \tau_{n+1}}{\partial s_1 \partial t_1}},$$

$$\left(hL_2^{\top k}h^{-1}\right)_{n,n+1} = \frac{p_{k-1}(-\tilde{\partial}_s)\tau_{n+2} \circ \tau_n}{\tau_{n+2}\tau_n} = \frac{\frac{\partial^2 \log \tau_{n+1}}{\partial t_1 \partial s_k}}{\frac{\partial^2 \log \tau_{n+1}}{\partial s_1 \partial t_1}}. \qquad (3.3.18)$$

Combining (3.3.18) with (3.3.17) for $k = 2$ yields

$$\left(L_1^2\right)_{n,n+1} = \frac{\frac{\partial^2}{\partial s_1 \partial t_2} \log \tau_{n+1}}{\frac{\partial^2}{\partial s_1 \partial t_1} \log \tau_{n+1}} = \frac{\partial}{\partial t_1} \log\left(-\frac{\tau_{n+2}}{\tau_n}\right) \qquad (3.3.19)$$

$$= \frac{\partial}{\partial t_1} \log\left(\left(\frac{\tau_{n+1}}{\tau_n}\right)^2 \frac{\partial^2}{\partial s_1 \partial t_1} \log \tau_{n+1}\right).$$

Then, subtracting $\partial/\partial s_1$ of (3.3.19) from $\partial/\partial t_1$ of the dual of the same equation (see remark at the end of Theorem 3.4) leads to (3.3.6).

3.4. The Toeplitz Lattice and Unitary Matrix Integrals

3.4.1. Toeplitz lattice, factorization of moment matrices and bi-orthogonal polynomials. Recall that a Toeplitz matrix is one whose (i,j)-th entry depends only on $i - j$. Consider the inner product

$$\langle f(z), g(z) \rangle_{t,s} := \oint_{S^1} \frac{dz}{2\pi i z} f(z) g(z^{-1}) e^{\sum_1^\infty (t_i z^i - s_i z^{-i})}, \quad t, s \in \mathbb{C}^\infty, \qquad (3.4.1)$$

where the integral is taken over the unit circle $S^1 \subset \mathbb{C}$ around the origin. It has the property

$$\langle z^k f, g \rangle_{t,s} = \langle f, z^{-k} g \rangle_{t,s}. \qquad (3.4.2)$$

The t, s-dependent semi-infinite moment matrix $m_\infty(t,s)$, where

$$m_n(t,s) := \left(\langle z^k, z^l \rangle_{t,s}\right)_{0 \leq k,l \leq n-1} = \left(\oint_{S^1} \frac{\rho(z)\,dz}{2\pi i z} z^{k-l} e^{\sum_1^\infty (t_i z^i - s_i z^{-i})}\right)_{0 \leq k,l \leq n-1}$$

$$= \text{Toeplitz matrix} \qquad (3.4.3)$$

satisfies the same differential equations as in (3.3.3):

$$\frac{\partial m_\infty}{\partial t_n} = \Lambda^n m_\infty \quad \text{and} \quad \frac{\partial m_\infty}{\partial s_n} = -m_\infty \Lambda^{\top n}. \quad \text{(2-Toda Lattice)} \qquad (3.4.4)$$

As before, define

$$\tau_n(t,s) := \det m_n(t,s).$$

Also, consider the factorization $m_\infty(t,s) = S_1^{-1}(t,s)S_2(t,s)$, as in (3.3.2), from which one defines $L_1 := S_1 \Lambda S_1^{-1}$ and $L_2 := S_2 \Lambda^\top S_2^{-1}$ and the bi-orthogonal polynomials $p_i^{(k)}(t,s;z)$ for $k = 1, 2$. Since m_∞ satisfies the same equations (3.3.3), the matrices L_1 and L_2 satisfy the 2-Toda lattice equations; the Toeplitz nature of m_∞ implies a peculiar "rank 2"-structure, with $h_i/h_{i-1} = 1 - x_i y_i$ and $x_0 = y_0 = 1$:

$$h^{-1}L_1 h = \begin{pmatrix} -x_1 y_0 & 1 - x_1 y_1 & 0 & 0 & \\ -x_2 y_0 & -x_2 y_1 & 1 - x_2 y_2 & 0 & \\ -x_3 y_0 & -x_3 y_1 & -x_3 y_2 & 1 - x_3 y_3 & \\ -x_4 y_0 & -x_4 y_1 & -x_4 y_2 & -x_4 y_3 & \\ & & & & \ddots \end{pmatrix}$$

and

$$L_2 = \begin{pmatrix} -x_0 y_1 & -x_0 y_2 & -x_0 y_3 & -x_0 y_4 & \\ 1 - x_1 y_1 & -x_1 y_2 & -x_1 y_3 & -x_1 y_4 & \\ 0 & 1 - x_2 y_2 & -x_2 y_3 & -x_2 y_4 & \\ 0 & 0 & 1 - x_3 y_3 & -x_3 y_4 & \\ & & & & \ddots \end{pmatrix}. \quad (3.4.5)$$

Some of the ideas in the next theorem are inspired by the work of Hisakado [37].

THEOREM 3.5 (Adler and van Moerbeke [7]). *The integrals* $I_n(t,s)$, *with* $I_0 = 1$,

$$\tau_n(t,s) = \det m_n = \frac{1}{n!} I_n := \frac{1}{n!} \int_{(S^1)^n} |\Delta_n(z)|^2 \prod_{k=1}^n \left(e^{\sum_1^\infty (t_i z_k^i - s_i z_k^{-i})} \frac{dz_k}{2\pi i z_k} \right)$$

$$= \int_{U(n)} e^{\sum_1^\infty \operatorname{Tr}(t_i M^i - s_i \bar{M}^i)} dM$$

$$= \sum_{\{\text{Young diagrams } \lambda | \hat{\lambda}_1 \leq n\}} \mathbf{s}_\lambda(t) \mathbf{s}_\lambda(-s), \quad (3.4.6)$$

satisfy the following relations:

(i) *An* $\mathrm{SL}(2,\mathbb{Z})$-*algebra of three* <u>Virasoro constraints</u> (2.3.2):

$$\mathbb{J}_{k,n}^{(2)}(t,n) - \mathbb{J}_{-k,n}^{(2)}(-s,n) - k\big(\theta \mathbb{J}_{k,n}^{(1)}(t,n) + (1-\theta)\mathbb{J}_{-k,n}^{(1)}(-s,n)\big) I_n(t,s) = 0,$$

$$\text{for } \begin{cases} k = -1, & \theta = 0, \\ k = 0, & \theta \text{ arbitrary}, \\ k = 1, & \theta = 1. \end{cases} \quad (3.4.7)$$

(ii) <u>2d-Toda identities</u>: *The matrices* L_1 *and* L_2 *defined above satisfy the 2-Toda lattice equations* (3.3.8); *in particular,*

$$\frac{\partial^2}{\partial s_1 \partial t_1} \log \tau_n = -\frac{\tau_{n-1} \tau_{n+1}}{\tau_n^2}.$$

and

$$\frac{\partial^2}{\partial s_2 \partial t_1} \log \tau_n = -2 \frac{\partial}{\partial s_1} \log \frac{\tau_n}{\tau_{n-1}} \cdot \frac{\partial^2}{\partial s_1 \partial t_1} \log \tau_n - \frac{\partial^3}{\partial s_1^2 \partial t_1} \log \tau_n, \quad (3.4.8)$$

the first being equivalent to the discrete sinh-Gordon equation

$$\frac{\partial^2 q_n}{\partial t_1 \partial s_1} = e^{q_n - q_{n-1}} - e^{q_{n+1} - q_n}, \quad \text{where } q_n = \log \frac{\tau_{n+1}}{\tau_n}.$$

(iii) <u>The Toeplitz lattice</u>: The 2-Toda lattice solution is a very special one — the matrices L_1 and L_2 have a "rank 2" structure, given by (3.4.5), whose x_n's and y_n's equal[25]

$$x_n(t,s) = \frac{1}{\tau_n} \int_{U(n)} \mathbf{s}_n \left(-\operatorname{Tr} M, -\tfrac{1}{2} \operatorname{Tr} M^2, -\tfrac{1}{3} \operatorname{Tr} M^3, \ldots \right) e^{\sum_1^\infty \operatorname{Tr}(t_i M^i - s_i \bar{M}^i)} dM$$

$$= \frac{\mathbf{s}_n(-\frac{\partial}{\partial t_1}, -\frac{1}{2}\frac{\partial}{\partial t_2}, -\frac{1}{3}\frac{\partial}{\partial t_3}, \ldots) \tau_n(t,s)}{\tau_n(t,s)} = p_n^{(1)}(t,s;0),$$

$$y_n(t,s) = \frac{1}{\tau_n} \int_{U(n)} \mathbf{s}_n \left(-\operatorname{Tr} \bar{M}, -\tfrac{1}{2} \operatorname{Tr} \bar{M}^2, -\tfrac{1}{3} \operatorname{Tr} \bar{M}^3, \ldots \right) e^{\sum_1^\infty \operatorname{Tr}(t_i M^i - s_i \bar{M}^i)} dM$$

$$= \frac{\mathbf{s}_n(\frac{\partial}{\partial t_1}, \frac{1}{2}\frac{\partial}{\partial t_2}, \frac{1}{3}\frac{\partial}{\partial t_3}, \ldots) \tau_n(t,s)}{\tau_n(t,s)} = p_n^{(2)}(t,s;0),$$

$$(3.4.9)$$

and satisfy the integrable Hamiltonian system

$$\frac{\partial x_n}{\partial t_i} = (1 - x_n y_n) \frac{\partial H_i^{(1)}}{\partial y_n}, \quad \frac{\partial y_n}{\partial t_i} = -(1 - x_n y_n) \frac{\partial H_i^{(1)}}{\partial x_n},$$

$$\frac{\partial x_n}{\partial s_i} = (1 - x_n y_n) \frac{\partial H_i^{(2)}}{\partial y_n}, \quad \frac{\partial y_n}{\partial s_i} = -(1 - x_n y_n) \frac{\partial H_i^{(2)}}{\partial x_n} \quad (3.4.10)$$

(**Toeplitz lattice**), with initial condition $x_n(0,0) = y_n(0,0) = 0$ for $n \geq 1$ and boundary condition $x_0(t,s) = y_0(t,s) = 1$. The traces

$$H_i^{(k)} = -\frac{1}{i} \operatorname{Tr} L_k^i, \quad i = 1, 2, 3, \ldots, \ k = 1, 2$$

of the matrices L_i in (3.4.5) are integrals in involution with regard to the symplectic structure $\omega := \sum_0^\infty (1 - x_k y_k)^{-1} dx_k \wedge dy_k$. The Toeplitz nature of m_∞ leads to identities between τ's, the simplest (due to Hisakado [37]) being:

$$\left(1 + \frac{\partial^2}{\partial s_1 \partial t_1} \log \tau_{n+1}\right)\left(1 + \frac{\partial^2}{\partial s_1 \partial t_1} \log \tau_n\right) = -\frac{\partial}{\partial t_1} \log \frac{\tau_{n+1}}{\tau_n} \frac{\partial}{\partial s_1} \log \frac{\tau_{n+1}}{\tau_n}.$$

$$(3.4.11)$$

---

[25] Remember that the $\mathbf{s}(t_1, t_2, \ldots)$ are elementary Schur polynomials.

REMARK. The first equation in the hierarchy above reads:

$$\frac{\partial x_n}{\partial t_1} = x_{n+1}(1 - x_n y_n), \quad \frac{\partial y_n}{\partial t_1} = -y_{n-1}(1 - x_n y_n),$$

$$\frac{\partial x_n}{\partial s_1} = x_{n-1}(1 - x_n y_n), \quad \frac{\partial y_n}{\partial s_1} = -y_{n+1}(1 - x_n y_n).$$

3.4.2. Sketch of Proof. The identity (3.4.6) between the determinant and the moment matrix uses again the Vandermonde identity (3.3.12),

$$\int_{U(n)} e^{\sum_1^\infty \operatorname{Tr}(t_i M^i - s_i \bar{M}^i)} dM$$

$$= \int_{(S^1)^n} |\Delta_n(z)|^2 \prod_{k=1}^n \left(e^{\sum_1^\infty (t_i z_k^i - s_i z_k^{-i})} \frac{dz_k}{2\pi i z_k} \right)$$

$$= \int_{(S^1)^n} \Delta_n(z) \Delta_n(\bar{z}) \prod_{k=1}^n \left(e^{\sum_1^\infty (t_i z_k^i - s_i z_k^{-i})} \frac{dz_k}{2\pi i z_k} \right)$$

$$= \int_{(S^1)^n} \sum_{\sigma \in S_n} \det \left(z_{\sigma(m)}^{l-1} \bar{z}_{\sigma(m)}^{m-1} \right)_{1 \leq l,m \leq n} \prod_{k=1}^n \left(e^{\sum_1^\infty (t_i z_k^i - s_i z_k^{-i})} \frac{dz_k}{2\pi i z_k} \right)$$

$$= \sum_{\sigma \in S_n} \det \left(\oint_{S^1} z_k^{l-1} \bar{z}_k^{m-1} e^{\sum_1^\infty (t_i z_k^i - s_i z_k^{-i})} \frac{dz_k}{2\pi i z_k} \right)_{1 \leq l,m \leq n}$$

$$= n! \det \left(\oint_{S^1} z^{l-m} e^{\sum_1^\infty (t_i z^i - s_i z^{-i})} \frac{dz}{2\pi i z} \right)_{1 \leq l,m \leq n}$$

$$= n! \det m_n(t,s) = n! \tau_n(t).$$

The last equation of (3.4.6) follows from the fact that the solution to equations (3.4.4) with initial condition $m_\infty(0,0)$ is given by

$$m_\infty(t,s) = e^{\sum_1^\infty t_i \Lambda^i} m_\infty(0,0) e^{-\sum_1^\infty s_i \Lambda^{\top i}}.$$

Since (3.4.3) gives the initial condition $m_\infty(0,0) = I_\infty$, the Cauchy–Binet formula implies

$$\tau_n(t,s) = \det m_n(t,s)$$
$$= \sum_{\substack{\lambda,\nu \\ \lambda_1,\nu_1 \leq n}} s_\lambda(t) s_\nu(-s) \det \left(\mu_{\lambda_i - i + n, \nu_j - j + n} \right)_{1 \leq i,j \leq n} = \sum_{\{\lambda \mid \tilde{\lambda} \leq n\}} \mathbf{s}_\lambda(t) \mathbf{s}_\lambda(-s),$$

establishing the last equation of (3.4.6); details can be found in [7].

Using the equality $z^{k\top} = z^{-k}$ (see (3.4.2)), one shows that the polynomials

$$p_{n+1}^{(1)}(z) - z p_n^{(1)}(z) \quad \text{and} \quad p_{n+1}^{(1)}(0) z^n p_n^{(2)}(z^{-1})$$

are perpendicular to the monomials $z^0, z^1, \ldots, z^n$ and that they have the same z^0-term; one makes a similar argument, by dualizing $1 \leftrightarrow 2$. Therefore, we have the Hisakado identities between the polynomials

$$p_{n+1}^{(1)}(z) - zp_n^{(1)}(z) = p_{n+1}^{(1)}(0)z^n p_n^{(2)}(z^{-1})$$
$$p_{n+1}^{(2)}(z) - zp_n^{(2)}(z) = p_{n+1}^{(2)}(0)z^n p_n^{(1)}(z^{-1}). \qquad (3.4.12)$$

The rank 2 structure (3.4.5) of L_1 and L_2, with $x_n = p_n^{(1)}(t, s; 0)$ and $y_n = p_n^{(2)}(t, s; 0)$, is obtained by taking the inner product of $p_{n+1}^{(1)}(z) - zp_n^{(1)}(z)$ with itself, for different n and m, and using the fact that $zp_n^{(1)}(z) = L_1 p_n^{(1)}(z)$.

To check the first equation in the hierarchy (see remark at the end of Theorem 3.5), consider, from (3.4.9),

$$\frac{\partial x_n}{\partial t_1} = \left.\frac{\partial p_n^{(1)}(t, s; z)}{\partial t_1}\right|_{z=0} = -\left((L_1)_- p^{(1)}\right)_n\Big|_{z=0} \quad \text{using (3.3.15)}$$

$$= h_n p_{n+1}^{(1)}(t, s; 0) \sum_{i=0}^{n-1} \frac{p_i^{(1)}(t, s; 0) p_i^{(2)}(t, s; 0)}{h_i} \quad \text{using (3.4.5)}$$

$$= h_n x_{n+1} \sum_{i=0}^{n-1} \frac{x_i y_i}{h_i}$$

$$= h_n x_{n+1} \sum_{i=0}^{n-1} \left(\frac{1}{h_i} - \frac{1}{h_{i-1}}\right) \quad \text{using } \frac{h_i}{h_{i-1}} = 1 - x_i y_i$$

$$= x_{n+1} \frac{h_n}{h_{n-1}} = x_{n+1}(1 - x_n y_n),$$

and similarly for the other coordinates. From (3.3.7) and (3.4.5), upon making the products of the corresponding diagonal entries of L_1 and $hL_2^\top h^{-1}$, one finds (3.4.11):

$$\frac{\partial}{\partial t_1} \log \frac{\tau_{n+1}}{\tau_n} \frac{\partial}{\partial s_1} \log \frac{\tau_{n+1}}{\tau_n} = -x_{n+1} y_n x_n y_{n+1} = -x_n y_n x_{n+1} y_{n+1}$$

$$= -\left(1 - \frac{h_n}{h_{n-1}}\right)\left(1 - \frac{h_{n+1}}{h_n}\right).$$

4. Ensembles of Finite Random Matrices

4.1. PDEs Defined by the Probabilities in Hermitian, Symmetric and Symplectic Random Ensembles.
As used earlier, the disjoint union $E = \bigcup_1^r [c_{2i-1}, c_{2i}] \subset \mathbb{R}$, and the weight $\rho(z) = e^{-V(z)}$, with $-\rho'/\rho = V' = g/f$ define an algebra of differential operators

$$\mathcal{B}_k = \sum_1^{2r} c_i^{k+1} f(c_i) \frac{\partial}{\partial c_i}, \quad k \in \mathbb{Z}.$$

The aim of this section is to find PDEs for the following probabilities in terms of the boundary points c_i of E (see (1.1.9), (1.1.11) and (1.1.18)), i.e.

$$P_n(E) := P_n(\text{ all spectral points of } M \in E)$$
$$= \frac{\int_{\mathcal{H}_n(E),\ \mathcal{S}_n(E)\ \text{or}\ \mathcal{T}_n(E)} e^{-\operatorname{tr} V(M)} dM}{\int_{\mathcal{H}_n(\mathbb{R}),\ \mathcal{S}_n(\mathbb{R})\ \text{or}\ \mathcal{T}_n(\mathbb{R})} e^{-\operatorname{tr} V(M)} dM} \qquad (4.1.1)$$
$$= \frac{\int_{E^n} |\Delta_n(z)|^\beta \prod_{k=1}^n e^{-V(z_k)} dz_k}{\int_{\mathbb{R}^n} |\Delta_n(z)|^\beta \prod_{k=1}^n e^{-V(z_k)} dz_k}, \quad \beta = 2, 1, 4 \text{ respectively,}$$

involving the classical weights below. In anticipation, the equations obtained in Theorems 4.1, 4.2 and 4.3 are closely related to three of the six Painlevé differential equations:

weight	$\rho(z)$	Painlevé
Gauss	e^{-bz^2}	IV
Laguerre	$z^a e^{-bz}$	V
Jacobi	$(1-z)^a(1+z)^b$	VI

For $\beta = 2$, the probabilities satisfy partial differential equations in the boundary points of E, whereas in the case $\beta = 1, 4$, the equations are inductive. Namely, for $\beta = 1$ (resp. $\beta = 4$), the probabilities P_{n+2} (resp. P_{n+1}) are given in terms of P_{n-2} (resp. P_{n-1}) and a differential operator acting on P_n. The weights above involve the parameters β, a, b and

$$\delta_{1,4}^\beta := 2\left(\left(\frac{\beta}{2}\right)^{1/2} - \left(\frac{\beta}{2}\right)^{-1/2}\right)^2 = \begin{cases} 0 & \text{for } \beta = 2, \\ 1 & \text{for } \beta = 1, 4. \end{cases}$$

As a consequence of the duality (2.1.12) between β-Virasoro generators under the map $\beta \mapsto 4/\beta$, and the equations (2.1.7), the PDEs obtained have a remarkable property: the coefficients Q and Q_i of the PDEs are functions in the variables n, β, a, b, having the invariance property under the map

$$n \to -2n, \quad a \to -\frac{a}{2}, \quad b \to -\frac{b}{2};$$

to be precise,

$$Q_i\left(-2n, \beta, -\frac{a}{2}, -\frac{b}{2}\right)\bigg|_{\beta=1} = Q_i(n, \beta, a, b)\big|_{\beta=4}. \qquad (4.1.2)$$

The results in this section are mainly due to Adler, Shiota, and van Moerbeke [11] for $\beta = 2$ and to Adler and van Moerbeke [6] for $\beta = 1, 4$. For more detailed references, see the end of Section 4.2.

4.1.1. Gaussian Hermitian, symmetric and symplectic ensembles.
Given the disjoint union E and the weight e^{-bz^2}, the differential operators $\mathcal{B}_k$ take on the form

$$\mathcal{B}_k = \sum_1^{2r} c_i^{k+1} \frac{\partial}{\partial c_i}.$$

Define the *invariant* polynomials

$$Q = 12b^2 n\left(n + 1 - \frac{2}{\beta}\right) \quad \text{and} \quad Q_2 = 4(1 + \delta_{1,4}^\beta) b\left(2n + \delta_{1,4}^\beta\left(1 - \frac{2}{\beta}\right)\right).$$

THEOREM 4.1. *The probabilities*

$$P_n(E) = \frac{\int_{E^n} |\Delta_n(z)|^\beta \prod_{k=1}^n e^{-bz_k^2}\, dz_k}{\int_{\mathbb{R}^n} |\Delta_n(z)|^\beta \prod_{k=1}^n e^{-bz_k^2}\, dz_k}, \qquad (4.1.3)$$

with $\beta = 2, 1, 4$, satisfy the following PDE, where we put $F := F_n = \log P_n$:

$$\delta_{1,4}^\beta Q\left(\frac{P_{n-\frac{2}{1}} P_{n+\frac{2}{1}}}{P_n^2} - 1\right) \qquad \text{with index} \begin{cases} 2 & \text{when } n \text{ even and } \beta = 1, \\ 1 & \text{when } n \text{ arbitrary and } \beta = 4 \end{cases}$$

$$= \left(\mathcal{B}_{-1}^4 + (Q_2 + 6\mathcal{B}_{-1}^2 F)\mathcal{B}_{-1}^2 + 4(2 - \delta_{1,4}^\beta)\frac{b^2}{\beta}(3\mathcal{B}_0^2 - 4\mathcal{B}_{-1}\mathcal{B}_1 + 6\mathcal{B}_0)\right) F. \qquad (4.1.4)$$

4.1.2. Laguerre Hermitian, symmetric and symplectic ensembles.
Given the disjoint union $E \subset \mathbb{R}^+$ and the weight $z^a e^{-bz}$, the $\mathcal{B}_k$ take on the form

$$\mathcal{B}_k = \sum_1^{2r} c_i^{k+2} \frac{\partial}{\partial c_i}.$$

Define the polynomials, also respecting the duality (4.1.2),

$$Q = \begin{cases} \frac{3}{4} n(n-1)(n+2a)(n+2a+1) & \text{for } \beta = 1, \\ \frac{3}{2} n(2n+1)(2n+a)(2n+a-1) & \text{for } \beta = 4, \end{cases}$$

$$Q_2 = \left(3\beta n^2 - \frac{a^2}{\beta} + 6an + 4\left(1 - \frac{\beta}{2}\right)a + 3\right)\delta_{1,4}^\beta + (1 - a^2)(1 - \delta_{1,4}^\beta)$$

$$Q_1 = \left(\beta n^2 + 2an + \left(1 - \frac{\beta}{2}\right)a\right), \quad Q_0 = b(2 - \delta_{1,4}^\beta)\left(n + \frac{a}{\beta}\right).$$

THEOREM 4.2. *The probabilities*

$$P_n(E) = \frac{\int_{E^n} |\Delta_n(z)|^\beta \prod_{k=1}^n z_k^a e^{-bz_k}\, dz_k}{\int_{\mathbb{R}_+^n} |\Delta_n(z)|^\beta \prod_{k=1}^n z_k^a e^{-bz_k}\, dz_k} \qquad (4.1.5)$$

satisfy the following PDE, where we put $F := F_n = \log P_n$ and use the same convention on the indices $n \pm 2$ and $n \pm 1$ as in (4.1.4):

$$\delta_{1,4}^{\beta} Q\left(\frac{P_{n-1^2} P_{n+1^2}}{P_n^2} - 1\right) = \left(\mathcal{B}_{-1}^4 - 2(\delta_{1,4}^{\beta} + 1)\mathcal{B}_{-1}^3\right.$$
$$+ (Q_2 + 6\mathcal{B}_{-1}^2 F - 4(\delta_{1,4}^{\beta} + 1)\mathcal{B}_{-1}F)\mathcal{B}_{-1}^2 - 3\delta_{1,4}^{\beta}(Q_1 - \mathcal{B}_{-1}F)\mathcal{B}_{-1}$$
$$+ \frac{b^2}{\beta}(2 - \delta_{1,4}^{\beta})(3\mathcal{B}_0^2 - 4\mathcal{B}_1\mathcal{B}_{-1} - 2\mathcal{B}_1) + Q_0(2\mathcal{B}_0\mathcal{B}_{-1} - \mathcal{B}_0)\Big) F. \quad (4.1.6)$$

4.1.3. Jacobi Hermitian, symmetric and symplectic ensembles. In terms of $E \subset [-1, +1]$ and the Jacobi weight $(1-z)^a(1+z)^b$, the differential operators $\mathcal{B}_k$ take on the form

$$\mathcal{B}_k = \sum_1^{2r} c_i^{k+1}(1-c_i^2)\frac{\partial}{\partial c_i}.$$

With $b_0 = a - b$, $b_1 = a + b$, introduce the variables

$$r = \frac{4}{\beta}(b_0^2 + (b_1 + 2 - \beta)^2), \quad s = \frac{4}{\beta}b_0(b_1 + 2 - \beta), \quad q_n = \frac{4}{\beta}(\beta n + b_1 + 2 - \beta)(\beta n + b_1),$$

which themselves have the invariance property (4.1.2). Introduce also the following *invariant* polynomials in q, r, s:

$$\begin{aligned}
Q &= \tfrac{3}{16}\left((s^2 - qr + q^2)^2 - 4(rs^2 - 4qs^2 - 4s^2 + q^2 r)\right), \\
Q_1 &= 3s^2 - 3qr - 6r + 2q^2 + 23q + 24 \\
Q_2 &= 3qs^2 + 9s^2 - 4q^2 r + 2qr + 4q^3 + 10q^2, \\
Q_3 &= 3qs^2 + 6s^2 - 3q^2 r + q^3 + 4q^2 \\
Q_4 &= 9s^2 - 3qr - 6r + q^2 + 22q + 24 = Q_1 + (6s^2 - q^2 - q). \quad (4.1.7)
\end{aligned}$$

THEOREM 4.3. *The probabilities*

$$P_n(E) = \frac{\int_{E^n} |\Delta_n(z)|^{\beta} \prod_{k=1}^n (1-z_k)^a(1+z_k)^b \, dz_k}{\int_{[-1,1]^n} |\Delta_n(z)|^{\beta} \prod_{k=1}^n (1-z_k)^a(1+z_k)^b \, dz_k} \quad (4.1.8)$$

satisfy the following PDE, where we put $F = F_n = \log P_n$:

for $\beta = 2$:

$$\left(2\mathcal{B}_{-1}^4 + (q-r+4)\mathcal{B}_{-1}^2 - (4\mathcal{B}_{-1}F - s)\mathcal{B}_{-1} + 3q\mathcal{B}_0^2 - 2q\mathcal{B}_0 + 8\mathcal{B}_0\mathcal{B}_{-1}^2\right.$$
$$- 4(q-1)\mathcal{B}_1\mathcal{B}_{-1} + (4\mathcal{B}_{-1}F - s)\mathcal{B}_1 + 2(4\mathcal{B}_{-1}F - s)\mathcal{B}_0\mathcal{B}_{-1} + 2q\mathcal{B}_2\Big)F$$
$$+ 4\mathcal{B}_{-1}^2 F\left(2\mathcal{B}_0 F + 3\mathcal{B}_{-1}^2 F\right) = 0 \quad (4.1.9)$$

for $\beta = 1, 4$:

$$Q\left(\frac{P_{n+\frac{2}{1}}P_{n-\frac{2}{1}}}{P_n^2} - 1\right)$$
$$= (q+1)\left(4q\mathcal{B}_{-1}^4 + 12(4\mathcal{B}_{-1}F - s)\mathcal{B}_{-1}^3 + 2(q+12)(4\mathcal{B}_{-1}F - s)\mathcal{B}_0\mathcal{B}_{-1}\right.$$
$$+ 3q^2\mathcal{B}_0^2 - 4(q-4)q\mathcal{B}_1\mathcal{B}_{-1} + q(4\mathcal{B}_{-1}F - s)\mathcal{B}_1 + 20q\mathcal{B}_0\mathcal{B}_{-1}^2 + 2q^2\mathcal{B}_2\big)F$$
$$+ \left(Q_2\mathcal{B}_{-1}^2 - sQ_1\mathcal{B}_{-1} + Q_3\mathcal{B}_0\right)F + 48(\mathcal{B}_{-1}F)^4 - 48s(\mathcal{B}_{-1}F)^3 + 2Q_4(\mathcal{B}_{-1}F)^2$$
$$+ 12q^2(\mathcal{B}_0F)^2 + 16q(2q-1)(\mathcal{B}_{-1}^2F)(\mathcal{B}_0F) + 24(q-1)q(\mathcal{B}_{-1}^2F)^2$$
$$+ 24(2\mathcal{B}_{-1}F - s)\left((q+2)\mathcal{B}_0F + (q+3)\mathcal{B}_{-1}^2F\right)\mathcal{B}_{-1}F. \tag{4.1.10}$$

The proof of these three theorems will be sketched in Sections 4.3, 4.4 and 4.5.

4.2. ODEs When E Has One Boundary Point. Assume the set E consists of one boundary point $c = x$, besides the boundary of the full range; thus, setting respectively $E = [-\infty, x]$, $E = [0, x]$, $E = [-1, x]$ in the PDEs (4.1.4), (4.1.6) and (4.1.9), (4.1.10), leads to the equations in x below. Notice that, for $\beta = 2$, the equations obtained are ODEs and, for $\beta = 1, 4$, these equations express P_{n+2} in terms of P_{n-2} and a differential operator acting on P_n:

(1) *Gauss ensemble* $(\beta = 2, 1, 4)$: $f_n(x) = \frac{d}{dx} \log P_n(\max_i \lambda_i \leq x)$ satisfies

$$\delta_{1,4}^\beta Q\left(\frac{P_{n-\frac{2}{1}}P_{n+\frac{2}{1}}}{P_n^2} - 1\right)$$
$$= f_n''' + 6f_n'^2 + \left(4\frac{b^2x^2}{\beta}(\delta_{1,4}^\beta - 2) + Q_2\right)f_n' - 4\frac{b^2x}{\beta}(\delta_{1,4}^\beta - 2)f_n. \tag{4.2.1}$$

(2) *Laguerre ensemble* $(\beta = 2, 1, 4)$: $f_n(x) = x\frac{d}{dx}\log P_n(\max_i \lambda_i \leq x)$ (with all eigenvalues $\lambda_i \geq 0$) satisfies:

$$\delta_{1,4}^\beta Q\left(\frac{P_{n-\frac{2}{1}}P_{n+\frac{2}{1}}}{P_n^2} - 1\right) - \left(3\delta_{1,4}^\beta f_n - \frac{b^2x^2}{\beta}(\delta_{1,4}^\beta - 2) - Q_0x - 3\delta_{1,4}^\beta Q_1\right)f_n$$
$$= x^3 f_n''' - (2\delta_{1,4}^\beta - 1)x^2 f_n'' + 6x^2 f_n'^2$$
$$- x\left(4(\delta_{1,4}^\beta + 1)f_n - \frac{b^2x^2}{\beta}(\delta_{1,4}^\beta - 2) - 2Q_0x - Q_2 + 2\delta_{1,4}^\beta + 1\right)f_n'. \tag{4.2.2}$$

(3) *Jacobi ensemble*: $f := f_n(x) - (1 - x^2)\frac{d}{dx}\log P_n(\max_i \lambda_i \leq x)$ (with all eigenvalues $-1 \leq \lambda_i \leq 1$) satisfies:

• for $\beta = 2$:

$$2(x^2 - 1)^2 f''' + 4(x^2 - 1)\left(xf'' - 3f'^2\right) + \left(16xf - q(x^2 - 1) - 2sx - r\right)f'$$
$$- f\left(4f - qx - s\right) = 0, \tag{4.2.3}$$

- for $\beta = 1, 4$:

$$Q\left(\frac{P_{n+\frac{2}{1}}P_{n-\frac{2}{1}}}{P_n^2} - 1\right)$$
$$= 4(q+1)(x^2-1)^2\left(-q(x^2-1)f''' + (12f - qx - 3s)f'' + 6q(q-1)f'^2\right)$$
$$-(x^2-1)f'\left(24f(q+3)(2f-s) + 8fq(5q-1)x - q(q+1)(qx^2+2sx+8) + Q_2\right)$$
$$+ f\left(48f^3 + 48f^2(qx+2x-s) + 2f(8q^2x^2 + 2qx^2 - 12qsx - 24sx + Q_4)\right.$$
$$\left. - q(q+1)x(3qx^2 + sx - 2qx - 3q) + Q_3x - Q_1s\right). \qquad (4.2.4)$$

For $\beta = 2$, the term containing the ratio $(P_{n+\frac{2}{1}}P_{n-\frac{2}{1}}/P_n^2) - 1$ on the left-hand side of (4.2.1), (4.2.2) and (4.2.4) vanishes, and one thus obtains the following ODEs:

- **Gauss:** $f_n(x) := \dfrac{d}{dx}\log P_n(\max_i \lambda_i \leq x)$ satisfies
$$f''' + 6f'^2 + 4b(2n - bx^2)f' + 4b^2xf = 0.$$

- **Laguerre:** $f_n(x) := x\dfrac{d}{dx}\log P_n(\max_i \lambda_i \leq x)$ satisfies
$$x^2f''' + xf'' + 6xf'^2 - 4ff' - ((a-bx)^2 - 4nbx)f' - b(2n + a - bx)f = 0.$$

- **Jacobi:** $f_n(x) = (1 - x^2)\dfrac{d}{dx}\log P_n(\max_i \lambda_i \leq x)$ satisfies
$$2(x^2-1)^2 f''' + 4(x^2-1)\left(xf'' - 3f'^2\right) + \left(16xf - q(x^2-1) - 2sx - r\right)f'$$
$$- f\left(4f - qx - s\right) = 0.$$

Each of these three equations is of the Chazy form (see the Appendix on Chazy classes)

$$f''' + \frac{P'}{P}f'' + \frac{6}{P}f'^2 - \frac{4P'}{P^2}ff' + \frac{P''}{P^2}f^2 + \frac{4Q}{P^2}f' - \frac{2Q'}{P^2}f + \frac{2R}{P^2} = 0, \qquad (4.2.5)$$

with $c = 0$ and P, Q, R having the following form:

Gauss	$P(x) = 1$	$4Q(x) = -4b^2x^2 + 8bn$	$R = 0$
Laguerre	$P(x) = x$	$4Q(x) = -(bx-a)^2 + 4bnx$	$R = 0$
Jacobi	$P(x) = 1 - x^2$	$4Q(x) = -\frac{1}{2}(q(x^2-1) + 2sx + r)$	$R = 0$

Cosgrove shows that such a third order equation (4.2.5) in $f(x)$, with $P(x)$, $Q(x)$, $R(x)$ of respective degrees 3, 2, 1, has a first integral (9.0.2), which is second order in f and quadratic in f'', with an integration constant c. Equation (9.0.2) is a master Painlevé equation, containing the 6 Painlevé equations. If $f(x)$ satisfies the equations above, then the new (renormalized) function $g(z)$ defined by

Gauss	$g(z) = b^{-1/2}f(zb^{-1/2}) + \frac{2}{3}nz$
Laguerre	$g(z) = f(z) + \frac{1}{4}b(2n+a)z + \frac{1}{4}a^2$
Jacobi	$g(z) := -\frac{1}{2}f(x)\big|_{x=2z-1} - \frac{1}{8}qz + \frac{1}{16}(q+s)$

satisfies the canonical equations, which then can be transformed into the standard Painlevé equations; these canonical equations are respectively

- $g''^2 = -4g'^3 + 4(zg' - g)^2 + A_1 g' + A_2,$ (**Painlevé IV**)
- $(zg'')^2 = (zg' - g)(-4g'^2 + A_1(zg' - g) + A_2) + A_3 g' + A_4,$ (**Painlevé V**)
- $(z(z-1)g'')^2 = (zg' - g)(4g'^2 - 4g'(zg' - g) + A_2) + A_1 g'^2 + A_3 g' + A_4,$ (**Painlevé VI**)

with respective coefficients

- $A_1 = 3\left(\frac{4}{3}n\right)^2,\ A_2 = -\left(\frac{4}{3}n\right)^3,$
- $A_1 = b^2,\ A_2 = b^2\left((n + \frac{1}{2}a)^2 + \frac{1}{2}a^2\right),\ A_3 = -a^2 b(n + \frac{1}{2}a),\ A_4 = \frac{1}{2}(ab)^2 \times \left((n + \frac{1}{2}a)^2 + \frac{1}{8}a^2\right),$
- $A_1 = \frac{1}{8}(2q + r),\ A_2 = \frac{1}{16}qs,\ A_3 = \frac{1}{64}\left((q - s)^2 + 2qr\right),\ A_4 = \frac{1}{512}q(2s^2 + qr).$

For $\beta = 1$ and 4, the inductive partial differential equations (4.1.4), (4.1.6), (4.1.10), and the derived differential equations (4.2.1), (4.2.2) and (4.2.4) are due to Adler and van Moerbeke [6]. For $\beta = 2$ and for general E, they were first computed by Adler, Shiota, and van Moerbeke [11], using the method of the present paper. For $\beta = 2$ and for E having one boundary point, the equations obtained here coincide with the ones first obtained by Tracy and Widom in [63], who recognized them to be Painlevé IV and V for the Gaussian and Laguerre distribution respectively. In his Louvain doctoral dissertation, J. P. Semengue, together with L. Haine [32], were lead to Painlevé VI for the Jacobi ensemble, for $\beta = 2$ and E having one boundary point, upon subtracting the Tracy and Widom differential equation [63] from the one computed with the method of Adler, Shiota, and van Moerbeke [11]. The classification by Cosgrove [23] and Cosgrove and Scoufis [24, (A.3)] leads directly to these results.

4.3. Proof of Theorems 4.1, 4.2 and 4.3

4.3.1. Gaussian and Laguerre ensembles. The three first Virasoro equations, as in (2.1.29) and (2.1.32), are differential equations, involving partials in $t \in \mathbb{C}^\infty$ and partials $\mathcal{D}_1, \mathcal{D}_2, \mathcal{D}_3$ in $c = (c_1, \ldots, c_{2r}) \in \mathbb{R}^{2r}$, for $F := F_n(t,c) = \log I_n$; they have the general form:

$$\mathcal{D}_k F = \frac{\partial F}{\partial t_k} + \sum_{-1 \leq j < k} \gamma_{kj} V_j(F) + \gamma_k + \delta_k t_1, \quad k = 1,2,3, \quad (4.3.1)$$

with first $V_j(F)$'s given by

$$V_j(F) = \sum_{i, i+j \geq 1} it_i \frac{\partial F}{\partial t_{i+j}} + \frac{\beta}{2} \delta_{2,j}\left(\frac{\partial^2 F}{\partial t_1^2} + \left(\frac{\partial F}{\partial t_1}\right)^2\right), \quad -1 \leq j \leq 2. \quad (4.3.2)$$

In (4.3.1) and (4.3.2), $\beta > 0, \gamma_{kj}, \gamma_k, \delta_k$ are arbitrary parameters; also $\delta_{2j} = 0$ for $j \neq 2$ and $= 1$ for $j = 2$. The claim is that the equations (4.3.1) enable one to express all partial derivatives,

$$\left.\frac{\partial^{i_1+\cdots+i_k} F(t,c)}{\partial t_1^{i_1}\cdots \partial t_k^{i_k}}\right|_{\mathcal{L}}, \quad \text{along } \mathcal{L} := \{\text{all } t_i = 0,\ c = (c_1,\ldots,c_{2r})\text{ arbitrary}\}, \tag{4.3.3}$$

uniquely in terms of polynomials in

$$\mathcal{D}_{j_1}\ldots\mathcal{D}_{j_r}F(0,c).$$

Indeed, the method consists of expressing $\partial F/\partial t_k|_{t=0}$ in terms of $\mathcal{D}_k f|_{t=0}$, using (4.3.1). Second derivatives are obtained by acting on $\mathcal{D}_k F$ with $\mathcal{D}_l$, by noting that $\mathcal{D}_l$ commutes with all t-derivatives, by using the equation for $\mathcal{D}_l F$, and by setting in the end $t=0$:

$$\begin{aligned}
\mathcal{D}_l \mathcal{D}_k F &= \mathcal{D}_l \frac{\partial F}{\partial t_k} + \sum_{-1\le j<k} \gamma_{kj} \mathcal{D}_l(V_j(F)) \\
&= \left(\frac{\partial}{\partial t_k} + \sum_{-1\le j<k} \gamma_{kj} V_j\right)\mathcal{D}_l(F), \quad \text{provided } V_j(F) \text{ does not contain nonlinear terms} \\
&= \left(\frac{\partial}{\partial t_k} + \sum_{-1\le j<k} \gamma_{kj} V_j\right)\left(\frac{\partial F}{\partial t_l} + \sum_{-1\le j<l} \gamma_{lj} V_j(F) + \delta_l t_1\right) \\
&= \frac{\partial^2 F}{\partial t_k \partial t_l} + \text{lower-weight terms.}
\end{aligned}$$

When the nonlinear term is present, it is taken care of as follows:

$$\mathcal{D}_l\left(\frac{\partial F}{\partial t_1}\right)^2 = 2\frac{\partial F}{\partial t_1}\mathcal{D}_l\frac{\partial F}{\partial t_1} = 2\frac{\partial F}{\partial t_1}\frac{\partial}{\partial t_1}\left(\frac{\partial F}{\partial t_l} + \sum_{-1\le j<l}\gamma_{lj}V_j(F) + \gamma_l + \delta_l t_1\right).$$

Higher derivatives are obtained in the same way. We only record here, for future use, the few partials appearing in the KP equation (3.1.6):

$$\left.\frac{\partial^2 F}{\partial t_1^2}\right|_{\mathcal{L}} = \left(\mathcal{D}_1^2 - \gamma_{10}\mathcal{D}_1\right)F + \gamma_{10}\gamma_1 - \delta_1$$

$$\left.\frac{\partial^4 F}{\partial t_1^4}\right|_{\mathcal{L}} = \left(\mathcal{D}_1^4 - 6\gamma_{10}\mathcal{D}_1^3 + 11\gamma_{10}^2\mathcal{D}_1^2 - 6\gamma_{10}^3\mathcal{D}_1\right)F - 6\gamma_{10}^2(\delta_1 - \gamma_1\gamma_{10})$$

$$\left.\frac{\partial^2 F}{\partial t_2^2}\right|_{\mathcal{L}} = \left(\mathcal{D}_2^2 - 2\gamma_{20}\mathcal{D}_2 + \beta\gamma_{21}\gamma_{32}\mathcal{D}_1^2 - ((2\gamma_1+\gamma_{10})\gamma_{21}\gamma_{32}\beta + 2\gamma_{2,-1})\mathcal{D}_1 - 2\gamma_{21}\mathcal{D}_3\right)F$$
$$+ \beta\gamma_{21}\gamma_{32}(\mathcal{D}_1 F)^2 + \beta\gamma_{21}\gamma_{32}(\gamma_1^2 + \gamma_{10}\gamma_1 - \delta_1) + 2(\gamma_{21}\gamma_3 + \gamma_{20}\gamma_2 + \gamma_1\gamma_{2,-1})$$

$$\left.\frac{\partial^2 F}{\partial t_1 \partial t_3}\right|_{\mathcal{L}} = \left(\mathcal{D}_1\mathcal{D}_3 - \frac{\beta}{2}\gamma_{32}\mathcal{D}_1^3 + \beta\gamma_{32}(\gamma_1+2\gamma_{10})\mathcal{D}_1^2 - \frac{3\beta}{2}\gamma_{10}\gamma_{32}(2\gamma_1+\gamma_{10})\mathcal{D}_1\right.$$
$$\left. - 3\gamma_{1,-1}\mathcal{D}_2 - 3\gamma_{10}\mathcal{D}_3\right)F + \frac{3\beta}{2}\gamma_{10}\gamma_{32}(\mathcal{D}_1 F)^2 - \beta\gamma_{32}(\mathcal{D}_1 F)(\mathcal{D}_1^2 F)$$
$$+ \tfrac{3}{2}\left(2\gamma_{10}\gamma_3 + \beta\gamma_{32}\gamma_{10}(\gamma_1^2 + \gamma_{10}\gamma_1 - \delta_1) + 2\gamma_{1,-1}\gamma_2\right).$$

4.3.2. Jacobi ensemble.
Here, from the Virasoro constraints (2.1.35), one proceeds in the same way as before, by forming $\mathcal{B}_i F|_{t=0}$, $\mathcal{B}_i \mathcal{B}_j F|_{t=0}$, etc., in terms of t_i partials. For example, from the expressions $\mathcal{B}_{-1} F|_{t=0}$, $\mathcal{B}_{-1}^2 F|_{t=0}$, $\mathcal{B}_0 F|_{t=0}$, one extracts

$$\frac{\partial F}{\partial t_1}\Big|_{t=0}, \quad \frac{\partial^2 F}{\partial t_1^2}\Big|_{t=0}, \quad \frac{\partial F}{\partial t_2}\Big|_{t=0}.$$

From the expressions $\mathcal{B}_{-1}^3 F|_{t=0}$, $\mathcal{B}_0 \mathcal{B}_{-1} F|_{t=0}$, $\mathcal{B}_1 F|_{t=0}$, and using the previous information, one extracts

$$\frac{\partial F}{\partial t_3}\Big|_{t=0}, \quad \frac{\partial^2 F}{\partial t_1^3}\Big|_{t=0}, \quad \frac{\partial^2 F}{\partial t_1 \partial t_2}\Big|_{t=0}.$$

Finally, from the expressions $\mathcal{B}_2 F|_{t=0}$, $\mathcal{B}_1 \mathcal{B}_{-1} F|_{t=0}$, $\mathcal{B}_0^2 F|_{t=0}$, $\mathcal{B}_0 \mathcal{B}_{-1}^2 F|_{t=0}$, $\mathcal{B}_{-1}^4 F|_{t=0}$, one deduces

$$\frac{\partial^4 F}{\partial t_1^4}\Big|_{t=0}, \quad \frac{\partial F}{\partial t_4}\Big|_{t=0}, \quad \frac{\partial^3 F}{\partial t_1^2 \partial t_2}\Big|_{t=0}, \quad \frac{\partial^2 F}{\partial t_1 \partial t_3}\Big|_{t=0}, \quad \frac{\partial^2 F}{\partial t_2^2}\Big|_{t=0}. \qquad (4.3.4)$$

This provides all the partials, appearing in the KP equation (3.1.6).

4.3.3. Inserting partials into the integrable equation.
From Theorem 3.3, the integrals $I_n(t, c)$, depending on $\beta = 2, 1, 4$, on $t = (t_1, t_2, \dots)$ and on the boundary points $c = (c_1, \dots, c_{2r})$ of E, relate to τ-functions, as follows:

$$I_n(t, c) = \int_{E^n} |\Delta_n(z)|^\beta \prod_{k=1}^n \left(e^{\sum_1^\infty t_i z_k^i} \rho(z_k)\, dz_k \right)$$

$$= \begin{cases} n!\, \tau_n(t, c), & n \text{ arbitrary}, \beta = 2, \\ n!\, \tau_n(t, c), & n \text{ even}, \beta = 1, \\ n!\, \tau_{2n}(t/2, c), & n \text{ arbitrary}, \beta = 4, \end{cases} \qquad (4.3.5)$$

where $\tau_n(t, c)$ satisfies the KP-like equation

$$12 \frac{\tau_{n-2}(t, c)\tau_{n+2}(t, c)}{\tau_n(t, c)^2} \delta_{1,4}^\beta = (KP)_t \log \tau_n(t, c), \quad \begin{cases} n \text{ arbitrary for } \beta = 2, \\ n \text{ even for } \beta = 1, 4, \end{cases} \qquad (4.3.6)$$

with

$$(KP)_t F := \left(\left(\frac{\partial}{\partial t_1}\right)^4 + 3\left(\frac{\partial}{\partial t_2}\right)^2 - 4 \frac{\partial^2}{\partial t_1 \partial t_3} \right) F + 6 \left(\frac{\partial^2}{\partial t_1^2} F \right)^2.$$

Evaluating the left hand side of (4.3.6): Here $I_n(t)$ will refer to the integral (4.3.5) over the full range. For $\beta = 2$, the left-hand side is zero. For $\beta = 1$, the left-hand side can be evaluated in terms of the probability $P_n(E)$, as follows: taking into account $P_n := P_n(E) = I_n(0, c)/I_n(0)$,

$$12 \frac{\tau_{n-2}(t, c)\tau_{n+2}(t, c)}{\tau_n(t, c)^2}\Big|_{t=0} = 12 \frac{(n!)^2}{(n-2)!(n+2)!} \frac{I_{n-2}(t, c) I_{n+2}(t, c)}{I_n(t, c)^2}\Big|_{t=0}$$

$$= 12\frac{n(n-1)}{(n+1)(n+2)}\frac{I_{n-2}(0)I_{n+2}(0)}{I_n(0)^2}\frac{P_{n-2}P_{n+2}}{P_n^2}$$

$$= 12b_n^{(1)}\frac{P_{n-2}(E)P_{n+2}(E)}{P_n^2(E)},$$

with $b_n^{(1)}$ given by[26]

$$b_n^{(1)} = \frac{n(n-1)}{(n+2)(n+1)}\frac{I_{n-2}(0)I_{n+2}(0)}{I_n(0)^2} = \begin{cases} \dfrac{n(n-1)}{16b^2} & \text{(Gauss)}, \\ \dfrac{n(n-1)(n+2a)(n+2a+1)}{16b^4} & \text{(Laguerre)}, \\ \dfrac{Q}{Q_6^\pm} & \text{(Jacobi)}. \end{cases}$$

(4.3.7)

For $\beta = 4$, we have

$$12\frac{\tau_{2n-2}(t/2,c)\tau_{2n+2}(t/2,c)}{\tau_{2n}(t/2,c)^2}\bigg|_{t=0} = 12\frac{(n!)^2}{(n-1)!(n+1)!}\frac{I_{n-1}(t,c)I_{n+1}(t,c)}{I_n(t,c)^2}\bigg|_{t=0}$$

$$= 12\frac{n}{(n+1)}\frac{I_{n-1}(0)I_{n+1}(0)}{I_n(0)^2}\frac{P_{n-1}P_{n+1}}{P_n^2}$$

$$= 12b_n^{(4)}\frac{P_{n-1}(E)P_{n+1}(E)}{P_n^2(E)},$$

with

$$b_n^{(4)} = \frac{(n!)^2}{(n-1)!(n+1)!}\frac{I_{n-1}(0)I_{n+1}(0)}{I_n^2(0)} = \begin{cases} \dfrac{2n(2n+1)}{4b^2} & \text{(Gauss)}, \\ \dfrac{2n(2n+1)(2n+a)(2n+a-1)}{b^4} & \text{(Laguerre)}, \\ \dfrac{Q}{Q_6^\pm} & \text{(Jacobi)}, \end{cases}$$

(4.3.8)

where Q is precisely the expression appearing in (4.1.7) and where

$$Q_6^\pm = 3q\,(q+1)\,(q-3)\,\left(q+4\pm 4\sqrt{q+1}\right)\quad \begin{cases} + \text{ for } \beta = 1, \\ - \text{ for } \beta = 4. \end{cases} \quad (4.3.9)$$

The exact formulae $b_n^{(4)}$ and $b_n^{(1)}$ show they satisfy the duality property (4.1.2):

$$b_n^{(4)}(a,b,n) = b_n^{(1)}\left(-\tfrac{1}{2}a, -\tfrac{1}{2}b, -2n\right).$$

---

[26]This calculation is based on Selberg's integrals: see Mehta [49, p. 340]. For instance, in the Jacobi case, one uses

$$I_n^{(\beta)} = \int_{[-1,1]^n} \Delta_n(x)^\beta \prod_{j=1}^n (1-x_j)^a (1+x_j)^b\, dx_j$$

$$= 2^{n(2a+2b+\beta(n-1)+2)/2} \prod_{j=0}^{n-1} \frac{\Gamma(a+j\beta/2+1)\Gamma(b+j\beta/2+1)\Gamma((j+1)\beta/2+1)}{\Gamma(\beta/2+1)\Gamma(a+b+(n+j-1)\beta/2+2)}.$$

Evaluating the right-hand side of (4.3.6): From Section 2.4, it also follows that $F_n(t;c) = \log I_n(t;c)$ satisfies Virasoro constraints, corresponding precisely to the situation (4.3.1), with

Gaussian ensemble:[27]

$$\gamma_{1,-1} = -\tfrac{1}{2},\ \gamma_{1,0} = \gamma_1 = 0, \delta_1 = -\tfrac{1}{2}n;$$
$$\gamma_{2,-1} = 0,\ \gamma_{2,0} = -\tfrac{1}{2},\ \gamma_{2,1} = 0, \gamma_2 = -\tfrac{1}{4}n\sigma_1,\ \delta_2 = 0;$$
$$\gamma_{3,-1} = -\tfrac{1}{4}\sigma_1,\ \gamma_{3,0} = 0,\ \gamma_{3,1} = -\tfrac{1}{2},\ \gamma_{3,2} = \gamma_3 = 0,\ \delta_3 = -\tfrac{1}{4}n\sigma_1.$$

Laguerre ensemble:[28] $\delta_1 = \delta_2 = \delta_3 = 0$, and

$$\gamma_{1,-1} = 0,\ \gamma_{1,0} = -1,\ \gamma_1 = -\tfrac{1}{2}n(\sigma_1 + a),$$
$$\gamma_{2,-1} = 0,\ \gamma_{2,0} = -\sigma_1,\ \gamma_{2,1} = -1,\ \gamma_2 = -\tfrac{1}{2}n\sigma_1(\sigma_1 + a);$$
$$\gamma_{3,-1} = 0,\ \gamma_{3,0} = -\sigma_1\sigma_2,\ \gamma_{3,1} = -\sigma_2,\ \gamma_{3,2} = -1,\ \gamma_3 = -\tfrac{1}{2}n\sigma_1\sigma_2(\sigma_1 + a).$$

Jacobi ensemble: see (4.3.4).

They lead to expressions for

$$\left.\frac{\partial^4 F}{\partial t_1^4}\right|_{t=0},\ \left.\frac{\partial^2 F}{\partial t_2^2}\right|_{t=0},\ \left.\frac{\partial^2 F}{\partial t_1 \partial t_3}\right|_{t=0},\ \left.\frac{\partial^2 F}{\partial t_1^2}\right|_{t=0},$$

in terms of $\mathcal{D}_k$ and $\mathcal{B}_k$, which substituted in the right-hand side of (4.3.6) — i.e., in the KP-expressions — leads to the right-hand side of (4.1.4),(4.1.6), (4.1.9) and (4.1.10). In the Jacobi case, the right-hand side of (4.3.6) contains the same coefficient $1/Q_6^\pm$ as in (4.3.9), which therefore cancels with the one appearing on the left-hand side; see the expression $b_n^{1,4}$ in (4.3.7) and (4.3.8).

5. Ensembles of Infinite Random Matrices: Fredholm Determinants As τ-Functions of the KdV Equation

Infinite Hermitian matrix ensembles typically relate to the Korteweg-de Vries hierarchy, itself a reduction of the KP hierarchy; a brief sketch will be necessary. The *KP-hierarchy* is given by t_n-deformations of a pseudo-differential operator[29] L: (commuting vector fields)

$$\frac{\partial L}{\partial t_n} = [(L^n)_+, L],\quad L = D + a_{-1}D^{-1} + \cdots,\quad \text{with } D = \frac{\partial}{\partial x}. \tag{5.0.1}$$

Wave and adjoint wave functions are eigenfunctions $\Psi^+(x,t;z)$ and $\Psi^-(x,t;z)$, depending on $x \in \mathbb{R}$, $t \in \mathbb{C}^\infty$, $z \in \mathbb{C}$, behaving asymptotically like (5.0.3) below

---

[27] Remember from Section 2.1 that $\sigma_1 = \beta(n-1) + 2$.

[28] Remember from Section 2.1 that $\sigma_1 = \beta(n-1) + a + 2$ and $\sigma_2 = \beta(n - \tfrac{3}{2}) + a + 3$.

[29] In this section, given P a pseudo-differential operator, P_+ and P_- denote the differential and the (strictly) smoothing part of P respectively.

and satisfying

$$z\Psi^+ = L\Psi^+, \quad \frac{\partial \Psi^+}{\partial t_n} = (L^n)_+\Psi^+, \quad z\Psi^- = L^\top \Psi^-, \quad \frac{\partial \Psi^-}{\partial t_n} = -(L^{\top n})_+\Psi^-. \quad (5.0.2)$$

According to Sato's theory, Ψ^+ and Ψ^- have the following representation in terms of a τ-function (see [25]):

$$\Psi^\pm(x,t;z) = e^{\pm(xz+\sum_1^\infty t_i z^i)} \frac{\tau(t \mp [z^{-1}])}{\tau(t)}$$
$$= e^{\pm(xz+\sum_1^\infty t_i z^i)}(1 + O(z^{-1})), \quad \text{for } z \nearrow \infty, \quad (5.0.3)$$

where τ satisfies and is characterized by the following bilinear relation

$$\oint e^{\sum_1^\infty (t_i - t_i') z^i} \tau(t - [z^{-1}])\tau(t' + [z^{-1}]) \, dz = 0 \quad \text{for all } t, t' \in \mathbb{C}^\infty; \quad (5.0.4)$$

the integral is taken over a small circle around $z = \infty$. From the bilinear relation, one derives the KP-hierarchy, already mentioned in Theorem 3.1, of which the first equation reads as in (3.1.6).

We consider the p-reduced KP hierarchy, i.e., the reduction to pseudo-differential L's such that $L^p = D^p + \cdots$ is a differential operator for some fixed $p \geq 2$. Then $(L^{kp})_+ = L^{kp}$ for all $k \geq 1$ and thus $\partial L/\partial t_{kp} = 0$, in view of the deformation equations (5.0.1) on L. Therefore the variables $t_p, t_{2p}, t_{3p}, \ldots$ are not active and can thus be set $= 0$. The case $p = 2$ is particularly interesting and leads to the KdV equation, upon setting all even $t_i = 0$.

For the time being, take the integer $p \geq 2$ arbitrary. The arbitrary linear combinations[30]

$$\Phi^\pm(x,t;z) := \sum_{\omega \in \zeta_p} a_\omega^\pm \Psi^\pm(x,t;\omega z) \quad (5.0.5)$$

are the most general solution of the spectral problems $L^p \Phi^+ = z^p \Phi^+$ and $L^{\top p} \Phi^- = z^p \Phi^-$ respectively, leading to the definition of the kernels:

$$k_{x,t}(y,z) := \int^x dx \, \Phi^-(x,t;y)\Phi^+(x,t;z),$$
$$k_{x,t}^E(y,z) := k_{x,t}(y,z) I_E(z), \quad (5.0.6)$$

where the integral is taken from a fixed, but arbitrary origin in $\mathbb{R}$. In the same way that $\Psi^\pm(x,t,z)$ has a τ-function representation, so also does $k_{x,t}^E(y,z)$ have a similar representation, involving the vertex operator

$$Y(x,t;y,z) := \sum_{\omega,\omega' \in \zeta_p} a_\omega^- a_{\omega'}^+ X(x,t;\omega y, \omega' z), \quad (5.0.7)$$

where (see [25; 11; 12])

$$X(x,t;y,z) := \frac{1}{z-y} e^{(z-y)x + \sum_1^\infty (z^i - y^i) t_i} e^{\sum_1^\infty (y^{-i} - z^{-i})\frac{1}{i}\frac{\partial}{\partial t_i}}. \quad (5.0.8)$$

---

[30] Here $\zeta_p := \{\omega \text{ such that } \omega^p = 1\}$.

A condition $\sum_{\omega \in \zeta_p} a_\omega^+ a_\omega^- / \omega = 0$ is needed to guarantee that the right-hand side of (5.0.7) is free of singularities in the positive quadrant $\{y_i \geq 0$ and $z_j \geq 0$ with $i, j = 1, \ldots, n\}$ and $\lim_{y \to z} Y(x, t; y, z)$ exists. Indeed, using Fay identities and higher degree Fay identities, one shows stepwise the following three statements, the last one being a statement about a Fredholm determinant:[31]

$$k_{x,t}(y, z) = \frac{1}{\tau(t)} Y(x, t; y, z) \tau(t),$$

$$\det\left(k_{x,t}(y_i, z_j)\right)_{1 \leq i,j \leq n} = \frac{1}{\tau} \prod_{i=1}^{k} Y(x, t; y_i, z_i) \tau,$$

$$\det(I - \lambda k_{x,t}^E) = \frac{1}{\tau} e^{-\lambda \int_E dz\, Y(x,t;z,z)} \tau =: \frac{\tau(t, E)}{\tau(t)}. \quad (5.0.9)$$

The kernel (5.0.12) at $t = 0$ will define the statistics of a random Hermitian ensemble, when the size $n \nearrow \infty$. The next theorem is precisely a statement about Fredholm determinants of kernels of the form (5.0.12); it will be identified at $t = 0$ with the probability that no eigenvalue belongs to a subset E; see Section 1.2. The initial condition that Virasoro annihilates τ_0, as in Sections 2.1.2 (Proof of Theorem 2.1), is now replaced by the *initial condition* (5.0.11) below.

THEOREM 5.1 (Adler, Shiota, and van Moerbeke [11; 12]). *Consider Virasoro generators $J_l^{(2)}$ satisfying*

$$\frac{\partial}{\partial z} z^{l+1} Y(x, t; z, z) = \left[\tfrac{1}{2} J_l^{(2)}(t),\ Y(x, t; z, z)\right], \quad (5.0.10)$$

where $Y(x, t; z, z)$ is defined in (5.0.7), and a τ-function satisfying the Virasoro constraint, with an arbitrary constant c_{kp}:

$$\left(J_{kp}^{(2)} - c_{kp}\right) \tau = 0 \quad \text{for a fixed } k \geq -1. \quad (5.0.11)$$

Then, given the disjoint union $E \subset \mathbb{R}^+$, the Fredholm determinant of

$$K_{x,t}^E(\lambda, \lambda') := \frac{1}{p} \frac{k_{x,t}(z, z')}{z^{(p-1)/2} z'^{(p-1)/2}} I_E(\lambda'), \quad \lambda = z^p,\ \lambda' = z'^p, \quad (5.0.12)$$

satisfies the following constraint for that same $k \geq -1$:

$$\left(-\sum_{i=1}^{2r} c_i^{k+1} \frac{\partial}{\partial c_i} + \frac{1}{2p}(J_{kp}^{(2)} - c_{kp})\right) \tau \det(I - \mu K_{x,t}^E) = 0. \quad (5.0.13)$$

---

[31] The Fredholm determinant of a kernel $A(y, z)$ is defined by

$$\det(I - \lambda A) = 1 + \sum_{m=1}^{\infty} (-\lambda)^m \int \cdots \int_{z_1 \leq \cdots \leq z_m} \det\left(A(z_i, z_j)\right)_{1 \leq i,j \leq m} dz_1 \ldots dz_m.$$

The generators $J_n^{(2)}$ take on the following precise form:

$$J_n^{(1)} := \frac{\partial}{\partial t_n} + (-n)t_{-n},$$

$$J_n^{(2)} := \sum_{i+j=n} :J_i^{(1)} J_j^{(1)}: - (n+1)J_n^{(1)} \qquad (5.0.14)$$

$$= \sum_{i+j=n} \frac{\partial^2}{\partial t_i \partial t_j} + 2\sum_{-i+j=n} it_i \frac{\partial}{\partial t_j} + \sum_{-i-j=n} it_i jt_j - (n+1)J_n^{(1)}.$$

REMARK. For KdV (i.e., $p = 2$), we have $(L^2)^\top = L^2 = D^2 - q(x)$, so the adjoint wave function has the simple expression $\Psi^-(x, t; z) = \Psi^+(x, t; -z)$. In the next two examples, which deal with KdV, set

$$\Psi(x, t; z) := \Psi^+(x, t; z).$$

Example 1: Eigenvalues of large random Hermitian matrices near the "soft edge" and the Airy kernel. Remember from Section 1.2, the spectrum of the Gaussian Hermitian matrix ensemble has, for large size n, its edge at $\pm\sqrt{2n}$, near which the scaling is given by $\sqrt{2}n^{1/6}$. Therefore, the eigenvalues in Theorem 5.2 must be expressed in that new scaling. Define the disjoint union $E = \bigcup_1^r [c_{2i-1}, c_{2i}]$, with c_{2r} possibly ∞.

THEOREM 5.2. *Given the spectrum $z_1 \geq z_2 \geq \cdots$ of the large random Hermitian matrix M, define the "eigenvalues" in the new scale:*

$$u_i = 2n^{2/3}\left(\frac{z_i}{\sqrt{2n}} - 1\right) \quad \text{for } n \nearrow \infty. \qquad (5.0.15)$$

The probability of the "eigenvalues"

$$P(E^c) := P(\text{all "eigenvalues" } u_i \in E^c) \qquad (5.0.16)$$

satisfies the partial differential equation (setting $\mathcal{B}_k := \sum_{i=1}^{2r} c_i^{k+1} \partial/\partial c_i$)[32]

$$(\mathcal{B}_{-1}^3 - 4(\mathcal{B}_0 - \tfrac{1}{2}))\mathcal{B}_{-1}\log P(E^c) + 6(\mathcal{B}_{-1}^2 \log P(E^c))^2 = 0. \qquad (5.0.17)$$

In particular, the statistics of the largest "eigenvalue" u_1 (in the new scale) is given by

$$P(u_1 \leq x) = \exp\left(-\int_x^\infty (\alpha - x)g^2(\alpha)\, d\alpha\right), \qquad (5.0.18)$$

with

$$\begin{cases} g'' = xg + 2g^3 & \textbf{(Painlevé II)} \\ g(x) \cong -\dfrac{e^{-(2/3)x^{3/2}}}{2\sqrt{\pi}x^{1/4}} & \text{for } x \nearrow \infty. \end{cases} \qquad (5.0.19)$$

---

[32] When $c_{2r} = \infty$, that term in $\mathcal{B}_k$ is absent.

The partial differential equation (5.0.17) is due to Adler, Shiota, and van Moerbeke [11; 12]. The equation (5.0.19) for the largest eigenvalue is a special case of (5.0.17), but was first derived by Tracy and Widom [63], by methods of functional analysis.

PROOF. Remember from Section 1.2 that the statistics of the eigenvalues is governed by the Fredholm determinant of the kernel (1.2.4), for the Hermite polynomials. In the limit,

$$\lim_{n \nearrow \infty} \frac{1}{\sqrt{2}n^{1/6}} K_n\left(\sqrt{2n} + \frac{u}{\sqrt{2}n^{1/6}}, \sqrt{2n} + \frac{v}{\sqrt{2}n^{1/6}}\right) = K(u,v),$$

where

$$K(u,v) = \int_0^\infty A(x+u)A(x+v)\,dx, \quad A(u) = \int_{-\infty}^\infty e^{iux - x^3/3}\,dx. \quad (5.0.20)$$

Then

$$P(E^c) := P(\text{all eigenvalues } u_i \in E^c) = \det\left(I - K(u,v)I_E(v)\right). \quad (5.0.21)$$

In order to compute the PDEs of this expression, with regard to the endpoints c_i of the disjoint union E, one proceeds as follows:

Consider the KdV wave function $\Psi(x,t;z)$, as in (5.0.2), with initial condition

$$\Psi(x,t_0;z) = z^{1/2}A(x+z^2) = e^{xz+(2/3)z^3}(1 + O(z^{-1})),$$
$$z \to \infty, \ t_0 = \left(0, 0, \tfrac{2}{3}, 0, \ldots\right), \quad (5.0.22)$$

in terms of the Airy function[33], which, by stationary phase, has the asymptotics

$$A(u) := \frac{1}{\sqrt{\pi}} \int_{-\infty}^\infty e^{-y^3/3 + yu}\,dy = u^{-1/4} e^{(2/3)u^{3/2}}\left(1 + O(u^{-3/2})\right).$$

The definition of $A(u)$ is slightly changed, compared to (5.0.20). $A(u)$ satisfies the differential equation $A(y)'' = yA(y)$, and thus the wave function $\Psi(x,t_0;z)$ satisfies $(D^2 - x)\Psi(x,t_0;z) = z^2\Psi(x,t_0;z)$. Therefore $L^2|_{t=t_0} = SD^2S^{-1}|_{t=t_0} = D^2 - x$, so that L^2 is a differential operator, and Ψ is a KdV wave function, with $\tau(t)$ satisfying[34] the Virasoro constraints (5.0.11) with $c_{2k} = -\tfrac{1}{4}\delta_{k0}$. The argument to prove these constraints is based on the fact that the linear span (a point in an infinite-dimensional Grassmannian)

$$\mathcal{W} = \mathrm{span}_{\mathbb{C}}\left\{\psi_n(z) := e^{-(2/3)z^3}\sqrt{z}\,\frac{\partial^n}{\partial u^n}A(u)\bigg|_{u=z^2},\ n = 0, 1, 2, \ldots\right\}$$

---

[33] The i in the definition of the Airy function is omitted here.

[34] Although not used here, the τ-function is Kontsevich's integral [44; 10]:

$$\tau(t) = \frac{\int_{\mathcal{H}} dY\, e^{-\mathrm{Tr}(Y^3/3 + Y^2 Z)}}{\int_{\mathcal{H}} dY\, e^{-\mathrm{Tr}\,Y^2 Z}} \quad \text{with } t_n = -\frac{1}{n}\mathrm{Tr}(Z^{-n}) + \frac{2}{3}\delta_{n,3},\ Z = \text{diagonal matrix}.$$

is invariant under multiplication by z^2 and under the operator

$$\frac{1}{2z}\left(\frac{\partial}{\partial z}+2z^2\right)-\frac{1}{4}z^{-2}.$$

Define, for $\lambda = z^2$ and $\lambda' = z'^2$, the kernel

$$K_t(\lambda,\lambda') := \frac{1}{2z^{1/2}z'^{1/2}}\int_0^\infty \Psi(x,t;z)\Psi(x,t;z')\,dx, \tag{5.0.23}$$

which flows off the Airy kernel, by (5.0.22),

$$K_{t_0}(\lambda,\lambda') = \frac{1}{2}\int_0^\infty A(x+\lambda)A(x+\lambda')\,dx.$$

Thus $\tau \det(I - K_{x,t}^E)$ satisfies (5.0.13), with that same constant c_{kp}, for $k = -1, 0, 1, \ldots$:

$$\left(-\sum_{i=1}^{2r} c_i^{k+1}\frac{\partial}{\partial c_i}+\frac{1}{4}J_{pk}^{(2)}+\frac{1}{16}\delta_{k,0}\right)\tau\det(I-K_t^E)=0. \tag{5.0.24}$$

Upon shifting $t_3 \mapsto t_3 + 2/3$, in view of (5.0.22), the two first Virasoro constraints for $k = -1$ and $k = 0$ read as follows, with $\mathcal{B}_k := \sum_{i=1}^{2r} c_i^{k+1}\partial/\partial c_i$:

$$\mathcal{B}_{-1}\log\tau(t,E) = \left(\frac{\partial}{\partial t_1}+\frac{1}{2}\sum_{i\geq 3}it_i\frac{\partial}{\partial t_{i-2}}\right)\log\tau(t,E)+\frac{t_1^2}{4}$$

$$\mathcal{B}_0\log\tau(t,E) = \left(\frac{\partial}{\partial t_3}+\frac{1}{2}\sum_{i\geq 1}it_i\frac{\partial}{\partial t_i}\right)\log\tau(t,E)+\frac{1}{16}. \tag{5.0.25}$$

The same method as in Section 4 enables one to express all the t-partials, appearing in the KdV equation,

$$\left(\frac{\partial^4}{\partial t_1^4}-4\frac{\partial^2}{\partial t_1\partial t_3}\right)\log\tau(t,E)+6\left(\frac{\partial^2}{\partial t_1^2}\log\tau(t,E)\right)^2=0,$$

in terms of c-partials, which upon substitution leads to the partial differential equation $\left(\mathcal{B}_{-1}^3-4(\mathcal{B}_0-\frac{1}{2})\right)f+6(\mathcal{B}_{-1}f)^2=0$ (announced in (5.0.17)) for

$$f := \mathcal{B}_{-1}\log P(E^c) = \sum_1^{2r}\frac{\partial}{\partial c_i}\log P(E^c), \text{ where } P(E^c)=\det(I-K^E)=\frac{\tau(t,E)}{\tau(t)}.$$

When $E = (-\infty, x)$, this PDE reduces to an ODE:

$$f''' - 4xf' + 2f + 6f'^2 = 0, \quad \text{with } f = \frac{d}{dx}\log P(\max_i \lambda_i \leq x). \tag{5.0.26}$$

According to Appendix on Chazy classes (section 9), this equation can be reduced to

$$f''^2 + 4f'(f'^2 - xf' + f) = 0, \qquad \textbf{(Painlevé II)} \tag{5.0.27}$$

which can be solved by setting

$$f' = -g^2 \text{ and } f = g'^2 - xg^2 - g^4.$$

An easy computation shows g satisfies the equation $g'' = 2g^3 + xg$ (Painlevé II), thus leading to (5.0.19). □

Example 2: Eigenvalues of large random Laguerre Hermitian matrices near the "hard edge" and the Bessel kernel. Consider the ensemble of $n \times n$ random matrices for the Laguerre probability distribution, thus corresponding to (1.1.9) with $\rho(dz) = z^{\nu/2} e^{-z/2} dz$. Remember from Section 1.2, the density of eigenvalues near the "hard edge" $z = 0$ is given by $4n$ for very large n. At this edge, the kernel (1.2.4) with Laguerre polynomials p_n tends to the Bessel kernel [52; 30]:

$$\lim_{n \nearrow \infty} \frac{1}{4n} K_n^{(\nu)}\left(\frac{u}{4n}, \frac{v}{4n}\right) = K^{(\nu)}(u,v) := \frac{1}{2} \int_0^1 x J_\nu(xu) J_\nu(xv)\, dx. \qquad (5.0.28)$$

Therefore, the eigenvalues in the theorem below will be expressed in that new scaling. Define, as before, the disjoint union $E = \bigcup_1^r [c_{2i-1}, c_{2i}]$.

THEOREM 5.3. *Given the spectrum $0 \leq z_1 \leq z_2 \leq \ldots$ of the large random Laguerre-distributed Hermitian matrix M, define the "eigenvalues" in the new scale:*

$$u_i = 4n z_i \quad \text{for } n \nearrow \infty. \qquad (5.0.29)$$

The statistics of the "eigenvalues"

$$P(E^c) := P(\text{all "eigenvalues" } u_i \in E^c) \qquad (5.0.30)$$

leads to the following PDE for $F = \log P(E^c)$, where $\mathcal{B}_k := \sum_{i=1}^{2r} c_i^{k+1} \partial/\partial c_i$:

$$(\mathcal{B}_0^4 - 2\mathcal{B}_0^3 + (1-\nu^2)\mathcal{B}_0^2 + \mathcal{B}_1(\mathcal{B}_0 - \tfrac{1}{2}))F - 4(\mathcal{B}_0 F)(\mathcal{B}_0^2 F) + 6(\mathcal{B}_0^2 F)^2 = 0. \qquad (5.0.31)$$

In particular, for very large n, the statistics of the smallest eigenvalue is governed by

$$P(u_1 \geq x) = \exp\left(-\int_0^x \frac{f(u)}{u} du\right), \quad u_1 \sim 4n z_1,$$

with f satisfying

$$(xf'')^2 - 4(xf' - f)f'^2 + ((x - \nu^2)f' - f)f' = 0. \quad \textbf{(Painlevé V)} \qquad (5.0.32)$$

Equation (5.0.32) for the smallest eigenvalue, first derived by Tracy and Widom [63], by methods of functional analysis, is a special case of the partial differential equation (5.0.31), originating in [11; 12].

REMARK. This same theorem would hold for the Jacobi ensemble, near the "hard edges" $z = \pm 1$.

PROOF. Define a wave function $\Psi(x, t; z)$, flowing off

$$\Psi(x, 0; z) = e^{xz} B(-xz) = e^{xz}\left(1 + O(z^{-1})\right),$$

where $B(z)$ is the Bessel function[35]

$$B(z) = \varepsilon\sqrt{z} e^z H_\nu(iz) = \frac{e^z 2^{\nu+1/2}}{\Gamma(-\nu+1/2)} \int_1^\infty \frac{z^{-\nu+1/2} e^{-uz}}{(u^2-1)^{\nu+1/2}} du = 1 + O(z^{-1}).$$

As the operator

$$L^2|_{t=0} = D^2 - \frac{\nu^2 - 1/4}{x^2}$$

is a differential operator, we are in the KdV situation; again one may assume $t_2 = t_4 = \ldots = 0$ and we have

$$\Psi^-(x, t; -z) = \Psi^+(x, t; z) = e^{xz + \sum t_i z^i} \frac{\tau(t - [z^{-1}])}{\tau(t)},$$

in terms of a τ-function[36] satisfying the Virasoro constraints

$$J_{2k}^{(2)} \tau = \left((2\nu)^2 - 1\right) \delta_{k0} \tau. \tag{5.0.33}$$

Set $p = 2$, $a_1^- = a_{-1}^+ = (1/4\pi) i e^{i\pi\nu/2}$ and $a_{-1}^- = a_1^+ = (1/4\pi) e^{-i\pi\nu/2}$ in (5.0.5); this defines the kernel (5.0.6) and so (5.0.12), which in terms of $\lambda = z^2$ and $\lambda' = z'^2$, takes on the form:

$$K_{x,t}^{(\nu)}(\lambda, \lambda') = \frac{1}{4\pi\sqrt{zz'}} \int^x \left(i e^{i\pi\nu/2} \Psi^*(x, t, z) + e^{-i\pi\nu/2} \Psi^*(x, t, -z)\right)$$
$$\cdot \left(e^{-i\pi\nu/2} \Psi(x, t, z') + i e^{i\pi\nu/2} \Psi(x, t, -z')\right) dx,$$

which flows off the *Bessel kernel*

$$K_{x,0}^{(\nu)}(\lambda, \lambda') = \frac{1}{2} \int_0^x x J_\nu(x\sqrt{\lambda}) J_\nu(x\sqrt{\lambda'}) dx.$$
$$= \frac{J_\nu(\sqrt{\lambda}) \sqrt{\lambda'} J_\nu'(\sqrt{\lambda'}) - J_\nu(\sqrt{\lambda'}) \sqrt{\lambda} J_\nu'(\sqrt{\lambda})}{2(\lambda - \lambda')} \quad \text{for } x = 1.$$

The Fredholm determinant satisfies for $E \subset \mathbb{R}_+$ and for $k = 0, 1, \ldots$:

$$\left(-\sum_1^{2r} c_i^{k+1} \frac{\partial}{\partial c_i} + \tfrac{1}{4} J_{2k}^{(2)} + \left(\tfrac{1}{4} - \nu^2\right) \delta_{k,0}\right) \left(\tau \det(I - K_{x,t}^{(\nu) E})\right) = 0. \tag{5.0.34}$$

---

[35] $\varepsilon = i\sqrt{\pi/2} e^{i\pi\nu/2}$, $-\tfrac{1}{2} < \nu < \tfrac{1}{2}$.

[36] $\tau(t)$ is given by the Adler–Morozov–Shiota–van Moerbeke double Laplace matrix transform, with t_n given in a similar way as in footnote 34 (see [10]):

$$\tau(t) = c(t) \int_{\mathcal{H}_N^+} dX \det X^{\nu - 1/2} e^{-\operatorname{Tr}(Z^2 X)} \int_{\mathcal{H}_N^+} dY S_0(Y) e^{-\operatorname{Tr}(XY^2)}.$$

Upon shifting $t_1 \mapsto t_1 + \sqrt{-1}$ and using the same $\mathcal{B}_i$ as in (5.0.25), the equations for $k=0$ and $k=1$ read

$$\mathcal{B}_0 \log \tau(t, E) = \frac{1}{2}\left(\sum_{i \geq 1} it_i \frac{\partial}{\partial t_i} + \sqrt{-1}\frac{\partial}{\partial t_1}\right) \log \tau(t, E) + \frac{1}{4}\left(\frac{1}{4} - \nu^2\right),$$

$$\mathcal{B}_1 \log \tau(t, E) = \frac{1}{2}\left(\sum_{i \geq 1} it_i \frac{\partial}{\partial t_{i+2}} + \frac{1}{2}\frac{\partial^2}{\partial t_1^2} + \sqrt{-1}\frac{\partial}{\partial t_3} + \frac{1}{2}\frac{\partial}{\partial t_1}\right) \log \tau(t, E). \quad (5.0.35)$$

Expressing the t-partials (5.0.20), appearing in the KdV-equation at $t=0$ (see formula just below (5.0.25)) in terms of the c-partials applied to $\log \tau(0, E)$, leads to the following PDE for $F = \log P(E^c)$:

$$\left(\mathcal{B}_0^4 - 2\mathcal{B}_0^3 + (1-\nu^2)\mathcal{B}_0^2 + \mathcal{B}_1(\mathcal{B}_0 - \tfrac{1}{2})\right) F - 4(\mathcal{B}_0 F)(\mathcal{B}_0^2 F) + 6(\mathcal{B}_0^2 F)^2 = 0. \quad (5.0.36)$$

Specializing this equation to the interval $E = (0, x)$ leads to an ODE for $f := -x \, \partial F / \partial x$, namely

$$f''' + \frac{1}{x}f'' - \frac{6}{x}f'^2 + \frac{4}{x^2}ff' + \frac{(x-\nu^2)}{x^2}f' - \frac{1}{2x^2}f = 0, \quad (5.0.37)$$

which is an equation of the type (9.0.1); changing $x \curvearrowright -x$ and $f \curvearrowright -f$ leads again to an equation of type (9.0.1), with $P(x) = x$, $4Q(x) = -x - \nu^2$ and $R = 0$. According to Cosgrove and Scoufis [24] (see the Appendix on Chazy classes), this equation can be reduced to the equation (9.0.2), with the same P, Q, R and with $c = 0$. Since $P(x) = x$, this equation is already in one of the canonical forms (9.0.3), which upon changing back x and f, leads to

$$(xf'')^2 + 4\bigl(-xf' + f\bigr)f'^2 + \bigl((x-\nu^2)f' - f\bigr)f' = 0. \quad \textbf{(Painlevé V)} \quad \square$$

Example 3: Eigenvalues of large random Gaussian Hermitian matrices in the bulk and the sine kernel. Setting $\nu = \pm\frac{1}{2}$ yields kernels related to the sine kernel:

$$K_{x,0}^{(+1/2)}(y^2, z^2) = \frac{1}{\pi}\int_0^x \frac{\sin xy \, \sin xz}{y^{1/2} z^{1/2}} dx = \frac{1}{2\pi}\left(\frac{\sin x(y-z)}{y-z} - \frac{\sin x(y+z)}{y+z}\right),$$

$$K_{x,0}^{(-1/2)}(y^2, z^2) = \frac{1}{\pi}\int_0^x \frac{\cos xy \, \cos xz}{y^{1/2} z^{1/2}} dx = \frac{1}{2\pi}\left(\frac{\sin x(y-z)}{y-z} + \frac{\sin x(y+z)}{y+z}\right).$$

Therefore the sine-kernel obtained in the context of the bulk-scaling limit (see (1.2.7)) is the sum $K_{x,0}^{(+1/2)} + K_{x,0}^{(-1/2)}$. Expressing the Fredholm determinant of this sum in terms of the Fredholm determinants of each of the parts, leads to the Painlevé V equation (1.2.8).

6. Coupled Random Hermitian Ensembles

Consider a product ensemble $(M_1, M_2) \in \mathcal{H}_n^2 := \mathcal{H}_n \times \mathcal{H}_n$ of $n \times n$ Hermitian matrices, equipped with a Gaussian probability measure,

$$c_n \, dM_1 \, dM_2 \, e^{-(1/2)\,\mathrm{Tr}(M_1^2 + M_2^2 - 2cM_1 M_2)}, \qquad (6.0.1)$$

where $dM_1 \, dM_2$ is Haar measure on the product $\mathcal{H}_n^2$, with each dM_i,

$$dM_1 = \Delta_n^2(x) \prod_1^n dx_i \, dU \quad \text{and} \quad dM_2 = \Delta_n^2(y) \prod_1^n dy_i \, dU \qquad (6.0.2)$$

decomposed into radial and angular parts. In terms of the coupling constant c, appearing in (6.0.1), and the boundary of the set

$$E = E_1 \times E_2 := \bigcup_{i=1}^r [a_{2i-1}, a_{2i}] \times \bigcup_{i=1}^s [b_{2i-1}, b_{2i}] \subset \mathbb{R}^2, \qquad (6.0.3)$$

define differential operators $\mathcal{A}_k, \mathcal{B}_k$ of "weight" k,

$$\mathcal{A}_1 = \frac{1}{c^2 - 1}\Big(\sum_1^r \frac{\partial}{\partial a_j} + c\sum_1^s \frac{\partial}{\partial b_j}\Big), \quad \mathcal{B}_1 = \frac{1}{1-c^2}\Big(c\sum_1^r \frac{\partial}{\partial a_j} + \sum_1^s \frac{\partial}{\partial b_j}\Big),$$

$$\mathcal{A}_2 = \sum_{j=1}^r a_j \frac{\partial}{\partial a_j} - c\frac{\partial}{\partial c}, \qquad \mathcal{B}_2 = \sum_{j=1}^s b_j \frac{\partial}{\partial b_j} - c\frac{\partial}{\partial c},$$

forming a closed Lie algebra.[37] The following theorem can be derived, via similar methods, from the Virasoro constraints (3.3.5) and the 2-Toda equation (3.3.6):

THEOREM 6.1 (Gaussian probability) (Adler and van Moerbeke [3]). *The joint statistics*

$$P_n(M \in \mathcal{H}_n^2(E_1 \times E_2)) = \frac{\iint_{\mathcal{H}_n^2(E_1 \times E_2)} dM_1 \, dM_2 \, e^{-\frac{1}{2}\mathrm{Tr}(M_1^2 + M_2^2 - 2cM_1 M_2)}}{\iint_{\mathcal{H}_n^2} dM_1 \, dM_2 \, e^{-\frac{1}{2}\mathrm{Tr}(M_1^2 + M_2^2 - 2cM_1 M_2)}}$$

$$= \frac{\iint_{E^n} \Delta_n(x)\Delta_n(y) \prod_{k=1}^n e^{-\frac{1}{2}(x_k^2 + y_k^2 - 2cx_k y_k)} \, dx_k dy_k}{\iint_{\mathbb{R}^{2n}} \Delta_n(x)\Delta_n(y) \prod_{k=1}^n e^{-\frac{1}{2}(x_k^2 + y_k^2 - 2cx_k y_k)} \, dx_k dy_k}$$

---

[37] We have

$$[\mathcal{A}_1, \mathcal{B}_1] = 0, \quad [\mathcal{A}_1, \mathcal{A}_2] = \frac{1+c^2}{1-c^2}\mathcal{A}_1, \quad [\mathcal{A}_2, \mathcal{B}_1] = \frac{2c}{1-c^2}\mathcal{A}_1,$$

$$[\mathcal{A}_2, \mathcal{B}_2] = 0, \quad [\mathcal{A}_1, \mathcal{B}_2] = \frac{-2c}{1-c^2}\mathcal{B}_1, \quad [\mathcal{B}_1, \mathcal{B}_2] = \frac{1+c^2}{1-c^2}\mathcal{B}_1.$$

satisfies the following nonlinear third-order partial differential equation[38] (independent of n), where $F_n := (1/n)\log P_n(E)$:

$$\left\{\mathcal{B}_2\mathcal{A}_1 F_n,\ \mathcal{B}_1\mathcal{A}_1 F_n + \frac{c}{c^2-1}\right\}_{\mathcal{A}_1} - \left\{\mathcal{A}_2\mathcal{B}_1 F_n,\ \mathcal{A}_1\mathcal{B}_1 F_n + \frac{c}{c^2-1}\right\}_{\mathcal{B}_1} = 0. \quad (6.0.4)$$

7. Random Permutations

The purpose of this section is to show that the generating function of the probability

$$P(L(\pi_n) \leq l) = \frac{1}{n!}\#\{\pi_n \in S_n \mid L(\pi_n) \leq l\}$$

is closely related to a special solution of the Painlevé V equation, with peculiar initial condition. Remember from Section 1.4 that $L(\pi_n)$ is the length of the longest increasing sequence in the permutation π_n.

THEOREM 7.1 (Tracy and Widom [66]). *For every $l \geq 0$,*

$$\sum_{n=0}^{\infty} \frac{x^n}{n!} P(L(\pi_n) \leq l) = \int_{U(l)} e^{\sqrt{x}\ \mathrm{tr}(M + \bar{M})}\, dM \qquad (7.0.1)$$

$$= \exp \int_0^x \log\left(\frac{x}{u}\right) g_l(u)\, du,$$

*with g_l satisfying the initial value problem for **Painlevé V**:*

$$\begin{cases} g'' - \dfrac{g'^2}{2}\left(\dfrac{1}{g-1} + \dfrac{1}{g}\right) + \dfrac{g'}{u} + \dfrac{2}{u}g(g-1) - \dfrac{l^2}{2u^2}\dfrac{g-1}{g} = 0, \\ g_l(u) = 1 - \dfrac{u^l}{l!^2} + O(u^{l+1}), \quad \text{near } u = 0. \end{cases} \qquad (7.0.2)$$

The systematic derivation below is due to Adler and van Moerbeke [7].

PROOF. The first identity in (7.0.1) follows from Proposition 1.1. Upon inserting $(t_1, t_2, \dots)$ and $(s_1, s_2, \dots)$ variables in the $U(n)$-integral (7.0.1), the integral

$$I_n(t,s) = \int_{U(n)} e^{\mathrm{Tr}\sum_1^\infty (t_i M^i - s_i \bar{M}^i)}\, dM \qquad (7.0.3)$$

$$= n!\det\left(\int_{S^1} z^{k-l} e^{\sum_1^\infty (t_i z^i - s_i z^{-i})} \frac{dz}{2\pi i z}\right)_{0 \leq k, l \leq n-1} = n!\, \tau_n(t,s)$$

puts us in the conditions of Theorem 3.5. It deals with semi-infinite matrices L_1 and $hL_2^\top h^{-1}$ of "rank 2", having diagonal elements

$$b_n := \frac{\partial}{\partial t_1} \log \frac{\tau_n}{\tau_{n-1}} = (L_1)_{n-1,n-1}, \quad b_n^* := -\frac{\partial}{\partial s_1} \log \frac{\tau_n}{\tau_{n-1}} = (hL_2^\top h^{-1})_{n-1,n-1}$$

To summarize Theorem 3.5, $I_n(t,s)$ satisfies three types of identities:

---

[38] in terms of the Wronskian $\{f,g\}_X = (Xf)g - f(Xg)$, with regard to a first order differential operator X.

(i) **Virasoro** (see (3.4.7)); we set $F := \log \tau_n$:

$$0 = \frac{\mathcal{V}_{-1}\tau_n}{\tau_n} = \left(\sum_{i\geq 1}(i+1)t_{i+1}\frac{\partial}{\partial t_i} - \sum_{i\geq 2}(i-1)s_{i-1}\frac{\partial}{\partial s_i} + n\frac{\partial}{\partial s_1}\right)F + nt_1,$$

$$0 = \frac{\mathcal{V}_0\tau_n}{\tau_n} = \sum_{i\geq 1}\left(it_i\frac{\partial}{\partial t_i} - is_i\frac{\partial}{\partial s_i}\right)F,$$

$$0 = \frac{\mathcal{V}_1\tau_n}{\tau_n} = \left(-\sum_{i\geq 1}(i+1)s_{i+1}\frac{\partial}{\partial s_i} + \sum_{i\geq 2}(i-1)t_{i-1}\frac{\partial}{\partial t_i}\frac{\partial}{\partial t_1}\right)F + ns_1,$$

$$0 = \frac{\partial}{\partial t_1}\frac{\mathcal{V}_{-1}\tau_n}{\tau_n}$$

$$= \left(\sum_{i\geq 1}(i+1)t_{i+1}\frac{\partial^2}{\partial t_1\partial t_i} - \sum_{i\geq 2}(i-1)s_{i-1}\frac{\partial^2}{\partial t_1\partial s_i} + n\frac{\partial^2}{\partial t_1\partial s_1}\right)F + n. \quad (7.0.4)$$

(ii) **two-Toda** (see (3.4.8)):

$$\frac{\partial^2 \log \tau_n}{\partial s_2 \partial t_1} = -2\frac{\partial}{\partial s_1}\log\frac{\tau_n}{\tau_{n-1}}\frac{\partial^2}{\partial s_1\partial t_1}\log\tau_n - \frac{\partial^3}{\partial s_1^2 \partial t_1}\log\tau_n$$

$$= 2b_n^* \frac{\partial^2}{\partial s_1\partial t_1}\log\tau_n - \frac{\partial^3}{\partial s_1^2\partial t_1}\log\tau_n. \quad (7.0.5)$$

(iii) **Toeplitz** (see (3.4.11)):

$$\mathfrak{T}(\tau)_n = \frac{\partial}{\partial t_1}\log\frac{\tau_n}{\tau_{n-1}}\frac{\partial}{\partial s_1}\log\frac{\tau_n}{\tau_{n-1}}$$

$$+ \left(1 + \frac{\partial^2}{\partial s_1\partial t_1}\log\tau_n\right)\left(1 + \frac{\partial^2}{\partial s_1\partial t_1}\log\tau_n - \frac{\partial}{\partial s_1}\left(\frac{\partial}{\partial t_1}\log\frac{\tau_n}{\tau_{n-1}}\right)\right)$$

$$= -b_n b_n^* + \left(1 + \frac{\partial^2}{\partial s_1\partial t_1}\log\tau_n\right)\left(1+\frac{\partial^2}{\partial s_1\partial t_1}\log\tau_n - \frac{\partial}{\partial s_1}b_n\right) = 0. \quad (7.0.6)$$

Defining the locus $\mathcal{L} = \{$ all $t_i = s_i = 0$, except $t_1, s_1 \neq 0\}$, and using the second relation (7.0.4), we have on $\mathcal{L}$

$$\left.\frac{\mathcal{V}_0\tau_n}{\tau_n}\right|_{\mathcal{L}} = \left(t_1\frac{\partial}{\partial t_1} - s_1\frac{\partial}{\partial s_1}\right)\log\tau_n\bigg|_{\mathcal{L}} = 0,$$

implying that $\tau_n(t,s)\big|_{\mathcal{L}}$ is a function of $x := -t_1 s_1$ only. Therefore we may write $\tau_n\big|_{\mathcal{L}} = \tau_n(x)$, and so, along $\mathcal{L}$, we have

$$\frac{\partial}{\partial t_1} = -s_1\frac{\partial}{\partial x}, \quad \frac{\partial}{\partial s_1} = -t_1\frac{\partial}{\partial x}, \quad \frac{\partial^2}{\partial t_1\partial s_1} = -\frac{\partial}{\partial x}x\frac{\partial}{\partial x}.$$

Setting

$$f_n(x) = \frac{\partial}{\partial x}x\frac{\partial}{\partial x}\log\tau_n(x) = -\frac{\partial^2}{\partial t_1\partial s_1}\log\tau_n(t,s)\bigg|_{\mathcal{L}}, \quad (7.0.7)$$

and using $x = -t_1 s_1$, the two-Toda relation (7.0.5) takes on the form

$$s_1 \frac{\partial^2 \log \tau_n}{\partial s_2 \partial t_1}\bigg|_{\mathcal{L}} = s_1 \left(2b_n^* \frac{\partial^2}{\partial s_1 \partial t_1} \log \tau_n - \frac{\partial}{\partial s_1}\left(\frac{\partial^2 \log \tau_n}{\partial s_1 \partial t_1}\right)\right) = x\left(2\frac{b_n^*}{t_1} f_n + f_n'\right).$$

Setting relation (7.0.7) into the Virasoro relations (7.0.4) yields

$$0 = \frac{\mathcal{V}_0 \tau_n}{\tau_n} - \frac{\mathcal{V}_0 \tau_{n-1}}{\tau_{n-1}}\bigg|_{\mathcal{L}} = \left(t_1 \frac{\partial}{\partial t_1} - s_1 \frac{\partial}{\partial s_1}\right) \log \frac{\tau_n}{\tau_{n-1}}\bigg|_{\mathcal{L}} = t_1 b_n + s_1 b_n^*$$

$$0 = \frac{\partial}{\partial t_1} \frac{\mathcal{V}_{-1} \tau_n}{\tau_n}\bigg|_{\mathcal{L}} = \left(-s_1 \frac{\partial^2}{\partial s_2 \partial t_1} + n \frac{\partial^2}{\partial t_1 \partial s_1}\right) \log \tau_n \bigg|_{\mathcal{L}} + n$$

$$= -x\left(2\frac{b_n^*}{t_1} f_n(x) + f_n'(x)\right) + n(-f_n(x) + 1).$$

This is a system of two linear relations in b_n and b_n^*, whose solution, together with its derivatives, are given by

$$\frac{b_n^*}{t_1} = -\frac{b_n}{s_1} = -\frac{n(f_n - 1) + x f_n'}{2x f_n},$$

$$\frac{\partial b_n}{\partial s_1} = \frac{\partial}{\partial x} x \frac{b_n}{s_1} = \frac{x(f_n f_n'' - f_n'^2) + (f_n + n)f_n'}{2 f_n^2}.$$

Substituting the result into the Toeplitz relation (7.0.6), namely

$$b_n b_n^* = (1 - f_n)\left(1 - f_n - \frac{\partial}{\partial s_1} b_n\right),$$

leads to f_n satisfying Painlevé equation (7.0.2), with $g = f_n$, as in (7.0.7) and $u = x$. Note, along the locus $\mathcal{L}$, we may set $t_1 = \sqrt{x}$ and $s_1 = -\sqrt{x}$, since it respects $t_1 s_1 = -x$. Thus, $I_n(t, s)|_{\mathcal{L}}$ equals (7.0.1).

The initial condition (7.0.2) follows from the fact that as long as $0 \leq n \leq l$, the inequality $L(\pi_n) \leq l$ is always verified, and so

$$\sum_0^\infty \frac{x^n}{(n!)^2} \#\{\pi \in S_n \mid L(\pi_n) \leq l\} = \sum_0^l \frac{x^n}{n!} + \frac{x^{l+1}}{(l+1)!^2}((l+1)! - 1) + O(x^{l+2})$$

$$= \exp\left(x - \frac{x^{l+1}}{(l+1)!^2} + O(x^{l+2})\right),$$

thus proving Theorem 7.1. □

REMARK. Setting

$$f_n(x) = \frac{g(x)}{g(x) - 1}$$

leads to standard Painlevé V, with $\alpha = \delta = 0$, $\beta = -n^2/2$, $\gamma = -2$.

8. Random Involutions

This section deals with a generating function for the distribution of the length of the longest increasing sequence of a fixed-point free random involution π_{2k}^0, with the uniform distribution:

$$P\left(L(\pi_{2k}^0) \leq l+1,\; \pi_{2k}^0 \in S_{2k}^0\right) = \frac{2^k k!}{(2k)!} \#\{\pi_{2k}^0 \in S_{2k}^0 \mid L(\pi_{2k}^0) \leq l+1\}.$$

PROPOSITION 8.1 (Adler and van Moerbeke [7]). *The generating function*

$$\begin{aligned}
2\sum_{k=0}^{\infty} \frac{(x^2/2)^k}{k!} & P(L(\pi_{2k}^0) \leq l+1) \\
&= E_{O(l+1)_-} e^{x\,\mathrm{Tr}\,M} + E_{O(l+1)_+} e^{x\,\mathrm{Tr}\,M} \\
&= \exp\left(\int_0^x \frac{f_l^-(u)}{u}\,du\right) + \exp\left(\int_0^x \frac{f_l^+(u)}{u}\,du\right),
\end{aligned} \qquad (8.0.1)$$

where $f = f_l^\pm$, satisfies the initial value problem for **Painlevé V**:

$$\begin{cases} f''' + \dfrac{1}{u}f'' + \dfrac{6}{u}f'^2 - \dfrac{4}{u^2}ff' - \dfrac{16u^2 + l^2}{u^2}f' + \dfrac{16}{u}f + \dfrac{2(l^2-1)}{u} = 0, \\ f_l^\pm(u) = u^2 \pm \dfrac{u^{l+1}}{l!} + O(u^{l+2}) \quad \text{near } u = 0. \end{cases} \qquad (8.0.2)$$

PROOF. The first equality in (8.0.1), due to Rains [57; 58], follows immediately from Proposition 1.1. The results of Section 1.3 lead to

$$\int_{O(2n+1)_\pm} e^{x\,\mathrm{Tr}\,M}\,dM = e^{\pm x}\int_{[-1,1]^n} \Delta_n(z)^2 \prod_{k=1}^n e^{2xz_k}(1-z_k)^a(1+z_k)^b dz_k, \qquad (8.0.3)$$

with $a = \pm\frac{1}{2}$, $b = \mp\frac{1}{2}$, (with corresponding signs). Inserting t_i's in the integral, the perturbed integral, with $e^{\pm x}$ removed and with $t_1 = 2x$, reads

$$I_n(t) = \int_{[-1,1]^n} \Delta_n(z)^2 \prod_{k=1}^n (1-z_k)^a(1+z_k)^b e^{\sum_1^\infty t_i z_k^i}\,dz_k = n!\,\tau_n(t); \qquad (8.0.4)$$

this is precisely integral (3.1.4) of Section 3.1.1 and thus it satisfies the Virasoro constraints (3.1.5), but without boundary contribution $\mathcal{B}_i F$. Explicit Virasoro expressions appear in (2.1.35), upon setting $\beta = 2$. Also, $\tau_n(t)$, as in (3.1.4), (see Theorem 3.1) satisfies the KP equation (3.1.6). Differentiating the Virasoro constraints in t_1 and t_2, and restricting to the locus

$$\mathcal{L} := \{t_1 = x,\; \text{all other } t_i = 0\},$$

lead to a linear system of five equations, with $b_0 = a - b$, $b_1 = a + b$,

$$\frac{1}{I_n}\left(\mathbb{J}^{(2)}_{k+2} - \mathbb{J}^{(2)}_k + b_0 \mathbb{J}^{(1)}_{k+1} + b_1 \mathbb{J}^{(1)}_{k+2}\right) I_n\bigg|_{\mathcal{L}} = 0 \quad \text{for } k = -1, 0,$$

$$\frac{\partial}{\partial t_1} \frac{1}{I_n}\left(\mathbb{J}^{(2)}_{k+2} - \mathbb{J}^{(2)}_k + b_0 \mathbb{J}^{(1)}_{k+1} + b_1 \mathbb{J}^{(1)}_{k+2}\right) I_n\bigg|_{\mathcal{L}} = 0 \quad \text{for } k = -1, 0,$$

$$\frac{\partial}{\partial t_2} \frac{1}{I_n}\left(\mathbb{J}^{(2)}_{k+2} - \mathbb{J}^{(2)}_k + b_0 \mathbb{J}^{(1)}_{k+1} + b_1 \mathbb{J}^{(1)}_{k+2}\right) I_n\bigg|_{\mathcal{L}} = 0 \quad \text{for } k = -1,$$

in five unknowns ($F_n = \log \tau_n$)

$$\frac{\partial F_n}{\partial t_2}\bigg|_{\mathcal{L}}, \quad \frac{\partial F_n}{\partial t_3}\bigg|_{\mathcal{L}}, \quad \frac{\partial^2 F_n}{\partial t_1 \partial t_2}\bigg|_{\mathcal{L}}, \quad \frac{\partial^2 F_n}{\partial t_1 \partial t_3}\bigg|_{\mathcal{L}}, \quad \frac{\partial^2 F_n}{\partial t_2^2}\bigg|_{\mathcal{L}}.$$

Setting $t_1 = x$ and $F'_n = \partial F_n / \partial x$, the solution is given by the expressions

$$\frac{\partial F_n}{\partial t_2}\bigg|_{\mathcal{L}} = -\frac{1}{x}\left((2n + b_1)F'_n + n(b_0 - x)\right),$$

$$\frac{\partial F_n}{\partial t_3}\bigg|_{\mathcal{L}} = -\frac{1}{x^2}\left(x\left(F''_n + F'^2_n + (b_0 - x)F'_n + n(n + b_1)\right)\right.$$
$$\left. - (2n + b_1)\left((2n + b_1)F'_n + b_0 n\right)\right),$$

$$\frac{\partial^2 F_n}{\partial t_1 \partial t_2}\bigg|_{\mathcal{L}} = -\frac{1}{x^2}\left((2n + b_1)(xF''_n - F'_n) - b_0 n\right),$$

$$\frac{\partial^2 F_n}{\partial t_1 \partial t_3}\bigg|_{\mathcal{L}} = -\frac{1}{x^3}\left(x^2(F'''_n + 2F'_n F''_n) - x\left((x^2 - b_0 x + 1)F''_n + F'^2_n + b_0 F'_n\right.\right.$$
$$\left.\left. + (2n + b_1)^2 F''_n + n(n + b_1)\right) + 2(2n + b_1)^2 F'_n + 2 b_0 n(2n + b_1)\right),$$

$$\frac{\partial^2 F_n}{\partial t_2^2}\bigg|_{\mathcal{L}} = \frac{1}{x^3}\left(x\left(2F'^2_n + 2 b_0 F'_n + ((2n + b_1)^2 + 2)F''_n + 2n(n + b_1)\right)\right.$$
$$\left. - 3(2n + b_1)^2 F'_n - 3 b_0 n(2n + b_1)\right).$$

Putting these expressions into KP and setting $t_1 = x$, one finds

$$0 = \left(\left(\frac{\partial}{\partial t_1}\right)^4 + 3\left(\frac{\partial}{\partial t_2}\right)^2 - 4\frac{\partial^2}{\partial t_1 \partial t_3}\right) F_n + 6\left(\frac{\partial^2}{\partial t_1^2} F_n\right)^2$$

$$= \frac{1}{x^3}\left(x^3 F'''' + 4x^2 F''' + x(-4x^2 + 4b_0 x + 2 - (2n + b_1)^2)F'' + 8x^2 F' F''\right.$$
$$\left. + 6x^3 F''^2 + 2x F'^2 + (2b_0 x - (2n + b_1)^2)F' + n(2x - b_0)(n + b_1) - b_0 n^2\right).$$

Finally, the function

$$H(x) := x \frac{d}{dx} F(x) = x \frac{d}{dx} \log \tau_n(x)$$

satisfies

$$x^2 H''' + x H'' + 6x H'^2 - \left(4H + 4x^2 - 4bx + (2n + a)^2\right) H' + (4x - 2b)H$$
$$+ 2n(n + a)x - bn(2n + a) = 0. \quad (8.0.5)$$

This third order equation is Cosgrove's [24; 23] equation, with $P = x$, $4Q = -4x^2 + 4bx - (2n + a)^2$, $2R = 2n(n + a)x - bn(2n + a)$. So, this third order equation can be transformed into the Painlevé V equation (9.0.3) in the appendix. The boundary condition $f(0) = 0$ follows from the definition of H above, whereas, after an elementary, but tedious computation, $f'(0) = f''(0) = 0$ follows from the differential equation (8.0.5) and the Aomoto extension [15] (see Mehta [49, p. 340]) of Selberg's integral:[39]

$$\frac{\int_0^1 \cdots \int_0^1 x_1 \ldots x_m |\Delta(x)|^\beta \prod_{j=1}^n x_j^\gamma (1-x_j)^\delta \, dx_1 \ldots dx_n}{\int_0^1 \cdots \int_0^1 |\Delta(x)|^\beta \prod_{j=1}^n x_j^\gamma (1-x_j)^\delta \, dx_1 \ldots dx_n}$$

$$= \prod_{j=1}^m \frac{\gamma + 1 + (n-j)\beta/2}{\gamma + \delta + 2 + (2n - j - 1)\beta/2}.$$

However, the initial condition (8.0.2) is a much stronger statement, again stemming from the fact that as long as $0 \leq n \leq l$, the inequality $L(\pi_n) \leq l$ is trivially verified, thus leading to

$$E_{O_\pm(l+1)} e^{x \operatorname{Tr} M} = \exp\left(\frac{x^2}{2} \pm \frac{x^{l+1}}{(l+1)!} + O(x^{l+2})\right),$$

ending the proof of Proposition 8.1. $\square$

Appendix: Chazy Classes

Most of the differential equations encountered in this survey belong to the general Chazy class

$$f''' = F(z, f, f', f''), \text{ where } F \text{ is rational in } f, f', f'' \text{ and locally analytic in } z,$$

subjected to the requirement that the general solution be free of movable branch points; the latter is a branch point whose location depends on the integration constants. In his classification Chazy found thirteen cases, the first of which is given by

$$f''' + \frac{P'}{P} f'' + \frac{6}{P} f'^2 - \frac{4P'}{P^2} ff' + \frac{P''}{P^2} f^2 + \frac{4Q}{P^2} f' - \frac{2Q'}{P^2} f + \frac{2R}{P^2} = 0 \quad (9.0.1)$$

with arbitrary polynomials $P(z), Q(z), R(z)$ of degree $3, 2, 1$ respectively. Cosgrove and Scoufis [24; 23, (A.3)] show that this third order equation has a first integral, which is second order in f and quadratic in f'',

$$f''^2 + \frac{4}{P^2}\left((Pf'^2 + Qf' + R)f' - (P'f'^2 + Q'f' + R')f\right.$$
$$\left. + \frac{1}{2}(P''f' + Q'')f^2 - \frac{1}{6}P'''f^3 + c\right) = 0; \quad (9.0.2)$$

c is the integration constant. Equations of the general form

$$f''^2 = G(x, f, f')$$

---

[39] Here $\operatorname{Re}\gamma, \operatorname{Re}\delta > -1$, and $\operatorname{Re}\beta > -2\min\left(\frac{1}{n}, \frac{\operatorname{Re}\gamma + 1}{n-1}, \frac{\operatorname{Re}\delta + 1}{n-1}\right).$

are invariant under the map

$$x \mapsto \frac{a_1 z + a_2}{a_3 z + a_4} \quad \text{and} \quad f \mapsto \frac{a_5 f + a_6 z + a_7}{a_3 z + a_4}.$$

Using this map, the polynomial $P(z)$ can be normalized to

$$P(z) = z(z-1), \ z, \ \text{or} \ 1.$$

In this way, Cosgrove shows (9.0.2) is a master Painlevé equation, containing the 6 Painlevé equations. In each of the cases, the canonical equations are respectively:

- $g''^2 = -4g'^3 - 2g'(zg' - g) + A_1$ (**Painlevé II**)
- $g''^2 = -4g'^3 + 4(zg' - g)^2 + A_1 g' + A_2$ (**Painlevé IV**)
- $(zg'')^2 = (zg' - g)(-4g'^2 + A_1(zg' - g) + A_2) + A_3 g' + A_4$ (**Painlevé V**)
- $(z(z-1)g'')^2 = (zg' - g)(4g'^2 - 4g'(zg' - g) + A_2) + A_1 g'^2 + A_3 g' + A_4$
(**Painlevé VI**)
(9.0.3)

The Painlevé II equation above can be solved by setting

$$g(z) = \tfrac{1}{2}(u')^2 - \tfrac{1}{2}(u^2 + \tfrac{1}{2}z)^2 - (\alpha + \tfrac{1}{2}\varepsilon_1)u,$$
$$g'(z) = -\tfrac{1}{2}\varepsilon_1 u' - \tfrac{1}{2}(u^2 + \tfrac{1}{2}z),$$
$$A_1 = \tfrac{1}{4}\big(\alpha + (u^2 + \tfrac{1}{2}z)^2 \varepsilon_1\big)^2 \quad (\varepsilon = \pm 1).$$

Then $u(z)$ satisfies yet another version of the Painlevé II equation

$$u'' = 2u^3 + zu + \alpha. \quad \textbf{(Painlevé II)}$$

Now, each of these Painlevé II, IV, V, VI equations can be transformed into the standard Painlevé equations, which are all differential equations of the form

$$f'' = F(z, f, f'), \quad \text{rational in } f \text{ and } f' \text{ and analytic in } z,$$

whose general solution has no movable critical points. Painlevé showed that this requirement leads to 50 types of equations, six of which cannot be reduced to known equations.

Acknowledgment

These lectures represent joint work especially with (but also inspired by) Mark Adler, Taka Shiota and Emil Horozov. Thanks also for many informative discussions with Jinho Baik, Pavel Bleher, Edward Frenkel, Alberto Grünbaum, Alexander Its, especially Craig Tracy and Harold Widom, and with other participants in the semester at MSRI.

I thank Pavel Bleher, David Eisenbud and Alexander Its for organizing a truly stimulating and enjoyable semester at MSRI.

References

[1] M. Adler and P. van Moerbeke: *Bäcklund transformations, Birkhoff strata and isospectral sets of differential operators*, Advances in Mathematics, **108**, 140–204 (1994).

[2] M. Adler and P. van Moerbeke: *Matrix integrals, Toda symmetries, Virasoro constraints and orthogonal polynomials*, Duke Math. J. **80**, 863–911 (1995).

[3] M. Adler and P. van Moerbeke: *The spectrum of coupled random matrices*, Annals of Mathematics, **149**, 921–976 (1999).

[4] M. Adler and P. van Moerbeke: *String orthogonal Polynomials, String Equations and two-Toda Symmetries*, Comm. Pure and Appl. Math., **50**, 241–290 (1997).

[5] M. Adler and P. van Moerbeke: *Vertex operator solutions to the discrete KP-hierarchy*, Comm. Math. Phys., **203**, 185–210 (1999).

[6] M. Adler and P. van Moerbeke: *Hermitian, symmetric and symplectic random ensembles: PDE's for the distribution of the spectrum*, Annals of Mathematics (2001). Preprint available at solv-int/9903009.

[7] M. Adler and P. van Moerbeke: *Integrals over classical groups, random permutations, Toda and Toeplitz lattices*, Comm. Pure Appl. Math. **53**, 1–53 (2000). Preprint available at math.CO/9912143.

[8] M. Adler and P. van Moerbeke: *The Pfaff lattice, matrix integrals and a map from Toda to Pfaff*, Duke Math J. (2001). Preprint available at solv-int/9912008.

[9] M. Adler, E. Horozov and P. van Moerbeke: *The Pfaff lattice and skew-orthogonal polynomials*, International Mathematics Research notices, **11**, 569–588 (1999).

[10] M. Adler, A. Morozov, T. Shiota and P. van Moerbeke: *A matrix integral solution to $[P,Q] = P$ and matrix Laplace transforms*, Comm. Math. Phys., **180**, 233–263 (1996).

[11] M. Adler, T. Shiota and P. van Moerbeke: *Random matrices, vertex operators and the Virasoro algebra*, Phys. Lett. **A 208**, 67–78, (1995).

[12] M. Adler, T. Shiota and P. van Moerbeke: *Random matrices, Virasoro algebras and non-commutative KP*, Duke Math. J. **94**, 379–431 (1998).

[13] M. Adler, T, Shiota and P. van Moerbeke: *Pfaff τ-functions*, Math. Annalen (2001). Preprint available at solv-int/9909010.

[14] D. Aldous and P. Diaconis: *Longest increasing subsequences: From patience sorting to the Baik-Deift-Johansson theorem*, Bull. Am. Math. Soc. (N.S.) **36** (4), 413–432 (1999).

[15] K. Aomoto: *Jacobi polynomials associated with Selberg integrals*, SIAM J. Math. Anal. **18**, 545–549 (1987).

[16] H. Awata, Y. Matsuo, S. Odake and J. Shiraishi: *Collective field theory, Calogero-Sutherland Model and generalized matrix models*, RIMS-997 reprint (1994). Available at hep-th/9411053.

[17] B. Baik, P. Deift and K. Johansson: *On the distribution of the length of the longest increasing subsequence of random permutations*, Journal Amer. Math. Soc. **12**, 1119–1178 (1999). Preprint available at math.CO/9810105.

[18] J. Baik and E. Rains: *Algebraic aspects of increasing subsequences*, preprint available at math.CO/9905083.

[19] P. Bleher, A. Its: *Semiclassical asymptotics of orthogonal polynomials, Riemann-Hilbert problem and universality in the matrix model*, Ann. of Math. **150** 1–81 (1999).

[20] N. Bourbaki: *Algèbre de Lie*, Hermann, Paris.

[21] Bowick and E. Brézin: *Universal scaling of the tail of the density of eigenvalues in random matrix models*, Phys. Letters **B 268**, 21–28 (1991).

[22] E. Brézin, H. Neuberger: *Multicritical points of unoriented random surfaces*, Nuclear Physics **B 350**, 513–553 (1991).

[23] C. M. Cosgrove: *Chazy classes IX-XII of third-order differential equations*, Stud. Appl. Math. **104**(3), 171–228 (2000).

[24] C. M. Cosgrove, G. Scoufis: *Painlevé classification of a class of differential equations of the second order and second degree*, Studies. Appl. Math. **88**, 25–87 (1993).

[25] E. Date, M. Jimbo, M. Kashiwara, T. Miwa: *Transformation groups for soliton equations*, pp. 39–119 in Proc. RIMS Symp. Nonlinear integrable systems — Classical and quantum theory (Kyoto 1981), World Scientific, 1983.

[26] P. Diaconis, M. Shashahani: *On the eigenvalues of random matrices* J. Appl. Prob., suppl. in honour of Takàcs **31**A, 49–61 (1994).

[27] F. Dyson: *Statistical theory of energy levels of complex systems*, I, II and III, J. Math Phys **3** 140–156, 157–165, 166–175 (1962).

[28] P. Erdös and G. Szekeres: *A combinatorial theorem in geometry*, Compositio Math., **2**, 463–470 (1935).

[29] A. S. Fokas, A. R. İts, A. V. Kitaev: *The isomonodromy approach to matrix models in 2d quantum gravity*, Comm. Math. Phys., **147**, 395–430 (1992).

[30] P. J. Forrester: *The spectrum edge of random matrix ensembles*, Nucl. Phys. B, **402**, 709–728 (1993).

[31] I. M. Gessel: *Symmetric functions and P-recursiveness*, J. of Comb. Theory, Ser A, **53**, 257–285 (1990).

[32] L. Haine, J. P. Semengue: *The Jacobi polynomial ensemble and the Painlevé VI equation*, J. of Math. Phys., **40**, 2117–2134 (1999).

[33] J. M. Hammersley: *A few seedlings of research*, Proc. Sixth. Berkeley Symp. Math. Statist. and Probability, Vol. 1, 345–394, University of California Press (1972).

[34] Harish-Chandra: *Differential operators on a semi-simple Lie algebra*, Amer. J. of Math., **79**, 87–120 (1957).

[35] S. Helgason: Groups and geometric analysis; integral geometry, invariant differential operators, and spherical functions, Academic Press 1984,

[36] S. Helgason: Differential geometry and symmetric spaces, Academic Press, 1962

[37] M. Hisakado: *Unitary matrix models and Painlevé III*, Mod. Phys. Letters, **A 11** 3001–3010 (1996).

[38] Cl. Itzykson, J.-B. Zuber: *The planar approximation, II*, J. Math. Phys. **21**, 411–421 (1980).

[39] M. Jimbo, T. Miwa, Y. Mori and M. Sato: *Density matrix of an impenetrable Bose gas and the fifth Painlevé transcendent*, Physica 1D, 80–158 (1980).

[40] K. Johansson: *On random matrices from the compact classical groups*, Ann. of Math., **145**, 519–545 (1997).

[41] K. Johansson: *The Longest increasing subsequence in a random permutation and a unitary random matrix model*, Math. Res. Lett., **5**(1–2), 63–82 (1998)

[42] V. G. Kac and J. van de Leur: *The geometry of spinors and the multicomponent BKP and DKP hierarchies*, pp. 159–202 in The bispectral problem (Montreal PQ, 1997), CRM Proc. Lecture notes **14**, AMS, Providence (1998).

[43] R. D. Kamien, H. D. Politzer, M. B. Wise: *Universality of random-matrix predictions for the statistics of energy levels* Phys. rev. letters **60**, 1995–1998 (1988).

[44] M. Kontsevich: *Intersection theory on the moduli space of curves and the matrix Airy function*, Comm. Math. Phys. **147**, 1–23 (1992).

[45] D. Knuth: "The art of computer programming, v. III: searching and sorting", 3rd edition, Addison-Wesley, Reading, MA, 1998.

[46] B. F. Logan and L. A. Shepp: *A variational problem for random Young tableaux*, Advances in Math., **26**, 206–222 (1977).

[47] I. G. MacDonald: "Symmetric functions and Hall polynomials", Clarendon Press, 1995.

[48] G. Mahoux, M. L. Mehta: *A method of integration over matrix variables: IV*, J. Phys. I (France) **1**, 1093–1108 (1991).

[49] M. L. Mehta: Random matrices, 2nd ed., Boston: Academic Press, 1991.

[50] M. L. Mehta: Matrix Theory, special topics and useful results, Les éditions de Physique, Les Ulis, France, 1989.

[51] Moore, G.: Matrix models of 2D gravity and isomonodromic deformations, Progr. Theor. Phys., Suppl. 102, 255–285 (1990).

[52] T. Nagao, M. Wadati: *Correlation functions of random matrix ensembles related to classical orthogonal polynomials*, J. Phys. Soc. of Japan, **60** 3298–3322 (1991).

[53] A. M. Odlyzko: *On the distribution of spacings between zeros of the zeta function*, Math. Comput. **48** 273–308 (1987).

[54] A. Okounkov: *Random matrices and random permutations*, preprint available at math.CO/9903176.

[55] L. A. Pastur: *On the universality of the level spacing distribution for some ensembles of random matrices*, Letters Math. Phys., **25** 259–265 (1992).

[56] C. E. Porter and N. Rosenzweig, *Statistical properties of atomic and nuclear spectra*, Ann. Acad. Sci. Fennicae, Serie A, VI Physica **44**, 1–66 (1960); *Repulsion of energy levels in complex atomic spectra*, Phys. Rev. **120**, 1698–1714 (1960).

[57] E. M. Rains: *Topics in probability on compact Lie groups*, Harvard University doctoral dissertation, (1995).

[58] E. M. Rains: *Increasing subsequences and the classical groups*, Elect. J. of Combinatorics, **5**, R12 (1998).

[59] P. Sarnak: *Arithmetic quantum chaos*, Israel Math. Conf. Proceedings, **8**, 183–236 (1995).

[60] A. Terras: "Harmonic analysis on Symmetric Spaces and Applications II", Springer, 1988.

[61] C. L. Terng: *Isoparametric submanifolds and their Coxeter groups*, J. Differential Geometry. **21**, 79–107 (1985).

[62] C. L. Terng, W. Y. Hsiang and R. S. Palais: *The topology of isoparametric submanifolds in Euclidean spaces*, J. of Diff. Geometry **27**, 423–460 (1988).

[63] C. A. Tracy and H. Widom: *Level-spacings distribution and the Airy kernel*, Commun. Math. Phys., **159**, 151–174 (1994).

[64] C. A. Tracy and H. Widom: *Level spacing distributions and the Bessel kernel*, Commun. Math. Phys., **161**, 289–309 (1994).

[65] C. A. Tracy and H. Widom: *On orthogonal and symplectic matrix ensembles*, Comm. Math. Phys. **177**, 103–130 (1996).

[66] C. A. Tracy and H. Widom: *Random unitary matrices, permutations and Painlevé*, preprint available at math.CO/9811154.

[67] K. Ueno and K. Takasaki: *Toda Lattice Hierarchy*, Adv. Studies in Pure Math. **4**, 1–95 (1984).

[68] S. M. Ulam: *Monte Carlo calculations in problems of mathematical physics*, pp. 261–281 in *Modern Mathematics for the Engineers*, E. F. Beckenbach ed., McGraw-Hill (1961).

[69] J. van de Leur: *Matrix integrals and geometry of spinors*, preprint available at solv-int/9909028.

[70] P. van Moerbeke: *The spectrum of random matrices and integrable systems*, pp. 835–852 in Physical applications and Mathematical aspects of Geometry, Groups and Algebras, Vol.II, Eds.: H.-D. Doebner, W. Scherer, C. Schulte, World Scientific, 1997.

[71] P. van Moerbeke: *Integrable foundations of string theory*, pp. 163–267 in Lectures on Integrable systems, Proceedings of the CIMPA-school, 1991, Ed.: O. Babelon, P. Cartier, Y. Kosmann-Schwarzbach, World Scientific, 1994.

[72] A. M. Vershik and S. V. Kerov: *Asymptotics of the Plancherel measure of the symmetric group and the limiting form of Young tables*, Soviet Math. Dokl., **18**, 527–531 (1977).

[73] H. Weyl: The classical groups, Princeton University Press, 1946.

[74] E. P. Wigner: *On the statistical distribution of the widths and spacings of nuclear resonance levels*, Proc. Cambr. Phil. Soc. **47** 790–798 (1951).

PIERRE VAN MOERBEKE

DEPARTMENT OF MATHEMATICS
UNIVERSITÉ DE LOUVAIN
1348 LOUVAIN-LA-NEUVE
BELGIUM

DEPARTMENT OF MATHEMATICS
BRANDEIS UNIVERSITY
WALTHAM, MA 02454
UNITED STATES
vanmoerbeke@geom.ucl.ac.be, vanmoerbeke@math.brandeis.edu

SL(2) and z-Measures

ANDREI OKOUNKOV

ABSTRACT. We provide a representation-theoretic derivation of the determinantal formula of Borodin and Olshanski for the correlation functions of z-measures in terms of the hypergeometric kernel.

1. Introduction

This paper is about z-measures, a remarkable two-parameter family of measures on partitions introduced by S. Kerov, G. Olshanski and A. Vershik [Kerov et al. 1993] in the context of harmonic analysis on the infinite symmetric group. In a series of papers, A. Borodin and Olshanski obtained fundamental results on z-measures; see the survey [Borodin and Olshanski 2001] in this volume and also [Borodin and Olshanski 1998]. The culmination of this development is an exact determinantal formula for the correlation functions of the z-measures in terms of the hypergeometric kernel [Borodin and Olshanski 2000]. We mention [Borodin et al. 2000] as one of the applications of this formula. The main result of this paper is a representation-theoretic derivation of the formula of Borodin and Olshanski.

In the early days of z-measures, it was already noticed that they have some mysterious connection to the representation theory of SL(2). For example, a z-measure is in fact positive if its two parameters z and z' are either complex conjugate $z' = \bar z$ or $z, z' \in (n, n+1)$ for some $n \in \mathbb{Z}$. In these cases $z - z'$ is either imaginary or lies in $(-1, 1)$, which is reminiscent of the principal and complementary series of representations of SL(2).

Later, Kerov (private communication) constructed an SL(2)-action on partitions for which the z-measures are certain matrix elements. Finally, Borodin and Olshanski computed the correlation functions of the z-measures is in terms of the Gauss hypergeometric function, which appears in matrix elements of representations of SL(2). The aim of this paper is to put these pieces together.

The constructions of this paper were subsequently generalized beyond SL(2) and z-measures in [Okounkov 1999].

2. z-measures, Kerov Operators, and Correlation Functions

2.1. Definition of z-Measures. Let $z, z' \in \mathbb{C}$ be two parameters and consider the following measure on the set of all partitions λ of n:

$$\mathcal{M}_n(\lambda) = \frac{n!}{(zz')_n} \prod_{\square \in \lambda} \frac{(z + c(\square))(z' + c(\square))}{h(\square)^2}, \qquad (2\text{--}1)$$

where

$$(x)_n = x(x+1)\ldots(x+n-1),$$

the product is over all squares $\square$ in the diagram of λ, $h(\square)$ is the length of the corresponding hook, and $c(\square)$ stands for the content of the square $\square$. Recall that, by definition, the content of $\square$ is

$$c(\square) = \text{column}(\square) - \text{row}(\square),$$

where $\text{column}(\square)$ denotes the column number of the square $\square$. See [Macdonald 1995] for general facts about partitions.

It is not immediately obvious from the definition (2–1) that

$$\sum_{|\lambda|=n} \mathcal{M}_n(\lambda) = 1. \qquad (2\text{--}2)$$

One possible proof of (2–2) uses the following operators on partitions, introduced by S. Kerov.

2.2. Kerov Operators. Consider the vector space with an orthonormal basis $\{\delta_\lambda\}$ indexed by all partitions of λ of any size. Introduce the operators

$$U\delta_\lambda = \sum_{\mu=\lambda+\square} (z + c(\square))\delta_\mu,$$

$$L\delta_\lambda = (zz' + 2|\lambda|)\delta_\lambda,$$

$$D\delta_\lambda = \sum_{\mu=\lambda-\square} (z' + c(\square))\delta_\mu,$$

where $\mu = \lambda + \square$ means that μ is obtained from λ by adding a square $\square$ and $c(\square)$ is the content of this square. The letters U and D here stand for "up" and "down".

These operators satisfy the commutation relations

$$[D, U] = L, \quad [L, U] = 2U, \quad [L, D] = -2D, \qquad (2\text{--}3)$$

as does the basis of $\mathfrak{sl}(2)$ given by

$$U = \begin{pmatrix} 0 & 1 \\ 0 & 0 \end{pmatrix}, \quad L = \begin{pmatrix} 1 & 0 \\ 0 & -1 \end{pmatrix}, \quad D = \begin{pmatrix} 0 & 0 \\ -1 & 0 \end{pmatrix}.$$

In particular, it is clear that if $|\lambda| = n$ then
$$(U^n \delta_\varnothing, \delta_\lambda) = \dim \lambda \prod_{\square \in \lambda} (z + c(\square)),$$
where
$$\dim \lambda = n! \prod_{\square \in \lambda} h(\square)^{-1}$$
is the number of standard tableaux on λ. It follows that
$$\mathcal{M}_n(\lambda) = \frac{1}{n!(zz')_n} (U^n \delta_\varnothing, \delta_\lambda)(L^n \delta_\lambda, \delta_\varnothing).$$

Using this presentation and the commutation relations (2–3) one proves (2–2) by induction on n.

2.3. The Measure $\mathcal{M}$ and Its Normalization. In a slightly different language, with induction on n replaced by the use of generating functions, this computation goes as follows.

As in [Borodin and Olshanski 2000], the sequence of the measures $\mathcal{M}_n$ can be conveniently assembled into one measure $\mathcal{M}$ on the set of all partitions of all numbers:
$$\mathcal{M} = (1-\xi)^{zz'} \sum_{n=0}^{\infty} \xi^n \frac{(zz')_n}{n!} \mathcal{M}_n,$$
where $\xi \in [0,1)$ is a new parameter. In other words, $\mathcal{M}$ is the mixture of the measures $\mathcal{M}_n$ by means of a negative binomial distribution on n with parameter ξ.

It is clear that (2–2) is now equivalent to $\mathcal{M}$ being a probability measure. It is also clear that
$$\mathcal{M}(\lambda) = (1-\xi)^{zz'} (e^{\sqrt{\xi}U} \delta_\varnothing, \delta_\lambda)(e^{\sqrt{\xi}D} \delta_\lambda, \delta_\varnothing). \tag{2-4}$$

Therefore
$$\sum_\lambda \mathcal{M}(\lambda) = (1-\xi)^{zz'} (e^{\sqrt{\xi}D} e^{\sqrt{\xi}U} \delta_\varnothing, \delta_\varnothing). \tag{2-5}$$

It follows from the definitions that
$$D\delta_\varnothing = 0, \quad L\delta_\varnothing = zz'\delta_\varnothing, \quad U^*\delta_\varnothing = 0, \tag{2-6}$$
where U^* is the operator adjoint to U. Therefore, in order to evaluate (2–5), it suffices to commute $e^{\sqrt{\xi}L}$ through $e^{\sqrt{\xi}U}$.

A computation in SL(2),
$$\begin{pmatrix} 1 & 0 \\ -\beta & 1 \end{pmatrix} \begin{pmatrix} 1 & \alpha \\ 0 & 1 \end{pmatrix} = \begin{pmatrix} 1 & \frac{\alpha}{1-\alpha\beta} \\ 0 & 1 \end{pmatrix} \begin{pmatrix} \frac{1}{1-\alpha\beta} & 0 \\ 0 & 1-\alpha\beta \end{pmatrix} \begin{pmatrix} 1 & 0 \\ -\frac{\beta}{1-\alpha\beta} & 1 \end{pmatrix},$$

implies that

$$\exp(\beta D)\exp(\alpha U) = \exp\left(\frac{\alpha}{1-\alpha\beta}U\right)(1-\alpha\beta)^{-L}\exp\left(\frac{\beta}{1-\alpha\beta}D\right), \quad (2\text{-}7)$$

provided $|\alpha\beta| < 1$. Therefore

$$\sum_\lambda \mathcal{M}(\lambda) = (1-\xi)^{zz'}\left(\exp\left(\frac{\sqrt{\xi}}{1-\xi}U\right)(1-\xi)^{-L}\exp\left(\frac{\sqrt{\xi}}{1-\xi}D\right)\delta_\varnothing,\delta_\varnothing\right)$$
$$= (1-\xi)^{zz'}\left((1-\xi)^{-L}\delta_\varnothing,\delta_\varnothing\right) = 1,$$

as was to be shown.

2.4. Correlation Functions. We now introduce coordinates on the set of partitions. To a partition λ we associate a subset

$$\mathfrak{S}(\lambda) = \{\lambda_i - i + \tfrac{1}{2}\} \subset \mathbb{Z} + \tfrac{1}{2}.$$

For example,

$$\mathfrak{S}(\varnothing) = \{-\tfrac{1}{2}, -\tfrac{3}{2}, -\tfrac{5}{2}, \dots\}$$

This set $\mathfrak{S}(\lambda)$ has a geometric interpretation. Take the diagram of λ and rotate it 135° thus:

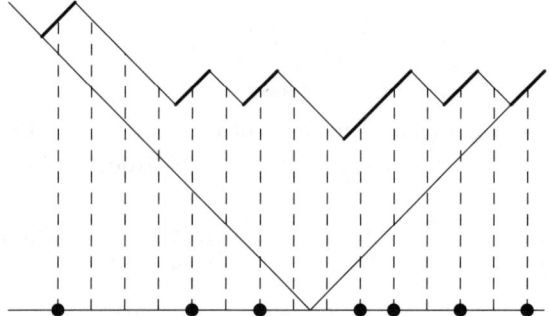

The positive direction of the axis points to the left in the figure. The boundary of λ forms a zigzag path and the elements of $\mathfrak{S}(\lambda)$, which are marked by •, correspond to moments when this zigzag goes up.

Subsets $S \subset \mathbb{Z} + \tfrac{1}{2}$ of the form $S = \mathfrak{S}(\lambda)$ can be characterized by

$$|S_+| = |S_-| < \infty,$$

where

$$S_+ = S \setminus \left(\mathbb{Z}_{\le 0} - \tfrac{1}{2}\right), \quad S_- = \left(\mathbb{Z}_{\le 0} - \tfrac{1}{2}\right) \setminus S.$$

The number $|\mathfrak{S}_+(\lambda)| = |\mathfrak{S}_-(\lambda)|$ is the number of squares in the diagonal of the diagram of λ and the finite set $\mathfrak{S}_+(\lambda) \cup \mathfrak{S}_-(\lambda) \subset \mathbb{Z} + \tfrac{1}{2}$ is known as the modified Frobenius coordinates of λ.

Given a finite subset $X \in \mathbb{Z} + \tfrac{1}{2}$, define the *correlation function* by

$$\rho(X) = \mathcal{M}(\{\lambda, X \subset \mathfrak{S}(\lambda)\}).$$

Borodin and Olshanski [2000] proved that

$$\rho(X) = \det \left[K(x_i, x_j) \right]_{x_i, x_j \in X}$$

where K the *hypergeometric kernel* introduced in [Borodin and Olshanski 2000]. This kernel involves the Gauss hypergeometric function and the explicit formula for K will be reproduced below.

It is our goal in the present paper to give a representation-theoretic derivation of the formula for correlation functions and, in particular, show how the kernel K arises from matrix elements of irreducible SL(2)-modules.

3. SL(2) and Correlation Functions

3.1. Matrix Elements of $\mathfrak{sl}(2)$-Modules and the Gauss Hypergeometric Function.
It is well known that the hypergeometric function arises as matrix coefficients of SL(2) modules. A standard way to see this is to use a functional realization of these modules; the computation of matrix elements leads then to an integral representation of the hypergeometric function, see for example how matrix elements of SL(2)-modules are treated in [Vilenkin 1968]. An alternative approach is to use explicit formulas for the action of the Lie algebra $\mathfrak{sl}(2)$ and it goes as follows.

Consider the $\mathfrak{sl}(2)$-module V with the basis v_k indexed by all half-integers $k \in \mathbb{Z} + \frac{1}{2}$, and the action of $\mathfrak{sl}(2)$ given by

$$\begin{aligned} U v_k &= \left(z + k + \tfrac{1}{2}\right) v_{k+1}, \\ L v_k &= \left(2k + z + z'\right) v_k, \\ D v_k &= \left(z' + k - \tfrac{1}{2}\right) v_{k-1}. \end{aligned} \qquad (3\text{-}1)$$

It is clear that

$$e^{\alpha U} v_k = \sum_{s=0}^{\infty} \frac{\alpha^s}{s!} \left(z + k + \tfrac{1}{2}\right)_s v_{k+s}.$$

Introduce the notation

$$(a)_{\downarrow s} = a(a-1)(a-2) \cdots (a-s+1),$$

so that

$$e^{\beta D} v_k = \sum_{s=0}^{\infty} \frac{\beta^s}{s!} \left(z' + k - \tfrac{1}{2}\right)_{\downarrow s} v_{k-s}.$$

Denote by $[i \to j]_{\alpha, \beta, z, z'}$ the coefficient of v_j in the expansion of $e^{\alpha U} e^{\beta D} v_i$:

$$e^{\alpha U} e^{\beta D} v_i = \sum_j [i \to j]_{\alpha, \beta, z, z'} v_j.$$

A direct computation yields

$$[i \to j]_{\alpha,\beta,z,z'} = \begin{cases} \dfrac{\alpha^{j-i}}{(j-i)!}(z+i+\tfrac12)_{j-i}\, F\!\left(\begin{matrix}-z-i+\tfrac12,\, -z'-i+\tfrac12\\ j-i+1\end{matrix};\alpha\beta\right) & \text{if } i \le j, \\[2ex] \dfrac{\beta^{i-j}}{(i-j)!}(z'+j+\tfrac12)_{i-j}\, F\!\left(\begin{matrix}-z-j+\tfrac12,\, -z'-j+\tfrac12\\ i-j+1\end{matrix};\alpha\beta\right) & \text{if } i \ge j, \end{cases}$$

(3-2)

where

$$F\!\left(\begin{matrix}a,\, b\\ c\end{matrix};z\right) = \sum_{k=0}^{\infty} \frac{(a)_k (b)_k}{(c)_k\, k!} z^k$$

is the Gauss hypergeometric function.

Now consider the dual module V^* spanned by functionals v_j^* such that

$$\langle v_i^*, v_j \rangle = \delta_{ij}$$

and equipped with the dual action of $\mathfrak{sl}(2)$:

$$U v_k^* = -\bigl(z+k-\tfrac12\bigr) v_{k-1}^*, \quad D v_k^* = -\bigl(z'+k+\tfrac12\bigr) v_{k+1}^*.$$

Denote by $[i \to j]^*_{\alpha,\beta,z,z'}$ the coefficient of v_j^* in the expansion of $e^{\alpha U} e^{\beta D} v_i^*$:

$$e^{\alpha U} e^{\beta D} v_i^* = \sum_j [i \to j]^*_{\alpha,\beta,z,z'}\, v_j^*.$$

We have

$$[i \to j]^*_{\alpha,\beta,z,z'} = \begin{cases} \dfrac{(-\beta)^{j-i}}{(j-i)!}(z'+i+\tfrac12)_{j-i}\, F\!\left(\begin{matrix}z+j+\tfrac12,\, z'+j+\tfrac12\\ j-i+1\end{matrix};\alpha\beta\right) & \text{if } i \le j, \\[2ex] \dfrac{(-\alpha)^{i-j}}{(i-j)!}(z+j+\tfrac12)_{i-j}\, F\!\left(\begin{matrix}z+i+\tfrac12,\, z'+i+\tfrac12\\ i-j+1\end{matrix};\alpha\beta\right) & \text{if } i \ge j. \end{cases}$$

(3-3)

3.2. Remarks

3.2.1. Periodicity. Observe that representations whose parameters z and z' are related by the transformation

$$(z, z') \mapsto (z+m, z'+m), \quad m \in \mathbb{Z},$$

are equivalent. The above transformation amounts to just a renumeration of the vectors v_k. Olshanski has pointed out that this periodicity in (z, z') is reflected in a similar periodicity of various asymptotic properties of z-measures; see [Borodin and Olshanski 1998, Sections 10 and 11].

3.2.2. Unitarity.
Recall that the z-measures are positive if either $z' = \bar z$ or $z, z' \in (n, n+1)$ for some n. By analogy with representation theory of SL(2), these cases were called the principal and the complementary series.

In these cases the representations above have a positive definite Hermitian form Q, invariant in the sense that

$$Q(Lu, v) = Q(u, Lv), \quad Q(Uu, v) = Q(u, Dv).$$

The form Q is given by

$$Q(v_k, v_k) = \begin{cases} 1 & \text{if } z' = \bar z, \\ \dfrac{\Gamma\left(z' + k + \frac{1}{2}\right)}{\Gamma\left(z + k + \frac{1}{2}\right)} & \text{if } z, z' \in (n, n+1), \end{cases}$$

and $Q(v_k, v_l) = 0$ if $k \ne l$. It follows that the operators

$$\frac{i}{2} L, \quad \frac{1}{2}(U - D), \quad \frac{i}{2}(U + D) \in \mathfrak{sl}(2),$$

which form a standard basis of $\mathfrak{su}(1,1)$, are skew-Hermitian; hence this representation of $\mathfrak{su}(1,1)$ can be integrated to a unitary representation of the universal covering group of SU(1, 1). This group SU(1, 1) is isomorphic to $SL(2, \mathbb{R})$ and the above representations correspond to the principal and complementary series of unitary representations of the universal covering of $SL(2, \mathbb{R})$; see [Pukánszky 1964].

3.3. The Infinite Wedge Module.
Consider the module $\Lambda^{\frac{\infty}{2}} V$, which is, by definition, spanned by vectors

$$\delta_S = v_{s_1} \wedge v_{s_2} \wedge v_{s_3} \wedge \cdots,$$

where $S = \{s_1 > s_2 > \ldots\} \subset \mathbb{Z} + \frac{1}{2}$ is a subset such that both sets

$$S_+ = S \setminus \left(\mathbb{Z}_{\le 0} - \tfrac{1}{2}\right), \quad S_- = \left(\mathbb{Z}_{\le 0} - \tfrac{1}{2}\right) \setminus S$$

are finite. We equip this module with the inner product in which the basis $\{\delta_S\}$ is orthonormal. Introduce the operators

$$\psi_k, \psi_k^* : \Lambda^{\frac{\infty}{2}} V \to \Lambda^{\frac{\infty}{2}} V.$$

The operator ψ_k is the exterior multiplication by v_k

$$\psi_k(f) = v_k \wedge f.$$

The operator ψ_k^* is the adjoint operator; it can be also given by the formula

$$\psi_k^*(v_{s_1} \wedge v_{s_2} \wedge v_{s_3}) = \sum_i (-1)^{i+1} \langle v_k^*, v_{s_i} \rangle v_{s_1} \wedge v_{s_2} \wedge \cdots \wedge \widehat{v_{s_i}} \wedge \cdots.$$

These operators satisfy the canonical anticommutation relations

$$\psi_k \psi_k^* + \psi_k^* \psi_k = 1,$$

all other anticommutators being equal to 0. It is clear that

$$\psi_k \psi_k^* \delta_S = \begin{cases} \delta_S & \text{if } k \in S, \\ 0 & \text{if } k \notin S. \end{cases} \tag{3-4}$$

A general reference on the infinite wedge space is [Kac 1990, Chapter 14].

The Lie algebra $\mathfrak{sl}(2)$ acts on $\Lambda^{\frac{\infty}{2}} V$. The actions of U and D are the obvious extensions of the action on V. In terms of the fermionic operators ψ_k and ψ_k^* they can be written as

$$U = \sum_{k \in \mathbb{Z}+\frac{1}{2}} (z+k+\tfrac{1}{2}) \psi_{k+1} \psi_k^*,$$

$$D = \sum_{k \in \mathbb{Z}+\frac{1}{2}} (z'+k+\tfrac{1}{2}) \psi_k \psi_{k+1}^*.$$

The easiest way to define the action of L is to set it equal to $[D, U]$ by definition. We obtain

$$L = 2H + (z+z')C + zz',$$

where H is the energy operator

$$H = \sum_{k>0} k \psi_k \psi_k^* - \sum_{k<0} k \psi_k^* \psi_k$$

and C is the charge

$$C = \sum_{k>0} \psi_k \psi_k^* - \sum_{k<0} \psi_k^* \psi_k.$$

It is clear that

$$C\delta_S = (|S_+| - |S_-|)\delta_S;$$

similarly,

$$H\delta_S = \left(\sum_{k \in S_+} k - \sum_{k \in S_-} k \right) \delta_S.$$

The charge is preserved by the $\mathfrak{sl}(2)$ action.

Consider the zero charge subspace, that is, the kernel of C:

$$\Lambda_0 \subset \Lambda^{\frac{\infty}{2}} V.$$

It is spanned by vectors which, abusing notation, we shall denote by

$$\delta_\lambda = \delta_{S(\lambda)}, \quad S(\lambda) = \{\lambda_1 - \tfrac{1}{2},\ \lambda_2 - \tfrac{3}{2},\ \lambda_3 - \tfrac{5}{2},\ \dots\},$$

where λ is a partition. One immediately sees that the action of $\mathfrak{sl}(2)$ on $\{\delta_\lambda\}$ is identical with Kerov operators.

3.4. Correlation Functions.
Recall that the correlation functions were defined by

$$\rho(X) = \mathcal{M}(\{\lambda, X \subset \mathfrak{S}(\lambda)\}),$$

where the finite set

$$X = \{x_1, \ldots, x_s\} \subset \mathbb{Z} + \tfrac{1}{2}$$

is arbitrary.

The important observation is that (2–4) and (3–4) imply the following expression for the correlation functions:

$$\rho(X) = (1-\xi)^{zz'}\left(e^{\sqrt{\xi}D}\prod_{x\in X}\psi_x\psi_x^* e^{\sqrt{\xi}U}\delta_\varnothing, \delta_\varnothing\right). \tag{3-5}$$

We apply to (3–5) the same strategy we applied to (2–5), which is to commute the operators $e^{\sqrt{\xi}D}$ and $e^{\sqrt{\xi}U}$ all the way to the right and left, respectively, and then use (2–6). From (2–7), we have for any operator A the identity

$$e^{\beta D} A e^{\alpha U} = e^{\frac{\alpha}{1-\alpha\beta}U}\left(e^{-\frac{\alpha}{1-\alpha\beta}U} e^{\beta D} A e^{-\beta D} e^{\frac{\alpha}{1-\alpha\beta}U}\right)(1-\alpha\beta)^{-L} e^{\frac{\beta}{1-\alpha\beta}D}.$$

We now apply this identity with $\alpha = \beta = \sqrt{\xi}$ and $A = \prod \psi_x \psi_x^*$ to obtain

$$\rho(X) = \left(G\prod_{x\in X}\psi_x\psi_x^* G^{-1}\delta_\varnothing,\ \delta_\varnothing\right), \tag{3-6}$$

where

$$G = \exp\left(\frac{\sqrt{\xi}}{\xi-1}U\right)\exp\left(\sqrt{\xi}D\right).$$

Consider the operators

$$\begin{aligned}\Psi_k &:= G\psi_k G^{-1} = \sum_i [k\to i]\,\psi_i, \\ \Psi_k^* &:= G\psi_k^* G^{-1} = \sum_i [k\to i]^*\,\psi_i^*,\end{aligned} \tag{3-7}$$

with the understanding that matrix elements without parameters stand for the choice of parameters

$$[k\to i] = [k\to i]_{\xi^{1/2}(\xi-1)^{-1},\xi^{1/2},z,z'}, \tag{3-8}$$

and with same choice of parameters for $[k\to i]^*$. The first equality on either line of (3–7) is a definition and the second equality follows from the definition of the operators ψ_i and the definition of the matrix coefficients $[i\to j]_{\alpha,\beta,z,z'}$.

From (3–6) we obtain

$$\rho(X) = \left(\prod_{x\in X}\Psi_x\Psi_x^*\delta_\varnothing,\ \delta_\varnothing\right).$$

Applying Wick's theorem to this equality, or simply unraveling the definitions on its right-hand side, we obtain:

THEOREM 3.1.
$$\rho(X) = \det \left[K(x_i, x_j) \right]_{1 \le i,j \le s},$$
where the kernel K is defined by
$$K(i,j) = \left(\Psi_i \Psi_j^* \delta_\varnothing, \delta_\varnothing \right).$$

Observe that
$$(\psi_l \psi_m^* \delta_\varnothing, \delta_\varnothing) = \begin{cases} 1 & \text{if } l = m < 0, \\ 0 & \text{otherwise} . \end{cases}$$
Therefore, applying formulas (3–7) we obtain:

THEOREM 3.2.
$$K(i,j) = \sum_{m=-1/2,-3/2,\ldots} [i \to m]\, [j \to m]^*, \qquad (3\text{--}9)$$
with the convention (3–8) about matrix elements without parameters.

Formula (3–9) is the analog of the [Borodin et al. 2000, Proposition 2.9] for the discrete Bessel kernel.

We conclude this section with a formula which, upon substitution of formulas (3–2) and (3–3) for matrix elements, becomes the formula of [Borodin and Olshanski 2000].

THEOREM 3.3.
$$K(i,j) = \frac{z'\sqrt{\xi}\,[i \to \tfrac{1}{2}]\,[j \to -\tfrac{1}{2}]^* - z\dfrac{\sqrt{\xi}}{(\xi-1)^2}\,[i \to -\tfrac{1}{2}]\,[j \to \tfrac{1}{2}]^*}{i-j}, \qquad (3\text{--}10)$$
where for $i = j$ the right-hand side is defined by continuity.

More generally, set
$$K(i,j)_{\alpha,\beta} = \left(\Psi_i(\alpha,\beta) \Psi_j^*(\alpha,\beta) \delta_\varnothing, \delta_\varnothing \right),$$
where
$$\Psi_k = e^{\alpha U} e^{\beta D} \psi_k e^{-\beta D} e^{-\alpha U} = \sum_i [k \to i]_{\alpha,\beta,z,z'}\, \psi_i$$
$$\Psi_k^* = e^{\alpha U} e^{\beta D} \psi_k^* e^{-\beta D} e^{-\alpha U} = \sum_i [k \to i]^*_{\alpha,\beta,z,z'}\, \psi_i^*.$$

We will prove that
$$K(i,j)_{\alpha,\beta} = \big(\beta z' \,[i \to \tfrac{1}{2}]_{\alpha,\beta,z,z'}\, [j \to -\tfrac{1}{2}]^*_{\alpha,\beta,z,z'} -$$
$$\alpha(\alpha\beta-1) z \, [i \to -\tfrac{1}{2}]_{\alpha,\beta,z,z'}\, [j \to \tfrac{1}{2}]^*_{\alpha,\beta,z,z'} \big)/(i-j). \qquad (3\text{--}11)$$

First we treat the case $i \ne j$, in which we can clear denominators in (3–11). Computing
$$\begin{pmatrix} 1 & \alpha \\ 0 & 1 \end{pmatrix} \begin{pmatrix} 1 & 0 \\ -\beta & 1 \end{pmatrix} \begin{pmatrix} 1 & 0 \\ 0 & -1 \end{pmatrix} \begin{pmatrix} 1 & 0 \\ \beta & 1 \end{pmatrix} \begin{pmatrix} 1 & -\alpha \\ 0 & 1 \end{pmatrix} = \begin{pmatrix} 1-2\alpha\beta & 2\alpha(\alpha\beta-1) \\ -2\beta & 2\alpha\beta-1 \end{pmatrix}$$

we conclude that
$$e^{\alpha U}e^{\beta D}Le^{-\beta D}e^{-\alpha U} = L+T,$$
where
$$T = -2\alpha\beta L + 2\beta D + 2\alpha(\alpha\beta - 1)U.$$
This can be rewritten as
$$\begin{aligned}[L, e^{\alpha U}e^{\beta D}] &= -Te^{\alpha U}e^{\beta D},\\ [L, e^{-\alpha U}e^{-\beta D}] &= e^{-\alpha U}e^{-\beta D}T.\end{aligned} \qquad (3\text{--}12)$$

From (3–12) and the equality $[L, \psi_i\psi_j^*] = 2(i-j)\psi_i\psi_j^*$ we get
$$\bigl[L,\, \Psi_i(\alpha,\beta)\,\Psi_j^*(\alpha,\beta)\bigr] = -\bigl[T,\, \Psi_i(\alpha,\beta)\,\Psi_j^*(\alpha,\beta)\bigr] + 2(i-j)\Psi_i(\alpha,\beta)\,\Psi_j^*(\alpha,\beta). \qquad (3\text{--}13)$$

Since $\delta_\varnothing$ is an eigenvector of L we have
$$\bigl([L,\, \Psi_i(\alpha,\beta)\,\Psi_j^*(\alpha,\beta)]\delta_\varnothing,\, \delta_\varnothing\bigr) = 0.$$

Expand this equality using (3–13) and the relations
$$\begin{aligned}T\delta_\varnothing &= -2\alpha\beta zz'\delta_\varnothing + 2\alpha(\alpha\beta-1)z\delta_\square,\\ T^*\delta_\varnothing &= -2\alpha\beta zz'\delta_\varnothing + 2\beta z'\delta_\square,\end{aligned}$$
where T^* is the operator adjoint to T and $\delta_\square$ is the vector corresponding to the partition $(1,0,0,\ldots)$. We obtain
$$\begin{aligned}(i-j)&K(i,j)_{\alpha,\beta}\\ &= \beta z'\bigl(\Psi_i(\alpha,\beta)\,\Psi_j^*(\alpha,\beta)\delta_\varnothing, \delta_\square\bigr) - \alpha(\alpha\beta - 1)z\bigl(\Psi_i(\alpha,\beta)\,\Psi_j^*(\alpha,\beta)\delta_\square, \delta_\varnothing\bigr).\end{aligned}$$

In order to obtain (3–11) for $i \neq j$, it now remains to observe that
$$(\psi_l\psi_m^*\delta_\varnothing, \delta_\square) = \begin{cases} 1 & \text{if } l = \tfrac{1}{2} \text{ and } m = -\tfrac{1}{2},\\ 0 & \text{otherwise,}\end{cases}$$
$$(\psi_l\psi_m^*\delta_\square, \delta_\varnothing) = \begin{cases} 1 & \text{if } l = -\tfrac{1}{2} \text{ and } m = \tfrac{1}{2},\\ 0 & \text{otherwise.}\end{cases}$$

In the case $i = j$ we argue by continuity. It is clear from (3–9) that $K(i,j)$ is an analytic function of i and j and so is the right-hand side of (3–10). The passage from (3–9) to (3–10) is based on the fact that the product i times $[i \to m]_{\alpha,\beta,z,z'}$ is a linear combination of $[i \to m]_{\alpha,\beta,z,z'}$ and $[i \to m \pm 1]_{\alpha,\beta,z,z'}$ with coefficients that are linear functions of m. Since the matrix coefficients are, essentially, the hypergeometric function, such a relation must hold for any i, not just half-integers. Hence, (3–9) and (3–10) are equal for any $i \neq j$, not necessarily half-integers. Therefore, they are equal for $i = j$.

3.5. Rim-Hook Analogs.
The same principles apply to rim-hook analogs of the z-measures, which were also considered by S. Kerov (private communication).

Recall that a rim hook of a diagram λ is, by definition, a skew diagram λ/μ that is connected and lies on the rim of λ. Here connected means that the squares have to be connected by common edges, not just common vertices. Rim hooks of a diagram λ are in the following 1-1 correspondence with the squares of λ: given a square $\square \in \lambda$, the corresponding rim hook consists of all squares on the rim of λ which are (weakly) to the right of and below $\square$. The length of this rim hook is equal to the hook-length of $\square$.

The entire discussion of the previous section applies to the more general operators

$$U_r v_k = \left(z + \tfrac{k}{r} + \tfrac{1}{2}\right) v_{k+r},$$
$$L_r v_k = \left(\tfrac{2k}{r} + z + z'\right) v_k,$$
$$D_r v_k = \left(z' + \tfrac{k}{r} - \tfrac{1}{2}\right) v_{k-r},$$

which satisfy the same $\mathfrak{sl}(2)$ commutation relations. The easiest way to check the commutation relations is to consider $\tfrac{k}{r}$ rather than k as the index of v_k; the above formulas then become precisely the formulas (3–1). The operator U_r acts on the basis $\{\delta_\lambda\}$ as follows

$$U_r \delta_\lambda = \sum_{\mu = \lambda + \text{rim hook}} (-1)^{\text{height}+1} \left(z + \frac{1}{r^2} \sum_{\square \in \text{rim hook}} c(\square) \right) \delta_\mu,$$

where the summation is over all partitions μ which can be obtained from λ by adding a rim hook of length r, height is the number of horizontal rows occupied by this rim hook and $c(\square)$ stands, as usual for the content of the square $\square$. Similarly, the operator D_r removes rim hooks of length r. These operators were considered by Kerov (private commutation).

It is clear that the action of the operators $e^{\alpha U_r}$ and $e^{\beta D_r}$ on a half-infinite wedge product like

$$v_{s_1} \wedge v_{s_2} \wedge v_{s_3} \wedge \cdots,$$

essentially (up to a sign which disappears in formulas like (3–5)) factors into the tensor product of r separate actions on

$$\bigwedge_{s_i \equiv k + \frac{1}{2} \bmod r} v_{s_i}, \quad k = 0, \ldots, r-1.$$

Consequently, the analogs of the correlation functions (3–5) have again a determinantal form with a certain kernel $K_r(i,j)$ which has the following structure. If $i \equiv j \bmod r$ then $K_r(i,j)$ is essentially the kernel $K(i,j)$ with rescaled arguments. Otherwise, $K_r(i,j) = 0$.

This factorization of the action on $\Lambda^{\frac{\infty}{2}} V$ is just one more way to understand the following well-known phenomenon. Let $\mathbb{Y}_r$ be the partial ordered set formed by partitions with respect to the following ordering: $\mu \leq_r \lambda$ if μ can be obtained from λ by removing a number of rim hooks with r squares. The minimal elements of $\mathbb{Y}_r$ are called the r-cores. The r-cores are precisely those partitions which do not have any hooks of length r. We have

$$\mathbb{Y}_r \cong \bigsqcup_{r\text{-cores}} (\mathbb{Y}_1)^r \tag{3-14}$$

as partially ordered sets. Here the Cartesian product $(\mathbb{Y}_1)^r$ is ordered as follows:

$$(\mu_1,\dots,\mu_r) \le (\lambda_1,\dots,\lambda_r) \iff \mu_i \le_1 \lambda_i \text{ for } i=1,\dots,r,$$

and the partitions corresponding to different r-cores are incomparable in the $\le_r$-order. Combinatorial algorithms which materialize the isomorphism (3–14) are discussed in [James and Kerber 1981, Section 2.7]. The r-core and the r-tuple of partitions which the isomorphism (3–14) associates to a partition λ are called the r-core of λ and the r-quotient of λ. Among more recent papers dealing with r-quotients let us mention [Fomin and Stanton 1997] where an approach similar to the use of $\Lambda^{\frac{\infty}{2}} V$ is employed, an analog of the Robinson–Schensted algorithm for $\mathbb{Y}_r$ is discussed, and further references are given.

Factorization (3–14) and the corresponding analog of the Robinson–Schensted algorithm play the central role in the recent paper [Borodin 1999]; see also [Rains 1998].

Acknowledgements

I want to thank A. Borodin, S. Kerov, G. Olshanski, and A. Vershik for numerous discussions of the z-measures. I also want to thank the organizers of the Random Matrices program at MSRI, especially P. Bleher, P. Deift, and A. Its. My research was supported by NSF under grant DMS-9801466.

References

[Borodin 1999] A. Borodin, "Longest increasing subsequences of random colored permutations", *Electronic Journal of Combinatorics* **6** (1999), #R13.

[Borodin and Olshanski 1998] A. Borodin and G. Olshanski, "Point processes and the infinite symmetric group", *Math. Research Lett.* **5** (1998), 799–816. Preprint at http://arXiv.org/abs/math/9810015.

[Borodin and Olshanski 2000] A. Borodin and G. Olshanski, "Distributions on partitions, point processes, and the hypergeometric kernel", *Comm. Math. Phys.* **211**:2 (2000), 335–358. Preprint at http://arXiv.org/abs/math/9904010.

[Borodin and Olshanski 2001] A. Borodin and G. Olshanski, "Z-measures on partitions, Robinson–Schensted–Knuth correspondence, and $\beta = 2$ random matrix ensembles", in *Random matrices and their applications*, edited by P. Bleher and A. Its, Math. Sci. Res. Inst. Publications **40**, Cambridge Univ. Press, New York, 2001.

[Borodin et al. 2000] A. Borodin, A. Okounkov, and G. Olshanski, "Asymptotics of Plancherel measures for symmetric groups", *J. Amer. Math. Soc.* **13** (2000), 491–515. Preprint at http://arXiv.org/abs/math/9905032.

[Fomin and Stanton 1997] S. Fomin and D. Stanton, "Rim hook lattices", *Algebra i Analiz* **9**:5 (1997), 140–150. In Russian; translated in *St. Petersburg Math. J.*, **9**:5, 1007–1016 (1998).

[James and Kerber 1981] G. James and A. Kerber, *The representation theory of the symmetric group*, Encyclopedia of Math. and its Appl. **16**, Addison-Wesley, Reading, MA, 1981.

[Kac 1990] V. G. Kac, *Infinite-dimensional Lie algebras*, 3rd ed., Cambridge University Press, Cambridge, 1990.

[Kerov et al. 1993] S. Kerov, G. Olshanski, and A. Vershik, "Harmonic analysis on the infinite symmetric group: a deformation of the regular representation", *Comptes Rend. Acad. Sci. Paris, Sér. I* **316** (1993), 773–778.

[Macdonald 1995] I. G. Macdonald, *Symmetric functions and Hall polynomials*, 2nd ed., Oxford University Press, 1995.

[Okounkov 1999] A. Okounkov, "Infinite wedge and measures on partitions", preprint, 1999. Available at http://arXiv.org/abs/math/9907127.

[Pukánszky 1964] L. Pukánszky, "The Plancherel formula for the universal covering group of $SL(R, 2)$", *Math. Ann.* **156** (1964), 96–143.

[Rains 1998] E. M. Rains, "Increasing subsequences and the classical groups", *Electron. J. Combin.* **5**:1 (1998), Research Paper 12, 9 pp. (electronic).

[Vilenkin 1968] N. J. Vilenkin, *Special functions and the theory of group representations*, Translations of Mathematical Monographs **22**, Amer. Math. Soc., Providence, 1968.

ANDREI OKOUNKOV
DEPARTMENT OF MATHEMATICS
UNIVERSITY OF CALIFORNIA AT BERKELEY
EVANS HALL #3840
BERKELEY, CA 94720-3840
UNITED STATES
okounkov@math.berkeley.edu

Some Matrix Integrals Related to Knots and Links

PAUL ZINN-JUSTIN

ABSTRACT. The study of a certain class of matrix integrals can be motivated by their interpretation as counting objects of knot theory such as alternating prime links, tangles or knots. The simplest such model is studied in detail and allows to rederive recent results of Sundberg and Thistlethwaite. The second nontrivial example turns out to be essentially the so-called *ABAB* model, though in this case the analysis has not yet been carried out completely. Further generalizations are discussed. This is a review of work done (in part) in collaboration with J.-B. Zuber.

1. Introduction

Using random matrices to count combinatorial objects is not a new idea. It stems from the pioneering work [Brézin et al. 1978], which showed how the perturbative expansion of a simple nongaussian matrix integral led, using standard Feynman diagram techniques, to the counting of discretized surfaces. It has resulted in many applications: from the physical side, it allowed to define a discretized version of 2D quantum gravity [Di Francesco et al. 1995] and to study various statistical models on random lattices [Kazakov 1986; Kostov 1989; Gaudin and Kostov 1989; Kostov and Staudacher 1992]. From the mathematical side, let us cite the Kontsevitch integral [Kontsevich 1991; Witten 1991; Itzykson and Zuber 1992], and the counting of meanders and foldings [Makeenko 1996; Di Francesco et al. 1997; 1998].

Here we shall try to apply this idea to the field of knot theory. Our basic aim will be to count knots or related objects. The next section defines these objects, and is followed by a brief overview of matrix models and how they can be related to knots. Section 4 explains the counting of alternating links, following [Zinn-Justin and Zuber 1999]; Section 5 presents a generalized model (the *ABAB* model of [Kazakov and Zinn-Justin 1999]), which leads to a digression and consideration of summations over Young diagrams; finally, Section 6 discusses further generalizations.

This work was supported in part by the DOE grant DE-FG02-96ER40559.

2. Knots, Links and Tangles

Let us recall basic definitions of knot theory (from a physicist's point of view; the reader is referred to the literature for more precise definitions). A *knot* is a smooth circle embedded in $\mathbb{R}^3$. A *link* is a collection of intertwined knots. Both kinds of objects are considered up to homeomorphisms of $\mathbb{R}^3$. Roughly speaking, a *tangle* is a knotted structure from which four strings emerge.

In the nineteenth century, Tait introduced the idea to represent such objects by their projection on the plane, with under/over-crossings at each double point and with *minimal number of such crossings* (Figure 1). We shall now consider

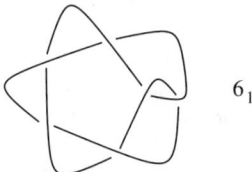

Figure 1. A reduced diagram with 6 crossings.

such *reduced* diagrams. To a given knot, there corresponds a finite number of (but not necessarily just one) reduced diagrams. We shall come back later to the problem of different reduced diagrams which correspond to the same knot (or link, or tangle).

To avoid redundancies, we can concentrate on *prime* links and tangles, whose diagrams cannot be decomposed as a connected sum of components (Figure 2).

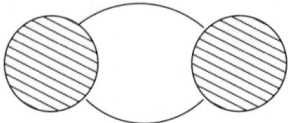

Figure 2. A nonprime diagram.

A diagram is called *alternating* if one meets alternatively under- and over-crossings as one travels along each loop. (Even though it may not seem obvious, there are knots that cannot be drawn in an alternating way—starting with eight crossings.) From now on, we shall concentrate on alternating diagrams only, since they are easier to count. There are two reasons for that.

The first reason is that there is a relatively simple way to characterize whether two reduced alternating diagrams correspond to the same knot or link (simpler than the general Reidemeister theorem [1932] for comparing any two knots). Indeed, a major result conjectured by Tait and proved by Menasco and Thistlethwaite [1991; 1993] is that two alternating reduced knot or link diagrams represent the same object if and only if they are related by a sequence of moves acting on tangles called "flypes" (Figure 3).

Figure 3. The flype of a tangle.

The second reason is that there is a correspondence between alternating diagrams and planar diagrams (see for example [Kauffman 1993]), which will be explained now as we discuss matrix integrals.

3. Matrix Integrals

We now start from a completely different angle and consider the matrix integral

$$Z^{(N)}(g) = \int dM \, e^{N \, \text{tr}\left(-\frac{1}{2}M^2 + \frac{g}{4}M^4\right)}, \qquad (3\text{-}1)$$

where M is a $N \times N$ hermitian matrix and g is a real parameter, which should be chosen negative to make the integral convergent.

As an application of Wick's theorem, the perturbative expansion of $Z^{(N)}$ in powers of g can be made using the following Feynman rules: one should count all diagrams made out of vertices $= gN\delta_{qi}\delta_{jk}$ and propagators $\langle M_{ij} M_{k\ell}^\dagger \rangle_0 = \frac{1}{N}\delta_{i\ell}\delta_{jk}$. Due to the double lines, these diagrams form so-called fat graphs which can be identified with triangulated surfaces. Each diagram has a weight

$$(gN)^{\text{V}} N^{-\text{E}} N^{\text{F}} \frac{1}{\text{symmetry factor}},$$

where V, E, F are the number of vertices, edges, faces of the triangulated surface (the factor N^{F} comes from the summation over internal indices). The symmetry factor (the order of the automorphism group of the diagram) is of little importance to us and we shall not discuss it any further. Note that the power of N is simply N^χ where χ is the Euler–Poincaré characteristic of the triangulated surface. If we take the logarithm, which amounts to considering only connected surfaces, we have the genus expansion

$$\log Z^{(N)}(g) = \sum_{h=0}^{\infty} F_h(g) N^{2-2h},$$

where F_h is the sum over surfaces of genus h. In particular, if we consider the large N limit, we see that

$$F(g) = \lim_{N \to \infty} \frac{\log Z^{(N)}(g)}{N^2} = \sum_{\text{planar graphs}} \frac{g^{\text{V}}}{\text{symmetry factor}}$$

is the sum over connected "planar" diagrams (i.e., with spherical topology). $F(g)$ is the quantity we are interested in. The formal power series $F(g) = \sum_p f_p g^p$

turns out to have, as is well-known, a finite radius of convergence (which allows to analytically continue it to positive values of g, as will be explained later). The position and nature of the closest singularity g_c determines the asymptotics of f_p as $p \to \infty$, that is, of the number of planar diagrams with large numbers of vertices.

In order to connect with knot theory, we take any planar diagram and do the following: starting from an arbitrary crossing, we decide it is a crossing of two strings (again there is an arbitrary choice of which is under/over-crossing). Once the first choice is made, we simply follow the strings and form alternating sequences of under- and over-crossings. The remarkable fact is that this can be done consistently (Figure 4). If we identify two alternating diagrams obtained

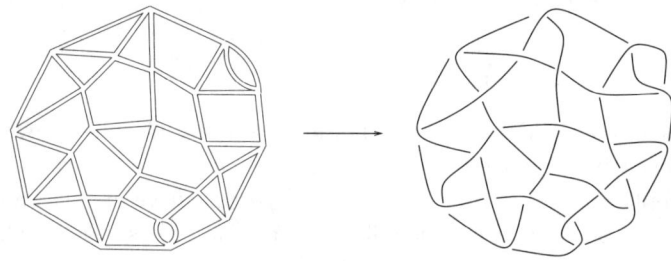

Figure 4. A planar diagram and the corresponding alternating link diagram.

from one another by inverting undercrossings and overcrossings, then there is a one-to-one correspondence between planar diagrams and alternating link diagrams. So the function $F(g)$ also counts alternating link diagrams with a given number of crossings.

A more detailed discussion of the properties of the resulting link diagrams will be made in the next section. For now, we shall address the question of the number of connected components of the link (as a 3-dimension object). Indeed, there is no reason for the diagram to represent a simple knot, and not several intertwined knots. In order to distinguish them, we introduce a more general model, which we shall call the intersecting loops $O(n)$ model. If n is a positive integer, consider the multi-matrix integral

$$Z^{(N)}(n,g) = \int \prod_{a=1}^{n} \mathrm{d}M_a \, \mathrm{e}^{N \, \mathrm{tr}\left(-\frac{1}{2}\sum_{a=1}^{n} M_a^2 + \frac{g}{4}\sum_{a,b=1}^{n} M_a M_b M_a M_b\right)} \quad (3\text{--}2)$$

and the corresponding free energy

$$F(n,g) = \lim_{N \to \infty} \frac{\log Z^{(N)}(n,g)}{N^2}. \quad (3\text{--}3)$$

This model has an $O(n)$-invariance where the M_a behave as a vector under $O(n)$. Its Feynman rules are a bit more complicated since we should draw the diagrams with n different colors. The colors "cross" each other at vertices just like strings

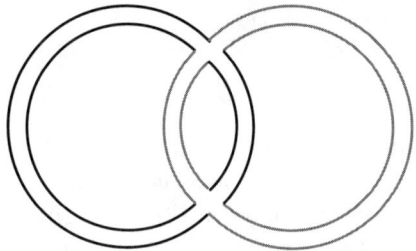

Figure 5. A planar diagram with 2 colors.

in links (Figure 5). So what we have done is allow each loop in the link to have n different colors. This is in itself an interesting generalization of the original counting problem. Indeed, we can write

$$F(n,g) = \sum_{k=1}^{\infty} F^k(g) n^k, \qquad (3\text{--}4)$$

where $F^k(g)$ is the sum over alternating link diagrams with exactly k intertwined knots. We see that the links are weighted differently according to their number of connected components.

But there is more. The expression (3–4) is an expansion of $F(n,g)$ as a function of n around 0; it provides a definition of $F(n,g)$ for noninteger values of n. In particular, we have the formal expression

$$F^1(g) = \frac{\partial F(n,g)}{\partial n} \bigg|_{n=0}$$

for the sum over alternating knots (this is the classical replica trick). Therefore, if one computed $F(n,g)$ for arbitrary (noninteger) values of n, one would have access to the generating function of the number of alternating knots. Of course, it might seem difficult to solve our model for all n; we shall discuss this again in the conclusion.

4. The One Matrix Model and the Counting of Links

Let us now come back to the one-matrix model and show how one can derive explicit formulae for the counting of prime alternating links. We recall the partition function

$$Z^{(N)}(g) = \int dM \, e^{N \operatorname{tr}\left(-\frac{1}{2}M^2 + \frac{g}{4}M^4\right)} \qquad (4\text{--}1)$$

and the corresponding free energy

$$F(g) = \lim_{N \to \infty} \frac{1}{N^2} \log Z^{(N)}(g). \qquad (4\text{--}2)$$

We also define the correlation functions

$$G_{2n}(g) = \lim_{N \to \infty} \left\langle \frac{1}{N} \operatorname{tr} M^{2n} \right\rangle. \tag{4-3}$$

Whereas the perturbative expansion of $F(g)$ generates closed diagrams (and therefore alternating links), the $G_{2n}(g)$ count diagrams with $2n$ external legs. In particular, we shall be interested later in $\Gamma(g) = G_4(g) - 2G_2(g)^2$ which counts connected diagrams with 4 legs, that is, alternating tangles.

There are various methods to compute all these quantities. We shall briefly recall the simplest one: the saddle point method.

4.1. Saddle Point Method for the One-Matrix Model. We start from Equation (4–1) and notice that the action and measure are $U(N)$-invariant; therefore we can go over to the eigenvalues λ_i of M:

$$Z(g) = \int \prod_{i=1}^{N} d\lambda_i \, \Delta[\lambda_i]^2 \, e^{N \sum_{i=1}^{N} \left(-\frac{1}{2}\lambda_i^2 + \frac{g}{4}\lambda_i^4\right)}, \tag{4-4}$$

up to an overall constant factor. Here $\Delta[\cdot]$ is the Vandermonde determinant:

$$\Delta[\lambda_i] = \det(\lambda_i^{j-1}) = \prod_{i<j}(\lambda_i - \lambda_j).$$

We are now interested in the large N limit of the N-uple integral (4–4). We would like to justify the fact that it is dominated by a saddle point in this limit. Of course this is not exactly the usual setting for a saddle point method, since not only does the integrand depend on N, but also the number of variables of integrations is equal to N. However, one notes the following: in (4–4), the "action" (that is, the log of the integrated function) is of order N^2 (essentially because the Vandermonde determinant is a product of $\sim N^2$ factors), whereas there are only N variables of integration. Since $N^2 \gg N$, a saddle point analysis does apply. It is easy to see that as $N \to \infty$, the eigenvalues λ_i will condense to form a continuous saddle point density $\rho(\lambda)$ whose support is an interval $[-2a, 2a]$. The density $\rho(\lambda)$ is defined by the property that $N\rho(\lambda)\,d\lambda$ eigenvalues lie in the interval $[\lambda; \lambda + d\lambda]$ (note the normalization, which is such that $\int \rho(\lambda)\,d\lambda = 1$). The saddle point equations are obtained by requiring the derivative with respect to the λ_i of the action of (4–4) to be zero:

$$\frac{2}{N} \sum_{j(\neq i)} \frac{1}{\lambda_i - \lambda_j} - \lambda_i + g\lambda_i^3 = 0 \quad \text{for all } i. \tag{4-5}$$

As $N \to \infty$, the sum in (4–5) tends to the principal part $\mathcal{P}P \int \frac{d\lambda'\, \rho(\lambda')}{\lambda_i - \lambda'}$; it is therefore convenient to introduce the resolvent

$$\omega(\lambda) = \lim_{N \to \infty} \left\langle \frac{1}{N} \operatorname{tr} \frac{1}{\lambda - M} \right\rangle = \int_{-2a}^{2a} d\lambda' \, \frac{\rho(\lambda')}{\lambda - \lambda'} \tag{4-6}$$

for any $\lambda \notin [-2a, 2a]$. Since the λ_i fill the interval $[-2a, 2a]$, the saddle point equations finally read

$$\omega(\lambda + i0) + \omega(\lambda - i0) - \lambda + g\lambda^3 = 0 \quad \text{for all } \lambda \in [-2a, 2a]. \quad (4\text{--}7)$$

This is a simple (scalar) Riemann–Hilbert problem which can be solved:

$$\omega(\lambda) = \tfrac{1}{2}\lambda - \tfrac{1}{2}g\lambda^3 - \left(-\tfrac{1}{2}g\lambda^2 + \tfrac{1}{2} - ga^2\right)\sqrt{\lambda^2 - 4a^2}, \quad (4\text{--}8)$$

where a is fixed, using the normalization condition $\int \rho(\lambda)\,d\lambda = 1$, to be $a^2 = (1 - \sqrt{1 - 12g})/(6g)$; equivalently,

$$\rho(\lambda) = \tfrac{1}{2}\pi i (\omega(\lambda - i0) - \omega(\lambda + i0)) = (-\tfrac{1}{2}g\lambda^2 + \tfrac{1}{2} - ga^2)\sqrt{4a^2 - \lambda^2}.$$

What we have found is a generalized semi-circle law (indeed, for $g = 0$ we recover the usual semi-circle law for the Gaussian Unitary Ensemble). From (4–6) it is clear that $\omega(\lambda)$ is a generating function of the G_{2n}, defined by (4–3); so that we can extract

$$G_2(g) = \;\text{—⬢—}\; = \tfrac{1}{3}a^2(4 - a^2),$$

$$\Gamma(g) = G_4(g) - 2G_2(g)^2 = \;\text{⨯⬢}\; = a^4(a^2 - 1)(2a^2 - 5).$$

Also, we find

$$F(g) = \tfrac{1}{2}\log a^2 - \tfrac{1}{24}(a^2 - 1)(9 - a^2).$$

All these expressions can now be analytically continued to $g > 0$ all the way to the singularity $g_c = 1/12$. This has a simple interpretation: changing the sign of $g > 0$ corresponds to making the potential in which the eigenvalues lie unstable; however, there is still a local minimum at the origin and since the large N limit is a *classical limit*, the eigenvalues cannot quantum tunnel to the unstable region and therefore remain in the valley (Figure 6). However, as g reaches its critical value g_c, the eigenvalues begin overflowing, which causes the singularity.

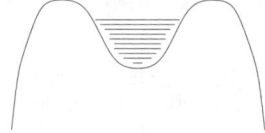

Figure 6. The potential for $g > 0$; analytic continuation is possible as long as the eigenvalues stay trapped inside the central valley.

Of course, $F(g)$ is not yet the counting function of prime alternating links. There are two separate problems to resolve:

1. Are the diagrams *reduced* (do they have minimal crossing number)? Do they correspond to *prime* links?

2. What about the flype equivalence? One should count only once different diagrams which are flype-equivalent.

We shall address them now.

4.2. Primality and Minimality. The diagrams obtained from the matrix model can have "nugatory" crossings or "nonprime" parts (Figure 7). However,

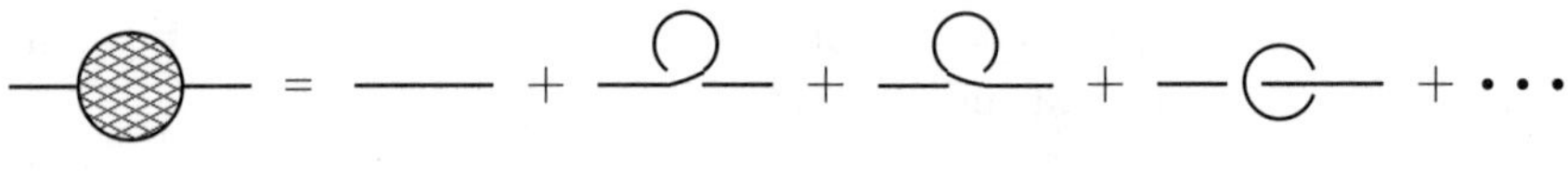

Figure 7. First terms in the perturbative expansion of the 2-point function.

all these unwanted features appear as part of the two-point function. Therefore, in order to remove them, we must simply set the two-point function equal to 1! This is achieved by introducing an additional parameter t in the action:

$$Z^{(N)}(t,g) = \int dM\, e^{N\,\mathrm{tr}\,\left(-\frac{t}{2}M^2 + \frac{g}{4}M^4\right)}. \tag{4-9}$$

Of course, t can be absorbed in a rescaling of M, so the model is essentially unchanged. However we can now ask that t be chosen as a function of g such that

$$G_2(t(g), g) = 1. \tag{4-10}$$

We can solve this equation; the auxiliary function $a(g)$ introduced earlier is now the solution of a third degree equation

$$27g = (a^2 - 1)(4 - a^2)^2, \tag{4-11}$$

equal to 1 when $g = 0$; and $t(g)$ is given by

$$t(g) = \tfrac{1}{3}a^2(g)\bigl(4 - a^2(g)\bigr). \tag{4-12}$$

The function $\Gamma(g) := \Gamma(t(g), g)$ is then the counting function for reduced alternating tangle diagrams. Similarly, $F(g)$ defined by $\frac{d}{dg}F(g) = \frac{1}{4}G_4(t(g),g)$ counts alternating link diagrams. We find in particular that the singularity of $F(g) = \sum_p f_p g^p$ (given by the equation $g_c/t^2(g_c) = \frac{1}{12}$) has moved to $g_c = \frac{4}{27}$; taking into account the power of the singularity, we find that the rate of growth of the number of alternating diagrams with p crossings is

$$f_p \stackrel{p\to\infty}{\sim} \mathrm{const}\ 6.75^p\, p^{-7/2}. \tag{4-13}$$

A similar result was found in [Tutte 1963].

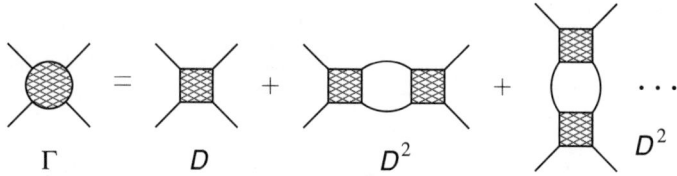

Figure 8. The whole set of diagrams Γ built out of the 2PI diagrams D.

4.3. Flype Equivalence.
The more serious problem we have to resolve is that we are not really counting links: we are counting diagrams, and links are flype-equivalence classes of diagrams. Here we shall follow [Sundberg and Thistlethwaite 1998].

Let us take a closer look at the action of a flype (Figure 3). The key remark is that it acts on tangles (four-point functions), but more precisely on *two-particle reducible* (2PR) tangles. This leads naturally to the idea of introducing *skeleton diagrams*: a general connected tangle can be created by putting *two-particle-irreducible* (2PI) diagrams in the "slots" of a *fully two-particle-reducible* skeleton diagram. We then expect that the 2PR skeleton will be modified by the flype-equivalence, whereas the 2PI pieces (or more precisely the corresponding skeletons, see below) will be unaffected. Let $\Gamma(g) = G_4(g) - 2G_2(g)^2$ be the counting function of connected tangles and $D(g)$ of 2PI tangles; then $\Gamma\{D\}$, that is the power series obtained by composing $\Gamma(g)$ and the inverse of $D(g)$, is the counting function of fully 2PR skeleton diagrams (Figure 8). It is easy to see from general combinatorial arguments that $D(g) = \Gamma(g)(1-\Gamma(g))/(1+\Gamma(g))$, and therefore

$$\Gamma\{D\} = \tfrac{1}{2}\bigl(1 - D - \sqrt{(1-D)^2 - 4D}\bigr). \tag{4-14}$$

Inversely if $D(g) = g + \zeta(g)$, then $\zeta[\Gamma]$ is the counting function of *fully 2PI skeletons diagrams* (Figure 9). From the solution of the one-matrix model, one

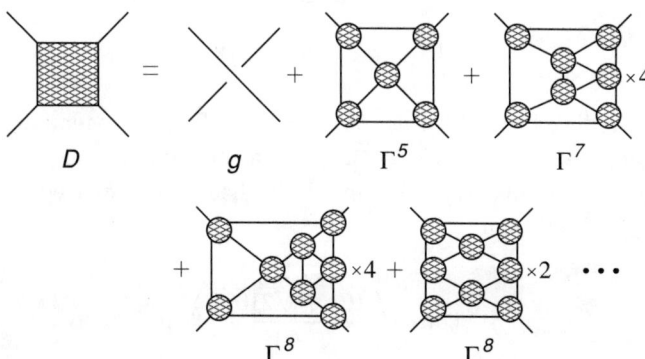

Figure 9. The set of 2PI diagrams built out of general diagrams Γ (plus the single crossing g).

obtains

$$\zeta[\Gamma] = -\frac{2}{1+\Gamma} + 2 - \Gamma - \frac{1}{2}\frac{1}{(\Gamma+2)^3}(1 + 10\Gamma - 2\Gamma^2 - (1-4\Gamma)^{3/2}). \quad (4\text{--}15)$$

As we mentioned earlier, after taking into account the flyping equivalence, Equation (4–14) will be modified, but not Equation (4–15). To demonstrate how it works, we show how the counting of Figure 8 is redone (Figure 10).

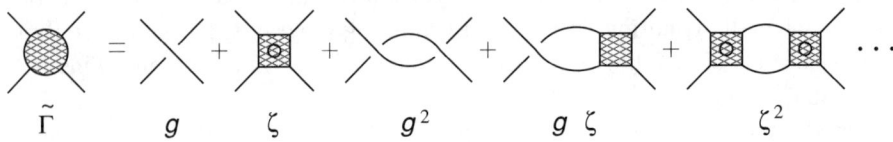

Figure 10. Taking into account the flyping equivalence forces us to distinguish simple crossings from nontrivial 2PI diagrams (marked with a circle). There is only one term $g\zeta$ because the other term is obtained by a flype.

More generally, we can redo the simple combinatorics to find the generating function of the 2PR skeletons, but this time taking into account the flype equivalence. We find

$$\tilde{\Gamma}\{g,\zeta\} = \frac{1}{2}\left((1+g-\zeta) - \sqrt{(1-g+\zeta)^2 - 8\zeta - 8\frac{g^2}{1-g}}\right). \quad (4\text{--}16)$$

This is to be combined with the (unaltered) matrix model data

$$\zeta[\Gamma] = -\frac{2}{1+\Gamma} + 2 - \Gamma - \frac{1}{2}\frac{1}{(\Gamma+2)^3}(1 + 10\Gamma - 2\Gamma^2 - (1-4\Gamma)^{3/2}). \quad (4\text{--}17)$$

In practice, this means that $\tilde{\Gamma}(g)$ is given by an implicit equation:

$$\tilde{\Gamma}(g) = \tilde{\Gamma}\{g, \zeta[\tilde{\Gamma}(g)]\}, \quad (4\text{--}18)$$

which can be reduced to a fifth degree equation. From the generating function of tangles $\tilde{\Gamma}(g)$ we can go back to the generating function of closed diagrams $\tilde{F}(g)$; we find in particular that the singularity has been displaced again, so that if $\tilde{F}(g) = \sum_{p=0}^{\infty} \tilde{f}_p g^p$, then

$$\tilde{f}_p \stackrel{p\to\infty}{\sim} \text{const}\left(\frac{101+\sqrt{21001}}{40}\right)^p p^{-7/2}, \quad (4\text{--}19)$$

where the quantity in parentheses equals approximately 6.14793. This result was first obtained in [Sundberg and Thistlethwaite 1998].

5. The $ABAB$ Model and Character Expansion

We shall now inspect the $n = 2$ case of the general $O(n)$ model (3–2). There are various reasons that this model is of particular interest, and we shall discover some of them along the way. Let us rewrite the partition function

$$Z^{(N)}(2,g) = \int dA\, dB\, e^{N\,\mathrm{tr}\left(-\frac{1}{2}(A^2+B^2)+\frac{g}{4}(A^4+B^4)+\frac{g}{2}(AB)^2\right)}. \quad (5\text{–}1)$$

We see that we could introduce two coupling constants α and β:

$$Z^{(N)}_{ABAB}(\alpha,\beta) = \int dA\, dB\, e^{N\,\mathrm{tr}\left(-\frac{1}{2}(A^2+B^2)+\frac{\alpha}{4}(A^4+B^4)+\frac{\beta}{2}(AB)^2\right)} \quad (5\text{–}2)$$

(which amounts to introducing "interaction" between the two colors of loops). For $\alpha = \beta = g$ we recover the $O(2)$ model. The more general model with α and β arbitrary is not necessary for the original counting problem, but since it turns out that we can solve it equally easily, we shall keep the two coupling constants. Note that when $\alpha \neq \beta$ the $O(2)$ symmetry of the model is broken. This is even more apparent if we make the change of variables $X = (A+iB)/\sqrt{2}$, $X^\dagger = (A - iB)/\sqrt{2}$:

$$Z^{(N)}_{8v}(b,c,d) = \int dX\, dX^\dagger\, e^{N\,\mathrm{tr}\left(-XX^\dagger+bX^2X^{\dagger 2}+\frac{c}{2}(XX^\dagger)^2+\frac{d}{4}(X^4+X^{\dagger 4})\right)}, \quad (5\text{–}3)$$

with $b = (\alpha + \beta)/2$ and $c = d = (\alpha - \beta)/2$. We recognize in (5–3) the partition function of the *8-vertex model* on random dynamical lattices (more precisely, a two-parameter slice of it, since c and d are not independent). A configuration of the model is defined by a quadr-angulated surface with arrows on the edges of the graph, such that each of the vertices displays one of the eight allowed configurations, which are weighted with the 3 constants b, c, d. For $\alpha = \beta$ the $U(1)$-breaking term $X^4 + X^{\dagger 4}$ vanishes and we recover the *6-vertex model*. (In this case note that the arrows "cross" each other just like strings in links, so that we manifestly recover our link model with the 2 orientations of the loops playing the same role as the 2 colors.)

We shall now show how to solve the model in the planar limit, i.e., compute the large N free energy.

5.1. Character Expansion.
All known matrix model solutions are (more or less implicitly) based on the fact that we can reduce the number of degrees of freedom from N^2 to N. Usually the N remaining degrees of freedom are the eigenvalues of the matrices. Unfortunately, from Equation (5–2) one cannot go directly to the eigenvalues of A and B: we do not know how to integrate over the relative angle between A and B. (Only one integral of this type is known exactly, the Harish-Chandra–Itzykson–Zuber integral [Harish-Chandra 1957; Itzykson and Zuber 1980], but it does not apply here.) Therefore, instead of working directly with (5–2), we expand the troublesome part $\exp\left(N\frac{\beta}{2}\,\mathrm{tr}(AB)^2\right)$

in *characters* of GL(N). Recalling that all class-functions can be expanded on the basis of characters, we write

$$e^{N\frac{\beta}{2}\operatorname{tr}(AB)^2} = \sum_{\{h\}} c_{\{h\}} \chi_{\{h\}}(AB), \tag{5-4}$$

where $\chi_{\{h\}}(AB)$ is the character taken at AB and $\{h\}$ is the set of shifted highest weights $h_i = m_i + N - i$ (m_i highest weights), $h_1 > h_2 > \cdots > h_n \geq 0$, which parameterize the GL(N) analytic irreducible representation. The coefficients of the expansion $c_{\{h\}}$ can be determined explicitly:

$$c_{\{h\}} = (N\beta/2)^{\#h/2} \frac{\Delta(h^{\text{even}}/2)\Delta((h^{\text{odd}}-1)/2)}{\prod_i \lfloor h_i/2 \rfloor!} \tag{5-5}$$

in terms of the set $\{h^{\text{even}}\}$ and $\{h^{\text{odd}}\}$ of even and odd h_i. The advantage of characters is that they satisfy orthogonality relations, so that we can now integrate over the relative angle between A and B:

$$\int_{U(N)} d\Omega \, \chi_{\{h\}}(A\Omega B \Omega^\dagger) = \frac{\chi_{\{h\}}(A)\chi_{\{h\}}(B)}{\chi_{\{h\}}(1)} \tag{5-6}$$

where the dimension $\chi_{\{h\}}(1)$ is up to an overall constant the Vandermonde determinant $\Delta[h_i]$.

Once Equations (5-4) and (5-6) are inserted into (5-2), we see that the integrand only depends on the eigenvalues of A and B:

$$Z^{(N)}_{ABAB}(\alpha,\beta) = \sum_{\{h\}} \frac{c_{\{h\}}}{\Delta[h_i]} \left(\int \prod_{i=1}^N d\lambda_i \, e^{N \sum_{i=1}^N \left(-\frac{1}{2}\lambda_i^2 + \frac{\alpha}{4}\lambda_i^4\right)} \Delta[\lambda_i] \det_{i,j}[\lambda_i^{h_j}] \right)^2 \tag{5-7}$$

The key observation here is that we still have an action of order N^2, but we have N highest weights h_i and N eigenvalues λ_i; therefore a saddle point analysis applies again.

5.2. Saddle Point on Young Diagrams. The notion of a saddle point on Young diagrams first appeared in [Vershik and Kerov 1977] in the context of the asymptotics of the Plancherel measure. It was rediscovered independently in the solution of large N 2D Yang–Mills [Douglas and Kazakov 1993], and was used to deal with character expansions in [Kazakov et al. 1996b; 1996a; 1996c; Kostov et al. 1998]. In the present calculation, the novelty is that we have to deal with a *double* saddle point equation on both eigenvalues and shifted highest weights (that is, the shape of the Young tableau) [Kazakov and Zinn-Justin 1999].

The idea here is to find an appropriate scaling ansatz for the shape of the dominant Young diagram in the large N limit. We find that the highest weights h_i scale as N (the Young diagrams become large both horizontally and vertically), so we can define a continuous density of rescaled h_i/N by

$$\rho(h) = \frac{1}{N}\sum_{i=1}^{N}\delta(h - h_i/N),$$

and the corresponding resolvent

$$H(h) = \int dh' \frac{\rho(h')}{h - h'}.$$

We also have a density of eigenvalues $\rho(\lambda)$ and the resolvent $\omega(\lambda)$.

If we introduce "slashed" functions by $\slashed{H}(h) := \frac{1}{2}(H(h+i0) + H(h-i0))$ (and similarly for the other functions), The saddle point equations now read:

$$\begin{cases} -\lambda + \alpha\lambda^3 + \slashed{\psi}(\lambda) + \slashed{h}(\lambda)/\lambda = 0, & \lambda \in [-\lambda_0, +\lambda_0], \\ \slashed{L}(h) - \dfrac{\slashed{H}(h)}{2} = \dfrac{1}{2}\log(h/\beta), & h \in [h_1, h_2]. \end{cases} \quad (5\text{-}8)$$

with $L(h) = \log \lambda^2(h)$. The new unknown functions $h(\lambda)$ and $\lambda(h)$ appear when taking the logarithmic derivative of $\det_{i,j}[\lambda_i^{h_j}]$; this type of functions was analyzed in [Zinn-Justin 1998], where it was shown that $\lambda(h)$ and $h(\lambda)$ are *functional inverses* of each other. Therefore, we have two saddle point equations which are connected by a functional inversion relation. This connection allows to solve them; skipping the details, one can show that one has a well-defined Riemann–Hilbert problem for the auxiliary function $D(h) := 2L(h) - H(h) - 3\log h + \log(h - h_1)$, whose solution can be expressed in terms of Θ functions in an appropriate elliptic parametrization $y(h)$ [Kazakov and Zinn-Justin 1999]:

$$D(h) = \log\frac{h - h_1}{-\alpha h^2} - \frac{\log(\beta/\alpha)}{K}y(h) + 2\log\frac{\Theta_2(x_0 - y(h))}{\Theta_2(x_0 + y(h))}.$$

5.3. Phase Diagram and Discussion. Just as the one-matrix model displayed a singularity at $g_c = \frac{1}{12}$, here the free energy and the various correlation functions have a line of singularities in the (α, β) plane, shown on Figure 11.

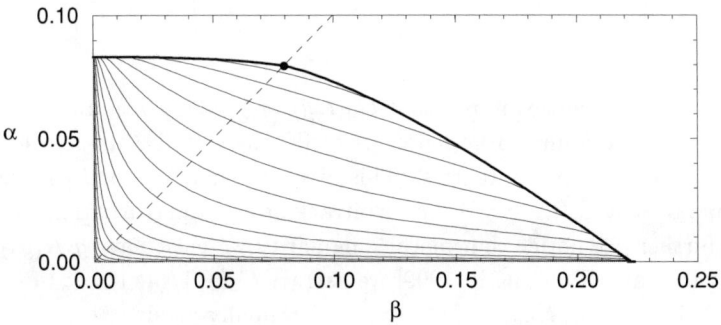

Figure 11. Phase diagram of the $ABAB$ model. The dashed line is the $\alpha = \beta$ line, the curves are equipotentials of the elliptic nome.

We recognize at $\alpha = \frac{1}{12}$, $\beta = 0$ the usual singularity of the one-matrix model. In fact, one can show that everywhere on the critical line except at the critical point $\alpha_c = \beta_c = \frac{1}{4\pi}$ the critical behavior is the same as the one of the one-matrix model ("pure gravity" behavior). This implies the following asymptotics: at fixed slope $s = \beta/\alpha \ne 1$, if the free energy $F(\alpha, \beta = s\alpha) = \sum_p f_p(s)\alpha^p$ then

$$f_p(s) \stackrel{p\to\infty}{\sim} \text{const } \alpha_c(s)^{-p} p^{-7/2}.$$

However, the point $\alpha_c = \beta_c = \frac{1}{4\pi}$, that is, the 6-vertex model point, is very special: it is the point where the elliptic functions degenerate into trigonometric functions, which implies logarithmic corrections:

$$f_p(1) \stackrel{p\to\infty}{\sim} \text{const } (4\pi)^p p^{-3} (\log p)^2.$$

This is characteristic of a $c = 1$ conformal field theory coupled to gravity.

5.4. Application to Reduced Alternating Diagrams.

We should remember that the $\alpha = \beta$ line is of special interest to us, since it is the intersecting loops $O(2)$ model (solving a certain counting problem for alternating links) we started from. In order to carry out the program that we have applied to the one-matrix model we should next address the two issues of primality/minimality and of the flype equivalence. We shall only consider the first issue; a discussion of the flype equivalence in the general $O(n)$ case will be made in the next section, and the corresponding calculation for $n = 2$ will appear in a future publication by the author and J.-B. Zuber.

Again, we introduce an additional parameter t in the action:

$$Z^{(N)}(2,t,g) = \int dA\, dB\, e^{N\,\text{tr}\left(-\frac{t}{2}(A^2 + B^2) + \frac{g}{4}(A^4 + B^4) + \frac{g}{2}(AB)^2\right)} \quad (5\text{--}9)$$

and we impose the condition that the 2-point function

$$G_2(t,g) = \lim_{N\to\infty} \langle \tfrac{1}{N} \text{tr } A^2 \rangle = \lim_{N\to\infty} \langle \tfrac{1}{N} \text{tr } B^2 \rangle$$

satisfy

$$G_2(t(g), g) = 1. \quad (5\text{--}10)$$

Obvious scaling properties imply that $G_2(t, g) = \frac{1}{t}G_2(1, g/t^2)$, and the formula for $G_2(1, g)$ can be found in appendix A of [Kazakov and Zinn-Justin 1999] in terms of complete elliptic integrals. This gives an equation for $t(g)$, which can in principle be solved (at least to an arbitrary order in perturbation theory).

To go further, we notice that at the singularity we must have $g_c/t(g_c)^2 = \frac{1}{4\pi}$; from [Kazakov and Zinn-Justin 1999] we extract $G_2(1, 1/(4\pi)) = \frac{\pi}{2}(4 - \pi)$, and therefore using (5–10), $t(g_c) = \frac{\pi}{2}(4 - \pi)$, which finally yields

$$g_c = \frac{\pi}{16}(\pi - 4)^2. \quad (5\text{--}11)$$

We conclude that the number f_p of reduced alternating link diagrams with 2 colors and p crossings has the asymptotics

$$f_p \overset{p \to \infty}{\sim} \text{const} \left(\frac{16}{\pi(\pi-4)^2}\right)^p p^{-3}(\log p)^2,$$

where the number $1/g_c = 6.91167\ldots$ is slightly larger than the value 6.75 obtained for only one color.

6. Further Generalizations and Prospects

We have already written a fairly general model, the intersecting loops $O(n)$ model (Equation (3–2)) which should contain in principle all information on the counting of alternating links and knots. We shall now show how this model is in fact not sufficient for our purposes.

Indeed, one should remember that the $O(n)$ model given above counts alternating link diagrams, and not alternating links. For the latter, one should address the problem of the flype equivalence. We have seen in the $n = 1$ case (Section 4) that we needed to do a little surgery on the four-point functions. One can convince oneself that what it amounts to, in more physical terms, is a *finite renormalization* which results in the appearance of quartic counterterms in the action. Generically these counterterms will have the most general form compatible with the symmetry. In our case, we find that there are two independent $O(n)$-symmetric tetravalent vertices, of the form $M_a M_b M_a M_b$ and $M_a M_a M_b M_b$. This results in a generalized $O(n)$ model:

$$Z^{(N)}(n,g,h) = \int \prod_{a=1}^{n} dM_a \, e^{N \operatorname{tr}\left(-\frac{1}{2}\sum_{a=1}^{n} M_a^2 + \frac{g}{4}\sum_{a,b=1}^{n}(M_a M_b)^2 + \frac{h}{2}\sum_{a,b=1}^{n} M_a^2 M_b^2\right)}, \quad (6\text{--}1)$$

where h will be given as a function of g by appropriate combinatorial relations of the same form as those of Section 4.

At the moment, the solution of this general model is unknown. We note, however, that for $g = 0$ this model is simply the usual (nonintersecting loops) $O(n)$ model, which has been completely solved [Kostov 1989; Gaudin and Kostov 1989; Kostov and Staudacher 1992]. It is tempting to speculate that there is no phase transition in the (g, h) plane as one moves away from the $g = 0$ line. (This is certainly true for $n = 2$, as shown by the study in [Dalley 1992] and the exact solution in [Zinn-Justin 2000].) Then, one can make predictions on *universal quantities* such as critical exponents of the model. For example, the number $\tilde{f}_p(n)$ of prime alternating links with n colors would have the asymptotics

$$\tilde{f}_p(n) \overset{?}{\sim} \text{const}(n) \, b(n)^p \, p^{-2-1/\nu}, \qquad n = -2\cos(\pi\nu), \ 0 < \nu < 1.$$

In particular the number $\tilde{f}_p$ of prime alternating knots would satisfy

$$\tilde{f}_p \stackrel{?}{\sim} \text{const } b(0)^p \, p^{-4}.$$

One interesting question is to determine the nonuniversal constant $b(0)$. This, of course, requires to really solve the $n = 0$ model (or more precisely to study the $n \to 0$ limit). An alternative option is to note that this model can be recast as a supersymmetric $Osp(2n|2n)$ model, the simplest $(n = 1)$ being a supersymmetrized version of the $O(2)$ model considered earlier; using bosonic and fermionic complex matrices X and Ψ, it can be written as

$$Z^{(N)}(g) = \int dX \, dX^\dagger \, d\Psi \, d\Psi^\dagger$$
$$\times e^{N \, \text{tr} \left(-XX^\dagger - \Psi\Psi^\dagger + g(XX^\dagger X^\dagger X + \Psi\Psi^\dagger \Psi^\dagger \Psi + \Psi X^\dagger \Psi^\dagger X + X^\dagger \Psi X \Psi^\dagger)\right)}.$$
(6–2)

Due to supersymmetry, this partition function is equal to 1, but nonsupersymmetric correlation functions such as $\langle \frac{1}{N} \text{tr}(XX^\dagger)^n \rangle$ are nontrivial and should contain the desired information. Whether this model is solvable or not is an open question.

Acknowledgements

I would like to thank J.B. Zuber for various discussions and with whom most of this work was done. I also want to thank Prof. D. Eisenbud for the hospitality of the MSRI, and the organizers of this semester on Random Matrices, P. Bleher and A. Its, for inviting me and giving me the opportunity to give seminars.

References

[Brézin et al. 1978] E. Brézin, C. Itzykson, G. Parisi, and J. B. Zuber, "Planar diagrams", *Comm. Math. Phys.* **59**:1 (1978), 35–51.

[Dalley 1992] S. Dalley, *Mod. Phys. Lett.* **A7** (1992), 1651–.

[Di Francesco et al. 1995] P. Di Francesco, P. Ginsparg, and J. Zinn-Justin, "2D gravity and random matrices", *Phys. Rep.* **254**:1-2 (1995), 133.

[Di Francesco et al. 1997] P. Di Francesco, O. Golinelli, and E. Guitter, "Meanders and the Temperley-Lieb algebra", *Comm. Math. Phys.* **186**:1 (1997), 1–59.

[Di Francesco et al. 1998] P. Di Francesco, B. Eynard, and E. Guitter, "Coloring random triangulations", *Nuclear Phys. B* **516**:3 (1998), 543–587.

[Douglas and Kazakov 1993] M. R. Douglas and V. A. Kazakov, *Phys. Lett.* **B319** (1993), 219–.

[Gaudin and Kostov 1989] M. Gaudin and I. Kostov, "$O(n)$ model on a fluctuating planar lattice. Some exact results", *Phys. Lett. B* **220**:1-2 (1989), 200–206.

[Harish-Chandra 1957] Harish-Chandra, "Differential operators on a semisimple Lie algebra", *Amer. J. Math.* **79** (1957), 87–120.

[Itzykson and Zuber 1980] C. Itzykson and J. B. Zuber, "The planar approximation, II", *J. Math. Phys.* **21**:3 (1980), 411–421.

[Itzykson and Zuber 1992] C. Itzykson and J.-B. Zuber, "Combinatorics of the modular group. II. The Kontsevich integrals", *Internat. J. Modern Phys. A* **7**:23 (1992), 5661–5705.

[Kauffman 1993] L. H. Kauffman, *Knots and physics*, 2nd ed., World Scientific, River Edge, NJ, 1993.

[Kazakov 1986] V. A. Kazakov, "Ising model on a dynamical planar random lattice: exact solution", *Phys. Lett. A* **119**:3 (1986), 140–144.

[Kazakov and Zinn-Justin 1999] V. A. Kazakov and P. Zinn-Justin, "Two-matrix model with $ABAB$ interaction", *Nuclear Phys. B* **546**:3 (1999), 647–668.

[Kazakov et al. 1996a] V. A. Kazakov, M. Staudacher, and T. Wynter, "Almost flat planar diagrams", *Comm. Math. Phys.* **179**:1 (1996), 235–256.

[Kazakov et al. 1996b] V. A. Kazakov, M. Staudacher, and T. Wynter, "Character expansion methods for matrix models of dually weighted graphs", *Comm. Math. Phys.* **177**:2 (1996), 451–468.

[Kazakov et al. 1996c] V. A. Kazakov, M. Staudacher, and T. Wynter, "Exact solution of discrete two-dimensional R^2 gravity", *Nuclear Phys. B* **471**:1-2 (1996), 309–333.

[Kontsevich 1991] M. L. Kontsevich, "Intersection theory on the moduli space of curves", *Funktsional. Anal. i Prilozhen.* **25**:2 (1991), 50–57, 96.

[Kostov 1989] I. K. Kostov, "$O(n)$ vector model on a planar random lattice: spectrum of anomalous dimensions", *Modern Phys. Lett. A* **4**:3 (1989), 217–226.

[Kostov and Staudacher 1992] I. K. Kostov and M. Staudacher, "Multicritical phases of the O(n) model on a random lattice", *Nuclear Phys. B* **384**:3 (1992), 459–483.

[Kostov et al. 1998] I. K. Kostov, M. Staudacher, and T. Wynter, "Complex matrix models and statistics of branched coverings of 2D surfaces", *Comm. Math. Phys.* **191**:2 (1998), 283–298.

[Makeenko 1996] Y. Makeenko, "Strings, matrix models, and meanders", *Nuclear Phys. B Proc. Suppl.* **49** (1996), 226–237. Theory of elementary particles (Buckow, 1995).

[Menasco and Thistlethwaite 1991] W. W. Menasco and M. B. Thistlethwaite, "The Tait flyping conjecture", *Bull. Amer. Math. Soc. (N.S.)* **25**:2 (1991), 403–412.

[Menasco and Thistlethwaite 1993] W. Menasco and M. Thistlethwaite, "The classification of alternating links", *Ann. of Math. (2)* **138**:1 (1993), 113–171.

[Reidemeister 1932] K. Reidemeister, *Knotentheorie*, Ergebnisse der Math. **1**, Springer, Berlin, 1932.

[Sundberg and Thistlethwaite 1998] C. Sundberg and M. Thistlethwaite, "The rate of growth of the number of prime alternating links and tangles", *Pacific J. Math.* **182**:2 (1998), 329–358.

[Tutte 1963] W. T. Tutte, "A census of planar maps", *Canad. J. Math.* **15** (1963), 249–271.

[Vershik and Kerov 1977] A. M. Vershik and S. V. Kerov, *Soviet. Math. Dokl.* **18** (1977), 527–.

[Witten 1991] E. Witten, "Two-dimensional gravity and intersection theory on moduli space", pp. 243–310 in *Surveys in differential geometry* (Cambridge, MA, 1990), Lehigh Univ., Bethlehem, PA, 1991.

[Zinn-Justin 1998] P. Zinn-Justin, "Universality of correlation functions of Hermitian random matrices in an external field", *Comm. Math. Phys.* **194**:3 (1998), 631–650. Erratum in **199**:3 (1999), 729.

[Zinn-Justin 2000] P. Zinn-Justin, "The six-vertex model on random lattices", *Europhys. Lett.* **50** (2000), 15–21.

[Zinn-Justin and Zuber 1999] P. Zinn-Justin and J.-B. Zuber, "Matrix integrals and the counting of tangles and links", in *Formal power series and algebraic combinatorics: 11th international conference, FPSAC'99* (Barcelona, 1999), 1999. Available at www.arxiv.org/abs/math-ph/9904019.

PAUL ZINN-JUSTIN
DEPARTMENT OF PHYSICS AND ASTRONOMY
RUTGERS UNIVERSITY
PISCATAWAY, NJ 08854-8019
UNITED STATES
pzinnstrings.rutgers.edu